Crustacea
and Arthropod Relationships

Crustacea and Arthropod Relationships

Stefan Koenemann
Institute for Animal Ecology and Cell Biology
University of Veterinary Medicine Hannover, Foundation
Hannover, Germany

Ronald A. Jenner
Section of Evolution & Ecology
University of California, Davis, U.S.A.

CRC Press
Taylor & Francis Group
Boca Raton London New York

CRC Press is an imprint of the
Taylor & Francis Group, an **informa** business

A TAYLOR & FRANCIS BOOK

CRC Press
Taylor & Francis Group
6000 Broken Sound Parkway NW, Suite 300
Boca Raton, FL 33487-2742

First issued in paperback 2019

© 2005 by Taylor & Francis Group, LLC
CRC Press is an imprint of Taylor & Francis Group, an Informa business

No claim to original U.S. Government works

ISBN-13: 978-0-8493-3498-6 (hbk)
ISBN-13: 978-0-367-39294-9 (pbk)

This book contains information obtained from authentic and highly regarded sources. Reasonable efforts have been made to publish reliable data and information, but the author and publisher cannot assume responsibility for the validity of all materials or the consequences of their use. The authors and publishers have attempted to trace the copyright holders of all material reproduced in this publication and apologize to copyright holders if permission to publish in this form has not been obtained. If any copyright material has not been acknowledged please write and let us know so we may rectify in any future reprint.

Except as permitted under U.S. Copyright Law, no part of this book may be reprinted, reproduced, transmitted, or utilized in any form by any electronic, mechanical, or other means, now known or hereafter invented, including photocopying, microfilming, and recording, or in any information storage or retrieval system, without written permission from the publishers.

For permission to photocopy or use material electronically from this work, please access www.copyright.com (http://www.copyright.com/) or contact the Copyright Clearance Center, Inc. (CCC), 222 Rosewood Drive, Danvers, MA 01923, 978-750-8400. CCC is a not-for-profit organization that provides licenses and registration for a variety of users. For organizations that have been granted a photocopy license by the CCC, a separate system of payment has been arranged.

Trademark Notice: Product or corporate names may be trademarks or registered trademarks, and are used only for identification and explanation without intent to infringe.

Library of Congress Cataloging-in-Publication Data

Catalog record is available from the Library of Congress

Visit the Taylor & Francis Web site at
http://www.taylorandfrancis.com

and the CRC Press Web site at
http://www.crcpress.com

Contents

Preface and dedication — vii

I *Introduction* — 1

Gould, Schram, and the paleontological perspective in evolutionary biology — 3
C. BARON & J.T. HØEG

II *Paleozoology* — 15

Decapod crustaceans, the K/P event, and Palaeocene recovery — 17
C.E. SCHWEITZER & R.M. FELDMANN

Oelandocaris oelandica and the stem lineage of Crustacea — 55
M. STEIN, D. WALOSZEK & A. MAAS

Early Palaeozoic non-lamellipedian arthropods — 73
J. BERGSTRÖM & X.-G. HOU

Comparative morphology and relationships of the Agnostida — 95
T.J. COTTON & R.A. FORTEY

III *Development and evolution* — 137

Heads, Hox and the phylogenetic position of trilobites — 139
G. SCHOLTZ & G.D. EDGECOMBE

Resolving arthropod relationships: Present and future insights from evo-devo studies — 167
S. HRYCAJ & A. POPADIĆ

IV *Comparative morphology* — 183

Evolution of eye structure and arthropod phylogeny — 185
C. BITSCH & J. BITSCH

Appendage loss and regeneration in arthropods: A comparative view — 215
D. MARUZZO, L. BONATO, C. BRENA, G. FUSCO & A. MINELLI

V   Arthropod phylogenetics — 247

What are Ostracoda? A cladistic analysis of the extant superfamilies of the subclasses Myodocopa and Podocopa (Crustacea: Ostracoda) — 249
D.J. HORNE, I. SCHÖN, R.J. SMITH & K. MARTENS

Relationships within the Pancrustacea: Examining the influence of additional Malacostracan 18S and 28S rDNA — 275
C.C. BABBIT & N.H. PATEL

Relationships between hexapods and crustaceans based on four mitochondrial genes — 295
A. CARAPELLI, F. NARDI, R. DALLAI, J.L. BOORE, P. LIÒ & F. FRATI

The position of crustaceans within the Arthropoda - Evidence from nine molecular loci and morphology — 307
G. GIRIBET, S. RICHTER, G.D. EDGECOMBE & W.C. WHEELER

VI   Metazoan phylogenetics — 353

Playing another round of metazoan phylogenetics: Historical epistemology, sensitivity analysis, and the position of Arthropoda within the Metazoa on the basis of morphology — 355
R.A. JENNER & G. SCHOLTZ

Appendix A - Publications of Frederick R. Schram — 387

Appendix B - Taxa erected by or in collaboration with F.R. Schram — 399

Contributors — 407

Index — 411

Previous series of *Crustacean Issues* — 425

Color insert

Preface and dedication

Crustacean Issues 16 is a special edition for two reasons. First, we would like to dedicate this *Festschrift* to the career of Frederick R. Schram, the founding editor of *Crustacean Issues* in 1983, in recognition of his many stimulating and wide-ranging contributions to the evolutionary biology of arthropods in general, and of crustaceans in particular. Second, this is the first volume of *Crustacean Issues* that is not exclusively 'all about' crustaceans. Although the previous 15 editions were all - more or less - narrowly concerned with crustaceans only, we hope that the wider range of topics covered in this book will be of interest to an even broader scientific audience.

However, the reader will notice that among the many topics covered herein, this book still is 'crustaceo-centric'. Only the focus has become more extended to include other groups of arthropods along with the Crustacea. This wider focus offers challenging opportunities to evaluate higher-level relationships within the Arthropoda from a carcinologic perspective. Compared to other arthropods, crustaceans are characterized by an unparalleled disparity of body plans. But although the specialization of arthropod body segments and appendages into distinct tagmata has served traditionally as a convenient basis for higher classification, many relationships within the phylum Arthropoda still remain controversial.

The lack of a detailed consensus on arthropod relationships has stimulated several fascinating recent investigations using novel approaches and modern techniques, as well as critical reappraisals of traditional evidence. A number of valuable summaries of this progress have been published during the last several years. Notable compendia are *Arthropod relationships* (Fortey & Thomas 1997), *Arthropod fossils and phylogeny* (Edgecombe 1998), and Deuve's (2001) *Origin of the Hexapoda*. These studies have aimed to solve some of the most intriguing puzzles of arthropod evolution.

Are the Crustacea at all a monophyletic group? And if so, which are their closest relatives within the Arthropoda? The answers to these and other questions will play a key role in understanding patterns and processes in arthropod evolution, for example, the 'disappearance' of given body plans from the fossil record, the transition(s) from aquatic to terrestrial environments, or the evolution of complete genomes and developmental pathways that control tagmatization.

Crustacean Issues 16 addresses the evolution and phylogenetic relationships of the Arthropoda based upon molecular, developmental, morphological, and paleontological evidence. Organized into six main sections, this volume comprises 14 chapters that shed light on *Crustacea and arthropod relationships* and offer an up-to-date summary of recent progress in several disciplines.

An introductory essay discusses the paleontological perspective in evolutionary biology. Four chapters in the *Paleozoology* section cover a range of topics concerned with the fossil record of arthropods. The first contribution addresses the impact of the Cretaceous/Paleogene (K/P) event on decapod extinction and recovery. The two following chapters deal with reconstructions of arthropod/crustacean stem lineages based on Orsten fossils, and on

Paleozoic non-lamellipedian arthropods, respectively. The last chapter in this section presents a phylogenetic analysis of agnostid trilobites. In the section on *Development and evolution*, two chapters use embryological and developmental genetic data to propose an innovative solution regarding the phylogenetic position of the trilobites, and to reevaluate the phylogeny of major arthropod groups. The two contributions in the *Comparative morphology* section critically appraise morphological evidence to discuss the evolution of arthropod eyes in a phylogenetic context, and to present a comprehensive overview of appendage loss and regeneration in Arthropoda. The section on *Arthropod phylogenetics* features four rigorous phylogenetic analyses based on different molecular loci and/or morphological data that investigate relationships within the Ostracoda, and among major groups of arthropods, with special emphasis on hexapods and crustaceans. The final section, *Metazoan phylogenetics*, presents one chapter on competing hypotheses for the placement of the arthropods in the animal kingdom on the basis of morphological evidence.

We hope that the selection of chapters written by internationally acknowledged experts will convey the fascination of crustacean and arthropod evolutionary biology to the reader, and we trust that the current work will be a valuable source of information for students and researchers alike.

Finally, we would like to thank all authors for their contributions to this *Festschrift*, and for their patient and constructive collaboration despite tight deadlines and our numerous editorial demands. We also wish to acknowledge the many reviewers, who contributed significantly to the quality of this volume, and thank Ronald Vonk for stimulating comments and discussions. John Sulzycki and Patricia Roberson from CRC Press provided reliable instructions and feedback throughout the preparation of this book. We are very grateful for their kind assistance.

Ronald A. Jenner, Davis, California
Stefan Koenemann, Amsterdam, Netherlands
November 2004

DEDICATION TO FREDERICK R. SCHRAM

It is fitting that this volume be dedicated to Dr. Frederick R. Schram on the occasion of the Sixth International Crustacean Conference, in Glasgow, Scotland, July 18-23, 2005. The year marks the 25th anniversary of the publication of the *Journal of Crustacean Biology*, the cornerstone of the Crustacean Society. Fred played a central role in the founding of the Society, serving on the Organizing Council from 1979-1982. He then served on the Board of Governors, as an Associate Editor of the Journal, and as President-Elect, becoming the third President of the Society in 1986. The record documents tireless energy and great foresight in recognizing the need for such an organization. His efforts have been rewarded by the growth and development of the Society and the stature of the Journal.

It is also noteworthy that the first article published in the *Journal of Crustacean Biology* was written by Fred. On the classification of Eumalacostraca has all the earmarks of a Schram publication, incorporating five aspects of research that are the hallmark of his remarkable career. The work documents Fred's extensive knowledge of the history of classification of Crustacea, builds on that structure by applying what, at the time, was groundbreaking work on the cladistic analysis of arthropods, utilizes an encyclopedic knowledge of the biology of Crustacea, applies data from the fossil record to an extent that few others do, and expresses his conclusions with crystal clarity and eloquence. This work, and his entire body of published research, sets a high standard and will continue to have profound impact on the research on Crustacea.

Rodney M. Feldmann, Department of Geology, Kent State University, Kent, OH 44242, U.S.A.

x   *Preface and dedication*

Frederick R. Schram. Keynote lecture at 4th European Crustacean Conference, Łódź, Poland; July 22-26, 2002.

"*When choosing between two evils, I always like to try the one I've never tried before.*"
   Mae West (1892-1980)

I INTRODUCTION

Gould, Schram and the paleontological perspective in evolutionary biology

CHRISTIAN BARON & JENS T. HØEG

Department of Cell Biology & Comparative Zoology, Biological Institute, University of Copenhagen, Denmark

ABSTRACT

In this essay, we compare the careers of Frederick R. Schram and Stephen Jay Gould and their views and influence on the importance of paleontological data in evolutionary (historical) biology. Both scientists have consistently advocated the importance of the fossil record, but both have also perceptibly changed their views and approaches throughout their careers and in very different directions. Gould initially advocated a model-based approach to historical biology in an apparent attempt to increase the status of an endangered discipline in a difficult political and financial environment. Later, he largely redirected his focus away from these model-based approaches and instead emphasized the 'contingency' perspective as of vital and overlooked importance in biological evolution. Schram only gradually moved from a 'Mantonian' view on arthropod evolution into being a convinced 'computer-based' cladist, but changed early enough to have immense influence on the development of phylogenetics of both arthropods and the Metazoa in general. We try to analyze how and why these changing views took place by plotting them against the socioscientific background of the two scientists.

1 INTRODUCTION

The theory of evolution has never had as strong a position in biology as during the last 10 years. Phylogeny or the establishment of trees of historical relationships between biological taxa is now seen as an integral part of evolutionary biology, and cladograms are now omnipresent, not only in morphology but also in ecology and molecular biology. This is contrary to the situation only 20 years ago, when phylogenetic research was commonly seen as a 'subjective', non-testable enterprise and largely dismissed as irrelevant in a modern scientific discipline. Both evolutionary biology and phylogenetic research attempt to analyze a sequence of historical events, and two eminent biologists, the late Stephen Jay Gould and Frederick R. Schram, have each in their way strongly advocated the importance of the paleontological perspective and the significance of fossil data. Yet, both these scientists have during their careers entertained both a 'traditionalist view' on paleontology and evolutionary biology and a physics-inspired ideal of a 'model-and-objectivity-based approach'. Interestingly, the change of views took place near simultaneously but in opposite directions. In this essay we will describe our personal impression of how this change took place. We wish to compare neither persons nor approaches in terms of what is correct or best. Instead we want to emphasize an old point from science studies, namely that the methods and corroborating arguments used by scientists are forged by their respective discipline, their personal background, and the sociological-scientific context in which they operate (for an example of this concerning a well-known scientist, see Desmond & Moore 1991).

2 PALEONTOLOGY AND THE PHYSICAL IDEAL

Already Darwin saw the virtual absence of intermediate forms as a problem for his theory of descent and he dedicated a whole chapter in *On the Origin of Species* to this problem, arguing that the available fossil remains were too insufficient to provide a reasonably accurate picture of biological evolution. Since then, this view has troubled paleontologists who attempted to join discussions in evolutionary theory (Darwin 1859: p. 279ff; Stanley 1979: p. 4).

Another stumbling block has been the classic physical ideal of science, i.e., on the nature of science and how to perform scientific research. (We are aware of the rather divergent meaning of the word 'science' in the Anglo-Saxon and the continental European tradition. Here, we use science as an equivalent to natural science, although we basically disagree that a clear distinction between this discipline and other scientific pursuits is either logical or sound.) There are several historical sources for this ideal, and it has perhaps been most clearly articulated by the philosopher Immanuel Kant in his *Kritik der Urteilskraft* (Kant 1790). To Kant, the physicist Isaac Newton epitomized the exemplary model describing how science should be performed. Kant, in his work on the metaphysical basis of the sciences, attempts to present his personal view of a Newtonian basis of natural science. He argues that a scientific insight must be based on a physical ontology in which the universe consists of matter and motion and can be perceived by means of mechanical causes formulated as mathematical laws of nature. Accordingly, Kant declares that a doctrine about the natural world can only contain true and valuable science to the extent that it contains mathematics (Kant 1790; Zammito 1992: pp. 191+207).

In the 19th and 20th century, the physical ideal can be found more or less explicitly in numerous papers and other forms of public statements from scientists and philosophers of science. It underlies the attempt of the classical positivist philosopher, Auguste Comte, to create a framework for a unified science, with the creation of general laws as the common goal, be it in physics or sociology (Comte 1856: p. 24). It appears again in the famous remark by the physicist Ernest Rutherford that there exists but one kind of science, this being physics, while the rest is "stamp collecting" (Mayr 1982: p. 33). It is, therefore, hardly surprising that in modern biological textbooks the physical ideal is still used to justify the claim that scientific explanations should be formulated in terms of mechanical causes (e.g., Vander et al. 1994: p. 2).

Thus, in the hierarchy of scientific disciplines, as seen by the traditional positivists, physics and chemistry are placed highest, while paleontology and geology are barely mentioned at all. In fact, in Comtes' own version of this hierarchy the sequence is physics, chemistry, biology, sociology. But there is hardly any doubt that the proponents of this view of science would place paleontology (and perhaps geology) very low on the list and definitely below biology. In the controversy about the great mass extinction at the Cretaceous-Tertiary boundary, the founder of the meteorite impact theory, the physicist Louis Alvarez, took every opportunity to use this hierarchy against his paleontologist opponents. In an interview with the *New York Times*, he accordingly characterized them as "stamp collectors" (Gould 1989: p. 281), obviously hinting that their arguments were not truly 'scientific'.

Lacking a prominent status as a discipline in its own right, paleontology has been compelled to see itself and function as a sub-discipline of geology, where it has especially contributed to the development of methods for correlating stratigraphic layers (Stanley 1979: p. 5). Since the late 1960s there has been a growing dissatisfaction with this situation among paleontologists and several of them have, therefore, tried to establish a stronger platform for their discipline in evolutionary biology. These efforts have primarily come from Anglo-Saxon paleontologists, initially American - but later also British, who launched a number of rather successful attempts to make paleontology more visible in the eyes of the general public, e.g., dinosaur research. Not surprisingly they correlated temporally with the financial cuts that hit American science under the Nixon administration in the 1970s and British natural sciences under Thatcher in the 1980s (F.R. Schram, pers. comm.). Paleontologists were already placed at the lower end of the scientific hierarchy, and before the blooming 'dinosaur industry' they had very limited commercial assets at their disposal. The threat of cutting positions and funds must accordingly have been most strongly felt in paleontology and was undoubtedly a motivating factor in trying to change the status of their discipline.

From a sociological perspective we can see these efforts as different political strategies in research policy to counter the threat of budget cuts. Compared to a traditional positivist or physically inspired ideal for science, they can be organized in two main groups. One strategy attempts to demonstrate that paleontology follows and lives up to the same values as do other disciplines in science such as objectivity and testability, and, like physics, strives for general principles and laws. The other strategy rejects the physical ideal as a universal measure for science and argues for the special nature of paleontology as a 'historical discipline'.

3 GOULD AND PALEONTOLOGY

Gould was an exponent of both these strategies and they are excellently illustrated in the way his views on evolutionary theory developed from the early 1970s onwards. Through the 1970s and into the early 1980s Gould strongly advocated a model-based approach to evolutionary history and argued that the 'new paleontology' had its own general historical principles or 'laws'. His and Eldredge's famous theory on *punctuated equilibria* was accordingly an attempt to propose exactly such a general principle for changes in the rate of evolution. The link between the theory of punctuated equilibria and the wish to bring a model-based approach into paleontology becomes evident when considering that Eldredge & Gould's paper appeared in the anthology *Models in Paleobiology* (Schopf 1972a). As the title may suggest, the anthology aimed to demonstrate the relevance of models in paleontological work and thereby counter the traditional and very empirically oriented tendency to merely collect and organize fossils into taxonomic groups without involving more general theoretical considerations. Whereas modern taxonomy of extant organisms is pervaded by a model-based cladistic approach, this empiricism is still very common in paleontology. The approach to paleobiology exemplified by (Schopf 1972a) clearly reveals an underlying ideal for scientific work inspired by biology, and the central position of model-based research forms part of the attempt to 'biologize' paleontology and liberate it from being only a technical method or subdiscipline of geology. *Models in Paleobiology* (Schopf 1972a) was first and foremost focused on invertebrate paleontologists, whose biological training is explicitly described as wanting in comparison with vertebrate paleontologists and paleobotanists (Schopf 1972b: p. 5). In the following decade Gould himself followed such a model-based approach in several papers (e.g., Gould 1980a+b).

The later Gould took the opposite direction in his attempts to argue for the special status of paleontology as a historical discipline of central importance for evolutionary biology. In this phase, and notably in the immensely popular book *Wonderful Life* (Gould 1989), he introduced the concept of 'contingency' to convey the idea that unique historical events are central for understanding evolutionary history (Gould 1989: p. 284). For the later Gould, the primary path to cognition in evolutionary biology is to be found in the historical 'narrative' that ties together particular single and unique events into a causal-historical sequence rather than searching for general principles or 'laws' (although such an approach was never entirely dismissed or even discontinued).

In this approach, the key contribution of paleontology to understanding evolution is to support exactly this narrative approach with documentation from historical biology. There is a historical irony in this. The later Gould, who had fought against the picture of paleontologists as 'stamp collectors', now openly embraced a position, of which the primary paleontological contribution to evolutionary biology lies exactly in such accurate historical biological documentation. In this view on the role of paleontology in understanding the theory of evolution, Gould joined a position that largely resembles the classical empirical approach in this discipline.

The principal difference between the traditional empiricism in paleontology and the approach taken by the later Gould's approach lies in the status each of them ascribes to research in natural history compared to a positivistically inspired ideal of how to perform science. The conventional view among paleontologists is apparently to accept the concept of being 'stamp collectors' and their discipline as having a low rank in the hierarchy of science (Stanley 1979). Opposed to this, both the early and the late Gould tries actively to remove this 'inferiority complex' by emphasizing that natural history has 'positive' program of its own, which is just as 'scientific' as the experiment-based disciplines in science (Gould 1989: p. 280). In this pursuit, he attempts to liberate paleontology from being the underdog in the hierarchy of scientific disciplines and to dissolve this hierarchy altogether as being without real substance. In this 'polished up natural history', the stamp collector suddenly becomes a positive metaphor, applicable to paleontologists who focus on details and, thereby, on revealing the nooks and corners in biological evolution. It is for this reason that the work of the 'Burgess Shale'-group, such as Derek Briggs and his meticulously detailed reconstructions of these fossils (e.g., Briggs 1978), fitted Gould's revised vision of paleontology (Gould 1989). In *Wonderful Life* and numerous later publications and debates, Gould advocates a view of a scientific 'stamp collector' wholly unlike the 'arrogant physicist or chemist'. This 'stamp collector' is quite aware that a universal theory will never explain everything in the natural world, but that real understanding of nature is first revealed in single concrete details.

Gould's revised view on historical biology is clearly revealed in the way he reacted to new methods and approaches such as cladistics, especially computer cladistics based on large data sets, which rapidly gained ground after the publication and widespread popularity of *Wonderful Life*. Throughout the 1970s and 1980s, cladistics had been largely an elitist approach, practiced by a few, although often very outspoken, biologists (Hull 1988). The latter half of the 1980s saw the simultaneous appearance of cheap but powerful desktop computers and the advent of PCR techniques. The entailed accumulation of molecular sequence seemed perfect for application in phylogenetics. Thus, the molecular biologists could present the data, the systematists the method of choice, and the 'silicon revolution' the computer power. The combined effect was that cladistics became all pervasive in evolutionary biology within less than a decade. Yet, Gould never saw cladistics as the real cornerstone with which to understand animal evolution, although he certainly accepted its analytical power. Several prominent zoologists have suggested to us that Gould failed to use cladistics, because he never really grasped what the method was all about, but we disagree with this view. We believe that Gould, like Ernst Mayer, fully comprehended the nature of cladistics. But he never used it himself, and we suspect this was because cladistics embodies exactly the kind of generalized theory that neglects the details in historical biology; details that Gould found all important. It is very interesting that following the publication of *Wonderful life* Gould entered a vigorous debate with Derek Briggs over the interpretation of the Burgess Shale fossils and the methods to use in historical biology (Briggs et al. 1992a+b; Foote & Gould 1992). This happened when Briggs began to use the enormous information compiled from the Burgess Shale studies in numerical cladistics and other quantified and 'testable' analyses of patterns and processes in Early Cambrian animal evolution (e.g., Briggs & Fortey 1989).

It is also noteworthy that Gould never embraced another detailed source of information from the Cambrian, viz., the fauna of 'Orsten' microfossils, which Waloszek (= Walossek) & Müller described in numerous monographs, several appearing before *Wonderful Life* was published and rivaling the Burgess Shale studies in their attention to detail (e.g., Müller & Walossek 1985a+b, 1986, 1987, 1988). While the Burgess Shale fossils appeared very difficult or impossible to classify into recent groups, Waloszek & Müller have repeatedly argued that many Orsten arthropods represent early members of well known modern groups, and this picture remains even in a very recent analysis (Schram & Koenemann 2004). So while the Orsten fauna was presented in a gradualist pattern, the Burgess Shale, at least in the original presentations, fitted perfectly with Gould's ideas of contingency and punctuated equilibria in biological evolution. We know that Waloszek met Gould face to face (Waloszek, pers. comm.), but the Orsten fossils may not have caught Gould's interest for the evolutionary problem at hand. In fact, Waloszek & Müller's gradualist presentation of the Orsten fauna could even be used as an argument against Gould's ideas.

4 SCHRAM, PHYLOGENETICS AND PALEONTOLOGY

The changes in Gould's attitudes toward historical biology, and the role of models in paleontology can be compared with the changes in F.R. Schram's view on almost the same issues. Schram's early career took place in a scientific environment that largely adhered to the Anglo-Saxon tradition of attention to morphological detail. Opposed to the influence of the continental and German schools, this tradition largely avoided the construction of phylogenies as being speculative and without any firm basis in empirical observations. In Schram's own area, arthropods, this view was strongly advocated by Sidnie Manton, whose dislike of speculative phylogenetic diagrams led to her highly influential claim that virtually all major arthropod groups have evolved independently from a 'pre-arthropod stock' (Manton 1977). His formal training aside, we see Schram not as a pure neontologist or a paleontologist but as a prominent advocate for the view that all data sources are necessary and important in evolutionary biology. Initially, Schram's contributions were primarily in the zoology of crustacean arthropods, with focus on the phylogeny of this group. He was one of the founding fathers, and an early president, of *The Crustacean Society,* and in the first issue of the society's *Journal of Crustacean Biology,* he presented a new classification of the Eumalacostraca. He illustrated his paper with diagrams that resembled cladistic trees but were in fact not (Schram 1981). A few years later he did in fact publish a fully fledged cladistic analysis of the Crustacea Malacostraca (Schram 1984), a whole book on Crustacea (Schram 1986) and the edited version of the famed 'Meglitsch textbook' *Invertebrate Zoology* (Meglitsch & Schram 1991). These two books are pioneering works in animal phylogeny. In *Crustacea* computer cladistic methods were used for the first time to reconstruct a phylogeny for this taxon, whilst *Invertebrate Zoology* presented the first use of this method for the phylogeny of the all Metazoa. Yet, in *Crustacea,* Schram's heritage from the Anglo-Saxon school of arthropod evolution is still very apparent. The approach in the chapters on specific taxa and in the concluding phylogeny section was innovative and very controversial to many colleagues (it still is even today), but in the introductory chapter *What are Crustaceans?,* Schram reveals himself as a follower of the tradition of Sidnie Manton, who considered arthropods, including Crustacea, to have evolved convergently. This is opposed to the cladists' adherence to the principle of parsimony: similarities between taxa must count as homologies until otherwise proven. We believe that the introductory chapter in *Crustacea* illustrates a brief period in Schram's scientific career. During this period he developed his new view on analyses of animal relationships in accordance with the physical and model-based ideal of science, while still carrying a 'heavy load' from the long British and American tradition that rejected exactly such an approach with respect to arthropods. One could argue that the methods of Manton and her followers are in fact more 'model-based' than cladistics, because their interpretations are founded in numerous more or less explicitly stated assumptions on how evolution proceeds. With 'model' we refer to the very strict methodology set out in especially computer cladistics as opposed to the approach of Manton, where theories are often based on a wealth of detailed information but not easily open to critique. A recent debate between a 'cladist' and a 'Mantonian' approach to crustacean (branchiopod) phylogeny illustrates this point very clearly (Fryer 1987a+b, 1999, 2001, 2002; Olesen 1998, 2001, 2002).

In the years following, Schram rapidly developed into a front figure in large scale phylogenetic research, and hardly any paper is published without reference to some of his works. Moreover, Schram is now one of the foremost advocates of a 'physical ideal'-based approach to these problems. At international symposia and in discussions he repeatedly argued against the practice of more traditional 'Hennigian' cladists, in which a comparatively small number of very accurately analyzed characters are used in creating phylogenetic diagrams. He very explicitly advocates a 'pattern before process' and 'total evidence approach', where all available data are used to arrive at the most parsimonious solution. This 'transformed', or 'pattern' cladism exemplified by Schram maintains an ideal of a purely data-based classification that closely approaches a 'Francis Baconian' empirism anchored in 'naked facts'. As with the pheneticists, we also here perceive the physical ideal, although transformed cladists extend it with a 'Popperian' view. By adhering to the quantity of data and the pursuit of a classification, transformed cladism directly supports the semi-automated, computerized analyses of data in current molecular-based cladistics, characterized by large data sets. It is hardly surprising that most 'transformed cladists', who often dislike being called by this name, are more positive towards computerized analyses of large data sets than traditional morphologists such as Waloszek, whose work is based on 'in-depth' analyses of comparatively few characters (e.g., Waloszek 2003; Walossek & Müller 1997, 1998). Those following the latter tradition consider it as a primary aim to analyze and explain the evolutionary history of single characters or character complexes, which are used to construct diagrams of relationship ('pattern with process' approach). In this research program, molecular data and large morphological matrices contribute comparatively little because they cannot be unambiguously coupled to single morphological structures under examination. But Schram does not reject or ignore the data compiled by 'Hennigian phylogeneticists'. He may not have agreed with the interpretations put forward by the 'Burgess Shale group' or the 'Orsten group', but he nevertheless used all their available data sets in his own large scale cladistic analyses (Schram & Hof 1998; Schram & Koenemann 2004). In a discussion concerning Schram and Hof (1998) he even said that the authors "decided to let the characters of the Orsten arthropods enter the data matrix as put forward by Waloszek himself just to see how the cards would play" (paraphrased from pers. comm. to J.T. Høeg). In doing so, rather than recode the characters of the Orsten fossils to their own preference, Schram and Hof (1998) undoubtedly felt that they would achieve a more 'objective' phylogenetic analysis. And here we see yet another angle to Schram's approach. While most proponents of the 'physical ideal' would maintain that their method provides a gradual approach towards some kind of 'truth' about the natural world, Schram states that all we can really do is "organize information". However, in doing this, we have to adhere to some generally accepted principles such as cladistic methodology and the total evidence approach.

5 CONCLUSION

We have portrayed Gould's change from advocating the 'physical ideal' into largely rejecting it in favour of a historical approach in evolutionary biology based on the principle of contingency. By comparison Schram developed from a supporter of the Mantonian school of 'phylogenetic grass' into one of the strongest advocates of modern cladistics, thereby embracing a scientific practice based on the physical ideal that was rejected by the later Gould. Yet, both scientists agree on the importance of the fossil data sets, however fractional and imperfect they may be. In this, Schram deviates from many other pattern cladists, since he always uses whatever fossil evidence is at hand as primary input for his analyses. He even extends this approach to model building. In his controversial Arthropod Pattern Theory (APT) Schram compares the recently found and very unique crustacean class Remipedia with the Carboniferous fossil *Tesnusocaris* (Emerson & Schram 1990, 1997) and arrives at a revolutionary interpretation of arthropod segmentation and limb homology. Here, Schram approaches Gould in not dismissing a hard-to-analyze fossil as largely irrelevant, but using it as a foundation for a new theory.

Still, Schram maintains his ideal of objectivity. When asked by us whether he still believes in his APT theory, he replied that "It has been put forward and it is there to use". He apparently sees little purpose in wasting energy on strongly defending past efforts and would rather press on to new frontiers of his science. Perhaps it is here that we see the deeper value of Schram's contributions to evolutionary biology. The 'unorthodox' inspiration he gives to scientists young and old with bold new ideas.

Finally, we wish to illustrate one surprising aspect where the attitudes of Gould and Schram approached each other. At the 4th International Crustacean Congress in Amsterdam (1998), the late Jack Sepkoski was a prominent keynote speaker, we surmise on the invitation of the chairman, F.R. Schram. Sepkoski, a student of Gould, has been a strong proponent of a quantitative, model-based approach to estimating extinction rates and their causes in animal evolution, and many of his papers have a very mathematical content (e.g., Raup & Sepkoski 1988; Sepkoski & Kendrick 1993; Sepkoski & Miller 1998; Plotnick & Sepkoski 2001). It is hardly surprising that Schram welcomed a talk by this model-and-objectivity oriented scientist, especially in a meeting that sported the subtitle *Crustacea and the Biodiversity Crisis*. Yet, on the occasion of Gould's visit to a meeting at the *Carlsberg Academy* in Copenhagen in the fall of 2001, one of us (J.T. Høeg) discussed Sepkoski's work with him and learned that Gould considered Sepkoski his best student ever. While the late Gould redirected his focus away from mathematically oriented approaches to the study of evolution, this evidently did not tarnish his appreciation of Sepkoski's contributions to paleontology.

In the last 10 to 15 years, both Gould and Schram have featured prominently in discussions on animal evolution. From the publication of *Wonderful Life* and onwards, Gould is more or less explicitly hostile to model-based, so-called objective approaches in historical biology, and he never in his own work embraced computer cladistics as the method of choice to analyze biological evolution. An informal discussion with one of us (J.T. Høeg) in late 2001 left the impression that Gould doubted that the diagrams resulting from computer cladistics gave any truly important insight into the evolutionary events in the Cambrian. In contrast, the early Gould was attempting to put paleontology higher in the hierarchy of science by formulating theories of evolution according to the physical ideal of science. Schram arrived fairly late at accepting and advocating computer cladistics. But contrary to entomology and vertebrate taxonomy, cladistics came very late to the field of crustacean systematics, where Schram in fact became a pioneer in using this approach. A turning point is illustrated in his book *Crustacea* (Schram 1986), the first publication to use computer cladistics, or cladistics in general, on that taxon as a whole. Nevertheless, he still, at least partly, adhered to the 'Mantonian' view of arthropod polyphyly, thereby in effect violating the principle of parsimony and objectivity, despite the massive amount of characters arguing against convergence. Schram & Koenemann (2004) discuss how such "straightjackets of the past" in terms of long standing assumptions can seriously bias even outwardly 'objective' analyses. In the last 10 years, few have rivaled Schram in advocating parsimony and computer cladistics, and he is openly suspicious of approaches in phylogeny and evolution that are overly focused on a few very detailed characters while missing the remaining data set. Thus Gould and Schram both have changed their approaches rather fundamentally, while, on the other hand, both scientists consistently concurred with the importance of paleontological data (e.g., Schram & Koenemann 2004). We venture that both of them would describe these changes as a logical development of their views and approaches to science. The battles to be won in paleontology and evolutionary biology differed in the 1970s and early 1980s from those of more recent years and therefore, also the scientific methods of choice. To us, this illustrates that natural science and the means of pursuing scientific research cannot be separated from the people pursuing it.

ACKNOWLEDGEMENTS

First and foremost we acknowledge both 'Fred' Schram and the late Stephen Jay Gould for lengthy and very open-minded, and most enjoyable discussions on their views on science. We also thank all members of the 'Crustacea Copenhagen' group for inspiring discussions and critique. Furthermore, the late Niels Møller Andersen, Niels Peder Kristensen, Niels Bonde, Claus Emmeche, Henrik Glenner, Reinhardt Møbjerg Kristensen and Claus Nielsen have each in their way contributed crucially to make zoology at the University of Copenhagen open to discussions on the history, philosophy and sociology of science. Finally, we acknowledge the financial support to Jens T. Høeg from the Danish Natural Science Research Council and the Carlsberg Foundation.

REFERENCES

Briggs, D.E.G. 1978. The morphology, mode of life, and affinities of *Canadaspis perfecta* (Crustacea, Phyllocarida), Middle Cambrian, Burgess Shale, British Columbia. *Phil. Trans. R. Soc. London, B,* 281: 439-487.
Briggs, D.E.G. & Fortey, R.A. 1989. The early radiation and relationships of the major arthropod groups. *Science* 246: 241-243.
Briggs, D.E.G., Fortey, R. & Wills, M.A. 1992a. Morphological Disparity in the Cambrian. *Science* 256: 1670-1673.
Briggs, D.E.G., Fortey, R. & Wills, M.A. 1992b. Cambrian and Recent Morphological Disparity [Reply to Foote & Gould 1992]. *Science* 258: 1817-1818.
Comte, A. 1856. *A General View of Positivism* [*Synthèse subjective ou système universel*]. Dubuque, Iowa: Brown Reprints, 1971 [translated from French].
Darwin, C. 1859. *On the Origin of Species*. London: John Murray.
Desmond, A. & Moore, J. 1991. *Darwin*. London: Penguin Books.
Eldredge, N. & Gould, S.J. 1972. Punctuated Equilibria: an alternative to phyletic gradualism. In: Schopf, T.J.M. (ed.), *Models in Paleobiology*: pp. 82-115. San Francisco: Freeman, Cooper & Company.
Emerson, M.J. & Schram, F.R. 1990. A novel hypothesis for the origin of biramous limbs in arthropods. In: Mikulic, D.G. (ed.), *Arthropod Paleobiology: Short Courses in Paleontology No 3*: pp. 157-76. Knoxville: Univ. Tennessee.

Emerson, M.J. & Schram, F.R. 1997. Theories, patterns and reality: game plan for arthropod phylogeny. In: Fortey, R.A. & Thomas, R.H. (eds.), *Arthropod Relationships*: pp. 67-86. London: Chapman & Hall.

Foote, M. & Gould, S.J. 1992. Cambrian and Recent Morphological Disparity [Reply to Briggs et al. 1992a]. *Science 258*: 1816.

Fryer, G. 1987a. A new classification of the branchiopod Crustacea. *Zool. J. Linn. Soc.* 91: 357-383.

Fryer, G. 1987b. Morphology and the classification of the so-called Cladocera. *Hydrobiologia* 145: 19-28.

Fryer, G. 1999. A comment on a recent phylogenetic analysis of certain orders of the branchiopod Crustacea. *Crustaceana* 72: 1039-1050.

Fryer, G. 2001. The elucidation of branchiopod phylogeny. *Crustaceana* 74: 105-114.

Fryer, G. 2002. Branchiopod phylogeny: Facing the facts. *Crustaceana* 75: 85-88.

Gould, S.J. 1980a. The promise of palaeobiology as a nomothetic discipline. *Paleobiology* 6: 96-118.

Gould, S. J. 1980b. Is a new and general theory of evolution emerging? *Paleobiology* 6: 119-130.

Gould, S.J. 1989. *Wonderful Life: The Burgess Shale and the Nature of History*. New York: W.W. Norton & Company.

Hull, D.L. 1988. *Science as a Process: An Evolutionary Account of the Social and Conceptual Development of Science*. London; Chicago: Univ. Chicago Press.

Kant, I. 1790. *The Critique of Judgment* [*Kritik der Urteilskraft*]. Oxford: Clarendon Press, 1988 [translated from German].

Manton, S.M. 1977. *The Arthropods: Habits, Functional Morphology, and Evolution*. Oxford: Clarendon Press.

Mayr, E. 1982. *The Growth of Biological Thought: Diversity, Evolution, and Inheritance*. Cambridge (MA), London: The Belknap Press of Harvard Univ. Press.

University Press

Meglitsch, P.A. & Schram, F.R. 1991. *Invertebrate Zoology*: pp. 1-623. New York: Oxford Univ. Press.

Müller, K.J. & Walossek, D. 1985a. Skaracarida, a new order of Crustacea from the Upper Cambrian of Västergötland, Sweden. *Fossils & Strata* 17: 1-65.

Müller, K.J. & Walossek, D. 1985b. A remarkable arthropod fauna from the upper Cambrian "Orsten" of Sweden. *Trans. R. Soc. Edinburgh, Earth Sci.,* 76: 161-172.

Müller, K.J. & Walossek, D. 1986. *Martinssonia elongata* gen. et sp. n., a crustacean-like euarthropod from the Upper Cambrian 'Orsten' of Sweden. *Zoologica Scripta* 15: 73-92.

Müller, K.J. & Walossek, D. 1987. Morphology, ontogeny, and life habit of *Agnostus pisiformis* from the Upper Cambrian of Sweden. *Fossils & Strata* 19: 1-124.

Müller, K.J. & Walossek, D. 1988. External morphology and larval development of the Upper Cambrian maxillopod *Bredocaris admirabilis*. *Fossils & Strata* 23: 1-70.

Olesen, J. 1998. A phylogenetic analysis of the Conchostraca and Cladocera (Crustacea, Branchiopoda, Diplostraca). *Zool. J. Linn. Soc.* 122: 491-536.

Olesen, J. 2000. An updated phylogeny of the Conchostraca-Cladocera clade (Branchiopoda, Diplostraca). *Crustaceana* 73: 869-886.

Olesen, J. 2002. Branchiopod phylogeny: Continued morphological support for higher taxa like the Diplostraca and Cladocera, and for paraphyly of 'Conchostraca' and 'Spinicaudata'. *Crustaceana* 75: 77-84.

Plotnick, R.E. & Sepkoski, J.J. 2001. A multiplicative multifractal model for originations and extinctions. *Paleobiology* 27 (1): 126-139.

Raup, D.M. & Sepkoski, J.J. 1988. Testing for periodicity of extinction. *Science* 241 (4861): 94-96.
Schopf, T.J.M. (ed.) 1972a. *Models in Paleobiology*. San Francisco: Freeman, Cooper & Company.
Schopf, T.J.M. 1972b. Introduction: About This Book. In: Schopf, T.J.M. (ed.), *Models in Paleobiology*: pp. 3-7. San Francisco: Freeman, Cooper & Company.
Schram, F.R. 1981. On the classification of Eumalacostraca. *J. Crust. Biol.* 1: 1-10.
Schram, F.R. 1984. Relationships within eumalacostracan crustaceans. *Trans. San Diego Soc. Nat. Hist.* 20: 301-312.
Schram, F.R. 1986. *Crustacea*: pp. 1-606. Oxford: Oxford Univ. Press
Schram, F.R. 1993. The British School: Calman, Cannon and Manton and their effect on carcinology in the English speaking world. In: Schram, F. & Truesdale, F. (eds.), *Crustacean Issues 8, History of Carcinology*: pp. 321-348. Rotterdam: A.A. Balkema.
Schram, F.R. & Hof, C.H.J. 1998. Fossils and the interrelationships of major crustacean groups. In: Edgecombe, G.D. (ed.), *Arthropod Fossils and Phylogeny*: pp. 233-302. New York: Columbia Univ. Press.
Schram, F.R. & Koenemann, S. 2004. Are crustaceans monophyletic? In: Cracraft, J. & Donoghue, M.J. (eds.), *Assembling the Tree of Life*: pp. 319-329. New York: Oxford Univ. Press.
Sepkoski, J.J. & Kendrick, D.C. 1993. Numerical experiments with model monophyletic and paraphyletic taxa. *Paleobiology* 19 (2): 168-184.
Sepkoski, J.J. & Miller, A.I. 1998. Analysing diversification through time. Trends in Ecology and Evolution 13 (4): 158-159.
Stanley, S.M. 1979. *Macroevolution: Pattern and Process*. New York: W.H. Freeman & Company.
Vander, A.J., Sherman, J.H. & Luciano, D.S. 1994. *Human Physiology*. New York: McGraw-Hill, Inc.
Waloszek, D. 2003. Cambrian "Orsten"-type preserved arthropods and the phylogeny of Crustacea. In: Legakis, A., Sfenthourakis, S., Polymeni, R. & Thessalou-Legaki, M. (eds.), *The New Panorama of Animal Evolution; Proc. XVIII Int. Congr. Zool.*: pp. 69-87. Sofia; Moscow: Pensoft Publ.
Walossek, D. & Müller, K.J. 1997. Cambrian "Orsten"-type arthropods and the Phylogeny of Crustacea. In: Fortey, R.A. & Thomas, R.H. (eds.), *Arthropod Relationships*: pp. 139-153. London: Chapman & Hall.
Walossek, D. & Müller, K.J. 1998. Early arthropod phylogeny in the light of the Cambrian "Orsten" fossils. In: Edgecombe, G.D. (ed.), *Arthropod Fossils and Phylogeny*: pp. 185-231. New York: Columbia Univ. Press.
Zammito, J.H. 1992. *The Genesis of Kant's Critique of Judgment*. Chicago: Univ. Chicago Press.

II PALEOZOOLOGY

Decapod crustaceans, the K/P event, and Palaeocene recovery

CARRIE E. SCHWEITZER[1] & RODNEY M. FELDMANN[2]

[1]*Department of Geology, Kent State University Stark Campus, Canton, Ohio, U.S.A.*
[2]*Department of Geology, Kent State University, Kent, Ohio, U.S.A.*

ABSTRACT

Recent work on fossil decapod Crustacea has provided a sufficiently large data set to facilitate a synthesis of the effect of the proposed end-Cretaceous extinction event(s) on the group in the North Pacific Ocean, the Central Americas, and the Southern Hemisphere, as well as on the global occurrences of eight families that have been recently revised by the authors. Of the 38 Late Cretaceous decapod families known from the data set as defined herein, 79% survived into the Palaeogene and 66% are extant. Only three of the families that became extinct during the Late Cretaceous have confirmed Maastrichtian records, and one of those is last known from lower Maastrichtian occurrences. Seventy percent of the Late Cretaceous genera in the data set became extinct during the Late Cretaceous; however, only about one-third had last occurrences during the Maastrichtian. Because of the high level of survivorship of families, these extinctions were pseudoextinctions; species within these families must have survived across the boundary. Of the survivors, some appear to have been preadapted survivors, ecological generalists, or refugium taxa, and some were protected by inhabiting buffered habitats in high-latitude areas. It is important to note that these features also confer survivability whether or not a mass extinction event occurred, hence the reason that many of those lineages are extant. Decapods inhabiting high-latitude areas survived preferentially, but notably, many genera survived into the Palaeogene in the tropical and subtropical Americas, proximal to the Chicxulub impact. Broad geographic range also conferred significant survival success on decapod genera. The Cretaceous, as well as the Eocene and Miocene, was a time of rapid evolutionary radiation within the Decapoda. Gouldian contingency may have played as significant a role in the survivorship of some decapod lineages as mass extinction events in causing selective extinctions.

1 INTRODUCTION

The effects of the Cretaceous/Palaeogene (K/P) event have been well-studied in a wide range of taxonomic groups (MacLeod & Keller 1996; Kauffman & Harries 1996; Harries et al. 1996; Harries 1999). Many recent works have summarized the K/P extinction event, its causes and effects, and the dynamics of the Cretaceous ocean and climate (among many others: MacLeod & Keller 1996; Zinsmeister & Feldmann 1996; Hart 1996; Barrera &

Johnson 1999; Culver & Rawson 2000). However, until recently, the macro-crustaceans have received relatively little attention with respect to this significant biotic event (see numerous works by the authors). Buckeridge (1999) described the history of barnacles across the Cretaceous/Palaeogene boundary in the Chatham Islands, New Zealand, and concluded that the balanomorph barnacles experienced a nearly explosive diversification following their appearance in the Palaeocene. Sepkoski (2000) provided a very broad study of crustacean biodiversity over the entire Phanerozoic, but did not extensively treat the Cretaceous-Palaeogene extinction, and Sheehan et al. (1996) suggested that arthropods have preferentially survived mass extinctions because of their detritus-feeding habit. Only recently have the Decapoda been examined for the effects of the end-Cretaceous event(s) that led to the extinction of many lineages of micro-organisms, vertebrates, invertebrates, and plants. These recent works have indicated that the end-Cretaceous extinctions among the Decapoda appear to have been less severe than in other groups (Collins & Jakobsen 1994; Feldmann et al. 1995; Schweitzer 2001; Schweitzer et al. 2002; Feldmann 2003; Fraaije 2003; Feldmann & Schweitzer accepted), especially at the family level, although generic level patterns are less clear. Herein, we synthesize the results of our previous studies to provide the first overview of the effects of the end-Cretaceous extinction event on the Decapoda. It should be noted that while we recognize that end-Cretaceous extinctions are well-documented among some organisms, it is far from clear whether the events or phenomena that precipitated those extinctions affected all organisms. It is clear that there is no demonstrated synchroneity of extinctions within or across taxonomic groups. Further, it is not clear that a bolide impact (Feldmann 1990); other global phenomena such as temperature changes, changes in ocean chemistry, etc.; or a combination of several events and/or phenomena caused the extinctions. In addition, we have tried to recognize and minimize the impact of the *a priori* assumption that an extinction 'event' occurred and to avoid attempting to see how well the Decapoda 'fit' this assumption.

2 DATA SET

The data set includes all families and genera in the regions previously summarized by us, including the Southern Hemisphere (Feldmann et al. 1995, 1997; Feldmann & Schweitzer accepted), the North Pacific Ocean (Schweitzer 2001; Schweitzer & Feldmann 2001b; Schweitzer et al. 2003), and the Central Americas as defined by Schweitzer et al. (2002: p. 39), which includes "the region of the southeastern United States, Mexico, the Caribbean, Central America, and northern South America". This includes the worldwide records of all those families and genera occurring in the geographic regions just mentioned. In addition, we have recently comprehensively treated or revised several families with Cretaceous records and have included all relevant fossil records in each; these include Galatheidae and Chirostylidae (Schweitzer & Feldmann 2000a), Hexapodidae (Schweitzer & Feldmann 2001a), Homolidae and Homolodromiidae (Collins 1997; Schweitzer et al. 2004), Necrocarcinidae and Orithopsidae (Schweitzer & Feldmann 2000b; Schweitzer et al. 2003), and Palaeoxanthopsidae (Schweitzer 2003). One or both of us have examined nearly every fossil decapod specimen from the Southern Hemisphere, most from the North Pacific and Central Americas, and many from Europe and the North Atlantic. We have examined the best specimens or illustrations obtainable for every taxon treated in family revisions. Thus, the data set is as tightly constrained and consistent as possible, because all generic and family designations have been confirmed by us. Generic occurrences (except for those in the families listed) in Europe, Asia exclusive of Japan, and North Africa that are not found in the geographic regions outlined above are not included. These records have not been personally confirmed by us, and we have purposely constrained this discussion only to those records personally studied intensively and confirmed by us.

Inherent problems with the data set exist. The record of Palaeocene decapods is not robust, which certainly limits our ability to determine whether or not genera crossed the K/P boundary. Some genera may in fact have survived across the boundary, but their record is unknown due to the paucity of Palaeocene fossils. Lack of records and, in some instances, lack of fine-scale temporal control on Palaeogene occurrences, also limit our ability to determine when during the Palaeogene genera arose. Some genera may have arisen directly in the wake of the K/P extinction due to selective pressures resulting from a stressed environment, but because we lack fine-scale temporal control on fossil occurrences, this is difficult or impossible to ascertain. In addition, many Eocene records are known, but some of those genera may have originated in the Palaeocene or even the Late Cretaceous. As more Palaeocene decapods are being recovered and described, the geologic range of genera and families previously known only from Eocene or younger rocks is being extended into the Palaeocene (for example, the Panopeidae, in Karasawa & Schweitzer, 2004). Nevertheless, many of the Palaeocene occurrences are known to have been Danian. These taxa appear to be the survivors that led to subsequent radiations among their respective lineages.

We acknowledge that fossil decapods are relatively rare as compared to other groups; however, it has become clear that the record is sufficiently robust and decapod systematics has matured sufficiently to make global conclusions about the decapod record possible (Feldmann 1986; Schweitzer 2001; Schweitzer et al. 2002; Feldmann 2003; Feldmann & Schweitzer accepted). We consider the Decapoda an excellent group to test the hypothesis that a mass extinction occurred at the end of the Cretaceous, because the group was reasonably diverse; geographically widespread; and its members inhabited a broad variety of niches and habitats including a range of substrates, depths, and latitudinal settings.

Among the Arthropoda, the Decapoda are certainly the best of the macroscopic organisms for this type of study. Because they are largely marine organisms, they have a more robust fossil record than, for example, the insects, chelicerates, and myriapods. The marine microscopic arthropods, including the ostracods, may have an excellent fossil record, but the methods of collecting and studying them require totally different techniques than those used for the larger invertebrates. Thus, this task must be reserved for micropaleontologists. In addition, we believe that because of their well-constrained fossil record, the K/P record of the Decapoda can serve as a useful model against which to test the response of other arthropod groups to the mass extinction events during this time interval.

Because the fossil record of the decapods is less robust than that for other groups, such as the Mollusca, we consider generic and family level taxa within the Decapoda, recognizing of course that evolution and therefore appearance of new taxa occurs at the individual level, which is observable in the fossil record as at the species level. Thus, throughout this work, we refer to taxa rather than species, for example, 'refugium taxa'. Other workers, when discussing extinction and recovery patterns, refer to "species", as in Kauffman & Harries (1996), who used the term "refugia species".

Tracking the history of genera and families, rather than species, is based upon the notion that these taxonomic levels are sufficiently constrained to serve as legitimate proxies for species. Because nearly every decapod occurrence referred to in this work has been personally verified, we feel confident that the results are valid. Two systematic placements have been updated or changed herein. *Palaeoxantho*, which Bishop (1986) had placed in the Xanthidae, is herein assigned to the Palaeoxanthopsidae, and *Caloxanthus* is placed within the Dynomenidae, contrary to Schweitzer et al. (2002). In addition, we concur with the restricted definition of *Xanthosia* Bell, 1863, given by Guinot & Tavares (2001), which limits the genus to only Cretaceous occurrences and renders its last occurrence as Cenomanian.

3 FAMILY-LEVEL K/P DATA

Thirty-eight decapod families have been recorded in the data set in rocks of Late Cretaceous age or older (Appendix 1, Tables 1-4). Of these, 79% (30 families) crossed the K/P boundary, and 66% (25 families) are represented by extant genera. Of the 21% (8 families) that became extinct by the end of the Cretaceous, six families survived into the Late Cretaceous and only three (8% of the total number of families) have confirmed Maastrichtian records (Appendix 1, Table 3). Of those with Maastrichtian records, two are known from the Central Americas, specifically, in the area known to have been affected by the Chicxulub impact (Bishop et al. 1998; Feldmann et al. 1998). The Mecochiridae is not known above the Campanian-Maastrichtian boundary in Antarctica (Feldmann et al. 1993), and thus appears to have disappeared well before the end of the Maastrichtian. The last known occurrence of the Etyidae *sensu* Guinot & Tavares (2001) was of Cenomanian age, and the sole record of the Retrorsichelidae occurred during the Campanian. Thus, at the family level, the extinctions during the Cretaceous were distributed throughout much of the Late Cretaceous. In addition, there does not appear to have been a major geographic bias in extinctions at the family level; family-level extinctions are about evenly distributed among high latitude, tropical, and subtropical families.

Almost all of the families that originated during the Cretaceous or earlier were macruran (shrimps and lobsters), anomuran (hermit crabs, ghost shrimps, mud shrimps, mole crabs, porcelain crabs, and others), or primitive brachyuran (true crabs) decapods. This is not a new observation; it has long been known that the lobsters have a fossil record extending into the Devonian (Schram et al. 1978) that was robust during the Mesozoic. Shrimps have a fossil record extending from the Permo-Triassic to Recent (Glaessner 1969) and are known with certainty from Lower Triassic rocks of Madagascar (Van Straelen 1933; Garassino & Pasini 2002; Garassino & Teruzzi 1993, 1995). Similarly, many anomuran groups have robust Jurassic records (Van Straelen 1925; Glaessner 1969). Primitive true crabs only make a confirmed appearance in the Jurassic (Glaessner 1969) although there are questionable Mississippian and Permian occurrences (Schram & Mapes 1984; Hotton et al. 2002). A reported Triassic occurrence of a eubrachyuran from New Mexico (Rinehart et al., 2003) appears not to be a decapod. A small number of derived brachyurans appeared by the Cretaceous (8 families in the data set); however, most brachyurans have fossil records beginning in the Palaeocene or Eocene. A similar pattern of appearance of primitive and derived brachyuran lineages previously has been observed for specific geographic areas (Feldmann et al. 1995; Schweitzer 2001; Schweitzer et al. 2002).

4 GENERIC-LEVEL K/P DATA

Genera within surviving families often did not survive the K/P boundary. Sepkoski (1989) reported worldwide extinction rates at the generic level between 63 and 77% at the K/P boundary. Of the decapod genera known from the entire Late Cretaceous, 70% became extinct. However, only about one-quarter of those genera (23%) had their last occurrence in the Maastrichtian, significantly lower than the percentages reported by Sepkoski (1989). Although the so-called Signor-Lipps Effect (Signor & Lipps 1982) of apparent step-wise loss of taxa in a modeled instantaneous extinction has application in local sections, the effect should be mitigated by examination of all known records of taxa worldwide. Documentation of global disappearance of a taxon, based upon all available records, must be interpreted as an extinction and the time of extinction must be best approximated by the timing of the last occurrence.

Only 30% of the genera originating before the end of the Cretaceous survived across the boundary. Often, entirely new genera within a surviving family appeared in the Palaeocene or later, sometimes not until the Recent. Because the temporal records of the decapods are not constrained at as fine a level as in other groups, it is usually not possible to know exactly when the first post-Cretaceous taxon appeared. Typically, resolution to the level of stage is as precise as is possible, and it would be unrealistic to assign absolute age values to events as is done with mollusks (Harries 1999; Stilwell 2003).

The macruran and anomuran decapods exhibit relatively high levels of survivorship of genera across the boundary (Appendix 1, Tables 4, 5). The Penaeidae (1 surviving genus), Glypheidae (2 surviving genera), Erymidae (1), Nephropidae (3), Callianassidae (3), Palinuridae (1), Ctenochelidae (1), Upogebiidae (1), and Galatheidae (3) all embrace genera that survived across the boundary. Most of the remainder of the surviving genera belong to the primitive crab lineages, within the Podotremata, and include genera within the Dynomenidae (2), Homolidae (1), Homolodromiidae (1), Prosopidae (1), and Raninidae (3). Only five genera from more derived brachyuran lineages survived across the K/P boundary, within the Retroplumidae (2), Palaeoxanthopsidae (2), and Hexapodidae (1). This, however, should not be surprising, because most of the more derived lineages within the Brachyura appeared post-Cretaceous, in the Palaeocene or Eocene. The end-Cretaceous events and subsequent recovery dynamics may well have opened niche-space and spurred their appearance.

Unfortunately, the record of Palaeocene decapods is poor. In the data set, only 19 genera are documented as originating in the Palaeocene (Appendix 1, Table 6). Compounding the problem of constraining times of origin, the ages of these occurrences are not well known. Some are known from the Danian, but it is generally not known exactly how long after the K/P event these genera appeared. Thus, it is difficult to classify some of these taxa in the terminology of Kauffman & Harries (1996) and Harries et al. (1996), because we do not know if the genera arose in the direct wake of and as a result of the end-Cretaceous events.

4.1 Preadapted survivors

Several lineages within the Decapoda exhibit patterns typical of those described for preadapted survivors. Kauffman & Harris (1996) described preadapted survivors as being reduced in abundance during an extinction event and immediately during its wake, but eventually showing a major expansion in numbers and geographic distribution during the recovery phase.

Within the Galatheidae, eleven genera are known from Cretaceous or earlier rocks. Only a few of these survived into the Late Cretaceous. The two genera that are known from the Upper Cretaceous crossed the boundary, and three new genera appeared in the Palaeocene (Appendix 1, Table 5). Ultimately, the family experienced a major radiation in the post-Cretaceous, exhibited robust records in the Eocene and later records, and is well represented in the modern oceans.

Within the Raninidae, seven genera have confirmed records in Cretaceous rocks. Of these, three survived into the Palaeogene. Three new genera evolved in the Palaeocene. Thus, there was considerable turnover among the genera within the family. The genera that crossed the boundary belong to an extant subfamily while those that became extinct do not. It should also be noted that numerous genera within the family appeared during the Eocene and later, often in high-latitude areas (Feldmann 1991) and that the family is extant (Tucker 1998). Members of the Raninidae may have been well-adapted to survive harsh conditions due to their burrowing, benthic, detritus-feeding or scavenging habit (Tucker 1995, 1998).

Within the Dynomenidae, there are nine genera known from the Cretaceous or earlier, only two of which survived into the Palaeocene. Two genera questionably assigned to the family, *Kierionopsis* and *Dromilites*, arose during the Palaeocene, and the next confirmed records of the family are Eocene in age. The end-Cretaceous survivors in all three of these families may have been preadapted survivors, giving rise to subsequent genera, including many extant members.

4.2 Possible ecological generalists

The Homolidae exhibits a strong faunal turnover across the K/P boundary. Eight genera within the family are known from Cretaceous or earlier rocks, and of these, only one genus survived into the Palaeocene. Subsequent occurrences of the family were Eocene and later, including Recent genera. Interestingly, the K/P-surviving genus, *Latheticocarcinus*, is morphologically very similar to extant *Homola* and other genera. It may have been the founder for the extant lineage within the family. The Homolodromiidae also exhibits high faunal turnover; three genera are known from Cretaceous or earlier rocks and only one genus crossed the boundary. Subsequent generic appearances were Oligocene or later in this extant group. *Homolodromia*, the one genus that crossed the K/P boundary and which is extant, was likely the founder for subsequent taxa.

Both Homolidae and Homolodromiidae exhibit features typical of the ecological generalists described by Kauffman & Harries (1996). They possess primitive features within their lineage, and each embrace a single taxon that survived into the Palaeogene which exhibited morphological features suggesting ancestral affinities for subsequent taxa.

The Homolidae could equally be a refugium or preadapted survivor taxon. It is known from shallow, often epicontinental, environments early in its history (Schweitzer et al. 2004) and has subsequently moved to deep-water settings, where members of the family are found today. It is possible that this family shifted habitat preference in response to stressed environmental conditions during the end-Cretaceous event(s) or had expanded its habitat preference towards deep water prior to that time. The Homolodromiidae, on the other hand, may have been protected from extinction by early habitat preferences. It has occurred in middle- to high-latitude, typically deep-water environments throughout most of its history, which may have contributed to a relatively effective protection during extinction events (Schweitzer et al. 2004; Feldmann & Wilson 1988; Feldmann 1993; Harries et al. 1996).

4.3 *Possible refugium taxa*

A large gap in the fossil record exists for many decapod families, due primarily to the paucity of Palaeocene decapod fossils, so-called Lazarus taxa (Jablonski 1986). The general lack of Palaeocene rocks makes it probable that many decapod taxa have interrupted stratigraphic ranges; however, this is a condition of the fossil record and has no necessary implications for the intrinsic nature of the organisms. For example, within the Chirostylidae, there is one Cretaceous genus; the next record of the family is from the Eocene. Similarly, the Ctenochelidae display a gap between Cretaceous and Eocene records. Within the Aeglidae, there are two Cretaceous records and in the Micheleidae, there is one Cretaceous record; these are the only fossil records for these two extant families. These families were either refugium taxa or were protected by inhabiting a buffered habitat (Harries et al. 1996), because all four are known from temperate to high-latitude locations during the Cretaceous. High latitudes are hypothesized as having been less affected by the extinction event (Jablonksi 1996; Kauffman & Harries 1996). This hypothesis appears to hold for the Late Cretaceous, although it is recognized that ocean temperatures decreased in the high-latitudes during that time (Barrera & Savin 1999). Temperature change probably had a less severe effect on high-latitude organisms already adapted to lower or seasonal variations in temperatures. However, the lack of occurrences of these families after the K/P event and prior to the Eocene makes it nearly impossible to determine where the descendants survived.

It is important to note that these very same characteristics: inhabiting a refugium, eurytypy, features permitting taxa to be preadapted survivors, etc., all confer survivability whether or not a mass extinction event occurred. These very same features are likely what permitted nearly every one of the decapod lineages at the family level and many of the genera discussed in those terms herein to survive into modern times. Kauffman & Harries (1996) and Harries et al. (1996) may in fact have been describing attributes of long-lived lineages, not attributes restricted uniquely to conferring survivorship of mass extinction events. Jablonski (1996) suggested that features enhancing survival would not have been effective during mass extinction events; however, the decapods indicate that in fact many of these very features did permit them to survive well into the Cenozoic and even into the Recent.

4.4 Danian problem

High-latitude regions with well-exposed K/P sections have historically shown low levels of extinction across the K/P boundary (Kauffman & Harries 1996: p. 18), which was termed the "historical Danian problem". Faunas in Antarctica (Zinsmeister & Macellari 1988; Zinsmeister et al. 1989) and Alaska (Marincovich 1993) each exhibit Late Cretaceous and Danian faunas that show low extinction levels. The same pattern is seen in at least two other high-latitude decapod localities. Ekdale & Bromley (1984) reported little change in abundance of ichnofossils attributed to arthropods across the K/P boundary in Denmark, and Collins & Jakobsen (1994) reported that 66% of the decapod genera in Denmark and Sweden survived from the Maastrichtian into the Danian. Fraaije (2003) reported that over half of the genera known from the Maastrichtian type area survived into the Palaeogene. Similarly, in Maastrichtian and Danian rocks of southern Argentina, all of the genera present in the Maastrichtian crossed the K/P boundary (Feldmann et al. 1995). Although many of the high-latitude molluscan faunas apparently did not persist long in the Danian (Jablonski 1996), at least some of the decapod K/P survivors did persist well into the Cenozoic, some into the Neogene (Feldmann & Schweitzer accepted).

Alternatively, the K/P event may not have been as catastrophic as previously stated, and the extinctions that define the end of the Cretaceous may have occurred over a long period of time. The cause-and-effect relationship between the Chicxulub impact and the extinction of taxa is still being debated, and it is not at all clear that it was the sole, or even the primary, cause of the pattern of extinctions in the Late Cretaceous. Clearly, an impact would have a devastating effect locally, or even regionally, but the global effect may well be overstated. For example, investigation of the biotic effects of the Manson impact in Iowa, which occurred 74 ma based upon melt-precipitated feldspars (Izett et al. 1993), has led to the conclusion that there were few long-term biotic consequences and no global extinctions (Anderson et al. 1996). Although the magnitude of the Chicxulub impact was much greater than that of Manson, such wide variation in the effect of the Chicxulub event on different taxa argues for detailed examination and interpretation of patterns within well-defined taxonomic units rather than attempting to generalize. Certainly, the widely different interpretation of patterns of extinction articulated by Jablonski (1996) and Sheehan et al. (1996, within the same volume), demonstrates that broad generalizations are premature.

4.5 Effect of broad geographic range

Decapod taxa with broad geographic ranges preferentially survived across the K/P boundary, as previously noted for other invertebrate groups (Jablonski 1996). Of the genera with Cretaceous or earlier records, those found in three or more of the geographic areas defined in this paper usually survived into the Palaeogene, (10 out of 13 genera). Clearly, the broad range enhanced survival.

Interestingly, many of the genera with broad geographic ranges survived into the Palaeogene in tropical and subtropical areas, often proximal to areas that would be expected to have been most severely affected by the Chicxulub impact event: the Central Americas, Western Interior, and parts of the north-central and central Atlantic. These taxa include *Enoploclytia, Hoploparia, Homarus* (extant), *Upogebia* (extant), *Latheticocarcinus, Pla-*

giophthalmus, and *Costacopluma*. In addition, many genera with restricted Cretaceous geographic ranges survived into the Palaeogene in this same area, including *Penaeus* (extant), *Galathea* (extant), *Palaeomunida*, *Caloxanthus*, *Dromiopsis*, *Ranina* (extant), *Lophoranina*, *Archaeopus*, and *Goniocypoda*. The decapods do appear to have experienced generic-level losses at the end of the Cretaceous, based upon records of last occurrence (Appendix 1, Table 6). However, many genera in the regions previously considered to have been most heavily impacted by the extinction event, i.e., the tropics (Kauffman & Harries 1996) and the Chicxulub impact area (Harries 1999), were not as severely affected as might be expected based on the record of other organisms (Harries 1999; Stilwell 2003).

4.6 Effect of high and temperate geographic range

Two of the Late Cretaceous genera were cosmopolitan, *Hoploparia* and *Linuparus* (extant), and each survived in high-latitude Southern Hemisphere regions in the Palaeogene. Several genera that had high-latitude occurrences in the Late Cretaceous survived into the Palaeogene in that region, perhaps as refugia taxa or because they were buffered by inhabiting higher latitudes, including *Metanephrops* (extant), *Glyphea*, *Trachysoma*, *Protocallianassa*, *Ctenocheles* (extant), *Munidopsis* (extant), and *Homolodromia* (extant). *Lobulata* had a temperate southern-latitude distribution during the Cretaceous and Palaeogene. Note that several of these high southern-latitude survivors are extant. At least two genera that were present during the Late Cretaceous in the high-temperate North Pacific realm survived into the Palaeogene; these include *Calliax* and *Neocallichirus*, both extant. Several temperate or high-latitude North Atlantic taxa also survived into the Palaeogene of that region, including *Palaeomunida*, *Hemioon*, and *Macroacaena*.

4.7 Extinction by Cretaceous Stage

The Late Cretaceous was a time of rapid evolution and extinction within the Decapoda. It has previously been noted that during the Late Cretaceous, a similar number of decapod genera arose as became extinct in the Central Americas and that the data for the Eocene showed a nearly identical pattern (Schweitzer et al. 2002). This suggests that these were time intervals characterized by rapid evolution and faunal turnover, which was at least as important as, or more important than, catastrophic events causing extinction.

When looking at the data set compiled herein, decapod taxa became extinct throughout much of the Late Cretaceous (Appendix 1, Table 7). A total of 70 genera are known from the Late Cretaceous, of which 51 (70%) became extinct by the end of the Cretaceous. Eight genera last appeared during the Cenomanian. Of those, most of the extinctions occurred in the Central Americas and the North Atlantic region (which during that time formed a single unit) and the fledgling North Atlantic Ocean and its environs. During the Turonian, five genera had last occurrences, all in the Central Americas and North Atlantic. Two Santonian taxa had their last and only occurrence in the North Pacific, and 13 genera had last occurrences in the Campanian, with no apparent geographic pattern. The Maastrichtian was the time of the largest number of last occurrences, with 23, or one-third of the total genera known from the Late Cretaceous. About half of the locations for these last occurrences were in the Central Americas, and all but three of the Central American genera belonged to two families that became extinct at the end of the Cretaceous, the Carcineretidae and the Dakoticancridae. Feldmann et al. (1998) have already suggested that the carcineretids may have been the victims of the Chicxulub impact event. The high southern latitudes also saw the disappearance of five genera. It seems clear from the Cretaceous data that the Central Americas and North Atlantic were the sites of the majority of the Late Cretaceous extinctions, which occurred throughout that time interval. This suggests that causality for this event a) predated the K/P boundary event(s), b) the causes for the extinctions must have been localized in, but not limited to, that region, and c) that mechanisms permitted many decapods to survive in the Central Americas and North Atlantic as well. Localized causes for extinctions in this region could include the 'Super Tethys' scenario (Johnson et al. 1996), marine regression, which would have especially impacted epicontinental areas, marine cooling, or anoxia (Hallam & Wignall 1997), each of which could have caused the demise of some lineages. Hallam & Wignall (1997) considered that several of these, probably operating together, could have been sufficient to cause a Late Cretaceous mass extinction even without invoking a large impact event.

5 CONCLUSIONS

The very high survival rate of decapod families across the K/P boundary suggests that many of the extinctions within the Decapoda were in fact pseudoextinctions. Very few lineages within the Decapoda at the family level (only 8, or 21%) were lost during the Late Cretaceous, and only three were known from the Maastrichtian, temporally proximal to the proposed K/P boundary event(s). Clearly, species must have survived the K/P boundary in order to carry on into the Palaeogene and even into the Recent in many instances. Thus, those lineages in fact did not become extinct.

Geographic distribution was unquestionably important for surviving into the Palaeogene among the Decapoda. Temperate- and high-latitude genera appear to have survived at a higher rate; many of these genera are extant. In addition, those genera with broad geographic ranges survived preferentially. However, many taxa survived in the subtropics and Chicxulub area, although those areas are typically considered to have been hard-hit by the Late Cretaceous events (Harries 1999). One possible explanation for the latter phenomenon is that the decapod groups present in the Cretaceous, and the Mesozoic in general, were usually representatives of primitive, ecologically generalized families. Even in the presence of environmental stress, possibly even related to the repercussions of the Chicxulub impact, these eurytopic groups could survive preferentially due to their being scavengers or omnivores (Van Valen 1994; Hallam & Wignall 1996; Sheehan et al. 1996) or due to their ability to survive on many substrates in broad depth ranges. Many of the more stenotopic, environment-specific groups, such as the Porcellanidae, adapted to living within reefs; many members of the Xanthoidea, specialized for reef-dwelling; and many families within the Thoracotremata, tending to be environmentally-specific, appeared in the Eocene or even later (Glaessner 1969; Schweitzer 2001). The lack of specialization in many Cretaceous lineages may have contributed to their survival success.

Harries et al. (1996) suggested that those organisms with organic skeletons, such as the chitinous exoskeleton of decapods, may have preferentially survived mass extinction events. This has primarily to do with fluctuations in the Carbonate Compensation Depth, making it difficult for organisms with carbonate shells, such as molluscs, to survive in carbonate-deficient waters (Harries et al. 1996). Barrera & Savin (1999) documented shallowing of the CCD during the latest Cretaceous. Thus, the exoskeletal makeup of the decapods may have conferred a significant benefit in surviving mass extinction events, at least the events causing significant changes in carbonate chemistry of the oceans. It is important to note that the CCD would need to have been very, very shallow, in the order of tens of meters, to serve as a plausible explanation for causing mass extinctions.

There is another possible explanation for the extinctions within the Decapoda during the Late Cretaceous. These extinctions have been perceived as mass extinctions or at least higher-than-background-level extinctions in other groups, but it is important to note that the rates of extinction and origination were nearly equivalent during the Late Cretaceous when using the maximum possible number of genera originating (Schweitzer et al. 2002). It is well recognized that the Decapoda experienced at least two major radiations, one in the Mesozoic and one during the Eocene (Glaessner 1969; Schram 1986; Schweitzer 2001; Schweitzer et al. 2002; Fraaije 2003; Feldmann 2003). The Miocene also appears to have been a time of rapid evolution within the Decapoda (Schweitzer 2001); most taxa appearing during that time are extant.

During the Cretaceous and Eocene radiations, many new forms appeared that do not seem to have been successful in the long term, such as is seen in the Middle Cambrian radiations recorded in the Burgess Shale fauna (Gould 1989). For example, the Eocene record of decapods is incredibly robust. Just among the taxa known from the Central Americas, 35 genera originated, over half of which were endemic to that region (Schweitzer et al. 2002). A similar pattern is seen among Eocene decapods of the North Pacific Ocean (Schweitzer 2001). Many unusual forms are known from the Eocene of Italy and Hungary (Müller & Collins 1991; De Angeli & Beschin 2001; Beschin et al. 2002; Beschin & De Angeli 2003; Busulini et al. 2003) that are difficult to place within a family and that do not seem to have closely related descendants. The rate of origination of decapod taxa during the Eocene is the same as that seen for the Late Cretaceous in the Central Americas and in both time intervals is the same as the rate of extinction (Schweitzer et al. 2002). Thus, it may be that these time intervals were simply times of major adaptive radiation within the Decapoda, perhaps controlled by environmental factors but also possibly controlled by intrinsic genetic and morphological factors. Within these major radiations, some forms were not successful beyond their time period or geographic region, hence, the high endemicity during the Eocene and relatively high endemicity during the Cretaceous in the Central Americas and the North Pacific (Schweitzer et al. 2002). Note also that the Late Cretaceous extinctions within the Decapoda were distributed over the entire time interval, with a notable spike in the Maastrichtian. Gouldian contingency may be as much a factor as any in explaining the patterns of extinction within the Decapoda (Gould 1989). The effects of contingency have already been proposed as a possible explanation for the rise of mammals at the expense of the dinosaurs (Gould 2001). It is clear that contingency as well as the phenomena of pseudoextinction, geographic patterns in lineages exhibiting extinctions, and differential impact of the proposed mass extinctions on various taxonomic groups and lineages within those groups need directed attention.

ACKNOWLEDGMENTS

This work has been supported by National Science Foundation (NSF) Grants OPP8715945, OPP 8915439, OPP 9417697, OPP 9526252, OPP 9909184, and National Geographic Society (NGS) Grant 4375 to R. Feldmann and NSF Grant INT 0003058 and NGS Grant 6265-98 to R. Feldmann and C. Schweitzer. Conversations with W.J. Zinsmeister (Purdue University), J.A. Crame (British Antarctic Survey), F. Schram (University of Amsterdam), and comments by E. Kauffman (Indiana University) and P. Harries (University of South Florida) all have contributed to this work. D.A. Waugh (Kent State University) read an earlier draft of the manuscript and improved it with his suggestions. The manuscript was substantially improved by the reviews of J.W.M. Jagt (Natuurhistorisch Museum Maastricht, Netherlands) and R.H.B. Fraaije (Oertijdmuseum de Groene Poort, Netherlands). K. Smith (Kent State University) assisted with formatting the manuscript.

APPENDIX 1

Table 1. Families originating in the Cretaceous or before and surviving into the Recent. Data are based upon families occurring in the Central Americas (CAM) as defined by Schweitzer et al. (2002); the temperate southern latitudes (TSL), the region lying between 30 and 60 degrees south latitude; the high southern latitudes (HSL), the region lying between 60 and 90 degrees south latitude; and the North Pacific (NPAC) as defined by Schweitzer (2001) as well as the families recently comprehensively treated or revised by us (Homolidae, Homolodromiidae, Orithopsidae, Necrocarcinidae, Palaeoxanthopsidae, Galatheidae, Chirostylidae, Hexapodidae). References under Geologic Range column direct the reader to recent changes in the geologic range of the family. * Currently, *Titanocarcinus* A. Milne Edwards, 1864 is referred to the Pilumnidae (Karasawa & Schweitzer, 2004) and embraces Cretaceous species (Glaessner, 1969). However, the status of all species currently referred to it is questionable and under study. The Cretaceous *Titanocarcinus* are the only possible Cretaceous members of the Pilumnidae.

Family	Geologic Range
Penaeidae Rafinesque, 1815	Triassic-Recent
Glypheidae von Zittel, 1885	Triassic-Recent
Palinuridae Latreille, 1802	Triassic-Recent; Garassino et al. 1996
Axiidae Huxley, 1879	Jurassic-Recent
Nephropidae Dana, 1852	Jurassic-Recent
Galatheidae Samouelle, 1819	Jurassic-Recent
Dynomenidae Ortmann, 1892a	Jurassic-Recent
Homolidae White, 1847	Jurassic-Recent
Homolodromiidae Alcock, 1900	Jurassic-Recent
Upogebiidae Borradaile, 1903	Jurassic-Recent
Polychelidae Wood-Mason, 1875	Jurassic-Recent
Scyllaridae Latreille, 1825	Lower Cretaceous-Recent
Aeglidae Dana, 1852	Lower Cretaceous-Recent
Dorippidae MacLeay, 1838	Lower Cretaceous-Recent
Raninidae de Haan, 1839	Lower Cretaceous-Recent
Paguridae *sensu lato* Latreille, 1802	Lower Cretaceous-Recent
Callianassidae Dana, 1852	Lower Cretaceous-Recent
Micheleidae Sakai, 1992	Upper Cretaceous-Recent
Ctenochelidae Manning & Felder, 1991	Upper Cretaceous-Recent
Chirostylidae Ortmann, 1892b	Upper Cretaceous-Recent
Poupiniidae Guinot, 1991	Upper Cretaceous-Recent
Retroplumidae Gill, 1894	Upper Cretaceous-Recent
Goneplacidae MacLeay, 1838	Upper Cretaceous-Recent
Hexapodidae Miers, 1886	Upper Cretaceous-Recent
Pilumnidae Samouelle, 1819*	Upper Cretaceous?-Recent

Table 2. Families originating in the Cretaceous or earlier and becoming extinct during the Palaeogene. Data records as defined in Table 1.

Family	Geologic Range
Erymidae Van Straelen, 1925	Triassic-Palaeocene
Prosopidae von Meyer, 1860	Jurassic-Palaeocene
Necrocarcinidae Förster, 1968	Lower Cretaceous-Eocene
Orithopsidae Schweitzer et al., 2003	Lower Cretaceous-Oligocene
Palaeoxanthopsidae Schweitzer, 2003	Upper Cretaceous-Eocene

Table 3. Families becoming extinct by the end-Cretaceous. Data records and abbreviations as defined in Table 1.

Family	Geologic Range	Geographic Position of Last Occurrence
Coleiidae Van Straelen, 1925	Triassic-Lower Cretaceous; Garassino & Teruzzi 1993	Tethyan
Mecochiridae Van Straelen, 1925	Triassic-Upper Cretaceous (early Maastrichtian); Garassino & Teruzzi 1993	HSL
Chilenophoberidae Tshudy & Babcock, 1997	Jurassic	HSL
Etyidae Guinot & Tavares, 2001	Lower-Upper Cretaceous	CAM
Carcineretidae Beurlen, 1930	Lower-Upper Cretaceous (Maastrichtian)	CAM
Torynommidae Glaessner, 1980	Lower-Upper Cretaceous (Maastrichtian)	HSL
Dakoticancridae Rathbun, 1917	Upper Cretaceous (Maastrichtian)	CAM
Retrorsichelidae Feldmann et al., 1993	Upper Cretaceous	HSL

Table 4. Geologic range of families and genera that appeared during or before the Cretaceous (all within the data set as defined in the text). Ranges are included for all genera within families crossing the K/P boundary as well as ranges for all relevant genera within the families we have recently revised. The heavy vertical line indicates the K/P boundary, and the lighter vertical line indicates the end of the range for those genera not reaching the end-Maastrichtian; thus, those genera becoming extinct before the Maastrichtian or during the early Maastrichtian have ranges terminating at the lighter vertical line. Families are indicated in all capital letters and their ranges in bold horizontal lines.

A question mark (?) within the range chart indicates a questionable occurrence in that time interval. Abbreviations: Pal. = Palaeocene; Eo. = Eocene; Oligo. = Oligocene; Mio. = Miocene; Pli. = Pliocene; Ple. = Pleistocene.

Taxon	Triassic	Jurassic	Cretaceous Early	Cretaceous Late	Pal.	Eo.	Oligo.	Mio.	Pli.	Ple.
PENAEIDAE	—	—	—	—	—	—	—	—	—	—
Ambilobeia		—								
Ifaya		—								
Penaeus				—	—	—	—	—	—	—
GLYPHEIDAE		—	—	—	—	—	—	—	—	—
Glyphea		—	—	—						
Trachysoma		—	—							
MECOCHIRIDAE		—	—	—						
Meyeria			—							
ERYMIDAE		—	—	—						
Eryma		—	—	—						
Erymastacus		—								
Palaeastacus		—	—	—						
Enoploclytia			—	—						
Phlyctisoma			—	—						
NEPHROPIDAE		—	—	—	—	—	—	—	—	—
Homarus			—	—	—	—	—	—	—	—
Hoploparia		—	—	—	—	—	—	—		
Tillocheles			—							
Metanephrops				—	—	—	—	—	—	—
PALINURIDAE		—	—	—	—	—	—	—	—	—
Astacodes		—	—							
Linuparus			—	—	—	—	—	—	—	—
Jasus				?	—	—	—	—	—	—
SCYLLARIDAE			—	—	—	—	—	—	—	—
Scyllarella					—					
POLYCHELIDAE		—	—	—	—	—	—	—	—	—
Antarcticheles		—								
AXIIDAE		—	—	—	—	—	—	—	—	—
Schlueteria		—	—	—						

	Triassic	Jurassic	Cretaceous		Palaeogene			Neogene		
			Early	Late	Pal.	Eo.	Oligo.	Mio.	Pli.	Ple.
MICHELEIDAE										
Paki										
CALLIANASSIDAE										
Calliax										
Neocallichirus										
Protocallianassa										
Callichirus										
CTENOCHELIDAE										
Ctenocheles										
Callianopsis										
CHIROSTYLIDAE										
Pristinaspina										
Eumunida										
UPOGEBIIDAE										
Upogebia										
GALATHEIDAE										
Munitheites										
Palaeomunidopsis										
Gastrosacus										
Mesogalathea										
Paragalathea										
Palaeomunida										
Eomunidopsis										
Galathea										
Brazilomunida										
Luisogalathea										
Munidopsis										
Protomunida										
Munida										
Faxegalathea										
Acanthogalathea										
Lessinigalathea										
Spathogalathea										
AEGLIDAE										
Protaegla										
Haumuriaegla										
RETRORSICHELIDAE										
Retrorsichela										
DYNOMENIDAE										
Cyphonotus										
Cyclothyreus										

	Triassic	Jurassic	Cretaceous		Palaeogene			Neogene		
			Early	Late	Pal.	Eo.	Oligo.	Mio.	Pli.	Ple.
Diaulax		■	■	■						
Graptocarcinus			■	■						
Maurimia			■							
Caloxanthus			■	■						
Acanthodiaulax				■						
Dromiopsis				■	■					
Dynomenopsis				■						
Dromilites					■					
Kierionopsis					■					
HOMOLIDAE		■	■	■	■	■	■	■	■	■
Gastrodorus		■								
Laeviprosopon		■								
Tithonohomola		■								
Lignihomola			■							
Homolopsis			■							
Hoplitocarcinus			■	■						
Latheticocarcinus			■	■	■					
Zygastrocarcinus			■							
Prohomola						■				
Homola						■	■	■	■	■
Dagnaudus							■	■	■	■
Paromolopsis								■	■	■
HOMOLODROMIIDAE		■	■	■	■	■	■	■	■	■
Eoprosopon		■								
Rhinodromia			■							
Homolodromia			■	■	■			■	■	■
Palehomola						■				
Antarctidromia								■		
Dicranodromia								■	■	■
PROSOPIDAE		■	■	■	■					
Pithonoton		■	■							
Oonoton			■							
Plagiophthalmus			■	■						
Ekalakia			■							
Rathbunopon			■							
Rugafarius			■							
POUPINIIDAE					■	■	■	■	■	■
Rhinopoupinia					■					
RANINIDAE		■	■	■	■	■	■	■	■	■
Cretacoranina			■	■						
Notopocorystes			■	■						

34 *Schweitzer & Feldmann*

Stratigraphic range chart showing the occurrences of raninoid and related decapod genera and families across the Triassic, Jurassic, Cretaceous (Early, Late), Palaeogene (Pal., Eo., Oligo.), and Neogene (Mio., Pli., Ple.) periods:

- *Eucorystes*
- *Cristafrons*
- *Hemioon*
- *Lophoranina*
- *Macroacaena*
- *Ranina*
- *Laeviranina*
- *Quasilaeviranina*
- *Raninella*
- *Rogueus*
- *Lyreidus*
- *Raninoides*
- ETYIIDAE
- *Etyus*
- *Xanthosia*
- *Feldmannia*
- DORIPPIDAE
- *Hillius*
- *Eodorippe*
- *Sodakus*
- DAKOTICANCRIDAE
- *Avitelmessus*
- *Dakoticancer*
- *Tetracarcinus*
- *Seorsus*
- NECROCARCINIDAE
- *Necrocarcinus*
- *Paranecrocarcinus*
- *Hasaracancer*
- *Cenomanocarcinus*
- *Pseudonecrocarcinus*
- *Campylostoma*
- ORITHOPSIDAE
- *Orithopsis*
- *Paradoxicarcinus*
- *Silvacarcinus*
- *Marycarcinus*
- *Goniochele*
- *Cherpiocarcinus*
- RETROPLUMIDAE
- *Costacopluma*

	Triassic	Jurassic	Cretaceous		Palaeogene			Neogene		
			Early	Late	Pal.	Eo.	Oligo.	Mio.	Pli.	Ple.
Archaeopus										
CARCINERETIDAE										
Lithophylax										
Woodbinax										
Cancrixantho										
Carcineretes										
Ophthalmoplax										
Longusorbis										
Branchiocarcinus										
Mascaranada										
TORYNOMMIDAE										
Dioratiopus										
Torynomma										
PORTUNIDAE										
Proterocarcinus										
PALAEOXANTHOPIDAE										
Palaeoxantho										
Palaeoxanthopsis										
Remia										
Lobulata										
Verrucoides										
Paraverrucoides										
PANOPEIDAE										
Glyphithyreus										
ZANTHOPSIDAE										
Zanthopsis										
HEXAPODIDAE										
Goniocypoda										
Palaeopinnixa										
Stevia										
GONEPLACIDAE										
Icriocarcinus										
Chirinocarcinus										
Tehuacana										
XANTHIDAE s. lat.										
Megaxantho										
Cyclocorystes										
PINNOTHERIDAE				?						
Pinnotheres ?				?						
Viapinnixa										

Table 5. Families from the data set as defined in the text with Cretaceous or earlier and post-Cretaceous records and their included relevant genera. Only those families with post-Cretaceous records are included here; their included genera may not have records extending into the Palaeogene. Number indicated for family is the number of genera within the family that survived from the Cretaceous into the Palaeogene. The indication "crosses" is given for genera that survived the K/P boundary. The Paguridae are not included because the efficacy of using extant generic names, as is typically done for fossil pagurids, is currently under consideration by the authors. * *Xanthilites* is assigned to a new genus by Schweitzer (in press). Data records and abbreviations as defined in Table 1.

Taxon	Geologic Range	Number of Genera Crossing K/P Boundary
Penaeidae Rafinesque, 1815	**Triassic-Recent**	**1**
Penaeus Fabricius, 1798	Upper Cretaceous-Recent	crosses
Ambilobeia Garassino & Pasini, 2002	Triassic	
Ifaya Garassino & Teruzzi, 1995	Triassic	
Glypheidae von Zittel, 1885	**Triassic-Recent**	**2**
Glyphea von Meyer, 1835	Triassic-Miocene	crosses
Trachysoma Bell, 1858	Jurassic-Eocene	crosses
Erymidae Van Straelen, 1925	**Triassic-Palaeocene**	**1**
Enoploclytia McCoy, 1849	Cretaceous-Palaeocene	crosses
Eryma von Meyer, 1840	Triassic-Upper Cretaceous	
Erymastacus Beurlen, 1928	Jurassic	
Palaeoastacus Bell, 1850	Jurassic-Upper Cretaceous	
Phlyctisoma Bell, 1863	Upper Cretaceous	
Nephropidae Dana, 1852	**Jurassic-Recent**	**3**
Hoploparia McCoy, 1849	Cretaceous-Miocene	crosses
Homarus Weber, 1795	Cretaceous-Recent	crosses
Metanephrops Jenkins, 1972	Upper Cretaceous-Recent	crosses
Tillocheles Woods, 1957	Lower Cretaceous	
Palinuridae Latreille, 1802	**Triassic-Recent**	**1**
Astacodes Bell, 1863	Jurassic-Upper Cretaceous	
Jasus Parker, 1883	Upper Cretaceous?-Recent	questionably crosses
Linuparus White, 1847	Lower Cretaceous-Recent	crosses
Polychelidae Wood-Mason, 1875	**Jurassic-Recent**	**none**
Antarcticheles Aguirre-Urreta et al., 1990	Jurassic	
Axiidae Huxley, 1879	**Jurassic-Recent**	
Schlueteria Fritsch (in Fritsch & Kafka 1887)	Jurassic-Upper Cretaceous	
Micheleidae Sakai, 1992	**Upper Cretaceous-Recent**	
Paki Karasawa & Hayakawa, 2000	Upper Cretaceous	

Table 5 continued.

Taxon	Geologic Range	Number of Genera Crossing K/P Boundary
Callianassidae Dana, 1852	**Lower Cretaceous-Recent**	**3**
Calliax de Saint Laurent, 1973	Upper Cretaceous-Recent	crosses
Neocallichirus Sakai, 1988	Upper Cretaceous-Recent	crosses
Protocallianassa Beurlen, 1930	Upper Cretaceous-Eocene	crosses
Callichirus? Stimpson, 1866	Eocene-Recent	
Ctenochelidae Manning & Felder, 1991	**Upper Cretaceous-Recent**	**1**
Callianopsis de Saint Laurent, 1973	Eocene-Recent	
Ctenocheles Kishinouye, 1926	Upper Cretaceous-Recent	crosses
Chirostylidae Ortmann, 1892b	**Upper Cretaceous-Recent**	**none**
Pristinaspina Schweitzer & Feldmann, 2000a	Upper Cretaceous	
Eumunida Smith, 1883	Eocene-Recent	
Upogebiidae Borradaile, 1903	**Jurassic-Recent**	**1**
Upogebia Leach, 1814	Jurassic-Recent	crosses
Galatheidae Samouelle, 1819	**Jurassic-Recent**	**3**
Acanthogalathea Müller & Collins, 1991	Eocene	
Brazilomunida Martins-Neto, 2001	Lower Cretaceous	
Eomunidopsis Via, 1981	Jurassic-Upper Cretaceous	
Faxegalathea Jakobsen & Collins, 1997	Palaeocene	
Galathea Fabricius, 1793	Lower Cretaceous-Recent	crosses
Gastrosacus von Meyer, 1851	Jurassic-Lower Cretaceous	
Lessinigalathea De Angeli & Garassino, 2002	Eocene	
Luisogalathea Karasawa & Hayakawa, 2000	Upper Cretaceous	
Mesogalathea Houša, 1963	Jurassic-Cretaceous	
Munida Leach, 1820	Palaeocene-Recent	
Munidopsis Whiteaves, 1874	Upper Cretaceous-Recent	crosses
Munitheites Lőrenthey (in Lőrenthey & Beurlen 1929)	Jurassic	
Palaeomunida Lőrenthey, 1902	Jurassic-Eocene	crosses
Palaeomunidopsis Van Straelen, 1925	Jurassic	
Paragalathea Patrulius, 1960	Jurassic-Cretaceous	
Protomunida Beurlen, 1930	Palaeocene-Eocene	
Spathagalathea De Angeli & Garassino, 2002	Eocene	

Table 5 continued.

Taxon	Geologic Range	Number of Genera Crossing K/P Boundary
Aeglidae Dana, 1852	**Lower Cretaceous-Recent**	**none**
Hamuriaegla Feldmann, 1984	Upper Cretaceous	
Protaegla Feldmann et al., 1998	Lower Cretaceous	
Dynomenidae Ortmann, 1892a	**Jurassic-Recent**	**2**
Acanthodiaulax Schweitzer et al., 2003	Upper Cretaceous	
Caloxanthus A. Milne Edwards, 1864	Lower Cretaceous-Palaeocene	crosses
Cyclothyreus Remeš, 1895	Jurassic-Upper Cretaceous	
Cyphonotus Bell, 1863	Jurassic-Lower Cretaceous	
Diaulax Bell, 1863	Jurassic-Upper Cretaceous	
Dromilites H. Milne Edwards, 1837	Palaeocene-Eocene	
Dromiopsis Reuss, 1859	Upper Cretaceous-Palaeocene	crosses
Dynomenopsis Secretan, 1972	Upper Cretaceous	
Graptocarcinus Roemer, 1887	Lower-Upper Cretaceous	
Kierionopsis Davidson, 1966	Palaeocene	
Maurimia Martins-Neto, 2001	Lower Cretaceous	
Homolidae de Haan, 1839	**Jurassic-Recent**	**1**
Dagnaudus Guinot & Richer de Forges, 1995	Oligocene-Recent	
Gastrodorus von Meyer, 1864	Jurassic	
Homola Leach, 1815	Eocene-Recent	
Homolopsis Bell, 1863	Lower-Upper Cretaceous	
Hoplitocarcinus Beurlen, 1928	Cretaceous	
Laeviprosopon Glaessner, 1933	Jurassic	
Latheticocarcinus Bishop, 1988c	Cretaceous-Palaeocene	crosses
Lignihomola Collins, 1997	Lower Cretaceous	
Paromolopsis Wood-Mason (in Wood-Mason & Alcock, 1891)	Miocene-Recent	
Prohomola Karasawa, 1992	Eocene	
Tithonohomola Glaessner, 1933	Jurassic	
Zygastrocarcinus Bishop, 1983b	Upper Cretaceous	
Homolodromiidae Alcock, 1900	**Jurassic-Recent**	**1**
Antarctidromia Förster et al., 1985	Miocene	
Dicranodromia A. Milne Edwards, 1880	Miocene-Recent	
Eoprosopon Förster, 1986	Jurassic	
Homolodromia A. Milne Edwards, 1880	Upper Cretaceous-Recent	crosses
Palehomola Rathbun, 1926	Oligocene	
Rhinodromia Schweitzer et al., 2004	Upper Cretaceous	

Table 5 continued.

Taxon	Geologic Range	Number of Genera Crossing K/P Boundary
Prosopidae von Meyer, 1860	**Jurassic-Palaeocene**	1
Ekalakia Bishop, 1976	Upper Cretaceous	
Oonoton Glaessner, 1980	Lower Cretaceous	
Pithonoton von Meyer, 1842	Jurassic-Upper Cretaceous	
Plagiophthalmus Bell, 1863	Lower Cretaceous-Palaeocene	crosses
Rathbunopon Stenzel, 1945	Upper Cretaceous	
Rugafarius Bishop, 1985	Upper Cretaceous	
Poupiniidae Guinot, 1991	**Upper Cretaceous-Recent**	
Rhinopoupinia Feldmann et al., 1993	Upper Cretaceous	
Raninidae de Haan, 1839	**Lower Cretaceous-Recent**	3, possibly 4
Cretacoranina Mertin, 1941	Lower-Upper Cretaceous	
Notopocorystes McCoy, 1849	Lower-Upper Cretaceous	
Eucorystes Bell, 1863	Lower-Upper Cretaceous	
Cristafrons Feldmann et al., 1993	Upper Cretaceous	
Ranina Lamarck, 1801	Upper Cretaceous?-Recent	questionably crosses
Lophoranina Fabiani, 1910	Upper Cretaceous-Eocene	crosses
Hemioon Bell, 1863	Upper Cretaceous-Palaeocene	crosses
Macroacaena Tucker, 1998	Upper Cretaceous-Miocene	crosses
Laeviranina Lőrenthey (in Lőrenthey & Beurlen 1929)	Palaeocene-Eocene	
Quasilaeviranina Tucker, 1998	Palaeocene-Eocene	
Raninella A. Milne Edwards, 1862	Palaeocene-Eocene	
Rogueus Berglund & Feldmann, 1989	Palaeocene-Eocene	
Lyreidus de Haan, 1841	Eocene-Recent	
Raninoides H. Milne Edwards, 1837	Eocene-Recent	
Dorippidae MacLeay, 1838	**Lower Cretaceous-Recent**	none
Eodorippe Glaessner, 1980	Upper Cretaceous	
Hillius Bishop, 1983a	Lower Cretaceous	
Sodakus Bishop, 1978	Upper Cretaceous	
Necrocarcinidae Förster, 1968	**Lower Cretaceous-Eocene**	none
Campylostoma Bell, 1858	Eocene	
Hasaracancer Jux, 1971	Upper Cretaceous	
Cenomanocarcinus Van Straelen, 1936	Upper Cretaceous	
Necrocarcinus Bell, 1863	Lower-Upper Cretaceous	
Paranecrocarcinus Van Straelen, 1936	Lower-Upper Cretaceous	

Table 5 continued.

Taxon	Geologic Range	Number of Genera Crossing the K/P Boundary
Pseudonecrocarcinus Förster, 1968	Upper Cretaceous	
Orithopsidae Schweitzer et al., 2003	**Lower Cretaceous-Oligocene**	**none**
Cherpiocarcinus Marangon & De Angeli, 1997	Oligocene	
Goniochele Bell, 1858	Eocene-Oligocene	
Marycarcinus Schweitzer et al., 2003	Eocene	
Orithopsis Carter, 1872	Lower-Upper Cretaceous	
Paradoxicarcinus Schweitzer et al., 2003	Upper Cretaceous	
Silvacarcinus Collins & Smith, 1992	Lower Eocene	
Retroplumidae Gill, 1894	**Upper Cretaceous-Recent**	**2**
Archaeopus Rathbun, 1908	Upper Cretaceous-Eocene	crosses
Costacopluma Collins & Morris, 1975	Upper Cretaceous-Palaeocene	crosses
Palaeoxanthopsidae Schweitzer, 2003	**Upper Cretaceous-Eocene**	**2**
Palaeoxantho Bishop, 1986	Upper Cretaceous	
Palaeoxanthopsis Beurlen, 1958	Upper Cretaceous	
Remia Schweitzer, 2003	Upper Cretaceous	
Lobulata Schweitzer et al., 2004	Upper Cretaceous-Palaeocene	crosses
Xanthilites * *gerthi* Glaessner, 1930	Upper Cretaceous-Palaeocene	crosses
Verrucoides Vega et al., 2001b	Palaeocene-Eocene	
Paraverrucoides Schweitzer, 2003	Eocene	
Hexapodidae Miers, 1886	**Upper Cretaceous-Recent**	**1**
Goniocypoda Woodward, 1867	Upper Cretaceous-Eocene	crosses
Palaeopinnixa Via, 1966	Palaeocene-Miocene	
Stevea Manning & Holthuis, 1981	Eocene-Recent	
Goneplacidae MacLeay, 1838	**Upper Cretaceous-Recent**	**none**
Icriocarcinus Bishop, 1988a	Upper Cretaceous	
Chirinocarcinus Karasawa & Schweitzer, 2004	Palaeocene	
Numerous other genera of Goneplacidae	Eocene origination	
Pilumnidae Samouelle, 1819	**Upper Cretaceous?-Recent**	**questionably 1**
Titanocarcinus A. Milne Edwards, 1864 (see note in Appendix 1, Table 1)	Upper Cretaceous-Miocene	crosses
Pilumnus Leach, 1815	Miocene-Recent	
Pinnotheridae de Haan, 1833	**Cretaceous?-Recent**	**questionably 1**
Pinnotheres? Bosc, 1802	Cretaceous?-Recent	
Viapinnixa Schweitzer & Feldmann, 2001a	Palaeocene-Eocene	
Asthenognathus Stimpson, 1858	Oligocene-Recent	

Table 6. Genera in the data set originating during the Palaeocene. Data records and abbreviations as defined in Table 1. Asterisk (*) = families originating during the Palaeocene. Double Asterisk (**) = questionable occurrence of the Pinnotheridae reported from the Cretaceous; the family was unequivocally present by the Palaeocene.

Genus	Genus Range	Family	Family Range
Scyllarella Rathbun, 1935	Palaeocene	Scyllaridae	Lower Cretaceous-Recent
Faxegalathea	Palaeocene (Danian)	Galatheidae	Lower Cretaceous-Recent
Munida	Palaeocene (Danian)-Recent	Galatheidae	Lower Cretaceous-Recent
Protomunida	Palaeocene-Eocene	Galatheidae	Lower Cretaceous-Recent
Laeviranina	Palaeocene-Eocene	Raninidae	Lower Cretaceous-Recent
Quasilaeviranina	Palaeocene-Eocene	Raninidae	Lower Cretaceous-Recent
Raninella	Palaeocene-Eocene	Raninidae	Lower Cretaceous-Recent
Rogueus Berglund & Feldmann, 1989	Palaeocene-Eocene	Raninidae	Lower Cretaceous-Recent
Dromilites	Palaeocene-Eocene	Dynomenidae?	Jurassic-Recent
Kierionopsis	Palaeocene	Dynomenidae?	Jurassic-Recent
Proterocarcinus Feldmann et al., 1995	Palaeocene (Danian)-Miocene	Portunidae*	Palaeocene-Recent
Palaeopinnixa	Palaeocene (Danian)-Miocene	Hexapodidae	Upper Cretaceous-Recent
Chirinocarcinus	Palaeocene (Danian)	Goneplacidae	Upper Cretaceous-Recent
Tehuacana Stenzel, 1944	Palaeocene (Danian)	Goneplacidae	Upper Cretaceous-Recent
Verrucoides	Palaeocene-Eocene	Palaeoxanthopsidae	Upper Cretaceous-Eocene
Glyphithyreus Reuss, 1859	Palaeocene-Eocene	Panopeidae*	Palaeocene-Recent
Cyclocorystes Bell, 1858	Palaeocene-Eocene	Xanthidae *s. lat.*	N/A
Zanthopsis McCoy, 1849	Palaeocene-Miocene	Zanthopsidae*	Palaeocene-Miocene
Viapinnixa	Palaeocene-Eocene	Pinnotheridae**	Palaeocene?-Recent

Table 7. Genera in the data set becoming extinct during the Late Cretaceous and the geographic location of last occurrence. Only those in which the last-known stage of occurrence is known are included; those known only as Late Cretaceous are not included in this list. Some of these genera do not appear in Table 5 because they belong to families which became extinct by the end of the Cretaceous, listed in Table 3. Data records and abbreviations as defined in Table 1. NATL = North Atlantic, WI = Western Interior of North America, TETH = Tethyan.

Stage	Taxon	Geographic Location of Last Occurrence
Cenomanian	*Diaulax*	NATL, NPAC
Cenomanian	*Dynomenopsis*	TSL
Cenomanian	*Graptocarcinus*	CAM, NATL
Cenomanian	*Lithophylax* A. Milne Edwards & Brocchi, 1879	NATL
Cenomanian	*Orithopsis*	NATL
Cenomanian	*Pithonoton*	NATL, NPAC
Cenomanian	*Rathbunopon*	CAM, NATL
Cenomanian	*Woodbinax* Stenzel, 1953	CAM
Turonian	*Cenomanocarcinus*	CAM, NPAC (questionable from WI Maastrichtian)
Turonian	*Etyus* Mantell, 1822	NATL
Turonian	*Ophthalmoplax* Rathbun, 1935	CAM
Turonian	*Paranecrocarcinus*	CAM
Turonian	*Feldmannia* Guinot & Tavares, 2001	CAM
Turonian	*Xanthosia* Bell, 1863	CAM, NATL
Santonian	*Acanthodiaulax*	NPAC
Santonian	*Longusorbis* Richards, 1975	NPAC
Campanian	*Astacodes* Bell, 1863	CAM
Campanian	*Cancrixantho* Van Straelen, 1934	NATL
Campanian	*Dioratiopus* Woods, 1953	WI
Campanian	*Eucorystes*	NPAC
Campanian	*Hasaracancer*	TETH
Campanian	*Notopocorystes*	TSL
Campanian	*Paki*	NPAC
Campanian	*Palaeastacus* Bell, 1850	NATL
Campanian	*Paradoxicarcinus*	NPAC
Campanian	*Retrorsichela* Feldmann et al., 1993	HSL
Campanian	*Rugafarius*	WI
Campanian	*Schlueteria*	TSL
Campanian	*Zygastrocarcinus*	WI
Maastrichtian	*Avitelmessus* Rathbun, 1923	CAM
Maastrichtian	*Branchiocarcinus* Vega et al., 1995	CAM
Maastrichtian	*Carcineretes* Withers, 1922	CAM
Maastrichtian	*Cristafrons*	HSL
Maastrichtian	*Dakoticancer* Rathbun, 1917	CAM, WI
Maastrichtian	*Eodorippe*	HSL
Maastrichtian	*Eomunidopsis*	WI (extinct by early Maastrichtian)
Maastrichtian	*Ekalakia* Bishop, 1976	WI
Maastrichtian	*Haumuriaegla*	HSL

Table 7 continued.

Stage	Taxon	Geographic Location of Last Occurrence
Maastrichtian	*Homolopsis*	TSL
Maastrichtian	*Luisogalathea*	NPAC
Maastrichtian	*Mascaranada* Vega & Feldmann, 1991	CAM
Maastrichtian	*Megaxantho* Vega et al., 2001a	CAM
Maastrichtian	*Meyeria* McCoy, 1849	HSL
Maastrichtian	*Necrocarcinus*	HSL
Maastrichtian	*Palaeoxantho*	WI
Maastrichtian	*Palaeoxanthopsis*	CAM
Maastrichtian	*Pristinaspina*	NPAC (youngest possible age)
Maastrichtian	*Remia* Schweitzer, 2003	TETH
Maastrichtian	*Seorsus* Bishop, 1988b	CAM
Maastrichtian	*Sodakus*	WI
Maastrichtian	*Tetracarcinus* Weller, 1905	CAM
Maastrichtian	*Torynomma* Glaessner, 1980	HSL

REFERENCES

Aguirre-Urreta, M.B., Buatois, L.A., Chernoglasov, G.C.B. & Medina, F.A. 1990. First Polychelidae (Crustacea, Palinura) from the Jurassic of Antarctica. *Antarctic Science* 2: 157-162.

Alcock, A. 1900. Materials for a carcinological fauna of India, 5: The Brachyura Primigenia or Dromiacea. *J. Asia. Soc. Bengal* 68 (II: 3): 123-169.

Anderson, R.R., Witzke, B.J. & Hartung, J.B. 1996. Impact materials recovered by research core drilling in the Manson Impact Structure, Iowa. In: Ryder, G., Fastovsky, D. & Gartner, S. (eds.), *The Cretaceous-Tertiary Event and Other Catastrophes in Earth History*: pp. 527-540. Boulder, Colorado: Geol. Soc. Amer. Spec. Pap. 307.

Barrera, E. & Johnson, C.C. (eds.) 1999. *Evolution of the Cretaceous Ocean-Climate System*: 445 pp. Boulder, Colorado: Geol. Soc. Amer. Spec. Pap. 332.

Barrera, E. & Savin, S.M. 1999. Evolution of late Campanian-Maastrichtian marine climates and oceans. In: Barrera, E. & Johnson, C.C. (eds.), *Evolution of the Cretaceous Ocean-Climate System*: pp. 245-282. Boulder, Colorado: Geol. Soc. Amer. Spec. Pap. 332.

Bell, T. 1850. Notes on the Crustacea of the Chalk Formation. In: Dixon, F. (ed.), *The Geology and Fossils of the Tertiary and Cretaceous Formations of Sussex*: pp. 344-345, pl. XXXVIII. London: Longman, Brown, Green & Longmans.

Bell, T. 1858. *A Monograph of the Fossil Malacostracous Crustacea of Great Britain; Pt. I, Crustacea of the London Clay*: 44 pp., 11 pls. London: Palaeontogr. Soc. Monogr.

Bell, T. 1863. *A Monograph of the Fossil Malacostracous Crustacea of Great Britain; Pt. II, Crustacea of the Gault and Greensand*: 40 pp., 11 pls. London: Palaeontogr. Soc. Monogr.

Berglund, R.E. & Feldmann, R.M. 1989. A new crab, *Rogueus orri* n. gen. and sp. (Decapoda: Brachyura), from the Lookingglass Formation (Ulatisian Stage: Lower Middle Eocene) of southwestern Oregon. *J. Paleontol.* 63: 69-73.

Beschin, C., Busulini, A., De Angeli, A. & Tessier, G. 2002. Aggiornamento ai crostacei eocenici di Cava "Main" di Arzignano (Vicenza-Italia settentrionale) (Crustacea, Decapoda). *Studi e Ricerche, Assoc. Amici Mus., Mus. civ. "G. Zannato", Montecchio Maggiore (Vicenza)*, 2002: 7-28.

Beschin, C. & De Angeli, A. 2003. *Spinipalicus italicus*, nuovo genere e specie di Palicidae (Crustacea, Decapoda) dell'Eocene del Vicentino (Italia settentrionale). *Studi e Ricerche, Assoc. Amici Mus., Mus. civ. "G. Zannato", Montecchio Maggiore (Vicenza)*, 2003: 7-12.

Beurlen, K. 1928. Die fossilen Dromiaceen und ihre Stammesgeschichte. *Paläontol. Zeitschr.* 10: 144-183.

Beurlen, K. 1930. Vergleichende Stammesgeschichte Grundlagen, Methoden, Probleme unter besonderer Berücksichtigung der höheren Krebse. *Fortschr. Geol. Paläontol.* 8: 317-586.

Beurlen, K. 1958. Contribução a paleontologia do Estado do Pará, Crustaceos decápodos da Formação Pirabas. *Bol. Mus. Par. Emílio Goeldi, N.S. (Geol.)*, 5: 2-48.

Bishop, G.A. 1976. *Ekalakia lamberti* n. gen., n. sp. (Crustacea, Decapoda) from the Upper Cretaceous Pierre Shale of eastern Montana. *J. Paleontol.* 50: 398-401.

Bishop, G.A. 1978. Two new crabs, *Sodakus tatankayotankaensis* n. gen., n. sp. and *Raninella oaheensis* n. sp. (Crustacea, Decapoda), from the Upper Cretaceous Pierre Shale of South Dakota. *J. Paleontol.* 52: 608-617.

Bishop, G.A. 1983a. Fossil decapod crustaceans from the Lower Cretaceous Glen Rose Limestone of Central Texas. *Trans. San Diego Soc. Nat. Hist.* 20: 27-55.

Bishop, G.A. 1983b. Two new species of crabs, *Notopocorystes (Eucorystes) eichhorni* and *Zygastrocarcinus griesi* (Decapoda: Brachyura) from the Bearpaw Shale (Campanian) of north-central Montana. *J. Paleontol.* 57: 900-910.

Bishop, G.A. 1985. Fossil decapod crustaceans from the Gammon Ferruginous Member, Pierre Shale (early Campanian), Black Hills, South Dakota. *J. Paleontol.* 59: 605-624.

Bishop, G.A. 1986. Two new crabs, *Parapaguristes tuberculatus* and *Palaeoxantho libertiensis*, from the Prairie Bluff Formation (middle Maastrichtian), Union County, Mississippi, U.S.A. *Proc. Biol. Soc. Wash.* 99: 604-611.

Bishop, G.A. 1988a. Two crabs, *Xandaros sternbergi* (Rathbun, 1926) n. gen., and *Icriocarcinus xestos* n. gen., n. sp., from the Late Cretaceous of San Diego County, California, USA, and Baja California Norte, Mexico. *Trans. San Diego Soc. Nat. Hist.* 21: 245-257.

Bishop, G.A. 1988b. A new crab, *Seorsus wadei*, from the Late Cretaceous Coon Creek Formation, Union County, Mississippi. *Proc. Biol. Soc. Wash.* 101: 72-78.

Bishop, G.A. 1988c. New fossil crabs, *Plagiophthalmus izetti, Latheticocarcinus shipiroi*, and *Sagittiformosus carabus* (Crustacea, Decapoda) from the Western Interior Cretaceous, U.S.A. *Proc. Biol. Soc. Wash.* 101: 375-381.

Bishop, G.A., Feldmann, R.M & Vega, F.J. 1998. The Dakoticancridae (Decapoda, Brachyura) from the Late Cretaceous of North America and Mexico. *Contrib. Zool.* 67: 237-255.

Borradaile, L. A. 1903. On the classification of the Thalassinidea. *Ann. Mag. Nat. Hist.* (7) 12: 534-551.

Bosc, L.A.G.1802. *Histoire naturelle des Crustacés, contenant leur description et leurs moeurs, avec figures dessinées d'après nature 1*: 1-258, pls. 1-8; *2*: 1-296, pls. 9-18. Paris: de Guilleminet.

Buckeridge, J. 1999. Post Cretaceous biotic recovery: a case study on (Crustacea: Cirripedia) from the Chatham Islands, New Zealand. *Rec. Canterb. Mus.* 13: 43-51.

Busulini, A., Tessier, G., Beschin, C. & De Angeli, A. 2003. *Boschettia giampietroi*, nuovo genere e specie di Portunidae (Crustacea, Decapoda) dell'Eocene medio della Valle del Chiampo (Vicenza, Italia settentrionale). *Studi e Ricerche, Assoc. Amici Mus., Mus. civ. "G. Zannato", Montecchio Maggiore (Vicenza), 2003*: 13-18.

Carter, J. 1872. On *Orithopsis Bonneyi*, a new fossil crustacean. *Geol. Mag.* 9: 529-532.

Collins, J.S.H. 1997. Fossil Homolidae (Crustacea; Decapoda). *Bull. Mizunami Fossil Mus.* 24: 51-71.

Collins, J.S.H. & Jakobsen, S.L. 1994. A synopsis of the biostratigraphic distribution of the crab genera (Crustacea, Decapoda) of the Danian (Palaeocene) of Denmark and Sweden. *Bull. Mizunami Fossil Mus.* 21: 35-46.

Collins, J.S.H. & Morris, S.F. 1975. A new crab, *Costacopluma concava*, from the Upper Cretaceous of Nigeria. *Palaeonology* 18: 823-829.

Collins, J.S.H. & Smith, R. 1992. Ypresian (Lower Eocene) crabs (Decapoda, Crustacea) from Belgium. *Bull. Inst. R. Sci. Nat. Belg., Sci. Terre,* 63: 261-270.

Culver, S.J. & Rawson, P.F. (eds.) 2000. *Biotic Response to Global Change: The Last 145 Million Years.* London: Nat. Hist. Mus. and Cambridge Univ. Press.

Dana, J.D. 1852. Crustacea, Pt. 1. In: *United States Exploring Expedition During the Years 1838, 1839, 1840, 1841, 1842; Under the Command of Charles Wilkes, U.S.N., Vol. 13:* 685 pp.

Davidson, E. 1966. A new Paleocene crab from Texas. *J. Paleontol.* 40: 211-213.

De Angeli, A. & Beschin, C. 2001. I crostacei fossili del territorio Vicentino. *Natura Vicentina* 5: 5-54.

De Angeli, A. & Garassino, A. 2002. Galatheid, chirostylid and porcellanid decapods (Crustacea, Decapoda, Anomura) from the Eocene and Oligocene of Vicenza (N Italy). *Mem. Soc. ital. Sci. nat. Mus. civ. St. nat. Milano* 30 (3): 1-40.

de Haan W. 1833-1850. Crustacea. In: Von Siebold, P.F. (ed.), *Fauna Japonica sive Descriptio Animalium, quae in Itinere per Japoniam, Jussu et Auspiciis Superiorum, qui summum in India Batava Imperium Tenent, Suscepto, Annis 1823-1830 Collegit, Notis, Observationibus et Adumbrationibus Illustravit:* i-xvii, i-xxxi, ix-xvi, 1-243, pls. A-J, L-Q, 1-55, circ. Tab. 2. Lugduni-Batavorum (= Leiden): J. Müller et Co.

de Saint Laurent, M. 1973. Sur la systématique et la phylogénie des Thalassinidea: définition des familles des Callianassidae et des Upogebiidae et diagnose de cinq genres nouveaux (Crustacea Decapoda). *Comptes Rendus hebdomad. Séances Acad. Sci., Paris (D),* 277: 513-516.

Ekdale, A.A. & Bromley, R.G. 1984. Sedimentology and ichnology of the Cretaceous-Tertiary boundary in Denmark: implications for the causes of the terminal Cretaceous extinction. *J. Sedim. Petrol.* 54: 681-703.

Fabiani, R. 1910. I crostacei terziarii del Vicentino. *Boll. Mus. civ. Vicenza* 1: 1-40.

Fabricius, J.C. 1793. *Entomologiae Systematica Emendata et Aucta, Secundum Classes, Ordines, Genera, Species, Adjectis Synonimis, Locis, Observationibus, Descriptionibus:* 519 pp. Hafniae (= Copenhagen): C.G. Proft et Storch.

Fabricius, J.C. 1798. *Supplementum Entomologiae Systematicae:* 572 pp. Hafniae (= Copenhagen): C.G. Proft et Storch.

Feldmann, R.M. 1984. *Haumuriaegla glaessneri* n. gen. and sp. (Decapoda; Anomura; Aeglidae) from Haumurian (Late Cretaceous) rocks near Cheviot, New Zealand. *New Zeal. J. Geol. Geophys.* 27: 379-385.

Feldmann, R.M. 1986. Paleobiogeography of two decapod crustacean taxa in the Southern Hemisphere: global conclusions with sparse data. In: Gore, R.H. & Heck, K.L. (eds.), *Crustacean Biogeography:* pp. 5-19. Rotterdam; Boston: A.A. Balkema.

Feldmann, R.M. 1990. On impacts and extinction: biological solutions to biological problems. *J. Paleontol.* 64: 151-154.

Feldmann, R.M. 1991. The genus *Lyreidus* de Haan, 1839 (Crustacea, Decapoda, Raninidae): systematics and biogeography. *J. Paleontol.* 66: 943-957.

Feldmann, R.M. 1993. Additions to the fossil decapod crustacean fauna of New Zealand. *New Zeal. J. Geol. Geophys.* 36: 201-211.

Feldmann, R.M. 2003. The Decapoda: new initiatives and novel approaches. *J. Paleontol.* 77: 1021-1039.

Feldmann, R.M., Aguirre-Urreta, M., Chirino-Gálvez, L. & Casadío, S. 1997. Paleobiogeography of Cretaceous and Tertiary decapod crustaceans from southern South America: the link with Antarctica. In: Ricci, C.A. (ed.), *The Antarctic Region: Geological Evolution and Processes*: pp. 1007-1016. Siena, Italy: Terra Antarctica Publ.

Feldmann, R.M., Casadío, S., Chirino-Galvez, L. & Aguirre-Urreta, M. 1995. Fossil decapod crustaceans from the Jagüel and Roca Formations (Maastrichtian-Danian) of the Neuquén Basin, Argentina. *Paleontol. Soc. Mem.* 43: ii + 1-22.

Feldmann, R.M. & Schweitzer, C.E. Accepted. Paleobiogeography of southern hemisphere decapod Crustacea. *J. Paleontol.* 80 (1).

Feldmann, R.M., Tshudy, D.M. & Thomson, M.R.A. 1993. Late Cretaceous and Paleocene decapod crustaceans from James Ross Basin, Antarctic Peninsula. *Paleontol. Soc. Mem.* 28: iv + 1-41.

Feldmann, R.M. & Wilson, M.T. 1988. Eocene decapod crustaceans from Antarctica. In: Feldmann, R.M. & Woodburne, M.O. (eds.), *Geology and Paleontology of Seymour Island, Antarctic Peninsula*: pp. 465-488. Boulder, Colorado: Geol. Soc. Amer. Memoir 169.

Feldmann, R.M., Vega, F.J., Applegate, S.P. & Bishop, G.A. 1998. Early Cretaceous arthropods from the Tlayúa Formation at Tepexi de Rodríguez, Puebla, México. *J. Paleontol.* 72: 79-90.

Förster, R. 1968. *Paranecrocarcinus libanoticus* n. sp. (Decapoda) und die Entwicklung der Calappidae in der Kreide. *Mitt. Bay. St.-Samml. Paläont. hist. Geol.* 8: 167-195.

Förster, R. 1986. Der erste Nachweis eines brachyuren Krebses aus dem Lias (oberes Pliensbach) Mitteleuropas. *Mitt. Bay. St.-Samml. Paläont. hist. Geol.* 26: 25-31.

Förster, R., Gaździcki, A. & Wrona, R. 1985. First record of a homolodromiid crab from a Lower Miocene glacio-marine sequence of West Antarctica. *Neues Jahrb. Geol. Paläont., Mh.,* 6: 340-348.

Fraaije, R.H. B. 2003. Evolution of reef-associated decapod crustaceans through time, with particular reference to the Maastrichtian type area. *Contrib. Zool.* 72: 119-130.

Fritsch, A. & Kafka, J. 1887. *Die Crustaceen der böhmischen Kreideformation*: iv + 53 pp. Praha: Selbstverlag.

Garassino, A. & Pasini, G. 2002. Studies on Permo-Trias of Madagascar; 5. *Ambilobeia karojoi* n. gen., n. sp. (Crustacea, Decapoda) from the Lower Triassic (Olenekian) of Ambilobé region (NW Madagascar). *Atti Soc. Ital. Sci. nat. Mus. civ. St. nat. Milano* 143: 95-104.

Garassino, A. & Teruzzi, G. 1993. A new decapod crustacean assemblage from the Upper Triassic of Lombardy (N. Italy). *Paleontol. Lombarda, N.S.,* 1: 27 pp.

Garassino, A. & Teruzzi, G. 1995. Studies on Permo-Trias of Madagascar; 3. The decapod crustaceans of the Ambilobé region (NW Madagascar). *Atti Soc. Ital. Sci. nat. Mus. civ. St. nat. Milano* 134: 85-113.

Garassino, A., Teruzzi,G. & Dalla Vecchia F. 1996. The macruran decapod crustaceans of the Dolomia di Forni (Norian, Upper Triassic) of Carnia (Udine, NE Italy). *Atti Soc. Ital. Sci. nat. Mus. civ. St. nat. Milano* 136: 15-60.

Gill, T. 1894. A new bassalian type of crabs. *Am. Nat.* 28: 1043-1045.

Glaessner, M.F. 1930. Neues Krebsreste aus der Kreide. *Jahrb. Preuss. Geol. Landesanst. Berlin* 51: 1-7.

Glaessner, M.F. 1933. Die Krabben der Juraformation. *Zentralbl. Mineral. Geol. Paläont., B*, 3: 178-191.

Glaessner, M.F. 1969. Decapoda. In: Moore, R.C. (ed.), *Treatise on Invertebrate Paleontology, Pt. R, Arthropoda, 4 (2)*: pp. R400-R533, R626-R628. Boulder; Lawrence: Geol. Soc. Amer. and Univ. Kansas Press.

Glaessner, M.F. 1980. New Cretaceous and Tertiary crabs (Crustacea: Brachyura) from Australia and New Zealand. *Trans. R. Soc. S. Austr.* 104: 171-192.

Gould, S.J. 1989. *Wonderful Life: the Burgess Shale and the Nature of History*. New York: W.W. Norton.

Gould, S.J. 2001. Contingency. In: Briggs, D.E.G. & Crowther, P.R. (eds.), *Palaeobiology II*: pp. 195-198. Oxford: Blackwell Science Ltd.

Guinot, D. 1991. Établissement de la famille des Poupiniidae pour *Poupina hirsuta* gen. nov., sp. nov. de Polynésie (Crustacea Decapoda Brachyura Homoloidea). *Bull. Mus. natn. Hist. nat., Paris, Sér.* 4, 12 (1990): 577-605.

Guinot, D. & Richer de Forges, B. 1995. Crustacea Decapoda Brachyura: Révision de la famille des Homolidae de Haan, 1839. In: Crosnier, A. (ed.), *Résultats des campagnes MUSORSTOM, Vol. 13*. Mém. Mus. natn. Hist. nat. 163: 283-517.

Guinot, D. & Tavares, M. 2001. Une nouvelle famille de Crabes du Crétacé, et la notion de Podotremata Guinot, 1977 (Crustacea, Decapoda, Brachyura). *Zoosystema* 23: 507-546.

Hallam, A. & Wignall, P.B. 1997. *Mass Extinctions and Their Aftermath*: 320 pp. Oxford: Oxford Univ. Press.

Harries, P.J. 1999. Repopulations from Cretaceous mass extinctions: environmental and/or evolutionary controls? In: Barrera, E & Johnson, C.C. (eds.), *Evolution of the Cretaceous Ocean-Climate System*: pp. 345-364. Boulder, Colorado: Geol. Soc. Amer. Spec. Pap. 332.

Harries, P.J., Kauffman, E.G. & Hansen, T.A. 1996. Models for biotic survival following mass extinction. In: Hart, M.B. (ed.), *Biotic Recovery from Mass Extinction Events*: pp. 41-60. London: Geol. Soc. Spec. Publ. 102.

Hart, M.B. (ed.). 1996. *Biotic Recovery from Mass Extinction Events*. London: Geol. Soc. Spec. Publ. 102.

Hotton III, N., Feldmann, R.M., Hook, R.W. & DiMichele, W.A. 2002. Crustacean-bearing continental deposits in the Petrolia Formation (Leonardian Series, Lower Permian) of North-Central Texas. *J. Paleontol.* 76: 486-494.

Houša, V. 1963. Parasites of Tithonian decapod crustaceans (Štramberk, Moravia). *Sborník Ústředniho Ústavu Geolog., Paleontol.* 28: 101-114.

Huxley, T.H. 1879. On the classification and the distribution of the crayfishes. *Proc. Sci. Meet. Zool. Soc. London* 1878: 752-788.

Izett, G.A., Cobban, W.A., Kunk, M.J. & Obradovich, J.D. 1993. The Manson impact structure; $^{40}Ar/^{39}Ar$ age and its distal impact ejecta in the Pierre Shale in southeastern South Dakota. *Science* 262 (5134): 729-732.

Jablonski, D. 1986. Causes and consequences of mass extinctions: a comparative approach. In: Elliott, D.K. (ed.), *Dynamics of Extinction*: pp. 183-229. New York: John Wiley & Sons.

Jablonski, D. 1996. Mass extinctions: Persistent problems and new directions. In: Ryder, G., Fastovsky, D. & Gartner, S. (eds.), *The Cretaceous-Tertiary Event and Other Catastrophes in Earth History*: pp. 1-9. Boulder, Colorado: Geol. Soc. Amer. Spec. Pap. 307.

Jakobsen, S.L. & Collins, J.S.H. 1997. New middle Danian species of anomuran and brachyuran crabs from Fakse, Denmark. *Bull. Geol. Soc. Denmark* 44: 89-100.

Jenkins, R.J.F. 1972. *Metanephrops*, a new genus of late Pliocene to Recent lobsters (Decapoda, Nephropidae). *Crustaceana* 22: 161-177.

Johnson, C.C., Barron, E.J., Kauffman, E.G., Arthur, M.A., Fawcett, P.J. & Yasuda, M.K. 1996. Middle Cretaceous reef collapse linked to ocean heat transport. *Geology* 24: 376-380.

Jux, U. 1971. Ein Brachyuren-Rest aus der Oberkreide Afghanistans. *Paläontol. Zeitschr.* 45: 154-166.

Karasawa, H. 1992. Fossil decapod crustaceans from the Manda Group (middle Eocene), Kyushu, Japan. *Trans. Proc. Palaeontol. Soc. Japan, N.S.*, 167: 1247-1258.

Karasawa, H. & Hayakawa, H. 2000. Additions to Cretaceous decapod crustaceans from Hokkaido, Japan; Part 1, Nephropidae, Micheleidae and Galatheidae. *Paleontol. Res.* 4: 139-145.

Karasawa, H. & Schweitzer, C.E. 2004. Revision of the genus *Glyphithyreus* Reuss, 1859 (Crustacea, Decapoda, Brachyura, Xanthoidea) and recognition of a new genus. *Paleontol. Res.* 8 (3): 143-154.

Kauffman, E.G. & Harries, P.J. 1996. The importance of crisis progenitors in recovery from mass extinction. In: Hart, M.B. (ed.), *Biotic Recovery from Mass Extinction Events*: pp. 15-39. London: Geol. Soc. Spec. Publ. 102.

Kishinouye, K. 1926. Two rare and remarkable forms of macrurous Crustacea from Japan. *Annot. Zool. Japon.* 11: 63-70.

Lamarck, J.B.P.A. de. 1801. *Systême des animaux vertèbres, ou tableau général des classes, des ordres et des genres de ces animaux; présentant leurs caractères essentiels et leur distribution, d'après la considération de leurs rapports naturels et leur organisation, et suivant l'arrangement établi dans les galeries du Muséum d'Histoire Naturelle, parmi leurs dépouilles conservées; précédé de discours d'ouverture du cours de zoologie, donné dans le Muséum national d'Histoire naturelle l'an 8 de la République*: 432 pp. Paris: Déterville.

Latreille, P.A. 1802-1803. *Histoire naturelle, générale et particulière, des crustacés et des insectes*, Vol. 3: xii + 467 pp. Paris: F. DuFart.

Latreille, P.A. 1825. In: *Genre de Crustacés; Encyclopedie méthodiqu; Histoire naturelle; Entomologie, ou histoire naturelle des Crustacés, des Arachnides et des Insectes* 10: 832 pp. Paris: Roret.

Leach, W.E. 1814. Crustaceology. In: Brewster, D. (ed.). *The Edinburgh Encylopaedia*, Vol. 7: pp. 383-437, pl. 221. Edinburgh: W. Blackwood & J. Waugh.

Leach, W.E. 1815. A tabular view of the external characters of four classes of Animals, which Linné arranged under Insecta: with the distribution of the genera composing three of these classes into orders, and descriptions of several new genera and species. *Trans. Linn. Soc. London* 11: 306-400.

Leach, W.E. 1820. Galatéadées. *Dictionnaire des Sciences Naturelles* 18: 49-56. Paris: F.G. Levreault.

Lőrenthey, E. 1902. Neue Beiträge zur Tertiären Dekapodenfauna Ungarns. *Math. Naturw. Ber. Ungarn* 18: 98-120.

Lőrenthey, E. & Beurlen, K. 1929. Die fossilen Dekapoden der Länder der Ungarischen Krone. *Geol. Hungar., Ser. Palaeontol.*, 3: 421 pp., 16 pls.

MacLeay, W.S. 1838. On the Brachyurous Decapod Crustacea brought from the Cape by Dr. Smith. In: Smith, A. (ed.), *Illustrations of the Annulosa of South Africa; consisting chiefly of Figures and Descriptions of the Objects of Natural History Collected during an Expedition into the Interior of South Africa, in the Years 1834, 1835, and 1836; fitted out by "The Cape of Good Hope Association for Exploring Central Africa"*: pp. 53-71, 2 pls. London: Smith, Elder & Company.

MacLeod, N. & Keller, G. (eds.). 1996. *Cretaceous-Tertiary Mass Extinctions: Biotic and Environmental Changes*. New York: W.W. Norton & Company.

Manning, R.B. & Felder, D.L. 1991. Revision of the American Callianassidae (Crustacea: Decapoda: Thalassinidea). *Proc. Biol. Soc. Wash.* 104: 764-792.

Manning, R.B. & Holthuis, L.B. 1981. West African brachyuran crabs (Crustacea: Decapoda). *Smithsonian Contrib. Zool.* 306: xiii + 1-379.

Mantell, G.A. 1822. *The Fossils of the South Downs; or Illustrations of the Geology of Sussex*: 327 pp., 42 pls. London: Lupton Relfe.

Marangon, S. & De Angeli, A. 1997. *Cherpiocarcinus*, nuovo genere di brachiuro (Decapoda) dell'Oligocene del Bacino Ligure-Piemontese (Italia settentrionale). *Lavori, Soc. Ven. Sci. Nat.*, 22: 97-106.

Marincovich, L. 1993. Danian mollusks from the Prince Creek Formation, northern Alaska, and implications for Arctic Ocean paleogeography. *Paleontol. Soc. Mem.* 67: iv + 35 pp.

Martins-Neto, R.G. 2001. Review of some Crustacea (Isopoda and Decapoda) from Brazilian deposits (Paleozoic, Mesozoic and Cenozoic) with descriptions of new taxa. *Acta Geol. Leopold.* 24 (52/53): 237-254.

McCoy, F. 1849. On the classification of some British fossil Crustacea with notices of new forms in the University Collection at Cambridge. *Ann. Mag. Nat. Hist.* 2 (4): 161-179, 330-335.

Mertin, H. 1941. Decapode Krebse aus dem subhercynen und Braunschweiger Emscher und Untersenon, sowie Bemerkungen über einige verwandte Formen in der Oberkreide. *Nov. Acta Leopold., N.F.,* 10: 149-262.

Miers, E. J. 1886. Report on the Brachyura collected by H.M.S. Challenger during the years 1873-1876. In: Wyville Thomson, C. & Murray, J. (eds.), *Report of the Scientific Results of the Voyage of H.M.S. Challenger during the years 1873-76, Zoology Vol. XVII*: pp. 1-362. Edinburgh: Challenger Office, published by Order of her Majesty's Government; first reprinting 1965, New York: Johnson Reprint Corp.

Milne Edwards, A. 1862. Sur l'existence de Crustacés de la famille des Raniniens pendant la période crétacée. *Comptes Rendus hebdomad. Séances Acad. Sci. Paris* 55: 492-494.

Milne Edwards, A. 1862-1865. Monographie des crustacés de la famille des cancériens. *Ann. Sci. Nat., Zool., Ser.* 4/18 (1862): 31-85, pls. 1-10; 20 (1863): 273-324, pls. 5-12; *Ser.* 5/1 (1864): 31-88, pls. 1-10; 3 (1865): 297-351, pls. 5-13.

Milne Edwards, A. 1880. Reports on the results of dredging, under the supervision of Alexander Agassiz, in the Gulf of Mexico and in the Caribbean Sea, 1877, 78, 79, by the United States Coast Survey Steamer "Blake". VIII. Études préliminaires sur les Crustacés, 1er partie. *Bull. Mus. Comp. Zool. Harvard Coll.* 8 (1): 1-68.

Milne Edwards, A. & Brocchi, P. 1879. Note sur quelques Crustacés fossiles appartenant au groupe des macrophthalmiens. *Bull. Soc. Philomath. Paris* 3: 113-117.

Milne Edwards, H. 1834-1840. *Histoire naturelle des Crustacés, comprenant l'anatomie, la physiologie, et la classification de ces animaux*; *Vol. 1,* 1834: xxxv + 468 pp.; *Vol. 2,* 1837: 532 pp.; *Vol. 3,* 1840: 638 pp.; *Atlas:* 32 pp., pls. 1-42.

Müller, P. & Collins, J.S.H. 1991. Late Eocene coral-associated decapods (Crustacea) from Hungary. *Contrib. Tert. Quatern. Geol.* 28: 47-92.

Ortmann, A. 1892a. Die Decapoden-Krebse des Strassburger Museums, V. Theil; Die Abtheilungen Hippidea, Dromiidea und Oxystomata. *Zool. Jahrb., Abth. Syst., Geogr. Biol. Thiere* 6: 532-588, pl. 26.

Ortmann, A. 1892b. Die Decapoden-Krebse des Strassburger Museums, IV. Theil; Die Abtheilungen Galatheidae und Paguridae. *Zool. Jahrb., Abth. Syst., Geogr. Biol. Thiere* 6: 241-326, pls. 11-12.

Parker, T.J. 1883. On the structure of the head in *Palinurus* with special reference to the classification of the genus. *Nature* 29: 189-190.

Patrulius, D. 1960. Contribution à la systématique des Décapodes néojurassiques. *Rev. Géol. Géogr.* 3: 249-257.

Rafinesque, C.S. 1815. *Analyse de la nature, ou tableau de l'univers et des corps organisés*: 224 pp. Palermo, Italy: J. Baravecchia.

Rathbun, M.J. 1908. Descriptions of fossil crabs from California. *Proc. US Nat. Mus.* 35: 341-349.

Rathbun, M.J. 1917. New species of South Dakota Cretaceous crabs. *Proc. US Nat. Mus.* 52: 385-391.

Rathbun, M.J. 1923. Decapod crustaceans from the Upper Cretaceous of North Carolina. *N. Carolina Geol. Sur.* 5: 403-407.

Rathbun, M.J. 1926. The fossil stalk-eyed Crustacea of the Pacific Slope of North America. *US Nat. Mus. Bull.* 138: viii + 1-155.

Rathbun, M.J. 1935. Fossil Crustacea of the Atlantic and Gulf Coastal Plain. *Geol. Soc. Amer. Spec. Pap.* 2: viii + 1-160.

Remeš, M. 1895. Beiträge zur Kenntnis der Crustaceen der Stramberger Schichten. *Bull. Intern. Acad. Sci. Bohème* 2: 200-204, pls. 1-3.

Reuss, A.E. 1859. Zur Kenntnis fossiler Krabben. *Denkschr. kaiserl. Akad. Wissensch. Wien* 17: 1-90, pls. 1-24.

Richards, B.C. 1975. *Longusorbis cuniculosus*: a new genus and species of Upper Cretaceous crab; with comments on Spray Formation at Shelter Point, Vancouver Island, British Columbia. *Can. J. Earth Sci.* 12: 1850-1863.

Rinehart, L.F., Lucas, S.G. & Heckert, A.B. 2003. An early eubrachyuran (Malacostraca: Decapoda) from the Upper Triassic Snyder Quarry, Petrified Forest Formation, North-central New Mexico. In: Zeigler, K.E., Heckert, A.B. & Lucas, S.G. (eds.), *Paleontology and Geology of the Snyder Quarry*: pp. 67-70. Albuquerque: New Mexico Mus. Nat. Hist. Sci. Bull. No. 24.

Roemer, F. A. 1887. *Graptocarcinus texanus*, ein Brachyure aus der oberen Kreide von Texas. *Leonhardt und Bronn's Neues Jahrb. Mineral., Geol., Paläontol.* 1: 173-176.

Sakai, K. 1988. A new genus and five new species of Callianassidae (Crustacea: Decapoda: Thalassinidea) from northern Australia. *The Beagle, Rec. N. Territ. Mus. Arts Sci.* 5: 51-69.

Sakai, K. 1992. The families Callianideidae and Thalassinidae with the description of two new subfamilies, one new genus and two new species (Decapoda: Thalassinidea). *Naturalists* 4: 1-33.

Samouelle, G. 1819. *The Entomologist's Useful Compendium, or An Introduction to the Knowledge of British Insects*: 496 pp. London: T. Boys.

Schram, F.R. 1986. *Crustacea*: 606 pp. Oxford: Oxford Univ. Press.

Schram, F.R., Feldmann, R.M. & Copeland, M.J. 1978. The Late Devonian Palaeopalaemonidae and the earliest decapod crustaceans. *J. Paleontol.* 52: 1375-1387.

Schram, F.R. & Mapes, R.H. 1984. *Imocaris tuberculata*, n. gen., n. sp. (Crustacea: Decapoda) from the upper Mississippian Imo Formation, Arkansas. *Trans. San Diego Soc. Nat. Hist.* 20: 165-168.

Schweitzer, C.E. 2001. Paleobiogeography of Cretaceous and Tertiary decapod crustaceans of the North Pacific Ocean. *J. Paleontol.* 75: 808-826.

Schweitzer, C.E. 2003. Utility of proxy characters for classification of fossils: an example from the fossil Xanthoidea (Crustacea: Decapoda: Brachyura). *J. Paleontol.* 77: 1107-1128.

Schweitzer, C.E. In press. The genus *Xanthilites* Bell, 1858 and a new xanthoid family (Crustacea: Decapoda: Brachyura: Xanthoidea): new hypotheses on the origin of the Xanthoidea MacLeay, 1838. *J. Paleontol.* 78.

Schweitzer, C.E. & Feldmann, R.M. 2000a. First notice of the Chirostylidae (Decapoda) in the fossil record and new Tertiary Galatheidae (Decapoda) from the Americas. *Bull. Mizunami Fossil Mus.* 27: 147-165.

Schweitzer, C.E. & Feldmann, R.M. 2000b. New species of calappid crabs from western North America and reconsideration of the Calappidae sensu lato. *J. Paleontol.* 74: 230-246.

Schweitzer, C.E. & Feldmann, R.M. 2001a. Differentiation of the fossil Hexapodidae Miers, 1886 (Decapoda: Brachyura) from similar forms. *J. Paleontol.* 75: 330-345.

Schweitzer, C.E. & Feldmann, R.M. 2001b. New Cretaceous and Tertiary decapod crustaceans from western North America. *Bull. Mizunami Fossil Mus.* 28: 173-210.

Schweitzer, C.E., Feldmann, R.M., Fam, J., Hessin, W. A., Hetrick, S. W., Nyborg, T. G. & Ross, R.L.M. 2003. *Cretaceous and Eocene Decapod Crustaceans from Eastern Vancouver Island, British Columbia, Canada*: 66 pp. Ottawa, Ontario: NRC Research Press.

Schweitzer, C.E., Feldmann, R.M. & Gingerich, P.D. 2004. New decapods (Crustacea) from the Eocene of Pakistan and a revision of *Lobonotus* A. Milne Edwards, 1864. *Univ. Michigan, Contrib. Mus. Paleontol.* 31 (4): 89-118.

Schweitzer, C.E., Feldmann, R.M., Gonzáles-Barba, G. & Vega, F.J. 2002. New crabs from the Eocene and Oligocene of Baja California Sur, Mexico and an assessment of the evolutionary and paleobiogeographic implications of Mexican fossil decapods. *Paleontol. Soc. Mem.* 59: 43 pp.

Schweitzer, C.E., Nyborg, T.G., Feldmann, R.M. & Ross, R.L.M. 2004. Homolidae de Haan, 1839 and Homolodromiidae Alcock, 1900 (Crustacea: Decapoda: Brachyura) from the Pacific Northwest of North America and a reassessment of their fossil records. *J. Paleontol.* 78: 133-149.

Secretan, S. 1972. Crustacés Décapodes nouveaux du Crétacé supérieur de Bolivie. *Bull. Mus. Nat. Hist. Nat.* 3 (49): 1-16.

Sepkoski Jr., J.J. 1989. Periodicity in extinction and the problem of catastrophism in the history of life. *J. Geol. Soc. London* 146: 7-19.

Sepkoski Jr., J.J. 2000. Crustacean biodiversity through the marine fossil record. *Contrib. Zool.* 69: 213-222.

Sheehan, P.M., Coorough, P.J. & Fastovsky, D.E. 1996. Biotic selectivity during the K/T and Late Ordovician extinction events. In: Ryder, G., Fastovsky, D. & Gartner, S. (eds.), *The Cretaceous-Tertiary Event and Other Catastrophes in Earth History*: pp. 477-489. Boulder, Colorado: Geol. Soc. Amer. Spec. Pap. 307.

Signor, P.W. & Lipps, J.H. 1982. Sampling bias, gradual extinction patterns, and catastrophes in the fossil record. In: Silver, L.T. (ed.), *Geological Implications of Large Asteroids and Comets on the Earth*: pp. 291-296. Boulder, Colorado: Geol. Soc. Amer. Spec. Pap. 190.

Smith, S.I. 1883. Preliminary report on the Brachyura and Anomura dredged in deep water off the south coast of New England by the United States Fish Commission in 1880, 1881, and 1882. *Proc. US Nat. Mus.* 6: 1-57, pls. 1-6.

Stenzel, H.B. 1944. A new Paleocene catometope crab from Texas, *Tehuacana tehuacana*. *J. Paleontol.* 18: 546-549.

Stenzel, H.B. 1945. Decapod crustaceans from the Cretaceous of Texas. *Univ. Texas Publ.* 4401: 401-477.

Stenzel, H.B. 1953. Decapod crustaceans from the Woodbine Formation of Texas. *US Geol. Survey Prof. Pap.* 242: 212-217.

Stilwell, J. 2003. Patterns of biodiversity and faunal rebound following the K-T boundary extinction event in Austral Palaeocene molluscan faunas. *Palaeogeogr., Palaeoclimat., Palaeoecol.* 195: 319-356.

Stimpson, W. 1858. Crustacea Ocypodoidea. *Proc. Acad. Nat. Sci. Philadel.* 1858: 93-110.

Stimpson, W. 1866. Descriptions of new genera and species of Macrurous Crustacea from the coasts of North America. *Proc. Chicago Acad. Sci.* 1: 46-48.

Tshudy, D.M. & Babcock, L.E. 1997. Morphology-based phylogenetic analysis of the clawed lobsters (family Nephropidae) and the new family Chilenophoberidae. *J. Crust. Biol.* 17: 253-263.

Tucker, A.B. 1995. *A systematic evaluation of fossil Raninidae from the Twin River Group, Olympic Peninsula, Washington, and a re-examination of the Raninidae.* Unpublished Ph.D. dissertation: Kent State Univ., Ohio.

Tucker, A.B. 1998. Systematics of the Raninidae (Crustacea: Decapoda: Brachyura), with accounts of three new genera and two new species. *Proc. Biol. Soc. Wash.* 111: 320-371.

Van Straelen, V. 1925. Contribution à l'étude des crustacés décapodes de la période jurassique. *Mém. Acad. R. Belg., Class. Sci.,* 2/7: 1-462, pls. 1-10.

Van Straelen, V. 1933. *Antrimpos madagascariensis* Crustace Décapode du Permotrias de Madagascar. *Bull. Mus. R. Hist. Nat. Belg.* 9 (15): 1-3.

Van Straelen, V. 1934. Contribution à l'étude des crustacés décapodes fossiles de la Catalogne. *Géol. Pays Catalans* 3 (25): 1-6.

Van Straelen, V. 1936. Crustacés Décapodes nouveaux ou peu connus de l'époque Crétacique. *Bull. Mus. R. Hist. Nat. Belg.* 12 (45): 1-50, 4 pls.

Van Valen, L.M. 1994. Concepts and the nature of selection by extinction: is generalization possible? In: Glen, W. (ed.), *The Mass-Extinction Debates: How Science Works in a Crisis*: pp. 200-216. Stanford, California: Stanford Univ. Press.

Vega, F.J., Cosma, T., Coutiño, M.A., Feldmann, R.M., Nyborg, T.G., Schweitzer, C.E. & Waugh, D.A. 2001b. New middle Eocene Decapods (Crustacea) from Chiapas, Mexico. *J. Paleontol.* 75: 929-946.

Vega, F.J. & Feldmann, R.M. 1991. Fossil crabs (Crustacea, Decapoda) from the Maastrichtian Difunta Group, northeastern Mexico. *Ann. Carnegie Mus.* 60: 163-177.

Vega, F.J., Feldmann, R.M., García-Barrera, P., Filkorn, H., Pimentel, F. & Avendaño, J. 2001a. Maastrichtian Crustacea (Brachyura: Decapoda) from the Ocozocuautla Formation in Chiapas, southeast Mexico. *J. Paleontol.* 75: 319-329.

Vega, F.J., Feldmann, R.M. & Sour-Tovar, F. 1995. Fossil crabs (Crustacea: Decapoda) from the Late Cretaceous Cárdenas Formation, East-Central Mexico. *J. Paleontol.* 69: 340-350.

Via, L. 1966. *Pinnixa (Palaeopinnixa) mytilicola*, nuevo braquiuro fósil, en el mioceno marino del Vallés (Barcelona). *Acta Geol. Hisp.* 1: 1-4.

Via, L. 1981. Les crustacés décapodes du Cénomanien de Navarra (Espagne): premiers résultats de l'étude des Galatheidae. *Geobios* 14: 247-251.

von Meyer, H. 1835. Briefliche Mitteilungen. In: *Leonhardt und Bronn's Neues Jahrb. Mineral., Geol., Paläontol.* Stuttgart: C.F. Winter.

von Meyer, H. 1840. Briefliche Mitteilungen. In: *Leonhardt und Bronn's Neues Jahrb. Mineral., Geol., Paläontol.*: p. 587. Stuttgart: C.F. Winter.

von Meyer, H. 1842. Briefliche Mitteilungen. In: *Leonhardt und Bronn's Neues Jahrb. Mineral., Geol., Paläontol.* Stuttgart: C.F. Winter.

von Meyer, H. 1851. Briefliche Mitteilungen. In: *Leonhardt und Bronn's Neues Jahrb. Mineral., Geol., Paläontol.* Stuttgart: C.F. Winter.

von Meyer, H. 1860. Die Prosoponiden oder die Familie der Maskenkrebse. *Palaeontographica* 7: 183-222, pl. 23.

von Meyer, H. 1864. Briefliche Mitteilungen. In: *Leonhardt und Bronn's Neues Jahrb. Mineral., Geol., Paläontol.* Stuttgart: C.F. Winter.

von Zittel, K.A. 1885. *Handbuch der Paläontologie, 2 (2), Molluska und Arthropoda*: pp. 679-721. München: R. Oldenbourg.

Weber, F. 1795. *Nomenclator entomologicus secundum Entomologiam Systematicam ill. Fabricii adjectis speciebus recens detectis et varietatibus*: 171 pp. Chilonii and Hamburgi: C.E. Bohn.

Weller, S. 1905. The fauna of the Cliffwood (N.J.) Clays. *J. Geol.* 13: 324-337.

White, A. 1847. Short descriptions of some new species of Crustacea in the collection of the British Museum. *Ann. Mag. Nat. Hist.* 20: 205-207.

Whiteaves, J.F. 1874. On recent deep-sea dredging operations in the Gulf of St. Lawrence. *Amer. J. Sci.* 3/7: 210-219.

Withers, T.H. 1922. On a new brachyurous crustacean from the Upper Cretaceous of Jamaica. *Ann. Mag. Nat. Hist.* 9/10: 534-541.

Wood-Mason, J. 1875. On the genus *Deidamia* Willemoes-Smith. *Ann. Mag. Nat. Hist.* 4/15: 131-135.

Wood-Mason, J. & Alcock, A. 1891. A note on the result of the last season's deep-sea dredging: natural history notes from H. M. Indian Marine Survey Steamer "Investigator": Commander R.F. Hoskyn, R.N. Commanding, no. 21. *Ann. Mag. Nat. Hist.* 6/7: 258-272.

Woods, J.T. 1953. Brachyura from the Cretaceous of central Queensland. *Mem. Queensl. Mus.* 13: 50-57.

Woods, J.T. 1957. Macrurous decapods from the Cretaceous of Queensland. *Mem. Queensl. Mus.* 13: 155-175.

Woodward, H. 1867. On a new genus of shore-crab, *Goniocypoda Edwardsi*, from the Lower Eocene of Hampshire. *Geol. Mag.* 4: 529-531.

Zinsmeister, W.J. & Feldmann, R.M. 1996. Late Cretaceous faunal changes in the high southern latitudes: a harbinger of global biotic catastrophe. In: MacLeod, N. & Keller, G. (eds.), *Cretaceous-Tertiary Mass Extinctions: Biotic and Environmental Changes*: pp. 303-325. New York: W.W. Norton and Company.

Zinsmeister, W.J., Feldmann, R.M., Woodburne, M.O. & Elliot, D.H. 1989. Latest Cretaceous/earliest Tertiary transition on Seymour Island, Antarctica. *J. Paleontol.* 63: 731-738.

Zinsmeister, W.J. & Macellari, C.E. 1988. Bivalvia (Mollusca) from the Late Cretaceous and earliest Tertiary of Seymour Island, Antarctic Peninsula. In: Feldmann, R.M. & Woodburne, M.O. (eds.), *Geology and Paleontology of Seymour Island*: pp. 253-284. Boulder, Colorado: Geol. Soc. Amer. Mem. 169.

Oelandocaris oelandica and the stem lineage of Crustacea

MARTIN STEIN[1], DIETER WALOSZEK[2] & ANDREAS MAAS[2]

[1] *Uppsala University, Department of Earth Sciences, Palaeobiology, Uppsala, Sweden*
[2] *Section for Biosystematic Documentation, University of Ulm, Ulm, Germany*

ABSTRACT

Oelandocaris oelandica Müller, 1983, was originally described from a single limbless fragment discovered in limestones from the Isle of Öland, Sweden. Subsequently, six, apparently conspecific specimens having preserved appendages were picked from the most productive Upper Cambrian 'Orsten' samples taken in Västergötland in southern Sweden. These new specimens permit a much more conclusive reconstruction and discussion of the systematic affinities of the species than was possible in 1983. The most significant feature of this approximately 1 mm long arthropod are its large antennulae (first pair of limbs), which are subdivided into three long, spine-bearing outgrowths (not rami). Most likely, all three anterior limbs were involved in food gathering, while the posterior limbs are, more or less, serially designed (large basipod with few median setae, endopod (four- or five-segmented?), and a large, leaf-shaped exopod with strong marginal setae) and may have served mainly for locomotion (swimming). There are remarkable similarities to the co-existing, very abundant Upper Cambrian *Agnostus pisiformis* (Wahlenberg 1818), traditionally interpreted as a diminutive trilobite. *O. oelandica* also shares features with some of the derivatives of the stem lineage of the Eucrustacea/Labrophora, particularly *Henningsmoenicaris scutula* Walossek & Müller, 1990. These similarities are used herein to evaluate the feeding apparatus of the head region of these taxa. Our aim is to get a better picture of the development of the feeding apparatus in the early stem lineage of Crustacea and its significance as a useful data set for phylogenetic analyses. *O. oelandica* seems to represent the earliest offshoot of the stem lineage of Crustacea.

We would like to thank Fred Schram for many stimulating discussions and, not least, for his book on the Crustacea, which may be disputable but still is the only available comprehensive textbook on this taxon.

1 INTRODUCTION

The first reports of minute, three-dimensionally preserved phosphatised arthropods etched from Upper Cambrian limestones found in southern Sweden have been produced by the finder, Klaus J. Müller, Bonn in 1979 (Müller 1979). More detailed descriptions of the small Phosphatocopina with their characteristic bivalved head shields followed (Müller 1982). Subsequently, Müller (1983) described a set of six very small arthropods in a three-dimensional preservation, which he assigned to the Crustacea. These are *Bredocaris admirabilis* Müller, 1983, *Oelandocaris oelandica* Müller, 1983, *Rehbachiella kinnekullensis* Müller, 1983, *Skara anulata* Müller, 1983, *Dala peilertae* Müller, 1983, and *Walossekia quinquespinosa* Müller, 1983. The assignment of these animals to the Crustacea was criticised by Lauterbach (1988), who suggested that, despite their apparent diversity, all forms in the 'Orsten' material should be regarded as "stem-group Mandibulata". Remarkably, but not unexpectedly, due to the size range of preserved specimens ranging from 100 µm to 1 mm the 'Orsten' material contains mainly larval or immature specimens. The adult state of the - still very small - largest specimens and eucrustacean affinities could be demonstrated at least for *Bredocaris admirabilis* (Müller & Walossek 1988), having five smaller instars. The two species of a eucrustacean, *Skara* (Müller & Walossek 1985 described a second species, *Skara minuta*, half the size of *S. anulata*), are represented by approximately 150 specimens each, all apparently representing adults. The approximately 90 specimens in the material now available of *Dala peilertae*, currently under investigation, also seem to represent only adults.

Rehbachiella kinnekullensis, another eucrustacean is known, however, only from larvae and pre-adult stages (Walossek 1993; large limb fragments indicate the presence of still later instars). The latest instar has 12 pairs of trunk limbs but 'still' lacks the subdivision of the abdominal region (interpreted by comparison with extant anostracans as close relatives of *Rehbachiella kinnekullensis*; Walossek 1993). A detailed study of *Walossekia quinquespinosa* is in progress. The material has also not yielded stages later than pre-adults, recognizable by the presence of limb buds. Eucrustacean affinities of *Walossekia* can be justified by numerous features shared with this taxon. Apart from the species mentioned, ventral soft-part morphology has become also known from juveniles of *Agnostus pisiformis*, traditionally interpreted as a miniaturized trilobite (Fortey & Theron 1994; Whittington et al. 1997; Fortey 2001) or of uncertain status (Walossek & Müller 1990). The calcified head and tail shields of the later instars and adults of these species are rock-forming (Müller & Walossek 1987). Moreover, the most abundant components in the 'Orsten' fauna, the Phosphatocopina, consist of various species with many specimens (material contains more than 50.000 specimens), but only the immature specimens showing ventral soft-part details (Maas et al. 2003).

Of the six arthropods described by Müller in 1983, only *Oelandocaris oelandica* was originally described from samples, which were collected on the Isle of Öland. The original description was based on a single, fragmentary specimen, of which only the limb bases were preserved. Subsequent investigations yielded five additional specimens in the progressively investigated Västergötland material. They show so much morphological correspondence to the holotype that they are regarded as conspecific specimens and now available for new and emended reconstructions of this species. The new data do not only include the morphology of the appendages, but also other details significant for a future emendation of the diagnosis and description of this animal that reaches a maximum size of no more than one mm. We also found a 400 µm long larva that we assigned to *O. oelandica* on the basis of shared features found only in this species. Both the new data and the similarities between *O. oelandica* and *Agnostus pisiformis* are used to re-evaluate the evolutionary path of the development of cephalic locomotory and feeding structures in the early stem lineage of Crustacea, as initially discussed by Walossek (1999), Maas et al. (2003) and Waloszek (2003a, b). We show that the differences between the taxa investigated are smaller than they appear and that the derivatives of the labrophoran stem lineage (cf. Fig. 7), together with the Phosphatocopina and modern crustaceans (Eucrustacea), share the functional specialization of all three anterior appendages for food accumulation, manipulation and locomotion (multipurpose function).

2 MATERIAL AND METHODS

The holotype of *Oelandocaris oelandica* Müller, 1983, UB 649, is from bituminous limestone nodules. It was collected by Klaus J. Müller near Grönhögen/Degerhamn on the eastern coast of southern Öland, Sweden. Its age has been dated to the *Olenus-gibbosus* Subzone (Zone 2a) of the Upper Cambrian Alum Shale succession. Our six additional specimens (UB W 260-265) are from Västergötland, Sweden. They have also been collected by Klaus Müller from the *Agnostus-pisiformis* Zone (Zone 1) near Gum, Västergötland. UB W 265 is apparently a larva with few limbs, the posterior of which appear less developed than those of the other specimens and more than the posterior limbs known from larval *Goticaris longispinosa* Walossek & Müller, 1990, and *Henningsmoenicaris scutula* (data unpublished). Its assignment to *Oelandocaris oelandica* is based, at least, on the outgrowths on the first appendage (antennula), a characteristic feature of this species.

The material was processed and mounted onto SEM stubs many years ago, hence could not be changed in its position any more. We used a Zeiss DSM 962 scanning electron microscope for our investigations. Digital images were adjusted and assembled in Adobe PhotoShop 7 (version 7.0.1) and CS (version 8.0.1). Line drawings were produced in the traditional way and adjusted to the SEM images using Adobe Illustrator 10 (version 10.0) and CS (version 11.0.0).

We adopted the terminology generally used for crustacean appendages and structures, mainly in the standardized and revised way of Walossek (1993) and Waloszek (2003a, b). For example, the term antennula refers to the first, antenna to the second and mandible to the third cephalic appendages. The hypostome is a ventral sclerotic plate between the antennulae and the antennae (with sclerotic extensions, named wings, which serve as holdfast for limb musculature), while the labrum is a soft, bulged outgrowth at the posterior end of the hypostome, only present in labrophoran Crustacea (most likely originating from the distal part of the mouth membrane, well developed in *Agnostus pisiformis*, Fig. 5A, B). Accordingly, both structures are not the same and cannot be synonymised, thus.

3 MERGING THE OLD AND NEW INFORMATION ON *OELANDOCARIS OELANDICA*

Summarising the original description of the species by Müller, the holotype (UB 649) had a moderately shallow head shield, tapering anteriorly into a pointed end that resembled a short rostrum. The lateral margins of the shield (Müller 1983 occasionally used the term 'carapace', which cannot be applied here, following Walossek 1993) were straight, turning

58   Stein et al.

sharply into the straight posterior rim and did not overhang the subsequent tergite. On the posterolateral angles of either side of the shield, a pair of extensions into spines is indicated by circular holes (Fig. 1A, arrow). The trunk consists of at least five segments. The tergites of the three to four anterior segments bore caudolaterally-pointing spines at the short, rounded pleural extensions, which are broken off as on the shield (Fig. 1A). The rest of the trunk region is so poorly preserved that no further ascertainment of its morphology is possible.

The shallow ventral area begins anteriorly with a weakly elevated hypostome, which was erroneously termed "labrum" by Müller (1983; see above). The anterior rim of the hypostome bears a pair of bean-shaped protuberances, possibly eyes (Figs. 3A, B). The post-hypostomal midventral body area is rather flat, except for a slight elevation in the region of the third pair of limbs = "lower lip" in Müller 1983).

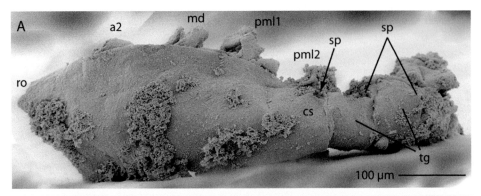

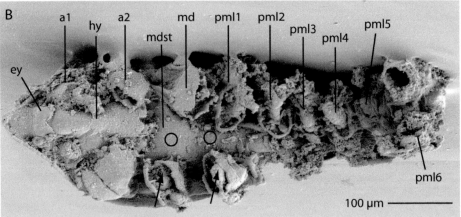

Figure 1. *Oelandocaris oelandica* Müller, 1983. Holotype specimen UB 649. (A) Dorsal view. The arrow points to a protuberance with a hole, possibly the area of breakage of an original spine, as on the caudolateral ends of the visible tergites (sp). (B) Ventral view. Circles mark the sternal parts of the mandibular and first post-mandibular segments, which together form one single unit. Arrows point to rod-like structures in the limb stumps. Abbreviations: a1 = antennula; a2 = antenna (or equivalent limb); cs = cephalic or head shield; ey = 'eyes'; hy = hypostome; md = mandible (or equivalent limb); mdst = mandibular sternite; pml1-5 = post-mandibular limbs; ro = rostrum (broken off distally); sp = spines; tg = tergite.

Oelandocaris and the stem lineage of Crustacea 59

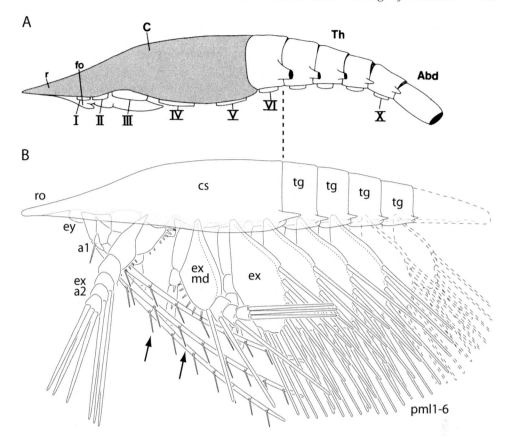

Figure 2. Reconstructions of *Oelandocaris oelandica* Müller, 1983. (A) Original drawing based on the fragmentary holotype specimen (Müller 1983, his fig. 12). (B) Reconstruction based largely on the limb-bearing new specimens. Note the tubular portions of the antennular outgrowths armed with spikes (arrows). The stippled line marks the posterior end of the head. Abbreviations: Abd = abdomen; C = head; ex = exopod; fo/ey = possible median eye; r/ro = rostrum; Th = thorax; I-X = appendages assumed by Müller; I is, however, a piece of residue, see text for details; other abbreviations as in Figure 1.

Little of the limbs is preserved in specimen UB 649, but at least the circular insertion areas of the antennulae and the anteroposteriorly flattened, but fragmentary and poorly preserved proximal parts of the subsequent cephalic appendages (following the posterior shield margin onto the ventral side) and three to four trunk limbs remained present. Only the second limb, the antenna, flanking the posterior edge of the hypostome, shows the basipod and part of the exopod. In some of the posterior limb stumps, a rod-like structure is preserved (indicated by arrows in Fig. 1B), possibly representing coagulated internal matter.

Our re-investigation of the holotype confirmed much of the original description. However, after having made more pictures from different angles of the holotype and of the new specimens, there are two major differences to be noted between the original description and our observations. These are mainly in the number of segments and in the arrangement of limbs. Obviously, the head limb insertions in Müller's original drawing were based on the misinterpretation of two humps near the 'eyes' (median eyes after Müller 1983; *fo* in Fig. 2A), which appear to be two pairs of limbs (I and II in Fig. 2A), while a third pair (III in Fig. 2A) flanked the hypostome. However, the anterior hump (I in Fig. 2A) exists only on the right hand side and is a piece of residue. Thus, the limb flanking the hypostome is in fact the second limb, the antenna and not the third. Consequently, Müller's fourth and fifth limbs are now the third and fourth ones. The actual fifth limb, the maxilla, is Müller's first trunk limb, possibly because a fold in the distorted shield (Fig. 1A) caused a seemingly separating off of this part from the head. In consequence the caudolateral spine of the shield (see new reconstruction in Fig. 2B) was shifted in Müller's drawing to the first trunk segment (Fig. 2A). In fact, shield and all tergites possess such a caudolateral spine (Fig. 2B). Furthermore, there are fewer tergite-bearing trunk segments than described by Müller (1983), most likely no more than four, and a posterior end (see above).

Apart from this, the new specimens (UB W 260-264) from Västergötland confirmed the general shape of the dorsal cuticle, particularly that the head shield was anteriorly extended into a shallow rostral spine, which might have been flat ventrally. New information is provided of the ventral region of the body and, in particular, of the appendages, and several cuticular details. The major features are described in the following.

The major feature of the ventral side of the head of *Oelandocaris oelandica* is the hypostome. It is about one third as long as the entire head and has a spatulate appearance. It ends bluntly in a softer region, where the mouth opens immediately proximal to the margin (Fig. 5D). Hence the mouth can be regarded as exposed, similar to that of *Agnostus pisiformis*, *Martinssonia elongata* Müller & Walossek, 1986, or *Henningsmoenicaris scutula*. The location of the mouth in a depression, the atrium oris, behind the bulged labrum at the rear of the hypostome is developed only in the Phosphatocopina and Eucrustacea (autapomorphy of the taxon Labrophora [Phosphatocopina + Eucrustacea], see Walossek & Müller 1990; Siveter et al. 2001, 2003; Maas et al. 2003; Walossek 2003a, b). The anterior region of the hypostome of *O. oelandica* bears a paired structure that consists of two medially connected plate-like protrusions, on each of which arises a depressed ovoid element (*ey* in Figs. 1B, 3A). Only the tip is somewhat free from the hypostome, while much of the proximal part seems to be fused to the surface of the hypostome. On the hypostomes of *A. pisiformis*, *H. scutula* (see also Fig. 5) and the Phosphatocopina are similar structures, which have been interpreted as median eyes (e.g., Müller & Walossek 1987; Maas et al. 2003). Such eyes belong to the ground pattern of Euarthropoda, as do the lateral eyes (compound eyes), which are, however, not associated with the hypostome. We therefore suggest that the structures in *O. oelandica* could well be median eyes, as already presumed by Müller (1983).

One of the striking features of *O. oelandica* is the morphology of the first appendage, the antennula. This appendage inserts posterior to the eyes in a circular area. The proximal article is a conical socket element (arrow 1 in Fig. 3B, C). It splits into a rod-like outgrowth (arrow 2 in Fig. 3B, C) and another conical article (arrow 3 in Fig. 3B, C). Again, terminally on this article a rod-like outgrowth (arrow 4 in Fig. 3C) and a conical socket element (arrow 5 in Fig. 3C) for another rod-like outgrowth (arrow 6 in Fig. 3C) insert. The long, rod-like outgrowths are subdivided into more or less equal, tubular portions, which must have been originally armed with rather prominent spikes (Fig. 2B), as can be deduced from the remaining pedestals on the surface. Since there is a distinctive joint above the socket, it is assumed that the antennulae could have swung far forward and back toward the ventral side, possibly being cleaned on the way back anteriorly by means of the various setae of the succeeding limbs. The entire length of the appendage was approximately at least as long as the entire head region (Figs. 2B, 3C). The other limbs reach at least one half or slightly more than half of the length of the head region, but did probably not exceed its entire length. The enormous size and design of this appendage suggest strongly its use in food gathering (raking) and not in sensation.

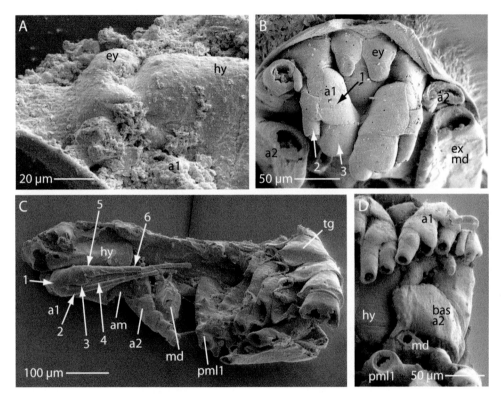

Figure 3. SEM images of *Oelandocaris oelandica* Müller, 1983. (A) Holotype UB 649 (see also Müller 1983: fig. 11 A, B). (B) UB W 260, ventral view of 'eyes' and antennulae; anterior is up. Compare Figure 3B and the text for explanation of the numbered arrows. (C) Oblique ventral view of UB W 261, with well preserved right antennula. The numbered arrows point to the portions of the antennula, see text for explanations. (D) UB W 262, oblique ventral view of proximal parts of antennulae and the basis of the antenna flanking the posterior end of the hypostome; anterior is up. Abbreviations: bas = basipod; am = arthrodial membrane; other abbreviations as in previous figures.

The second limb, the antenna, inserts lateral and slightly anterior to the posterior edge of the hypostome in a subtriangular area. The arthrodial membrane of the antenna is extensive (Fig. 3C), which implies a wide mobility range of this limb. The anteroposteriorly compressed basipod has a distinct armature of setae on its median edge pointing toward the mouth (Figs. 3D, 4B). The endopod arises from the mediodistal rim of the basipod and is four-segmented. The exopod arises from the lateral slanting edge of the basipod. The lateral margin of the basipod extends further down, almost to the body surface. The exopod has a long proximal part, from which four rod-shaped segments arise. The proximal part has fine setae along its inner edge, while each of the distal segments bears a large rigid seta mediodistally. Apparently, this outer ramus had a locomotory as well as a food-gathering and grooming function. A separate sternite for the antennal segment could not be found.

The mandible and all post-mandibular limbs are also anteroposteriorly compressed, representing a plesiomorphic design retained from the ground pattern of the Euarthropoda (cf. Maas et al. 2004). The mandibular basipod is long and angular, bearing more than 10 setae and spines of different size along its inner margin (Fig. 4A, B). A small separate sclerotic element located within the arthrodial membrane medioproximally of the basipod (Fig. 4B) is interpreted as a proximal endite (see below for further discussion of this structure). Its shape is slightly oval, the long axis in proximodistal orientation of the limb. Mediodistally, the element bears a small seta (size approximated from its socket; Fig. 4C). The mandibular endopod is unknown, but may have been similar to that of the antenna. The exopod inserts on the slanting outer edge of the basipod (Fig. 4D). In contrast to the morphology of other known arthropods, it has a long proximal paddle-shaped part (anteroposteriorly flattened) bearing a curved distal margin, from which the three-segmented distal part arises (Fig. 4E). This distal part of the exopod is bent strongly outwards (laterally) due to the special shape of its portions. The proximal two bear long setae mediodistally, and the distal one bears a set of three long spines - similar to as in the post-mandibular limbs of *A. pisiformis* (Fig. 6C; see also Müller & Walossek 1987). The morphology of the mandible of *A. pisiformis* is, however, different (see below). As in the antenna, the inner rim of the proximal part of the exopod of *O. oelandica* is lined some way up by fine setae. This limb may also have served several different functions including the cleaning of the antennular outgrowths.

The remaining limbs, the two posterior cephalic and the trunk limbs, are similar to each other (Fig. 4F, G), becoming progressively smaller towards the posterior trunk (Fig. 4F). Their basipods resemble that of the mandible, but seem to become progressively shorter towards the posterior trunk end and the setation is progressively reduced and more distally located (Fig. 4F). It also seems that the limbs arise progressively closer together medially, so that the median food path was more or less restricted to the anterior part of the ventral surface behind the mouth. Only one to two proximal podomeres of the post-mandibular endopods are preserved (Fig. 4F), but those preserved are significantly larger than those of the antennal endopod. In contrast to the antenna and mandible, all exopods of the post-mandibular limbs are huge oval flaps with very robust and possibly long outer marginal setae (Fig. 4E). It is noteworthy, that they are joined not only with the basipod, but also with the first endopodal podomere (Fig. 4G). Similarly large exopods with robust setae are known only from *Henningsmoenicaris scutula*, while a connection between endopod and exopod occurs in this species and *Agnostus pisiformis* (Fig. 6C), possibly also in the Cambrian naraoiids (Hou & Bergström 1997, their figs. 41, 43). If so, this would reflect a plesiomorphic feature, but has to be investigated in more detail also for other early euarthropods. The size of the exopods by far exceeds the height of the body proper, and it is most likely that *Oelandocaris oelandica* was capable of swimming with these structures rather than walking or crawling along the bottom.

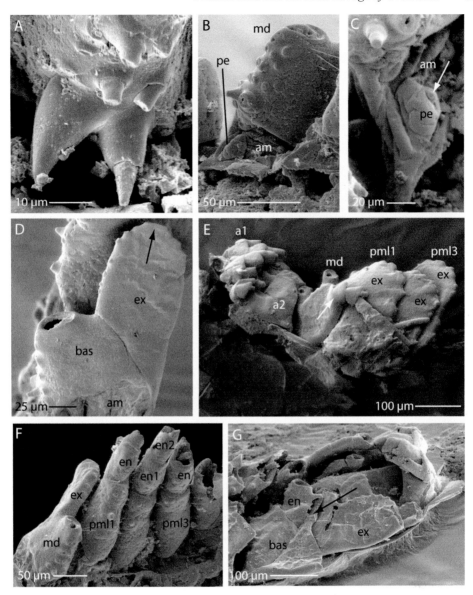

Figure 4. Details of *Oelandocaris oelandica* Müller, 1983. (A) UB W261, close-up of mandibular armature. (B) UB W263, image flipped horizontally, close-up of proximal part of the mandible, displaying the proximal endite. (C) UB W263, image flipped horizontally, close-up of proximal endite showing its mediodistal seta (arrow). (D) UB W263, right mandible with preserved proximal part of the exopod; missing distal part marked by arrow. (E) UB W262, image flipped horizontally, view of the distal part of the left mandibular exopod (arrow) with partly preserved setation. (F) UB W263, image flipped horizontally, median view of mandible and post-mandibular limbs. (G) UB W260, lateroventral view of posterior thorax, displaying a post-mandibular limb. The arrow points to the connection of the proximal region of endopod and exopod. Abbreviations: pe = proximal endite; other abbreviations as in previous figures.

The ventral region posterior to the mouth up to the third post-antennal segment is a single sclerotised plate, the sternum, formed by the sternites of the mandibular and the two subsequent segments. The mandibular part of the sternum is slightly elevated but not produced into a pair of paragnath humps (Figs. 1B, 5D).

4 EVOLUTION OF THE CEPHALIC FEEDING SYSTEM

The new evidence indicates that, as a rare life style among the known 'Orsten' arthropods, *Oelandocaris oelandica* was a swimmer using its specialised three anterior limbs, particularly the antennulae, for food gathering. Food transport by sweeping movements of the exopods and manipulation by endopods and the basipod in particular was achieved by the subsequent two limbs. Such special design and function of these anterior three cephalic appendages was most likely not present in the ground pattern of Euarthropoda. As exemplified by early Arthropoda *s. str.*, such as *Fuxianhuia protensa* Hou, 1987, early representatives of the Chelicerata, and the trilobites and related Palaeozoic arthropods, only the antennulae were different from the rest of appendages and uniramous - either appendage-like or grasping organs (Chen et al. 2004; Maas et al. 2004), so possibly mainly used for food gathering alone, or finely annulated and feeler-like sense organs, as in the trilobites and their allied. The post-antennular limbs of euarthropod taxa were 'still' serially similar, as has to be reconstructed for the ground pattern of Arthropoda *s. str.* (Maas et al. 2004), but comprised a rigid basipod for food transport, an endopod for this and locomotion, and a flap-shaped exopod for locomotion too (early stem lineage derivatives had only a multi-annulated limb rod and an exopod).

More detailed similarities exist between the anterior appendages of *Oelandocaris oelandica* and *Agnostus pisiformis* and also the known derivatives of the stem lineage of Crustacea, particularly *Henningsmoenicaris scutula*, to some degree also those of the Phosphatocopina and some Eucrustacea. For example, as does *O. oelandica*, *A. pisiformis* clearly has food-gathering antennulae, raking food into its shell and toward the mouth behind the hypostome (cf. Fig. 6A). The large antennulae even had to be folded and crossed when closing the head and tail shields (Müller & Walossek, their fig. 25, pl. 16.3). Yet, the antennae of *A. pisiformis* comprised 15 segments, with every second segment bearing mediodistal spines, most likely a plesiomorphic feature (we presume that the original antennula was not a sense organ [feeler] but acting more like an appendage; see Chen et al. 2004; Maas et al. 2004). A feature shared between *O. oelandica*, *A. pisiformis*, and the Crustacea is that the exopods of the anterior two post-antennular appendages, the antennae and mandibles in crustacean terminology, are annulated and swimming devices, which can or could also be used to guide food currents ventrally by sweeping movements, while their basipods and the smaller endopods were more involved in food intake in the proximity of the mouth. Yet, though in the same position and of similar general shape in *A. pisiformis* and *O. oelandica*, the exopodal setae insert laterally in the former (Fig. 6B) and medially in the latter (cf. Fig. 5B, D, F). This also holds true for the known stem lineage derivatives of Crustacea and the Phosphatocopina and Eucrustacea (e.g., Walossek et al. 1996 for extant rhizocephalan nauplii). Again, while *A. pisiformis* has 'only' four appendage-bearing head segments and its sternites are separate (cf. Fig. 5B), five appendage-bearing segments are included in the head in *O. oelandica* and, at least, the labrophoran Crustacea. Furthermore, the mandibular part of the sternal region of the Labrophora bears a pair of humps, the paragnaths (Maas et al. 2003, their pls. 3A, E, F, 25B), while in *O. oelandica* (Figs. 1B, 5D) the according area is only slightly elevated, and in *A. pisiformis* the same sternite is as shallow as all other separate post-oral sternites. The situation is as yet uncertain for other stem lineage derivatives, which still have to be restudied, and it cannot be excluded that the specimens known from these forms are immature and, thus, show deviations from the adult morphology.

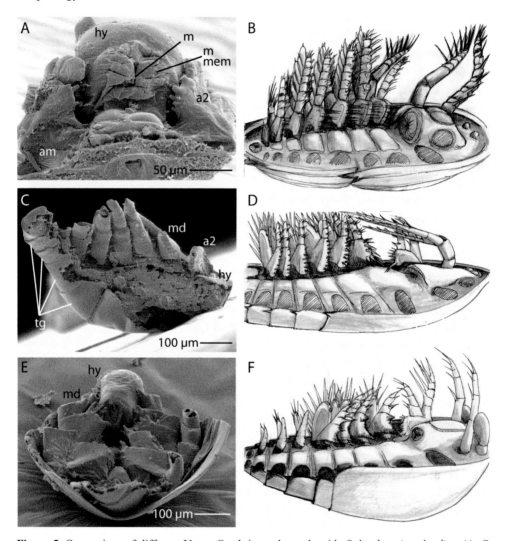

Figure 5. Comparison of different Upper Cambrian arthropods with *Oelandocaris oelandica*. (A, C, E) SEM pictures. (B, D, F) Reconstructions of the cephalic feeding area, appendages of left side omitted. (A, B) *Agnostus pisiformis* (Wahlenberg, 1818). (A) UB 845; see also Müller & Walossek 1987: pls. 12.5, 14.1+2, 16.1. (C, D) *Oelandocaris oelandica* Müller, 1983; (C) UB W263. (E, F) *Henningsmoenicaris scutula* (Walossek & Müller, 1990); (E) UB W266, destroyed. Abbreviations: m = mouth; m mem = mouth membrane; other abbreviations as in previous figures.

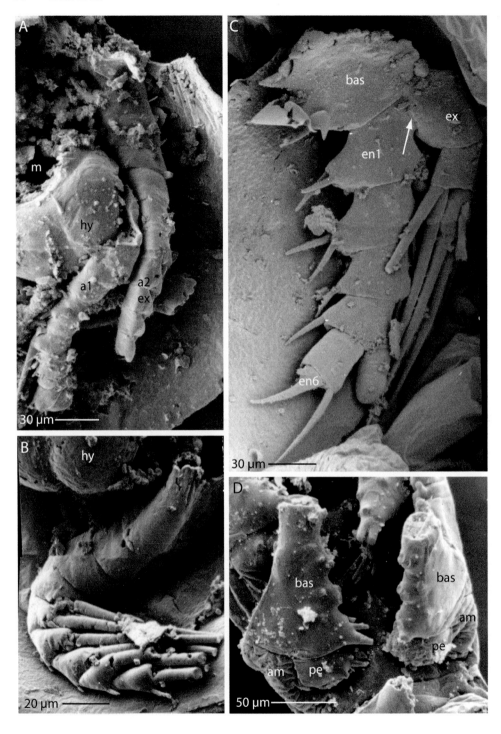

One characteristic feature found in all hitherto described Upper Cambrian arthropods recognised as Crustacea - and one of the autapomorphies of this taxon (Fig. 7: apomorphies set 3) - is a small setiferous lobate endite at the inner proximal edge of the basipod, better proximal to it within the arthrodial membrane, of all post-mandibular appendages, the 'proximal endite' (see, e.g., Walossek & Müller 1990; Walossek & Szaniawski 1991; Fig. 6D for *Martinssonia elongata*). Remarkably, in *Oelandocaris oelandica* this endite could be found only on the mandible, being small and only having a single median seta (Fig. 4B, C).

We regard this 'proximal endite' to be of crucial importance in the re-organisation of the cephalic feeding apparatus in the evolution of the Crustacea. This endite is a first sign of significant changes in the feeding process by permitting to de-couple the locomotory function of the distal limb parts from the food transport function of the endite proximal and close to the ventral body proper. Before this, as exemplified by *Agnostus pisiformis* (Fig. 5A, B), all basipods with their median setae transported food toward the mouth in an exposed position at the posterior end of the hypostome, and this was directly coupled with locomotion since basipod and rami could, more or less, not operate separately from each other. Now, food transport was possible by using the 'proximal endite', as exemplified by the stem lineage derivatives of Crustacea (Walossek & Müller 1990; Fig. 5E-F). In a further step, the mouth became recessed into an atrium oris, being overhung by a bulged soft labrum – with internal slime glands and posterior openings –, the cephalic sternites fused to a single sternal plate (sternum), and fine setulae occurred on the parts of all associated appendages, the flanks of the labrum, and the sternum (see also Walossek 1999, his fig. 8). Moreover, we suggest that the 'proximal endite' modified into a large spine-bearing proximal part of the limb stem of antenna and mandible, the coxa, as exemplified by the Labrophora (autapomorphy of this taxon; Fig. 7: set 4; see, e.g., Walossek 1999; Maas et al. 2003; Waloszek 2003a, b). The mandibular coxa gained a median protrusion with an oblique, flattened grinding edge (gnathobase). Even more, food transport was supported by paired humps, the paragnaths, immediately behind and below the coxal gnathobases (Walossek 1999, his fig. 8).

Figure 6. Details of other 'Orsten' species. (A-C) *Agnostus pisiformis* (Wahlenberg, 1818). (A) UB 857, ventral view of head. Most distal podomeres of antennula broken off (cf. Müller & Walossek 1987: pl. 17.3). (B) UB 832, curved exopod of antenna with outwardly directed setae (cf. Müller & Walossek 1987: pl. 18.1). (C) UB 858, rather early larva (not all endopod segments present), first post-mandibular appendage. Note the connection between proximal parts of endopod and exopod (arrow; cf. Müller & Walossek 1987: pl. 20.2). (D) *Martinssonia elongata* Müller & Walossek, 1986. UB 750, third postantennular limb pair, exhibiting a comprehensive basipod (bas) and proximal endite (pe) within arthrodial membrane (am; cf. Müller & Walossek 1986). See previous figures for abbreviations.

5 CONCLUSIONS

The different taxa preserved in the 'Orsten' demonstrate, in our view, the evolutionary transition of the cephalic feeding and locomotory system from the euarthropod ground pattern into the crustacean lineage: Initially, the cephalic nutritive and locomotory system involved only a food-gathering or sensorial antennula and serial limbs comprising a basipod, endopod and flap-shaped exopod. In the lineage toward Crustacea, evolution was largely triggered by the progressive specialisation of the anterior head limbs and later also more of the ventral structures.

The first step is indicated in the morphology of *Agnostus pisiformis* with the exopods of the second and third limbs engaged in sweeping. The next step may be exemplified by *Oelandocaris oelandica* by an inward orientation of the exopodal setation (Figs. 5C, D, 7: set 2). A third step - development of a prominent 'proximal endite' on *all* post-antennular limbs (Figs. 6D, 7: set 3) - is present in 'Orsten' stem lineage crustaceans such as *Henningsmoenicaris scutula* (Fig. 5E, F). The most significant evolutionary changes in the cephalic feeding system become apparent in the ground pattern of the Labrophora. They include:

- labrum as the enlarged distal part of mouth membrane;
- recession of the mouth behind the labrum into an atrium oris;
- sternum with mandibular paragnath humps;
- coxa on the second and third head limbs (enlarged proximal endites of these limbs);
- fine setulae on sternum, labral sides and on the setae themselves (Fig. 7: set 4).

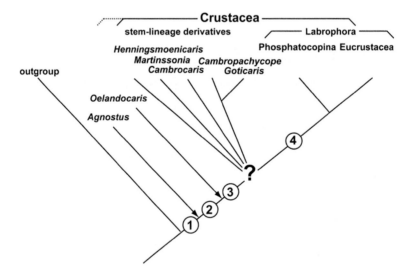

Figure 7. Suggested relationship of the taxa discussed in the text. Numbers refer to autapomorphies. Out-group taxa: 'great-appendage' arthropods (Chen et al. 2004), trilobites, *Fuxianhuia protensa* Hou, 1987 (see also Maas et al. 2004). Set (1) Exopods of second and third limbs employed for sweeping. Set (2) Inward orientation of exopodal setation, proximal endite on mandible. Set (3) Prominent proximal endite on antenna and post-mandibular limbs. Set (4) Recession of the mouth behind the labrum; sternum with mandibular paragnath humps, coxa on antenna and mandible present; fine setulation on sternum; labrum and limb setae present.

This large amount of features newly present suggests that there should be even more taxa having split off earlier and having possessed only part of this character set. Moreover, Eucrustacea specialised their fourth head limb as a food-intake device, the maxillula, while the second maxilla appeared only within special eucrustacean in-groups.

Indeed, *O. oelandica* shares some of the crustacean features - e.g., a five-segmented head - but has a hypostome and only one proximal endite. Hence, we interpret *O. oelandica* as the phylogenetically earliest stem lineage representative of the Crustacea. Taking the new data on *O. oelandica* into account, this interpretation has two consequences:

(1) The fact that *O. oelandica* possesses only part of the autapomorphies that characterises the taxon Crustacea is a combination of features, which requires either re-evaluation of the ground pattern composition of Crustacea or consideration that *Oelandocaris* is the sister taxon to Crustacea *sensu* Walossek 1999.
(2) Given that the morphology of *Agnostus pisiformis* can be used as a model for the ground pattern of Agnostida, the latter cannot be in-group trilobites, as traditionally understood, but represent derivatives of the early evolutionary lineage towards the Crustacea.

Traditional placement of the Agnostida within the Trilobita relies heavily on features of the calcified parts of the cuticle (for the most recent attempt to unravel agnostid relationships by dorsal features alone, see Jell 2003). Already such 'conventional' approaches (e.g., Fortey & Theron 1994; Wills et al. 1998) have rendered assignment controversial (see also the review in Fortey 2001), mostly because it omits the most informative morphological data of the ventral side and particularly the appendages from the analysis. Again, various of the characters mentioned are either ambiguous, clearly symplesiomorphies, can be shown to be convergences, or are disputable. For examples features such as trilobation, presence of a marginal rim, a cephalon incorporating three post-antennal segments, presence of the exopods on post-antennular appendages, a ventral anus in the terminal division of the body and presence of gut diverticula are very likely symplesiomorphies dating back to the ground patterns of Euarthropoda and Arthropoda *s. str.* (Maas et al. 2004). Some of the shared characters may be less ambiguous, such as the mineralised cuticle with calcium carbonate (only later instars of *Agnostus pisiformis*), the presence of a dorsally enhanced glabella, an occipital width exceeding that of the glabella in front (which is coupled to the presence of a dorsally elevated glabella) or the presence of a 'rolled' cephalic border (or marginal rim) as well as a narrow cephalic 'doublure' (margin tilted inwards). Yet, at least some of these features can also be found in other arthropods, which makes clear decisions of the particular character state difficult. A significant difference between agnostids and the Trilobita and their possible Cambrian relatives (e.g., naraoiids), however, is the development of a feeler-type of antennula ('antenna' in trilobite terminology) in the latter group.

Agnostus pisiformis clearly shares the appendage-like design of the antennula present in the ground patterns of Arthropoda *s. str.*, Euarthropoda, and apparently being retained in the stem lineage of the Crustacea as a plesiomorphy. Furthermore, the connection between basipod and exopod in the more posterior limbs may rather be a plesiomorphy. The resemblance between *Oelandocaris oelandica* and *A. pisiformis* in general shape and usage of the anterior three appendages is, at present, our only but in our view strong feature to suggest closer relationships of *Agnostus* - and the agnostids - with the Crustacea (Fig. 7: set 1) than with the trilobites.

ACKNOWLEDGEMENTS

We thank Jason Dunlop, Berlin, and David Siveter, Leicester, for critically reading different versions of our manuscript and making useful comments particularly with regard to linguistic improvements. Thanks are also due to three anonymous referees and particularly Stefan Koeneman, who made the manuscript much more consistent and easily readable. We are however fully responsible for all final reductions of language quality.

REFERENCES

Chen Junyuan, Waloszek, D. & Maas, A. 2004. A new "great appendage" arthropod from the Lower Cambrian of China and the Phylogeny of Chelicerata. *Lethaia* 37: 3-20.

Fortey, R.A. 2001. Trilobite systematics: The last 75 years. *J. Paleontol.* 75: 1141-1151.

Fortey, R.A. & Theron, J.N. 1994. A new Ordovician arthropod *Soomaspis*, and the agnostid problem. *Palaeontology* 37: 841-861.

Jell, P.A. 2003. Phylogeny of early Cambrian trilobites. In: Lane, P.D., Siveter, D.J. & Fortey, R.A. (eds.), *Trilobites and their Relatives. Contrib. 3rd Int. Conf., Oxford 2001*. Spec. Pap. Palaeontol. 70: 45-58.

Lauterbach, K.-E. 1988. Zur Position angeblicher Crustacea aus dem Ober-Kambrium im Phylogenetischen System der Mandibulata (Arthropoda). *Verh. naturwiss. Ver. Hamburg, N.F.*, 30: 409-467.

Maas, A., Waloszek, D., Chen Junyuan, Braun, A., Wang Xiquiang & Huang Diying 2004. Phylogeny and life habits of Early Arthropoda - Predation in the Early Cambria Sea. *Progr. Nat. Sci.* 14: 1-9.

Maas, A., Waloszek, D. & Müller, K.J. 2003. Morphology, Ontogeny and Phylogeny of the Phosphatocopina (Crustacea) from the Upper Cambrian "Orsten" of Sweden. *Fossils & Strata* 49: 1-238.

Müller, K.J. 1979. Phosphatocopine ostracodes with preserved appendages from the Upper Cambrian of Sweden. *Lethaia* 12: 1-27

Müller, K.J. 1982. *Hesslandona unisulcata* sp. nov. (Ostracoda) with phosphatised appendages from Upper Cambrian 'Orsten' of Sweden. In: Bate, R.H., Robinson, E. & Sheppard, L.M. (eds.), *Fossil and Recent Ostracods*: 276-304. Chichester: Ellis Horwood.

Müller, K.J. 1983. Crustacea with preserved soft parts from the Upper Cambrian of Sweden. *Lethaia* 16: 93-109.

Müller, K.J. & Walossek, D. 1985. Skaracarida, a new order of Crustacea from the Upper Cambrian of Västergötland, Sweden. *Fossils & Strata* 17: 1-65.

Müller, K.J. & Walossek, D. 1986. *Martinssonia elongata* gen. et sp. n., a crustacean-like euarthropod from the Upper Cambrian 'Orsten' of Sweden. *Zool. Scri.* 15 (1): 73-92.

Müller, K.J. & Walossek, D. 1987. Morphology, ontogeny, and life-habit of *Agnostus pisiformis* (Linnaeus, 1757) from the Upper Cambrian of Sweden. *Fossils & Strata* 19: 1-124.

Müller, K.J. & Walossek, D. 1988. External morphology and larval development of the Upper Cambrian maxillopod *Bredocaris admirabilis*. *Fossils & Strata* 23: 1-70.

Schram, F.R. 1986. *Crustacea*: 1-606. New York, Oxford: Oxford Univ. Press.

Siveter, D.J., Williams, M. & Walossek, D. 2001. A phosphatocopid crustacean with appendages from the lower Cambrian. *Science* 293: 479-481.

Siveter, D.J., Waloszek, D. & Williams, M. 2003. An Early Cambrian Phosphatocopid Crustacean with Three-dimensionally Preserved Soft Parts from Shropshire, England. In: Lane, P.D., Siveter, D.J. & Fortey, R.A. (eds.), *Trilobites and their Relatives. Contrib. 3rd Int. Conf., Oxford 2001.* Spec. Pap. Palaeontol. 70: 9-30.

Walossek, D. 1993. The Upper Cambrian *Rehbachiella kinnekullensis* Müller, 1983, and the phylogeny of Branchiopoda and Crustacea. *Fossils & Strata* 32: 1-202.

Walossek, D. 1999. On the Cambrian diversity of Crustacea. In: Schram, F.R. & von Vaupel Klein, J.C. (eds.), *Crustaceans and the Biodiversity Crisis. Proc. 4th Int. Crust. Congr., Amsterdam 1998, Vol. 1*: 3-27. Leiden: Brill Acad. Publ.

Walossek, D., Høeg, J.T. & Shirley, T.C. 1996. Larval development of the rhizocephalan cirripede *Briarosaccus tenellus* (Maxillopoda: Thecostraca) reared in the laboratory: A scanning electron microscopy study. *Hydrobiology* 328: 9-47.

Walossek, D. & Müller, K.J. 1990. Upper Cambrian stem-lineage crustaceans and their bearing upon the monophyletic origin of Crustacea and the position of *Agnostus*. *Lethaia* 23: 409-427.

Walossek, D. & Szaniawski, H. 1991. *Cambrocaris baltica* n. gen. n. sp., a possible stem-lineage crustacean from the Upper Cambrian of Poland. *Lethaia* 24: 363-378.

Waloszek, D. 2003a. Cambrian 'Orsten'-type Arthropods and the Phylogeny of Crustacea. In: Legakis, A., Sfenthourakis, S., Polymeni, R. & Thessalou-Legakis, M. (eds.), *The New Panorama of Animal Evolution. Proc. 18th Int. Congr. Zoology*: 69-87. Sofia; Moscow: Pensoft Publ.

Waloszek, D. 2003b. The 'Orsten' Window - A three-dimensionally preserved Upper Cambrian Meiofauna and its Contribution to our Understanding of the Evolution of Arthropoda. *Paleontol. Res.* 7: 71-88.

Whittington, H.B., Chang, W.T., Dean, W.T., Fortey, R.A., Jell, P.E., Laurie, J.A., Palmer, A.R., Repina, L.N., Rushton, A.W.A. & Shergold, J.H. 1997. Systematic descriptions of the class Trilobita. In: Kaesler, R.L. (ed.), *Treatise on Invertebrate Paleontology, Part O, Arthropoda 1, Trilobita Revised, Vol. 1*: 331-530. Boulder; Lawrence: Geol. Soc. Amer. and Univ. Kansas Press.

Wills, M.A., Briggs, D.E.G., Fortey, R.A., Wilkinson, M. & Sneath, H.A. 1998. An arthropod phylogeny based on fossil and recent taxa. In: Edgecombe, G.D. (ed.), *Arthropod Fossils and Phylogeny*: 33-105. New York: Columbia Univ. Press.

Early Palaeozoic non-lamellipedian arthropods

JAN BERGSTRÖM[1] & XIAN-GUANG HOU[2]

[1] *Department of Palaeozoology, Swedish Museum of Natural History, Stockholm, Sweden*
[2] *Yunnan Research Center for Chengjiang Biota, Yunnan University, Kunming, People's Republic of China*

ABSTRACT

Evolutionary steps leading to the basal crustaceomorphs are traced. In the most plesiomorphic Cambrian arthropods, all appendages except the first antenna were similarly shaped throughout the body. The endopod was a simple stem, with some 15-20 identical short segments and a conical tip. The exopod was a thin rounded flap devoid of setae. No appendages were adapted to assist in feeding, and substrate sediment with its inclusions was simply engulfed for food. Successive steps leading to a basal crustaceomorph level included endopod segment reduction and diversification, formation of exopod setae and segments, and later on, specialization of the first three limbs for swimming. With the agnostids, we enter the evolutionary stairway to modern crustaceans previously traced from fossil evidence in Upper Cambrian 'Orsten' concretions. In a number of side lineages, the 1st or 2nd appendage evolved into a clumsy grasping organ, indicating the need for limb specialization when sediment feeding was succeeded by other modes of feeding. However, in contrast, ancestors of the crustaceomorphs stuck to a fairly plesiomorphic morphology, which preserved an evolutionary flexibility.

This contribution is dedicated to Frederick R. Schram for his devoted work on fossil crustaceans.

1 THE PALAEOZOIC DIVERSITY

Late Palaeozoic crustaceans have been dealt with in many publications (for instance, Briggs & Clarkson 1990; Brooks 1962; Jenner et al. 1998; Schram & Hof 1998; Taylor et al. 1998). Klaus Müller, Dieter Walossek and others (for instance, Müller & Walossek 1988; Walossek 1993; Walossek & Müller 1998a, b) have contributed greatly to the understanding of Late Cambrian crustaceans and how crustaceans came into being in the Cambrian. The latter authors have also described the evolution from the "agnostid level" up to the crown-group Crustacea (for instance, Walossek & Müller 1998b). However, there is much less understanding of the evolution preceding the agnostid level.

The dominating arthropods in World collections from Lower Palaeozoic strata are the trilobites, with the number of species estimated to be more than 20,000. This high number results in part from the strongly calcified exoskeleton that was easily preserved. Other groups, each represented by many species, include taxa originally interpreted as ostracodes. Such 'pseudo-ostracodes' include the Cambrian bradoriids and phosphatocopids, both with phosphatic valves. In the Ordovician, there are forms with calcified valves: the leperditicopids, palaeocopids, podocopids and myodocopids. At least the podocopids and myodocopids are true ostracodes, still surviving today.

Other groups of arthropods are rare in the Lower Palaeozoic. Although there are a number of trilobite-like groups (cf. Bergström & Hou 2003b), for example, marrellomorphs, nectaspidids (including *Naraoia*), helmetiids, xandarellids, emeraldellids, aglaspidids, limulavids, and a few others, there are altogether only some 45 Cambrian and Ordovician species. All of these, as well as the trilobites, have more or less laterally directed legs and exopods with large, distinctly flattened setae arranged in a straight line, like the teeth of a comb, and they do not play a role in the origin of crustaceans. Collectively, they are known as lamellipedians. They may lead off to the chelicerate line, but they still have an antenna and biramous limbs along the entire body. Chelicerates are rare in the Cambrian and Ordovician.

The remaining Cambrian groups are more interesting in the search for the origin of the crustacean lineage. Again, these are small groups, each with a handful of known species or less. Here, we find the fuxianhuiids, canadaspidids, leanchoiliids, yohoiids, occacaridids, burgessiids, and some others.

Recently, the anomalocaridids have been suggested (e.g., by Chen et al. 2004) to be early, ancestral arthropods, but there are good arguments to conclude that they have achieved arthropod-like characters in parallel with the arthropods (Bergström & Hou 2003b). The most obvious, general argument is that the sequence of steps in the 'arthropodization' process differs considerably between arthropods and anomalocaridids. For instance, anomalocaridids developed segmented appendages after hard mouthparts, whereas arthropods developed - entirely different - mouthparts from appendages long after they had become segmented and further modified.

The most important evidence for our research comes from well-preserved *Lagerstätten* material. The *Lagerstätten* providing most of the evidence is the Lower Cambrian Chengjiang deposit in Yunnan, south China, the Middle Cambrian Burgess Shale deposit in British Columbia, and the Upper Cambrian Orsten nodules in Sweden. The approximate age of the Chengjiang fauna is about 520 million years, that of the Burgess Shale fauna 510, and that of the Orsten fauna 500 million years, as far as we now understand the situation. The beginning of the Cambrian is now set 543 million years back in time. The Chengjiang fauna therefore lived some 20-25 million years after the beginning of the Cambrian. It appears to represent a late stage in a second major phase of the 'Cambrian explosion', with the rise and first development of anatomies characteristic of modern phyla. The first step saw the evolution of 'wormy' animals, of which we have little evidence other than strange 'small shelly fossils' (for instance, Qian & Bengtson 1989).

1.1 *Before setae*

Fuxianhuia (Figs. 2B, 3A; Hou & Bergström 1997; Hou et al. 2004: 104-105) from the Chengjiang fauna at first sight has the appearance of a scorpion, with anterior fangs, a 'prosoma' with legs, and a limbless abdomen with a terminal spine. At closer view, this description is far from accurate. It is the leg morphology that strikes the observer as looking uniquely 'primitive'. The leg is simply sausage-like; there are some twenty (!) short segments, all of which are identical. The distal end is a rounded tip, devoid of claws. The leg can be understood as an endopod, since there is also a much shorter, thin and oval outer flap, also seemingly devoid of any structure except for a slightly thicker margin. The branching between endopod and flap has not been observed. Adding to the oddness is a mismatch between the dorsal and ventral body sides: there are many more pairs of limbs than there are dorsal tergites. The related *Chengjiangocaris* appears not to have this mismatch (Hou et al. 2004: 106-107).

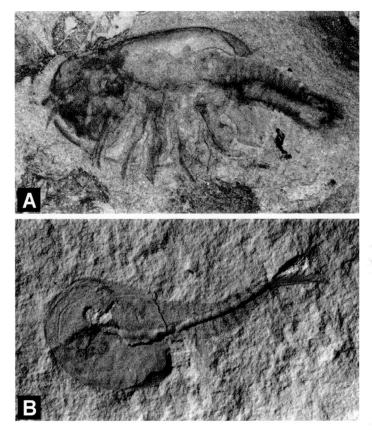

Figure 1. Examples of Chengjiang arthropods showing gut filled with sediment that consists of fine-grained quartz and muscovite/illite. (A) *Canadaspis laevigata*, NIGPAS 115361, from Maotianshan, Chengjiang, Yunnan; lateral view, length 27 mm. (B) *Clypecaris pteroidea*, NIGPAS 115413, from Xiaolantian, Chengjiang Yunnan; dorsolateral view, length 11.7 mm.

The tagma that looks like a prosoma has some composite parts. The anterior end, perhaps an acron, extends in front of the main part of the head. It bears a pair of eyes. The successive part is covered by a 'carapace'; on the ventral side, there is a pair of uniramous antennae succeeded by a pair of stout, uniramous, and pointed appendages. The 'carapace' hangs over about three more posterior segments with simple body appendages. Posterior to the carapace, there are 16 fairly wide segments carrying legs, followed by ca. 15 narrower, rounded segments without legs. On both sides of the terminal spine, there is a pair of smaller spines or blades.

Fuxianhuia is certainly not an entirely non-specialized arthropod. However, based on leg morphology, there is reason to believe that we have a rather early, plesiomorphic arthropod feature. The appendage type is simple enough to allow the rise of every possible limb specialization we see in modern arthropods.

Fortiforceps (Figs. 2D, 3C herein; Hou & Bergström 1997: figs. 31-35; Hou et al. 2004: 128-130) from the Chengjiang fauna represents a very different kind of 'great appendage arthropod'. It has a short head shield succeeded by approximately 20 segments, all provided with appendages of uniform appearance. The body ends with a complex, shrimp-like fluke. The endopod has up to about 15 segments of uniform shape, which are decreasing in size from proximal to distal. The exopod is a simple flap similar to that of *Canadaspis*, but with short hairs or setules along the margin. The great appendage is notably stout. The distal element and single long fingers from each of the three preceding segments form complex pincers. A separate appendage with bulbous tip was first interpreted as a possible first antenna, but it may have been an appendage belonging to another animal. Anteriorly the head bears a pair of stalked eyes.

Canadaspis, apparently somewhat more advanced than *Fuxianhuia* as far as the biramous limbs are concerned, has a completely different habitus (Figs. 2A, 3B; Hou & Bergström 1997; Hou et al. 2004: 112-113). It is deceivingly similar to malacostracan crustaceans in its possession of a 'carapace' embracing the anterior part of the body, a tagmosis into cephalothorax and abdomen, and a terminal furca. It should not surprise that it was interpreted as a crustacean before differently preserved specimens gave a better knowledge of the structural organization.

Except for the simple antennae, all appendages have a stout endopod with a terminal claw. It has ca. 12 short segments, notably narrower toward the distal end, and each with a strong enditic spine. The exopod is a large oval flap. It has no setae, but a peripheral field is partly ribbed in a way that may give the impression of long setae. The abdomen is devoid of limbs. At the posterior end, there is a pair of ventral spines.

Similarities in exopod morphology, including shape and radiating structures possibly for support, may indicate a relationship between the otherwise dissimilar *Canadaspis* and *Fortiforceps*.

Despite the presence of large grasping appendages in *Fuxianhuia* and the absence of apparent feeding specializations in *Canadaspis*, specimens of both fossils are occasionally found with part of the intestinal canal filled with sediment (Fig. 1 herein; for instance, Hou & Bergström 1997: figs. 9F, 18B+D, 19) consisting of fine quartz particles and clay. This appears also to be also the case with *Perspicaris*, a probable relative of *Canadaspis*, but the different kind of embedding sediment makes the interpretation more difficult (Briggs 1977: 603; Briggs & Whittington 1985: 152). It would be unusual to encounter sediment feeding in Recent arthropods. However, without any limb specialization for feeding, it seems logical that early arthropods ingested sediment together with enclosed organic matter and occasional burrowers. *Fuxianhuia* had a pair of grasping appendages, but of very simple and probably inefficient construction, e.g., without elongated fingers. It still ingested sediment, but occasionally we also recognize organic remains in the gut, particularly in large specimens.

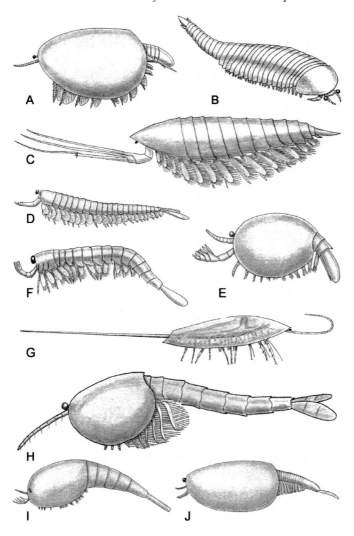

Figure 2. Reconstructions of some Cambrian arthropods from the Lower Cambrian Chengjiang fauna (A-D, G, I-J) and the Middle Cambrian Burgess Shale fauna (E-F, H). (A) *Canadaspis laevigata*. (B) *Fuxianhuia protensa*. (C) *Leanchoilia illecebrosa*. (D) *Fortiforceps foliosa*. (E) *Occacaris oviformis* (abdomen poorly known). (F) *Yohoia tenuis*. (G) *Burgessia bella*. (H) *Waptia fieldensis*. (I) *Ovalocephalus mirabilis*. (J) The branchiopod crustacean *Pectocaris eurypetala*. Some of the species have seemingly non-specialized features which may be inherited from a common ancestor, such as limbs with simple multisegmented endopod, and simple, flap-shaped exopod, without setae, and also a somewhat myriapod-like body shape (B, D). Early attempts to adapt to particle-feeding include the development of a sub-chela or chela from the antenna (C-D, F) or the first post-antennal appendage (B, E, I). Another attempt in the same direction led to the development of a 2nd antenna and mandible in crustaceans (J: note absence of 2nd antenna in A+H). Some animals have a marked tagmosis (A, B, F, H, I). Superficially carapace-like tergal expansions evolved many times (for instance, in A-B, E, G-J; E+I have no posterior fold). Altogether, evolution resulted in a distinct mosaic pattern. Drawings: JB.

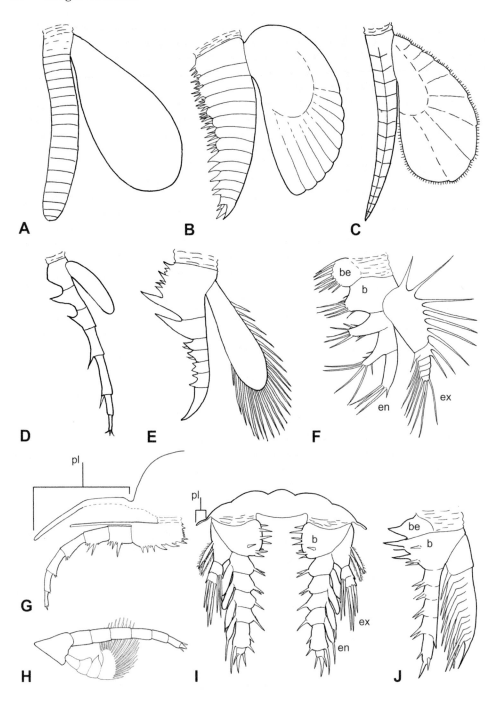

1.2 Arthropods with setae

All arthropods dealt below have exopods provided with setae of varied shapes. A major group of early arthropods, the lamellipedians, have laterally flattened setae. They constitute a major sideline and are not considered herein. Other groups have setae with round cross-section.

There are, in the Burgess Shale and Chengjiang deposits, a number of arthropods with a pair of enlarged and extended anterior appendages that have a more advanced morphology than the simple hooks in *Fuxianhuia*. These fossils have been summarized as "great appendage arthropods", but represent quite different types of arthropods. *Fortiforceps* and *Fuxianhuia* (see above) lack setae (although the former has short hair-like setules). The enlarged appendage is the first anterior appendage in some taxa (which, thus, are devoid of an ordinary, simple antenna), the second in others. It carries strong pincers in some groups, but antenna-like, filiform branches in others.

The significance of the enlarged appendage may be disputed. It is possible that the development of a single strong, anterior pair of appendages is a first primitive attempt at limb specialization, parallel to that of *Fuxianhuia* and its kin. The differences described above hint at the possibility that the types dealt with here are also the result of parallel evolution. Still another, independent case is the development of a 'great appendage' in the anomalocaridids.

1.2.1 Occacaris *and* Ovalicephalus

In *Occacaris*, it is the second appendage that is developed as a 'great appendage' (Fig. 2E herein; Hou 1999: figs 1.1, 2; Hou et al. 2004: 130-131). It is a bivalved arthropod with two abdominal segments and the telson exposed behind the 'carapace'. At the other end, a pair of stalked eyes, a pair of uniramous antennae and parts of a few endopods are visible, similar to the antenna but shorter. More posteriorly are the tips of setiferous structures, presumably exopods. The 'great appendage' has three (or four) segments prolonged into paired fingers, which appear to form complex pincers together with the last segment. In its position, this appendage apparently corresponds with the 2nd antenna in crustaceans.

Figure 3. Body appendages of representative Cambrian arthropods. The basal portion including the attachment of the exopod is uncertain except in the Orsten species (F, I, J). (A) *Fuxianhuia protensa*. (B) *Canadaspis laevigata*. (C) *Fortiforceps foliosa* (A-C modified from Hou & Bergström 1997). (D) *Burgessia bella* (modified from Hughes 1975). (E) *Alalcomenaeus cambricus* (original). (F) *Hesslandona unisulcata* (modified from Müller 1982). (G) The trilobite *Olenoides serratus* (modified from Whittington 1980), cross section. (H) *Ceraurus pleurexanthemus* (original). (I) *Agnostus pisiformis* (original reconstruction of body, limbs redrawn from Müller & Walossek 1987). (J) *Martinssonia elongata* (redrawn from Walossek & Müller 1990). A-C are from the early Cambrian Chengjiang fauna, D-E and G from the middle Cambrian Burgess Shale fauna, F and I-J from the late Cambrian Orsten fauna, and H from the Ordovician Utica Shale, New York. Note the immense difference between the trilobite (G-H) and *Agnostus* (I) on one hand, and the similarity between *Agnostus* and pancrustaceans (F, J) on the other. Abbreviations: b = basis; be = basal endite (corresponding to the coxal segment in crustaceans); en = endopod; ex = exopod; pf = pleural field. Drawings: JB.

Ovalicephalus (Fig. 2I) is similar to *Occacaris* in having both (1st) antenna and 'great appendage'. Similarities also include the possession of a carapace-like structure. This is attached to the body at its posterior end, and is therefore more like a shield with a wide pleural fold.

1.2.2 *Megacheirans: yohoiids and leanchoiliids*

Yohoia (Fig. 2F; Whittington 1974) from the Burgess Shale belongs to a major group of 'great appendage arthropods', the megacheirans. On the whole, it is similar to *Fortiforceps*, but has only 13 trunk segments behind the head, the telson is a single, flattened plate, and the exopods are bordered by strong setae. *Jianfengia* is a yohoiid from the Chengjiang fauna, differing from *Yohoia* and *Fortiforceps* in having a slender body with as many as 22 segments posterior to the head. A 2 mm long larva of *Jianfengia* already has the characters of the adult. The five longer anterior appendages suggest the possibility that the first larval stage was composed of five segments bearing pairs of appendages.

Leanchoilia, found both in the Burgess Shale and in the Chengjiang beds, represents another type of megacheiran (Fig. 2C herein; Bruton & Whittington 1983; Hou & Bergström 1997: figs. 22-30; Hou et al. 2004: 122-123). *Leanchoilia* has a head shield succeeded by 11 segments and a pointed telson. Posterior to the 'great appendage', all segments carry appendages of uniform shape. As in the other yohoiids and leanchoiliids, the endopod has at least ten segments. The exopod has a shaft and an oval distal blade. The latter is bordered by very strong, needle-shaped setae. The 'great appendage' has a fairly strong proximal portion. It divides into three very long, stiff extensions, probably the distal prolongation and branches from two successive segments. The terminal segment ends in pincers in a Burgess Shale representative, while the proximal two segments end in filiform threads. In a Chengjiang species, all three segments end in filiform threads. An ordinary antenna is apparently lacking. Eyes are rarely observed, however, they are definitely present in at least the Chengjiang representative.

Yohoiids and leanchoiliids share some features, including the development of the first appendage as pincers with successive branching and a flat tail plate (Fig. 2C, F). The most obvious difference is that in leanchoiliids, but not in yohoiids, three 'fingers' of the pincers are extended into extremely long, jointed and antenna-like structures. This might be interpreted as serially multiplied remains of the original antennal structure, and therefore as an indication that this appendage corresponds with the (1st) antenna.

1.2.3 *Some additional types*

The Burgess Shale animals *Habelia* and *Molaria* (Whittington 1981) both have a caudal spine. The appendages seem to be fairly simple, and there is no obvious specialization of the 'head', apart from the simple antennae. These fossils are, thus, among the few early arthropods which are demonstrably not far off a generalized pattern.

The Burgess Shale genus *Sanctacaris* (Briggs & Collins 1988), nicknamed Santa Claws, is similar to the yohoiids in having a broad tail plate, but the head shield comprises probably six segments rather than (three to) four. Also, five pairs of anterior legs are narrowly spaced out and apparently specialized for grasping, as compared with the single pair of grasping limbs in yohoiids. In this animal, we can speak of a comprehensive, functional anterior tagma, a head or prosoma.

A superficially myriapod-like creature is the Chengjiang *Pseudoiulia* (Hou & Bergström 1998; Hou et al. 2004: 134-135). The body has at least 31 segments. Unfortunately, the head end is unknown. All that is known of the appendages appear to be the exopods with strong setae. There is a different group of somewhat myriapod-like, but much shorter arthropods, both in the Chengjiang and the Burgess Shale faunas. It includes genera such as *Urokodia* (Hou et al. 2004: 102-103), *Mollisonia* (Walcott 1912) and *Thelxiope* (Simonetta & Delle Cave 1975). In these fossils, it seemed often difficult to tell tail and head ends apart. Unfortunately, no appendages have been found.

Burgessia is distinctly different from other Cambrian arthropods by having a large anterior carapace and a long pointed tail (Figs. 2G, 3D). It is known only from the Burgess Shale, represented by a single species, *B. bella* (Walcott 1931; Simonetta & Delle Cave 1975; Hughes 1975). The exopod lacks setae, and in this respect *Burgessia* is similar to *Canadaspis* and *Fortiforceps*. In other respects, it is quite different from these two genera. The body is slender. Walcott indicated the presence of a pair of small eyes, but this has not been confirmed by later observations. There is a weakly defined head tagma with a long antenna and probably three post-antennal appendages. The exopod may consist of a long filamentous structure. The endopod is fairly sturdy and has ca. six segments distal to the basis. The proximal segments have endites. The trunk is equipped with perhaps seven pairs of biramous limbs. The endopods are like those on the head, whereas the exopods are slender and oval, fairly short blades with smooth margins. An eighth trunk segment bears a dwarfed limb. The body ends with a slender spine, which is longer than the rest of the body. The animal carries a rounded carapace. A pair of intestinal diverticula enters the carapace at the boundary between head and trunk. Each diverticulum splits up into finer canals toward the lateral margins. It leaves the gut at the base of the carapace fold. Similar diverticula are known from agnostids and lamellipedian naraoiids. The morphology of *Burgessia* is completely unique and, apart from the exopods, there are hardly any features shared with other arthropods that could indicate phylogenetic relationships.

1.2.4 *Agnostid 'trilobites'*

Agnostids (Fig. 3I) have usually been, and are generally still recognized as trilobites (Kaesler 1997). The main reason to separate them from typical trilobites has been the lack of facial sutures (Jaekel 1909). After Müller & Walossek (1987) had described superbly preserved appendages in *Agnostus pisiformis*, it became clear that these limbs were distinctly different from those in trilobites, naraoiids, marrellids, helmetiids, emeraldellids and all other lamellipedians. Instead, agnostid limbs are more similar to those in crustacean-like arthropods. The significance of this similarity has not yet been generally appreciated, however. For instance, in discussing the position of agnostids within or outside the Trilobita, Fortey (1997: 294-295) did not mention the differences in limb morphology revealed by Müller & Walossek (1987). There is obviously reason to repeat this argument as well as add a series of others. Therefore, we list below important character differences between trilobites and agnostids. Numbers 1-9 and 12-14 are also arguments against affinities to other lamellipedians.

(1) The appendage morphology is strikingly different in various respects. First, all lamellipedians have limbs directed more or less laterally (cf. Figs. 3G-H herein; also for instance, *Emeraldella* in Bruton & Whittington 1983; different trilobites in Müller & Walossek 1987: fig. 27A-E; various trilobites and other lamellipedians in Whittington

1997: figs. 83, 86, 92, 93, 95, 96, 132). The legs of *Agnostus*, on the other hand, have a pendent posture like in the crustacean lineage (Fig. 3I herein; see also Müller & Walossek 1987: fig. 27F; or Whittington 1997: fig. 82).

(2) Lamellipedians have flat setae forming a lamellar comb (for instance, Hou & Bergström 1979: fig. 44 and pp. 42-88; cf. Fig. 3H herein), whereas *Agnostus* has notably different types of setae, with round cross sections, similar to those in crustaceans (Müller & Walossek 1987: e.g., fig. 9; see also Whittington 1997: figs. 78, 79, 82; cf. Fig. 3I herein).

(3) In *Agnostus*, some setal rows are double (Fig. 4A-B; Müller & Walossek 1987: fig. 6B, C). In lamellipedians including trilobites, the rows are always single (for instance, *Ceraurus*, Fig. 3H herein).

(4) Soft setae are present on the exopods of *Agnostus* (Fig. 4A-B) and *Henningsmoenicaris*. This type of setae is unknown in trilobites and other lamellipedians.

(5) The lamellipedian endopod has a terminal claw with associated telotarsal spurs, whereas *Agnostus* has slender spines with bristles. The agnostid endopod is very similar to that of the stem-lineage crustacean *Martinssonia*, except that it has an extra podomere and club-shaped outgrowths (cf. Fig. 3I and J).

(6) The lamellipedian antenna is a simple sensory structure, whereas the antenna as known from larvae in *Agnostus* and in 'stem-lineage crustaceans' is a locomotory and feeding organ (Müller & Walossek 1987, Walossek & Müller 1990: 420).

(7) In lamellipedians, post-antennal appendages are typically all similar (for instance, Whittington 1997: figs. 83, 86, 91, 93), or arranged in two or three homogeneous tagmata (for instance, *Emeraldella* in Bruton & Whittington 1983; Whittington 1992: fig. 9, showing tagmosis in segmental muscles in *Lonchodomas*). In *Agnostus*, as in crustaceans, there is a strong morphological change throughout the cephalon, the first three appendage pairs belonging to a functional head. It may be noted that the crustacean nauplius has three pairs of appendages. The fourth cephalic pair is similar to the thoracic and pygidial pairs. The cephalic shield is probably merely an adult reminiscence of the shield in the first larva, that is, a memory of the number of functional segments.

(8) Lamellipedians have the appendages situated under the axial lobe (Fig. 3G). In *Agnostus*, the appendages are situated in part outside the axial lobe (Figs. 3I, 4B-C; Müller & Walossek 1987; Whittington 1997: fig. 80).

(9) In lamellipedians, notably in trilobites, we find the bulk of the soft parts in the central lobe; the lateral lobes are thin pleural folds (Olenoides, Fig. 3G herein; Bergström & Hou 2003b: fig. 2). In *Agnostus*, the lateral lobes contain body parts such as muscles from the appendages attached to the lateral lobes. The agnostid lateral lobes are obviously not just pleural folds, but voluminous parts of the main body (Fig. 3I herein; Müller & Walossek 1987: figs. 13.1, 14.3, 27.1, 28.1-3). Thus, the lobes and axial furrows of lamellipedians do not correspond with comparable agnostid structures (Fig. 3G, I).

(10) In *Agnostus*, the lateral lobes contain thick alimentary diverticula, which did not anastomose (Öpik 1961; Bergström 1973: fig. 3B). There is no evidence of such structures in the lateral lobes of trilobites. Instead, where alimentary structures are known they are confined to the axial lobe (Bergström 1973: fig. 3D; Whittington 1997: fig.

94). Similar diverticula are known from other arthropods, including arachnids, the Burgess Shale *Burgessia*, and naraoiid lamellipedians (cf. Bergström 1973: fig. 3).

(11) Trilobites have an anastomosing vascular network in their lateral lobes (pleural areas; for instance, Harrington 1959: fig. 73A-C; Bergström 1973: fig. 2A; Jell 1978). Such a network is a characteristic of gill areas (Jell 1978 for trilobites; cf. for instance Shu et al. 2000: fig. 9 for a crustacean). Nothing similar is known from *Agnostus*, which is instead thought to have the gill function concentrated to club-shaped structures on the endopods (Fig. 3I).

(12) All trilobites and probably all other lamellipedians have a ventral plate, the hypostome, extending below the mouth and behind it as a fold (Whittington 1997: figs. 4, 5, 26, 121). A somewhat similarly positioned structure in agnostids (Robison 1972; Müller & Walossek 1987, e.g., fig. 5; see also Whittington 1997: figs. 27.1b+c, 78, 80, 82), misleadingly called a hypostome, is a pre-oral support for the oesophagus-stomach-brain region, has no posterior fold, and is not directly comparable with a trilobite hypostome. Instead, *Agnostus* is close to the stem-group crustaceans *Henningsmoenicaris* and *Martinssonia* in having the mouth at the tip of the 'hypostome'.

(13) In trilobites and other lamellipedians, the antenna extends to the side of the hypostome, whereas in *Agnostus* the antenna emerges through a hole in the 'hypostome'.

(14) Trilobites have dorsal compound eyes, except when secondarily lost. Paired lobes in windows of the ventral support structure ('hypostome') in *Agnostus* (Müller & Walossek 1987; Whittington 1997: fig. 80) are reasonably interpreted as compound eyes. One reason for such an interpretation is that many Palaeozoic arthropods have ventrally positioned eyes, and that in this position they must have been close to the visual centre of the brain.

(15) Trilobites have a skeletonised doublure, whereas agnostids lack this structure. As a consequence, agnostids also lack the submarginal ecdysial suture, which is characteristic for trilobites (usually, but not always and not originally, in combination with facial sutures; see Whittington 1997: figs. 27-37).

(16) The trilobite exoskeleton has a complex layered structure, whereas the agnostid counterpart consists of a simple prismatic layer (Wilmot 1990).

It is, thus, difficult to find even a single genuine similarity between *Agnostus* and trilobites. As in the case of the superficially similar ostracodes and phosphatocopids (Maas et al. 2003: 15), however, it should not come as a surprise if agnostids and trilobites are placed together in phylogenetic analyses based on the use of formalistic characters (exoskeleton calcified, exoskeleton trilobed, presence of hypostome, and so fort). However, in several of the characters listed above, *Agnostus* instead exposes a close similarity to the stem-line crustaceans. In particular, there is a most striking similarity in the design of the basis plus endopod between *Agnostus* (Fig. 3I), *Martinssonia*, (Fig. 3J) and also *Cambrocaris*, although the latter lacks endites (Walossek & Müller 1990: fig. 5C4, 5E5; Walossek & Szaniawski 1991: fig. 8). The animals form a successive series regarding the number of endopod podomeres, namely 6, 5 and 4, respectively.

A 1st antenna with prolonged segments and a swimming and/or feeding function may be the first indication for the rise of crustaceans (see below). This is seen already in *Agnostus*. Another early indication is the development of a caudal furca, although it was probably not yet present in the shared ancestor of agnostids and crustaceans.

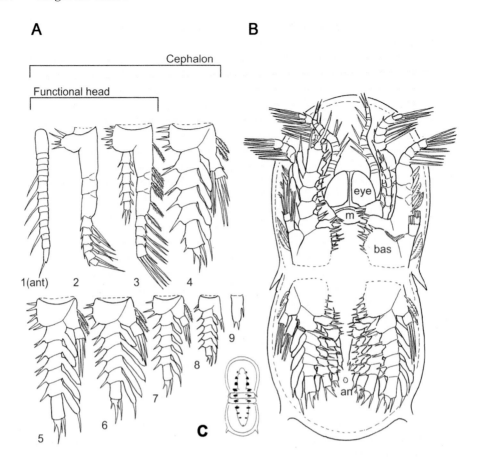

Figure 4. *Agnostus pisiformis*. (A) All appendages of a large larva, meraspis stage 2c, with only one of the ultimately two thoracic segments separated from the protopygidium. The cephalon has 4 pairs of appendages. The last one of these is similar to the appendages in the thorax and pygidium. The anterior three pairs in the cephalon are specialised for feeding and swimming. Therefore, they define the length of a functional head. (B) Ventral view of the same larva, with cephalic appendages folded forwards, post-cephalic legs folded backwards. The last cephalic limb on the right side (as seen) lacks the endopod and much of the exopodal setae for a better view of the more anterior limbs. Similarly, the first post-cephalic leg on the left side lacks much of its endopod. The narrow border of the body is the pleural field. Abbreviations: an = anus; bas = basis; eye = possible compound eye; m = mouth. (C) Young adult with two thoracic segments. Muscle scars (black) show a morphological distinction between the three most anterior pairs of scars and the more posterior pairs (A-C modified from Müller & Walossek 1987).

Yet another early event was the development of a proximal endite on the basal part of endopods that later evolved into the crustacean coxa. *Agnostus* has no coxal endite, but there is one in *Henningsmoenicaris*. It was pointed out that these genera share two "unique and noteworthy" features. These are "the fusion of the proximal parts of the two rami only in the fourth and subsequent limbs, and the soft setae on the outer edges of the limbs" (Walossek & Müller 1990: p. 422, fig. 5D.5-6, 5E.4-5; their reference to figures is erroneous). These two genera and *Martinssonia* (which also has a proximal endite) share also the unique ventral plate, erroneously referred to as a hypostome. Despite these apparent synapomorphies, the authors hesitated to separate agnostids from trilobites and shift their position to the crustacean side (Walossek & Müller 1990: 422-424, fig. 7). However, one problem remains. If the structures mentioned as shared are true synapomorphies, why does *Agnostus* have no coxal endite? If it had not been lost, it should mean that either the coxal endite or the morphology of the 4th and successive limbs (and the 'hypostome'?) had evolved twice.

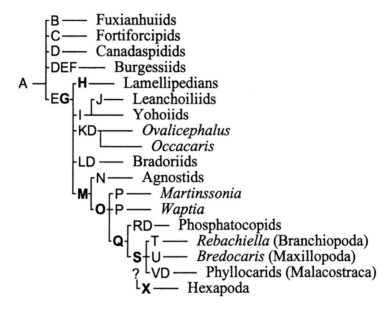

Figure 5. One hypothesis for the evolutionary path leading to the Crustacea. Many groups are excluded because of insufficient understanding. Important steps are marked with bold letters. [A] (1) Specialization of (1st) antenna; (2) biramous appendages with simple, blade-like exopod. [B] 2nd appendage hook-like grappler. [C] (1) 1st appendage with strong pincers; (2) crustacean-like tail fluke. [D] Large carapace. [E] Endopod segments fewer than 10. [F] (1) Alimentary caeca in carapace; (2) filiform exopods in head. **[G] Exopod with cylindrical setae. [H] Lamellipedia: (1) lamellipedian setae; (2) endopod with terminal claw; (3) limbs laterally deflexed; (4) hypostome.** [I] 1st appendage with strong pincers. [J] Each pincer finger with long filiform annulated extension. [K] 2nd appendage with strong pincers. [L] (1) Three first endopods with strong enditic spines; (2) trunk reduced to four segments. **[M] Crustaceomorpha: (1) larval (1st) antenna locomotory and feeding organ; (2) six endopodal podomeres in post-antennal limbs.** [N] Agnostida: (1) limbs 2-3 for swimming, with endopod small or lost; (2) tergum calcified; (3) dorsal morphology superficially trilobite-like. **[O] (1) Proximal endite; (2) setae directed toward endopod in limbs 2 and 3; (3) five endopodal podomeres in post-antennal limbs.** [P] Reduction of posterior appendages. **[Q] Labrophora: (1) 2nd antenna; (2) mandible with coxa; (3) labrum; (4) atrium oris; (5) sternum; (6) paragnaths; (7) furca.** [R] Phosphatic exoskeleton. **[S] Crustacea (Eucrustacea): (1) nauplius larva with three pairs of appendages and related larval and adult specializations in feeding and locomotion; (2) 1st maxilla included in functional head.** [T] Branchiopoda: limb tagmosis formula 4+12. [U] Maxillopoda: limb tagmosis formula 4+13+7. [V] Malacostraca: limb tagmosis formula 5+8+7. **[X] Hexapoda: (1) loss of 2nd antenna; (2) only three pairs of post-oral limbs left as legs.**

In conclusion, it seems extremely difficult to accept agnostids as trilobites. They are not built according to a lamellipedian body plan. Surprisingly, Müller & Walossek (1987) did not realize this when they described *Agnostus*. Even a decade later, they presented an agnostid limb as a trilobitoid limb (Walossek & Müller 1998b: fig. 5.5a). We concur with Shergold (1988) that agnostids should not be placed with the trilobites, and we suggest that they are close to the lineage leading to the crustaceans.

Eodiscids, usually thought of as being closely related to agnostids, are typical trilobites. Like all other trilobites (the so-called polymerids), they have a number of features typical to trilobites. Most notably, some eodiscids have laterally directed eyes looking through an eye-slit, an arrangement exclusively found in trilobites (including the olenellids).

So, where do the trilobites and other lamellipedians belong? Since they share mostly only basal arthropod characters with other Cambrian arthropods, they must have diverged quite early (Bergström & Hou 2003a, b), perhaps even before the needle-shaped crustacean-type setae evolved. This would probably be a level where we can now place canadaspidids, fortiforcipids and burgessiids (Figs. 2G, 3D). Presumably because the radiation was on such a low level, the main resulting groups are extremely divergent - it is hardly possible to recognise any shared derived characters. The flat lamellipedian setae forming lamellae in a 'comb' is a unique feature.

1.2.5 *From the agnostid level to the true crustaceans*

The evolution beyond the agnostids has been well analyzed by Walossek & Müller (1990, text, diagram in Fig. 7; Walossek & Müller 1998b; Maas et al. 2003: 182-191), and there is little need or opportunity to add much of relevance. Some of the chronology may be questioned, however. For example, the shift, in the larvae, in function of the (1st) antenna from a sensory flagellum to a limb with locomotory and feeding functions. We see that this shift had occurred already in *Agnostus*. In the decrease in the number of endopodal podomeres from 7 to 5, *Agnostus* is intermediary in having 6-7. The re-orientation of exopodal setae from outward to inward is initiated in *Agnostus* behind the functional head, that is, in the fourth cephalic appendage. Probably the most significant single result produced by Müller and Walossek in their studies of stem-lineage crustaceans is that we now understand how the coxal limb segment evolved through the formation of a proximal endite that expanded until it embraced the cormus, thereby forming a new segment, the coxa. The last step in the lineage leading to the crustaceans involves a shift from four to three limb pairs in the first larva. These limbs receive specific swimming and food collecting functions in the larva, which from now on is a nauplius larva. The strong development of the 2nd larval appendage resulted in a modification into the adult 2nd antenna. Similarly, the larval modification of the 3rd limb was perpetuated in the adult as a strong masticatory coxa and ultimately as a true jaw. The morphology of the adult can therefore tell us whether or not a crustacean-like extinct animal had a nauplius larva. The superficially ostracode-like Cambrian phosphatocopids are recognized as the closest relatives of Recent crustaceans. They still had a larva with four limb pairs (Maas et al. 2003: fig. 23 and p. 158-159). It was ostracode-like, without similarity to a nauplius larva. However, the phosphatocopids are the only extinct forms that had a crustacean-like labrum, which is interpreted as a shared apomorphy.

Shu et al. (2000) suggested that the bivalved bradoriids, formerly thought to be ostracodes related to phosphatocopids, could be another group of 'stem-group crustaceans', but stressed that we need more information, particularly on the possible coxal precursor for a reliable judgment. They may very well be right.

It may be relevant to comment on the position of the Burgess Shale and Chengjiang waptiids of the genus *Waptia* (Fig. 2H). They are notably shrimp-like in their habitus. They are also similar to the crustaceans in having exopodal setae directed medially, a condition typical of the crustacean lineage after the splitting from the agnostids. The (1st) antenna has long segments, as typical for crustaceomorphs (such as defined in Fig. 5). According to early studies, an anterior head tagma has (1st) antennae and about five appendages possibly lacking exopods, a thoracic tagma may have eight segments with bifid appendages, and the abdomen with six segments is limbless. The eyes are stalked. This sounds almost like the description of a crustacean, but there is no 2nd antenna and we do not know the details of the appendages. Therefore, it seems likely that the waptiids belong to the stem-group crustaceans, but further evidence is needed for a more reliable judgement.

Maas et al. (2003: 184) presented a list of crustacean features that were supposed to be plesiomorphically retained from the ground pattern of the Euarthropoda. If we define the Euarthropoda as including the first ancestor with body and leg segmentation, we include all arthropods discussed herein - and Euarthropoda becomes a synonym of the Arthropoda in its classical meaning. Any other definition would introduce great difficulties in the classification of the Cambrian arthropods. Regardless of the definition, there is no evidence that the four-segmented "head larva" of Walossek & Müller (1998a: fig. 12.8) was an inherited character. In their illustration, such a four-segmented larva is shared only with trilobites. For trilobites, this number is based on both larval and adult dorsal morphology. However, there are Cambrian examples with five pairs of muscle or tendon furrows and six axial lobes (for instance, species of zacanthoidids and oryctocephalids), indicating that the dorsal morphology may be unreliable or, alternatively, that there is some variation within the Trilobita. Whatever the explanation, the larva consisted virtually of the future cephalon. A similar segmental correspondence between the first larva and the length of a cephalic shield (thus not a functional head) is present in at least some of the 'stem-lineage crustaceans' (Walossek & Müller 1990: fig. 5). It is quite possible that the length of the cephalic shield marks the number of functional segments in the first larva in many early arthropods. If so, it is likely that the number of functional segments in the first larva was 4 in *Burgessia*, *Yohoia* and *Naraoia*, 4 and/or 5 in trilobites, and 5 or 6 in *Sanctacaris*. Xandarellids with a long cephalon could have had some 6 to 11 functional larval segments. From these numbers alone, no conclusions can be made on more ancestral numbers.

It seems possible that the hexapods belong phyletically either to the stem-group crustaceans or to the true crustaceans. If coxa and mandibles are structures genuinely shared with crustaceans, hexapods must have branched off after the phosphatocopids (Figs. 3F, 4). The 2nd antennae, having no function in a terrestrial habitat, would then have to be reduced together with the segment that bears them.

2 THE LAST SHARED ANCESTOR - THE UR-ARTHROPOD

With recent growth of understanding of Cambrian arthropods, it is becoming possible to discern the main steps in the evolution from the earliest arthropods to the crustaceans. By removing all the modifications caused by evolutionary trends and unique evolutionary steps we can sort out primitive characters and get a glimpse of the last common ancestor of all arthropods (Fig. 6). We arrive at an animal in which all post-antennal segments were alike, with individual dorsal sclerites, although it is possible that there was a dorsal cephalic shield covering the segments that carried functional legs in the first larva - a protaspis or 'cephalon larva'. The legs had a multi-segmented endopod and a simple exopodal flap. The antenna was similar to the endopod of the successive limbs. The mouth was situated on the ventral side behind the antennae, where it could reach the substrate for feeding; there was no hypostome or labrum, and no assisting feeding appendages. In fact, the most striking new understanding is, that before appendage specialization, arthropods seem to have indiscriminately ingested sediment with its contents for food (Hou & Bergström 1997; Bergström & Hou 2003a). The eyes were anteroventral. From such a beginning, the evolutionary trends easily - and in short time - could have given rise to the mosaic of morphologies that we see in the Cambrian.

A striking result of the search for the last shared ancestor is its similarity to the first crustacean, the urcrustacean, as viewed by Hessler & Newman (1975: figs. 5-7, 9, 11). The latter differs from the urarthropod in having a large protopod with endites, an epipod, setae on the exopod, a labrum, and a telson fork. Hessler & Newman claim that the crustaceans evolved from trilobitomorphs, but their concept of these includes a wide variety of Cambrian arthropods (trilobites, *Naraoia*, and non-trilobitomorphs such as *Burgessia*, *Waptia* and *Yohoia*). What they mean is therefore that the urcrustacean evolved from less advanced early arthropods as indicated in our Fig. 5. Fryer (1992: fig. 1), on the other hand, derived the urcrustacean from a segmented worm without passing through a stage of a generalized urarthropod or indeed any non-crustaceomorph arthropod. Thus, his first stage with arthropod legs (his Fig. 1c) already has a specialized naupliar set of three anterior appendages and a limbless abdomen.

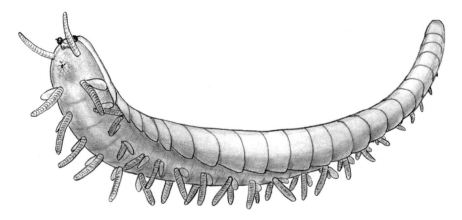

Figure 6. Reconstruction of the last shared ancestor of the arthropods.

The arthropod lineage could have existed well before the Cambrian, but there is no evidence for arthropod organization or arthropod radiation older than the Cambrian. Among modern crustacean groups, only branchiopods have been identified from the Lower Cambrian (Hou et al. 2004). Starting from our last shared arthropod ancestor, it seems that the time from the beginning of the Cambrian to the time of the Chengjiang fauna, an interval of approximately 15 million years, would be sufficient for the evolution of the primitive Chengjiang arthropods.

This conclusion may appear to be in conflict with molecular results suggesting a divergence between metazoan phyla at least some 40 million years before the Cambrian (Bromham et al. 1998; Bromham & Hendy 2000), but this is merely another hypothesis. Lineages could have been separated for this long, but the radiation within phyla of sizeable, skeletonised arthropods must have occurred in the Cambrian. Despite well-preserved sediments, there are no older body or trace fossils. Whether or not the lineages are of pre-Cambrian origin, there was a notable radiation in the Cambrian: the Cambrian explosion.

3 SUMMARIZING CONCLUSIONS

We can see two general types of patterns in arthropod evolution. First, there are the more or less general trends, the frequently repeated changes, with some variation in detail. Starting from an early arthropod with strict serial similarity along the segmental body, several new features could have been developed by changes in the *Hox* gene control of embryological growth (e.g., Schram & Koenemann 2004). This is particularly the case when the new character is locally restricted to one or a set of segments. The result may have been an evolutionary mosaic with notable morphological differences involved, and also that genetic changes outside the *Hox* genes may have been minimal. Trends often involving a high degree of parallel evolution include the following:

(1) A decrease in the number of leg podomeres (segments) from 15-20 to less than 10.
(2) A shape differentiation between podomeres.
(3) A formation of a sub-chelate to multi-chelate grasping appendage ('great appendage') in the antennal or first post-antennal segment.
(4) Abandonment of the primeval type of sediment-eating, partly in association with the development of grasping appendages.
(5) A functional differentiation along the body, with the formation of distinct tagmata.
(6) A formation of a larger functional head by the specialization of post-antennal appendages for feeding. This is independent of the segmental composition of the cephalon, since the cephalic shield tends to be the adult derivative of the shield of the first larva.
(7) A formation of a carapace fold. It is possible that there was also the opposite trend in groups with a carapace.
(8) A split of the carapace into a bivalved shape.
(9) A shift of the eye position up onto the head shield. This has happened among the crustaceans, but perhaps more often among the lamellipedian arthropods (Bergström & Hou 2003a).

Second, there are evolutionary changes of seemingly more unique characters.
(1) Except for a few groups with little advancement in limb morphology, all others have well-developed setae.
(2) Setae usually have round cross sections, but all the trilobite-like groups, the lamellipedians, have setae that are more or less flat. These groups also have other characteris-

tics, such as a hypostome and a wide pleural fold, and seem to represent a major side lineage.

(3) The modification of the antenna for locomotory purposes appears to be unique to the crustacean lineage.

(4) The formation of a head in agnostids is unique: the 2nd and 3rd appendages become specialized for swimming.

(5) The evolutionary steps in stem-lineage crustaceans described by Walossek and Müller (1990, 1998b): Among others, they include the formation of a "proximal endite", which was later enlarged to form a new segment, the coxa. The coxal setae were a prerequisite for the formation of a labrum. The nauplius larva embodies a revolutionary division of labour involving the first three appendages, which results in the adult 1st and 2nd antennae and mandibles.

The interpretation of 'great appendages' in many early arthropods is problematic (Budd 2002). It is either associated with a well-developed antenna (*Fuxianhuia*, *Occacaris*, *Ovalicephalus*), or it represents the most anterior appendage (*Leanchoilia*, *Actaeus*, *Yohoia*, *Jianfengia*, *Tanglangia*, probably *Fortiforceps*). Therefore, it is possible that it is a modified 2nd appendage in some forms, but a modified (1st) antenna in others. The generalized swimming and food-collecting function of the antenna in *Agnostus* indicates the ability of the antenna to develop into a food-catching organ (Müller & Walossek 1987). This casts severe doubts on the idea (Chen et al. 2004) that anomalocaridids, 'great appendage' arthropods and chelicerates form a monophyletic group.

Alternatively, the antenna may be reduced in certain taxa having a 'great appendage' as the most frontal appendage. In any case, the 'great appendage' would not have been developed in the lineage leading to the crustaceans. Many Cambrian arthropods are found with sediment-stuffed guts (e.g., *Canadaspis* and *Clypecaris*, Fig. 1 herein; Hou & Bergström 1997). In Chengjiang fossils, the contained genuine sediment consists of microscopic quartz grains and muscovite flakes (and possibly illite) (radiographic and mineralogical study, Ulf Hålenius, Swedish Museum of Natural History, Stockholm, pers. comm.). The nutritive contents would have been dead organic matter, micro-organisms, or burrowing 'worms'. These arthropods usually had no 'great appendage'. Therefore, the 'great appendage' seems to have been the first attempt to shape a tool for dealing with prey, when arthropods started to utilize other food sources than the sedimentary substrate. The distribution of a 'great appendage' among arthropods indicates that it was the result of independent evolution in several lineages.

ACKNOWLEDGEMENTS

We are grateful to Professor Ulf Hålenius, the Swedish Museum of Natural History, for analyzing the mineral composition of gut contents of Chengjiang arthropods, and to Mr. Magnus Hellbom for assistance with digital images. We are also in great debt to Dr. Gregory D. Edgecombe and an anonymous colleague for most valuable and welcome reviews with suggestions for improvement.

REFERENCES

Bergström, J. 1973. Organization, life, and systematics of trilobites. *Fossils & Strata* 2: 1-69.
Bergström, J. & Hou, X.-G. 2003a. Cambrian arthropods: a lesson in convergent evolution. In: Legakis, A., Sfenthourakis, S., Polymeni, R. & Thessalou-Legaki, M. (eds.), *The New Panorama of Animal Evolution*: 89-96. Sofia-Moscow: Pensoft Publ..
Bergström, J. & Hou, X.-G. 2003b. Arthropod origins. *Bull. Geosci.* 78: 323-334.
Briggs, D.E.G. 1977. Bivalved arthropods from the Middle Cambrian Burgess Shale of British Columbia. *Palaeontology* 20: 595-621.
Briggs, D.E.G. & Clarkson, E.N.K. 1990. The late Palaeozoic radiation of malacostracan crustaceans. In: Taylor, P.D. & Larwood, G.P. (eds.), *Major Evolutionary Radiations. Syst. Assoc. Spec. Vol.* 42: 165-186. Oxford: Clarendon Press.
Briggs, D.E.G. & Collins, D. 1988. A Middle Cambrian chelicerate from Mount Stephen, British Columbia. *Palaeontology* 31: 779-798.
Briggs, D.E.G. & Whittington, H.B. 1985. Modes of life of arthropods from the Burgess Shale, British Columbia. *Trans. R. Soc. Edinburgh* 76: 149-160.
Bromham, D.L. & Hendy, M.D. 2000. Can fast early rates reconcile molecular dates with the Cambrian explosion? *Proc. Royal Soc. London, Ser. B,* 267: 1041-1047.
Bromham, D.L., Rambaut, A., Fortey, R., Cooper, A. & Penny, D. 1998. Testing the Cambrian explosion hypothesis by using a molecular dating technique. *Proc. Nat. Acad. Sci. USA* 95: 12386-12389.
Brooks, H.K. 1962. The Paleozoic Eumalacostraca of North America. *Bull. Amer. Paleontol.* 44: 163-338.
Bruton, D.L. & Whittington, H.B. 1983. *Emeraldella* and *Leanchoilia*, two arthropods from the Burgess Shale, Middle Cambrian, British Columbia. *Phil. Trans. R. Soc., Ser. B,* 300: 553-585.
Budd, G.E. 2002. A palaeontological solution to the arthropod head problem. *Nature* 417: 271-275.
Chen, J.-Y., Waloszek, D. & Maas, A. 2004. A new 'great-appendage' arthropod from the Lower Cambrian of China and homology of chelicerate chelicerae and raptorial antero-ventral appendages. *Lethaia* 2004: 3-20.
Fortey, R.A. 1997. Classification. In: Kaesler, R.L. (ed.), *Treatise on Invertebrate Paleontology, Part O, Arthropoda 1, Trilobita, Vol. 1* (revised): 289-302. Boulder; Lawrence: Geol. Soc. Amer. and Univ. Kansas Press.
Fryer, G. 1992. The origin of the Crustacea. *Acta Zoologica* 73: 273-386.
Harrington, H.J. 1959. General description of Trilobita. In: Moore, R.C. (ed.), *Treatise on Invertebrate Paleontology, Arthropoda I, Part O*: 38-118. Boulder; Lawrence, Kansas: Geol. Soc. Amer. and Univ. Kansas Press.
Hessler, R.R. & Newman, W.A. 1975. A trilobitomorph origin for the Crustacea. *Fossils & Strata* 4: 437-459.
Hou, X.-G. 1999. New rare bivalved arthropods from the Lower Cambrian Chengjiang fauna, Yunnan, China. *J. Paleontol.* 73: 102-116.
Hou, X.-G., Aldridge, R.J., Bergström, J., Siveter, D.J., Siveter, D.J. & Feng, X.-H. 2004. *The Cambrian Fossils of Chengjiang, China; The Flowering of Early Animal Life*: 228 pp. Oxford: Blackwell.
Hou, X.-G. & Bergström, J. 1997. Arthropods of the Lower Cambrian Chengjiang fauna, southwest China. *Fossils & Strata* 45: 116 pp.

Hou, X.-G. & Bergström, J. 1998. Three additional arthropods from the Early Cambrian Chengjiang fauna, Yunnan, southwest China. *Acta Palaeont. Sinica* 37 (4): 395-401.

Hou, X.-G., Bergström, J. & Xu, G.-H. 2004. The Lower Cambrian crustacean *Pectocaris* from the Chengjiang biota, Yunnan, China. *J. Paleont.* 78 (4): 700-708.

Hou, X.-G., Siveter, D.J., Williams, M., Walossek, D. & Bergström, J. 1996. Appendages of the arthropod *Kunmingella* from the early Cambrian of China: its bearing on the systematic position of the Bradoriida and the fossil record of the Ostracoda. *Phil. Trans. R. Soc. London, Ser. B*, 351: 1131-1145.

Hughes, C.P. 1975. Redescription of *Burgessia bella* from the Burgess Shale, Middle Cambrian, British Columbia. *Fossils & Strata* 4: 415-435.

Jaekel, O. 1909. Ueber die Agnostiden. *Zeitschr. Deutsch. Geol. Ges.* 61: 380-401.

Jell, P.A. 1978. Trilobite respiration and genal caeca. *Alcheringa* 2: 251-260.

Jenner, R.A., Hof, C.H.J. & Schram, F.R. 1998. Palaeo- and archaeo-stomatopods (Hoplocarida, Crustacea) from the Bear Gulch Limestone, Mississippian (Namurian), of central Montana. *Contrib. Zool.* 67 (3): 155-185.

Kaesler, R.L. (ed.) 1997. *Treatise on Invertebrate Paleontology, Part O, Arthropoda 1, Trilobita, Vol. 1* (revised): xxiv + 530 pp. Boulder; Lawrence: Geol. Soc. Amer. and Univ. Kansas Press.

Maas, A., Waloszek, D. & Müller, K. 2003. Morphology, ontogeny and phylogeny of the Phosphatocopina (Crustacea) from the Upper Cambrian "Orsten" of Sweden. *Fossils & Strata* 49: 238 pp.

Müller, K.J. 1982. *Hesslandona unisulcata* sp. nov. (Ostracoda) with phosphatised appendages from Upper Cambrian 'Orsten' of Sweden. In: Bate, R.H., Robison, E. & Sheppard, L.M. (eds.), *A Research Manual of Fossil and Recent Ostracods*: 276-304. Chichester: Ellis Horwood.

Müller, K.J. & Walossek, D. 1987. Morphology, ontogeny, and life habit of *Agnostus pisiformis* from the Upper Cambrian of Sweden. *Fossils & Strata* 19: 124 pp.

Müller, K.J. & Walossek, D. 1988. External morphology and larval development of the Upper Cambrian maxillopod *Bredocaris admirabilis*. *Fossils & Strata* 23: 70 pp.

Öpik, A.A. 1961. Alimentary caeca of agnostids and other trilobites. *Palaeontology* 3: 410-438.

Qian, Y. & Bengtson, S. 1989. Palaeontology and biostratigraphy of the Early Cambrian Meishucunian Stage in Yunnan Province, South China. *Fossils & Strata* 24: 156 pp.

Robison, R.A. 1972. Hypostoma of agnostid trilobites. *Lethaia* 5: 239-248.

Schram, F.R. & Koenemann, S. 2004: Developmental genetics and arthropod evolution: On body regions of Crustacea. In: Scholtz, G. (ed.), *Crustacean Issues 15, Evolutionary Developmental Biology of Crustacea*: 75-92. Lisse: Balkema.

Schram, F.R. & Hof, C.H.J. 1998. Fossils and the interrelationships of major crustacean groups. In: Edgecombe, G.D. (ed.), *Arthropod Fossils and Phylogeny*: 233-302. Columbia Univ. Press.

Shergold, J.H. 1988. Review of the *Agnostus* paper of Müller & Walossek (1987). *Nomen Nudum* 17: 21-25.

Shu D.-G., Vannier, J., Luo, H.-L., Chen, L., Zhang, X.-L. & Hu, S.-X. 2000 (for 1999). Anatomy and lifestyle of *Kunmingella* (Arthropoda, Bradoriida) from the Chengjiang fossil Lagerstätte (lower Cambrian; Southwest China). *Lethaia* 32: 279-298.

Simonetta, A.M. & Delle Cave, L. 1975. The Cambrian non-trilobite arthropods from the Burgess Shale of British Columbia. A study of their comparative morphology, taxonomy and evolutionary significance. *Palaeontographica Italica (N.S. 39)* 69: 1-37.

Taylor, R.S., Shen, Y.-B. & Schram, F.R. 1998. New pygocephalomorph crustaceans from the Permian of China and their phylogenetic relationships. *Palaeontology* 41 (5): 815-834.

Walcott, C.D. 1912: Cambrian geology and paleontology; II, 6; Middle Cambrian Branchiopoda, Malacostraca, Trilobita and Merostomata. *Smiths. Misc. Coll.* 57: 145-228.

Walcott, C.D., 1931. Addenda to descriptions of Burgess Shale fossils. *Smiths. Misc. Coll.* 85: 1-46.

Walossek, D. 1993. The Upper Cambrian *Rehbachiella* and the phylogeny of Branchiopoda and Crustacea. *Fossils & Strata* 32: 202 pp.

Walossek, D. & Müller, K.J. 1990. Upper Cambrian stem-lineage crustaceans and their bearing upon the monophyletic origin of the Crustacea and the position of *Agnostus*. *Lethaia* 23: 409-427.

Walossek, D. & Müller, K.J. 1998a. Cambrian 'Orsten'-type arthropods and the phylogeny of Crustacea. In: Fortey, R.A. & Thomas, R.H. (eds.), *Arthropod Relationships*. Syst. Assoc. Spec. Vol. Ser. 55: 139-153.

Walossek, D. & Müller, K.J. 1998b. Early arthropod phylogeny in light of the Cambrian "Orsten" fossils. Pp. 185-231. In: Edgecombe, G.D. (ed.), *Arthropod Fossils and Phylogeny*: 347 pp. Columbia Univ. Press.

Walossek, D. & Szaniawski, H. 1991. *Cambrocaris baltica* n. gen. n. sp., a possible stem-lineage crustacean from the Upper Cambrian of Poland. *Lethaia* 24: 363-378.

Whittington, H.B. 1974. *Yohoia* Walcott and *Plenocaris* n. gen., arthropods from the Burgess Shale, Middle Cambrian, British Columbia. *Geol. Surv. Canada, Bull.* 231: 1-21.

Whittington, H.B. 1997. Morphology of the exoskeleton. In: Kaesler, R.L. (ed.), *Treatise on Invertebrate Paleontology, Part O, Arthropoda 1, Trilobita, Vol. 1* (revised): 1-85. Boulder; Lawrence: Geol. Soc. Amer. and Univ. Kansas Press.

Whittington, H.B. 1980. Exoskeleton, moult stage, appendage morphology and habits of the Middle Cambrian trilobite *Olenoides serratus*. *Palaeontology* 23: 171-204.

Whittington, H.B. 1981. Rare arthropods from the Burgess Shale, Middle Cambrian, British Columbia. *Phil. Trans. R. Soc., Ser. B*, 292: 329-357.

Whittington, H.B. 1992. *Trilobites*: xi + 145 pp + 120 pls. Woodbridge, Suffolk and Rochester, N.Y: The Boydell Press.

Wilmot, N.V. 1990. Cuticular structure of the agnostine trilobite *Homagnostus obesus*. *Lethaia* 23: 87-92.

Comparative morphology and relationships of the Agnostida

TREVOR J. COTTON[1] & RICHARD A. FORTEY[2]

[1] *1 Colewood Drive, Higham, Rochester, Kent, U.K.*
[2] *Department of Palaeontology, The Natural History Museum, London, U.K.*

ABSTRACT

The relationships of the agnostid trilobites have been controversial. They have been claimed as highly derived, neotenously-derived trilobites on the one hand, or as arthropods close to the stem group of crustaceans on the other. On the former view, their distinctive autapomorphies are regarded as specialisations related to their mode of life; on the latter, emphasis is placed on the differences from polymerid trilobites as indicating a separate derivation. We consider it necessary to include Eodiscina in any assessment of agnostid relationships. Previously suggested synapomorphies uniting crustaceans and agnostids to the exclusion of trilobites are reviewed, and the majority is rejected. A number of new synapomorphies uniting agnostids with all or some eodiscinids are described. These characters are tested in the context of a cladistic analysis of 79 eodiscinid and 3 agnostid taxa. We conclude that agnostids are deeply nested within the eodiscinids. The eodiscinid family Weymouthiidae is strongly supported as sister taxon to the Agnostina. Several supposed agnostid autapomorphies are found to be synapomorphies shared with one or more eodiscinids. At least as far as exoskeletal characters are concerned, the supposed distinctiveness of agnostids has been over-emphasised. To assert non-trilobite relationships of agnostids would require implausible levels of homoplasy.

1 INTRODUCTION

Fred Schram has often had to evaluate the phylogenetic position of anomalous crustacean groups or stem-group fossils in his many contributions to arthropod evolution. However, few groups have been as contentious as the agnostids (suborder Agnostina Salter, 1864), a large and varied Cambrian-Ordovician group of arthropods - universally recognised as a clade - which have traditionally been regarded as trilobites, but have been considered to be stem-group crustaceans by some authors.

As a major component of Cambrian arthropod diversity, agnostids are valuable biostratigraphic indices (e.g., Peng & Robison 2000), since they evolved rapidly and most genera and many species are cosmopolitan in their distribution (Robison 1984). Agnostids originated early in the Cambrian (Rasetti & Theokritoff 1967; Blaker & Peel 1997) and survived until the late Ordovician (e.g., Shergold & Laurie 1997; Shergold et al. 1990). Agnostid morphology is remarkably conservative throughout this range, despite their high diversity (123 genera and subgenera, according to Shergold & Laurie 1997: p. 332). For example, Öpik (1967: p. 65) argued that due to the "combinative nature" of agnostid taxa "all agnostoids within their suborder are relatively close to each other", a point of view that has been supported more recently by Shergold & Laurie (*op. cit.*), who suggested that "Agnostina reiterate morphological conditions at different times in different family groups". Furthermore, there has been a universal consensus that agnostids represent a highly distinctive and specialised body plan that differs morphologically from any other arthropod (Fortey 1997a: p. 294), e.g., "It is not in question that Agnostina were highly specialized arthropods, with a whole series of autapomorphies" (Fortey & Theron 1994: p. 851). This specialised morphology has fuelled the proliferation of discussions about agnostid life habits in the literature, which have spanned "almost the whole range of possibilities open to marine arthropods" (Fortey 1985: p. 3).

The distinctive morphology of agnostids has also led to intense debate about their relationships (e.g., see Fortey & Theron 1994; Fortey 1997a; Shergold 1991). They are conventionally regarded as trilobites closely related to eodiscinids (suborder Eodiscina Raymond, 1913). This hypothesis is reflected in the long history of uniting agnostids and eodiscinids to the exclusion of other trilobite groups (for which the term polymerid is used here) in the Isopygia Gürich, 1907, Miomera Jaekel, 1909, or the order Agnostida Salter, 1864. However, some authorities have considered this classification to not reflect relationships. Instead, many authors have suggested that agnostids are not trilobites, but constitute a distantly related group with a separate arthropod origin, usually considered to be close to crustaceans (Resser 1938; Walossek & Müller 1990; Shergold 1991).

Here, we briefly review previous work on agnostid relationships and, in particular, reassess the phylogenetic significance of characters that have previously been cited as supporting the exclusion of agnostids from the Trilobita. Secondly, the comparative morphology of agnostids and eodiscinids is revised within the context of polymerid trilobite morphology and potential synapomorphies uniting the Agnostina and some or all eodiscinids are identified and illustrated. These hypotheses of homology are tested in a preliminary cladistic analysis of 79 eodiscinids taxa and 3 agnostids coded for 123 characters, which provides a well-supported hypothesis of eodiscinid phylogeny and agnostid origins.

Throughout this work informal names refer to taxa employed in the recent *Treatise on Invertebrate Paleontology* (Kaesler 1997). Agnostid refers to the Agnostina, i.e., both Agnostoidea and Condylopygoidea, which are called agnostoids and condylopygoids, respectively. The term eodiscinid is used for the Eodiscina. No informal name is used for the Agnostida - the group including agnostoids, condylopygoids and eodiscinids.

2 PREVIOUS CONTRIBUTIONS TO THE AGNOSTID PROBLEM

Arguments against assigning agnostids to the Trilobita have largely relied on the distinctiveness of agnostid morphology, rather than the identification of derived similarities shared between agnostids and non-trilobites. Müller & Walossek (1987), for example, discussed the appendage morphology of the Upper Cambrian *Agnostus pisiformis*, which differs considerably from that of other trilobites (see, e.g., Whittington 1997b), and Shergold (1991) has highlighted the lack of a protaspis larval stage in agnostids. Peng & Robison (2000: p. 11) identified four suites of characters which they considered unique to the Agnostina: the modification of the cephalo-thoracic articulation, the protuberant hypostome, the lack of segmentation of the pleural lobes of the pygidium and the presence of triangular basal lobes on the cephalic axis, and they excluded both condylopygoids and eodiscinids from the Agnostida. On the other hand, those who favour cladistic reasoning regard unique features (autapomorphies) as of no consequence in determining relationships.

Of all the characters that have been discussed in the debate over agnostid relationships, the only potential synapomorphies uniting agnostids and non-trilobite arthropods described to date are features of the appendages. Agnostid appendages are known only from exceptionally preserved specimens of *Agnostus pisiformis* from the Upper Cambrian Orsten deposits of Sweden (Müller & Walossek 1987). These appendages show a number of similarities, which have been considered derived, to those of supposed stem-group crustaceans from the same deposits (Walossek & Müller 1990; Bergström 1992; Hou & Bergström 1997). This evidence for crustacean affinities is ambiguous. Firstly, the primitive condition of the euarthropod exopod is unclear. Secondly, all of the Orsten material is of sub-adult individuals and the appendages of many arthropods, both fossil (including *Agnostus* and the supposed stem-group crustaceans) and living (e.g., Olesen & Walossek 2000, Schram & Koenemann 2001), show a remarkable degree of ontogenetic variation. All other described trilobite appendages are from large adult specimens (Whittington 1997b; Chatterton & Speyer 1997: p. 200) - it remains possible that the appendages of larval polymerids resembled those of agnostids (e.g., Speyer & Chatterton 1989), since no larval appendages of the latter have been described.

In contrast, a number of unambiguous synapomorphies uniting agnostids and other trilobites have been recognised (Fortey 1990, 1997a; Fortey & Theron 1994). These include the trilobation (particularly possession of a glabella) and calcification of the exoskeleton, the presence of a cephalic border, and the method of thoracic articulation. The inclusion of agnostids within a trilobite clade has been confirmed in wider analyses of arthropod phylogeny (e.g., Briggs et al. 1992, 1993; Wills et al. 1994). Agnostids therefore provide a suitable test case on the application of cladistics as opposed to other methods of analysis of relationships.

On the basis of a cladistic analysis of 26 characters coded for six eodiscinid genera, six genera of polymeroids, two condylopygoids, three agnostoids and the nektaspid *Naraoia*, Babcock (1994: pp. 112-114) concluded that the Eodiscina was polyphyletic and that "available evidence does not suggest that agnostoids and eodiscids shared a close common ancestor". This conclusion was accepted by Blaker & Peel (1997). Babcock's study, however, is flawed in terms of data, methodology and interpretation.

Many characters of potential phylogenetic importance were excluded from Babcock's analysis. For example, the monophyly of the Corynexochida, from which many of Babcock's polymeroid taxa are drawn (*Olenoides*, *Thoracocare* and *Tonkinella*), is supported by the fusion of the hypostome to the rostral plate (Fortey 1990). Given the small number of characters actually used, excluding this character may have had a considerable impact on the analysis. Only one, calcification of the exoskeleton, of the long list of trilobite synapomorphies presented by Fortey & Whittington (1989) was included. Inclusion of the other characters would result in the exclusion of *Naraoia* from the Trilobita rather than its placement deep within the trilobite clade. Babcock's choice of taxa may also have significantly biased his results since a number of eodiscinid taxa, notably *Tannudiscus* and *Chelediscus*, which have been compared to agnostids by previous authors (Rushton 1966:

p. 10; Jell 1975: p. 14) were ignored. *Serrodiscus* is included in the published matrix but does not appear in the tree supposedly derived from it. No explanation for this is given.

In addition to these problems with the matrix, re-analysis of the published data generated quite different results from those reported by Babcock (1994). Many more equally parsimonious trees exist than the five that he reported. Treating all characters as unordered yielded 3204 equally parsimonious trees 59 steps long and treating them as ordered, 52 trees 67 steps long (the omission of *Serrodiscus* resulted in 1068 trees 57 steps long or 13 trees of 80 steps, respectively). The majority-rule consensus trees for these analyses are broadly similar to Babcock's result, but significantly less well resolved. Babcock's finding that the agnostoids form a clade to the exclusion of the condylopygoids, polymerids and *Naraoia* is supported by our re-analysis, but the sister-group relationship between *Naraoia* and the agnostoids to the exclusion of condylopygoids is not. In the four different analyses we carried out, a closer relationship between *Naraoia* and agnostoids than between condylopygoids and agnostoids was only supported when multistate characters were treated as unordered and then only in 534 of 1068 most parsimonious trees (MPTs) with *Serrodiscus* excluded, and in 534 of 3204 MPTs with *Serrodiscus* included. The *Naraoia*-agnostoid clade in Babcock's (1994: fig. 27) majority-rule tree therefore seems to represent an error in either his analysis or calculation of the majority-rule consensus tree, and is not supported by the published data.

Finally, Babcock's (1994) results do not support his conclusions. Babcock claims that his results suggest that the eodiscids are "polyphyletic from trilobites of the order Polymerida". In fact, on Babcock's tree, the eodiscinids are paraphyletic with respect to a clade including *Naraoia* and agnostids. With the exception of *Naraoia*, this is exactly the cladistic pattern that would be expected if, as argued here, the agnostids are derived from eodiscinid ancestors and this clade from polymeroids. The placement of *Naraoia* is influenced both by the limitations in the choice of characters and analytical errors described above, and is almost certainly not a trilobite *sensu stricto* (see, e.g., Edgecombe & Ramsköld 1999; Cotton & Braddy 2004).

3 COMPARATIVE MORPHOLOGY OF AGNOSTIDS AND EODISCINIDS

The distinctive morphology of agnostids has resulted in the adoption of a specialised descriptive terminology and a tendency for trilobite researchers to specialize in either agnostids or polymerids. This is illustrated by the separate description of agnostid and co-occurring polymerid and eodiscinid faunas by different authors - an approach rarely applied to other trilobite groups. For example, the polymerid faunas of the Middle Cambrian Henson Gletscher and Cap Stanton formations of North Greenland were described by Babcock (1994) and the accompanying agnostids by Robison (1988, 1994). Both the persistence of a distinct terminology and the paucity of researchers familiar with both agnostids and polymerids have in turn led to a neglect of the comparative morphology of agnostids and other trilobites. Other factors, such as the largely stratigraphic, as opposed to biological, interests of the majority of Cambrian trilobite workers interested in agnostids, have also contributed to this. There is currently a considerable lack of clarity concerning the homology of many features of agnostids with those of other trilobites. For example, the axial segmentation of agnostids has been considered too distinctive to allow homologies with that of other trilobites to be drawn (e.g., Rushton 1966).

General similarities between agnostids and eodiscinid trilobites have long been recognised. For example, the two groups were united in Jaekel's (1909) Miomera or Kobayashi's (1939, 1944) Agnostida on the basis of a small number of thoracic segments, isopygy, and the loss of eyes and facial sutures. Since then a number of authors have identified other general similarities between agnostids and eodiscinids as a whole, including discussions of ontogeny by Rushton (1966: p. 10) and Jell (1970) and similarities in the mechanism of enrollment discussed by Müller & Walossek (1987: p. 52). These similarities have been put into a cladistic context by Fortey (1990; Fortey & Theron 1994). However, in recent years, no general attempt has been made to systematically compare the morphology of the two groups.

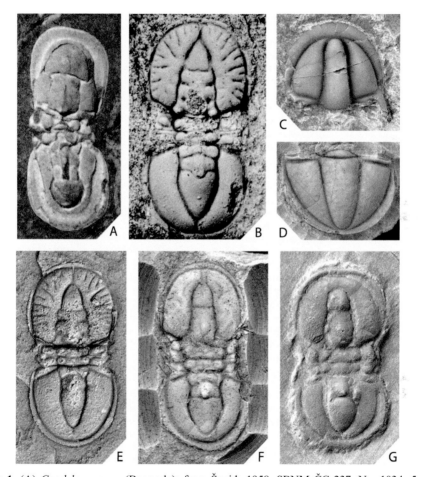

Figure 1. (A) *Condylopyge rex* (Barrande); from Šnajdr 1958, SBNM ČC 237, No. 1034; 5x. (B) *Goniagnostus nathorsti* (Brøgger); from Öpik 1979, CPC 14324; 10x. (C) *Litometopus longispinus* Rasetti; cephalon, NHM It 3762; 3x. (D) *Litometopus longispinus* Rasetti; pygidium, NHM It 3762; 2.5x. (E) Unidentified ptychagnostid; NHM In 19943; 9x. (F) *Triplagnostus burgessensis* Rasetti; NHM It 3564; 7x. (G) *Peronopsis interstricta* (White); NHM It 21015; 6x.

Similarities between agnostids and particular eodiscinid taxa have received even less attention, although the eodiscinid genera *Chelediscus* Rushton, 1966 and *Tannudiscus* Pokrovskaya, 1959 have been considered to share a number of features with agnostids (Rushton 1966; Jell 1975, 1997). Both these taxa are here included in the family Weymouthiidae (cf. Jell 1997). *Chelediscus* differs from the other taxa included in the Calodiscidae by Jell (1997) in having a pointed glabella, a larger number of pygidial segments, genal spines and an occipital furrow that slope backwards dorsally in lateral view, among other characters. These features are all found in a number of weymouthiids.

Here, the homology of 11 character complexes in agnostids, eodiscids and polymerids is discussed and their possible phylogenetic significance inferred. In particular, the morphology of weymouthiid eodiscinids is compared to that of agnostids. The choice of characters for discussion is based primarily on previous work, although the significance of a number of characters is newly identified. All characters that have previously been claimed as agnostid synapomorphies are re-assessed.

3.1 *Blindness*

All agnostids lack eyes and facial sutures, as do a number of eodiscinid taxa. It is widely accepted, following Jell (1975) and Öpik (1975), that blindness arose polyphyletically amongst eodiscinids. This may or may not be correlated with the loss of the facial sutures, since some eodiscinids possess eyes but lack facial sutures (e.g., *Yukonia intermedia* Palmer, 1968: pl. 2, fig. 14; *Helepagetia bitruncula* Jell, 1975: pl. 29, fig. 9). A similar observation has been used (Fortey 1990: p. 563) to suggest different modes of eye loss in various clades formerly included in the polyphyletic ptychopariid family Conocoryphidae (see Cotton 2001). Jell (1975, 1997) recognised at least three independent origins of blindness within the eodiscinids, within his families Calodiscidae and Eodiscidae and in the origin of the family Weymouthiidae. Other lineages may also have lost their sight convergently (e.g., Jell in Bengtson et al. 1990: p. 258). Blindness may have evolved independently a fourth time in the origin of the Agnostina from eodiscinid ancestors but, unless other evidence suggests otherwise, it is most parsimonious to assume that the agnostids share a close common ancestor with a blind, sutureless eodiscinid.

3.2 *Cephalic outline*

Kobayashi (1943: p. 45; 1944: p. 10) differentiated the agnostids and eodiscinids partly on the basis of the difference in cephalic outline. However, as Fortey & Theron (1994) recognised, certain eodiscinids have a cephalic outline more similar to agnostids than to either other eodiscinids or typical polymeroid trilobites. In most eodiscinids and polymeroids, the cephalon is considerably wider (transversely) than long (sagittally) and is widest at the posterior margin (Figs. 1C, 2I). In agnostids (e.g., Fig. 1A, B, E-G) and a number of weymouthiid taxa including *Chelediscus acifer* Rushton, 1966 (Fig. 2A), *Jinghediscus numularius* Xiang & Zhang, 1985, and *Tannudiscus balanus* Rushton, 1966, the cephalon is as long or longer than it is wide, and is widest at a point well anterior of the posterior margin. If the Agnostida are regarded as having polymerid ancestors, the situation in these eodiscinids and agnostids is derived. This is also supported by the distribution of cephalic outline amongst eodiscinids. In the earliest occurring species and those that have been considered primitive, such as *Tsunyidiscus*, *Sinodiscus* and *Calodiscus*, the cephalon is particularly wide compared to its length. In other eodiscinids this ratio is higher. This change in the shape of the cephalon therefore appears to represent a valid synapomorphy uniting at least some of Jell's Weymouthiidae with the Agnostina.

Examination of the ontogeny of polymeroid trilobites suggests that this character may reflect the probable progenetic origin of agnostids, since the cephalic outline is highly variable in the early ontogeny of many basal trilobites. The meraspid cephalon of most polymeroids has an outline more similar to agnostids than that of the typical holaspid cephalon. This seems to be particularly true of primitive polymeroids, as illustrated by Chatterton & Speyer (1997: figs. 168-169) and Zhang & Pratt (1999).

Isopygy has long been recognised as a distinctive feature of agnostids (see Fig. 1A, B, E-G) and has most recently been discussed by Fortey (1990; Fortey & Theron 1994). The similarity in outline between the cephalon and pygidium is, however, also a feature of all eodiscinids where the pygidium is known (e.g., Fig. 1C-D), and represents a potential synapomorphy at the level of Agnostida.

3.3 Genal spines

Fortey (1990: text-fig. 14) used the character 'genal spines reduced or absent' as a synapomorphy uniting agnostids and eodiscinids to the exclusion of other trilobites but provided no discussion of this character. The best known eodiscinid taxa (i.e., Eodiscidae, e.g., see Jell 1975: pls. 17-19; fig. 2I) and many agnostids do indeed lack, or have much reduced, genal spines.

Öpik (1979) suggested that agnostids lack genal spines and that the spines near the genal angles of agnostids instead represent 'fulcral spines' or 'fulcral prongs'. Müller & Walossek (1987) also employed this terminology. However, neither Öpik nor Müller & Walossek present a detailed argument in support of this view. It is true that the short agnostid spines are directly dorsal to the sockets against which the first thoracic segment articulates, whereas in most trilobites a distinct fulcral point lies adaxial to the genal spines on each side of the cephalon. In adult agnostids the posterior cephalic spines are positioned considerably dorsal to the lateral cephalic margin in lateral view. In the majority of trilobites, the genal spines lie on the same plane dorsoventrally as the lateral cephalic margin. The short, triangular agnostid 'prongs' are also structurally distinct from the typically long and slender genal spines of other trilobite groups. These observations could be interpreted as supporting the non-homology of agnostid posterior cephalic spines with the genal spines of polymeroids.

However, during the ontogeny of *Agnostus pisiformis*, the 'fulcral prongs' can be seen to migrate medially and dorsally with respect to the lateral cephalic margin (Müller & Walossek 1987: figs. 10, 12). In the youngest growth stages observed they are at the ventral extreme of the cephalon in lateral view and close to the lateral extreme in dorsal view - essentially the same position as conventional polymeroid genal spines. The strong geniculation between the lateral margin and the genal spines develops gradually through ontogeny. Similarly, the spines are relatively much longer in early growth stages than in later ones.

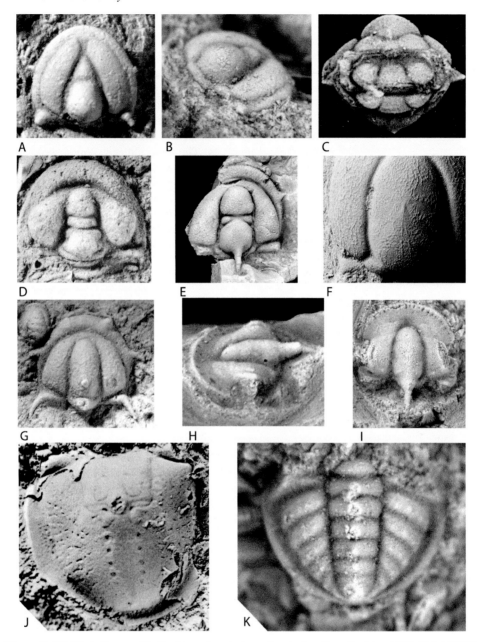

Figure 2. (A) *Cheledisucs acifer* Rushton; dorsal view of cephalon, from Rushton 1966, SM A57107; 11x. (B) *Cheledisucs acifer* Rushton; posterior lateral oblique view of cephalon, NHM; 19x. (C) *Geragnostus clusus* Whittington; from Whittington 1965, GSC 16171; 14x. (D) *Meniscuchus menetus* Öpik, 1975; 6x. (E) *Acimetopus bilobatus* Rasetti; NHM It 3742, 3x. (F) *Arthrorhachis latelimbata* (Ji); from Fortey 1997b, NHM It 25510; 20x. (G) *Acidiscus theristes* Rushton; from Rushton 1966, SM A57084; 7x. (H) *Pagetia resseri* Kobayashi; lateral oblique view of cranidium, NHM It 3630; 18x. (I) *Pagetia resseri* Kobayashi; NHM It 3631; 12.5x. (J) *Xestagnostus legirupa* Öpik; from Öpik 1967, CPC 5923; 8x. (K) *Pagetia resseri* Kobayashi; dorsal view of pygidium, NHM It 3632; 30x.

Within the eodiscinids, taxa with genal spines similar to those of polymeroids and taxa with agnostid 'fulcral spines' are known. This was recognised by Öpik (1975: table 6), who considered that *Eodiscus* possessed genal spines, but *Pagetia* fulcral spines. These taxa are generally considered to be very closely related. Within the Weymouthiidae, *Litometopus longispinus* has long genal spines that are distinctly abaxial to the fulcral points (Fig. 1C herein; Rasetti 1967: pl. 3, fig. 3; pl. 8, figs. 1, 4), whereas *Bathydiscus dolichometopus* has very short spines directly dorsal to the fulcra and well inside the lateral margins of the cephalon (Rasetti 1967: pl. 1, fig. 3; pl. 9, figs. 1, 4).

A number of eodiscinids, in particular weymouthiids, and agnostids have long genal spines of typical trilobite type (e.g., Figs. 1C, 2A, G). Genal spines are universally present in condylopygoids, which are generally considered to be the sister-group of agnostoids. It may be that genal spines are convergently reduced in agnostoids and eodiscinids, and that both primitive agnostids and primitive eodiscinids possessed genal spines comparable to those of polymeroids. When the phylogeny of the eodiscinids is better known, the significance of the variation in genal spine morphology may become clear. However, spines of very similar morphology to the posterolateral cephalic spines of agnostids are found in eodiscinids, and there seems to be no reason to regard agnostid spines as fundamentally different from those of polymeroids.

3.4 *Glabellar segmentation*

A distinct terminology has been applied to the glabellar segmentation of agnostoids, following Robison (e.g., 1964, 1982). The glabella is divided by a complete trans-glabellar furrow into an anteroglabella and posteroglabella. The posterior portion of the posteroglabella projects posteriorly, between more or less triangular basal lobes, as the glabellar culmination (*sensu* Whittington 1997a). This may be rounded or angular and bears a distinct small node in many taxa. The basal lobes and the narrow (sag. and trans.) occipital band which runs between them demarcated by shallow basal furrows, are excluded in this scheme from the glabella (see Fig. 2C herein). The lateral furrows of the glabella are numbered F1 to F3 from the posterior forward, F3 being the trans-glabellar furrow, and the axial lobes defined by these furrows at their anterior margin, are numbered M1 to M3 respectively (see Fig. 1B). This terminology has not been applied to the Eodiscina. Jell (1975), for example, regarded the basal lobe of the eodiscinid cephalic axis as occipital, and used the notation S and L for glabellar furrows and lobes, as applied to other trilobites.

Despite the use of a distinct terminology, it seems that most authors have considered the agnostid cephalic axis to be directly homologous with the polymeroid glabella, the basal lobes and occipital band making up a modified occipital ring and the trans-glabellar furrow a modified pair of lateral glabellar furrows. Fortey & Theron (1994: table 1), Hunt (1967), Robison (1984: p. 9) and Whittington (1965), for example, all regarded the basal lobes as occipital. Whilst other authors (Peng & Robison 2000: 11; Müller & Walossek 1987: 51) expressed doubt that the basal lobes represent the occipital ring, no coherent alternative has been proposed. The suggestion that the occipital band and basal lobes make up the segmental homologue of the axial rings of the thorax (Öpik 1979: p. 30; Müller & Walossek 1987: p. 51) can be applied equally to the polymeroid occipital ring.

Most trilobites possess four pairs of lateral furrows anterior to the occipital furrow. It is therefore unclear to which polymeroid furrows the three pairs in agnostids are homologous. In particular, the trans-glabellar furrow may represent S3 or S4 and the homologues of the polymeroid S1, S2 or S4 furrows may be missing in agnostids. If either the S1 or S2 furrows are missing then the posteroglabella represents L1 through L4 and the anteroglabella L5. If, on the other hand, the S4 furrows are missing then the anteroglabella would represent both the L4 and L5 lobes. The latter scenario seems more likely given that the S4 furrows are weak or effaced in many polymeroid trilobites.

Agnostid specimens showing presumed muscle attachment sites on the ventral surface of the exoskeleton support the homology of the agnostid trans-glabellar furrow with the polymeroid S3 furrows. Specimens of *Galbagnostus galba* (see Whittington 1965: pl. 3, fig. 7, 15; pl. 3, fig. 9) and *Arthrorhachis latelimbata* (Fig. 2F herein), clearly show four pairs of smooth areas anterior to the basal lobes. Judging from their distribution along the glabella, two of these pairs are on the anteroglabella and two on the posteroglabella. This suggests that the two main divisions of the agnostid cephalic axis each consist of two segments.

The insertion of the genal caecae, in taxa where these are well developed, also suggest that it is the homologue of S4 that is missing in most agnostids. In polymeroids a similar caecal network underlies the dorsal exoskeleton of the genae and inserts into the glabella at the same point as the eye ridges (e.g., *Meneviella venulosa*, see Cotton 2001: text-fig. 1a). This branch of the caecal network is commonly overlain by the eye ridges but is clearly visible, for example, in blind polymeroid taxa (Cotton 2001: pp. 173-174). In all polymeroids the eye ridges (and hence the caecal network) insert into the glabella anterior to the S3 furrows but posterior to the S4 furrows. In many agnostids, where the caecae are well developed, they seem to insert just anterior to the trans-glabellar furrow (Fig. 1B). This is indicated by a shallowing of the axial furrows at the point of insertion - just as the insertion of the caecal network in blind polymeroids is indicated by a shallowing of the axial furrows (see Cotton 2001: p. 188; *Pseudatops reticulatus*: pl. 2, fig. 1; *Alacephalus contortus*: pl. 1, fig. 6). This assumes that the agnostid caecae are homologous with those of polymeroids.

The trans-glabellar S3 furrow is also a general feature of eodiscinids (e.g., Fig. 2A, D-E). Taxa with trans-glabellar furrows are known from all eodiscinid families and in most species lacking the trans-glabellar furrow all lateral furrows are effaced. Few species (such as *Discomesites fragum* Öpik, 1975: pl. 5, figs. 1-4) possess strong, divided S3 furrows. The trans-glabellar S3 furrow is likely to be a distinctive synapomorphy uniting the majority of agnostids and eodiscinids. Even in eodiscinids where the frontal lobe is very long and the lateral furrows well impressed (such as *Serrodiscus daedalus*, see Blaker & Peel 1997: fig. 25.3+9-11), no furrows are present anterior to the trans-glabellar furrow. Complete effacement of the S4 furrow therefore also characterises eodiscinids and agnostids although, as mentioned above, this character has a wide distribution amongst trilobites.

3.5 *Form of the occipital ring*

The structure of the cephalic axis in eodiscinids is highly variable. In some taxa, such as *Sinodiscus* (Zhang et al. 1980: pl. 4, figs. 12, 18-19, 21) and *Korobovia ocellata* (Jell in Bengtson et al. 1990: fig. 177A-F), it differs from the primitive polymeroid condition only by the loss of the S4 furrows and presence of a trans-glabellar S3. In such cases, the occipital furrow is approximately straight or slightly curved posteriorly across the axis (in dorsal view). In lateral view the furrow is directed dorsally approximately perpendicular to the plane of the cephalic margin. The occipital ring is either of uniform width or slightly wider (sagittally) dorsally than ventrally with the result that, in lateral view, the posterior margin of the ring is angled backwards.

The occipital ring is highly modified in most eodiscinids. The occipital ring is more or less strongly angled backwards, with the result that the occipital furrow is angled backwards dorsally in lateral view and the occipital ring is lower (dorsoventrally) than the posterior lobe of the glabella (e.g., Fig. 2E). In some taxa (e.g., *Bathydiscus dolichometopus* Rasetti, 1966: pl. 9, fig. 3) this feature is present without any modification of the glabella anterior of the occipital furrow. Usually, however, the medial part of the occipital ring is completely covered dorsally by a posterior projection of the glabella. In these cases the occipital ring, in dorsal view, consists of two sub-triangular lateral lobes connected by a (sag.) narrow band behind or underneath the glabellar projection. This situation is strongly reminiscent of the agnostid basal lobes and occipital band, and there is no reason not to regard these modifications of the occipital ring and posterior glabella as homologous.

The form of the glabellar projection differs between eodiscinid groups. In weymouthiids it is posteriorly rounded (Fig. 2A-B, D), as in the majority of agnostids, but in most members of the eodiscinid families Yukoniidae and Eodiscidae the glabella is extended into a long posterodorsally directed spine (e.g., Fig. 2H-I), and the occipital ring is divided. This 'cranidial spine' therefore consists of a glabellar expansion and part of the occipital ring (Jell 1975: p. 4) and is not closely comparable to the situation in weymouthiids and agnostids. The backward expansion of the glabella may therefore not be homologous between agnostids and all eodiscinids, but the rounded expansion over a complete occipital furrow in agnostids and weymouthiids seems a convincing synapomorphy. Division of the occipital ring into an occipital band and basal lobes by the band furrow (*sensu* Whittington & Kelly 1997: p. 315) in agnostids is also shared by the species of the eodiscinid genus *Chelediscus* (Fig. 2A-B herein; Rushton 1966: p. 20).

The backward displacement of the occipital ring is likely to have had a function during enrollment. It is likely that, when fully extended, the thorax of many eodiscinids would have been angled ventrally with respect to the plane of the cephalic border due to the angle of the occipital ring, as discussed for *Agnostus pisiformis* by Müller & Walossek (1987). This decreases the degree of flexion necessary between the cephalon and the thoracic segments during complete enrollment of a trilobite with a short thorax.

Finally, the presence of a spine on the occipital band has been recognised as a distinctive feature of condylopygoids (Rushton 1966: p. 29). Some weymouthiid eodiscinids also have an occipital spine (Fig. 2G; *Leptochilodiscus punctulatus* Rasetti, 1967: pl. 3, figs. 18-20) but this is unknown in any agnostoid. This character may therefore not represent a valid synapomorphy of the Condylopygoidea, since it may be inherited from an eodiscinid ancestor. In this case, the loss of the occipital spine may instead be a synapomorphy of the Agnostoidea. The phylogenetic significance of this character will only be resolved when the sister-taxon to the agnostids within the Weymouthiidae is more precisely identified.

3.6 Median glabellar node

The median glabellar node on the dorsal midline of the posteroglabella has generally been considered a distinctive agnostid character. The position of this node is somewhat variable. In some taxa (e.g., *Goniagnostus fumicola* Öpik, 1961: pl. 20, figs. 14-17; *Ptychagnostus atavus*, see Peng & Robison 2000: fig. 52.1-3), it is at the level of the S1 furrows, in others (e.g., *Oidalagnostus trispinifer*, see Peng & Robison 2000: fig. 42.10-11) it is at the level of the S2 furrows. It seems clear, however, that the node belongs to the L2 glabellar lobe in all cases. The variation in the position of the node compared to the lateral furrows in dorsal view can be explained by changes in the dorsoventral orientation of L2.

Accepting the furrow homology scheme discussed above, in a number of weymouthiid taxa L2 bears a dorsally directed spine that is likely to be homologous to the median glabellar node of agnostids. In some cases the glabella is so strongly expanded posterodorsally (e.g., *Acimetopus bilobatus* Rasetti, 1966: pl. 4, figs. 3-4) and/or the lateral furrows are effaced (e.g., *Serrodiscus ctenoa* Rushton, 1966: p. 15), so that the position of the spine in terms of glabellar lobes is impossible to determine. In such cases, however, it seems most likely that the spine is homologous with those that are clearly on L2 in other species such as *Acidiscus theristes* (see Rushton, 1966: text-fig. 4; Fig. 2G herein) and *Bolboparia canadensis* (see Rasetti 1966: pl. 5, fig. 13). In some other weymouthiids only a low node is preserved, and it is unclear whether or not this was the base of a spine originally (e.g., *Tannudiscus balanus* Rushton, 1966: pl. 3, figs. 9a, 10).

3.7 Sagittal pre-glabellar furrow

A few polymeroid trilobites have a furrow running from the anterior of the glabella to the anterior cephalic border furrow. The sagittal pre-glabellar furrow, however, has rather a wide distribution within the eodiscinids and agnostids. It seems likely to have been acquired (or lost) convergently in many lineages in both groups. In the Eodiscina, *Natalina* (see, e.g., *Natalina incita* Repina & Romanenko, 1978: pl. 6, fig. 15) in the Hebediscidae, *Chelediscus* (e.g., *Chelediscus chathamensis* Rasetti, 1967: pl. 3, figs. 14-15; *Chelediscus acifer*, Fig. 2A herein) in the Weymouthiidae and all genera of the Eodiscidae (e.g., *Pagetia fluitata* Jell, 1975: pl. 8, figs. 9, 12; *Helepagetia bitruncula* Jell, 1975: pl. 29, figs. 1, 5-6) have such a furrow. Similarly, the pre-glabellar furrow may be present or absent in agnostoid taxa thought to be closely related. For example, within the Ammagnostidae (*sensu* Peng & Robison 2000) the genus *Nahannagnostus* (e.g., *N. nganasanicus* Rozova, 1964; see Peng & Robison, fig. 16) has a long, well-developed pre-glabellar furrow but *Kormagnostus* (e.g., *K. minutus*, see Peng & Robison: fig. 24) lacks the furrow altogether. Given this distribution, the phylogenetic importance of this character is unclear. It seems likely that this distribution is a function of the expression or effacement of the furrow rather than the presence or absence of a significant underlying structure.

3.8 Hypostome

The hypostome of *Agnostus pisiformis* described by Müller & Walossek (1987) is strikingly different from that of any known polymeroid. This has led to some authors rejecting the homology of agnostid and polymeroid hypostomes (Ramsköld & Edgecombe 1991). However, a range of hypostomal morphology is known from the agnostids. The hypostome of *Oidalagnostus trispinifer* (see Robison 1988: fig. 9) is very similar to those of many polymeroid hypostomes and that of *Peronopsis interstricta* (see Robison 1972: figs. 1c, d, 4a-c) is somewhat intermediate between that of *O. trispinifer* and the *Agnostus* hypostome (Müller & Walossek, 1987: fig. 26). Few eodiscinid hypostomes have been identified, and nothing is known of the ventral morphology of weymouthiids. The hypostome of *Pagetia ocellata* (see Jell 1975: pl. 28, figs. 1-2) closely resembles that of ptychopariid polymeroids (Fortey 1990). Agnostid hypostome morphology provides no support for a non-trilobite origin of the group and in most taxa is not fundamentally different from that of polymeroids.

3.9   *Thoracic segments*

The presence of only two or three segments in the thorax of agnostid and eodiscinids was historically the basis for the division of the Trilobita into the subclasses Miomera and Polymera. More recently, this criterion has been rejected by some authors following the discovery of other taxa with few thoracic segments in the Corynexochida (*Thoracocare*, see Robison & Campbell 1974) and Raphiophoridae (see Zhang 1980). Such discoveries, however, have no bearing on the status of this character as a potential synapomorphy for the Agnostida, albeit not a unique one. Further reduction of the number of segments from three to two is likely to be a synapomorphy for a narrower clade including agnostids and some eodiscinids. However, like the loss of eyes and sutures, this may well have occurred polyphyletically - many of Jell's eodiscinid families include taxa with two segments and those with three segments. Within the Weymouthiidae, only *Cheledisus acifer* Rushton, 1966 is known to have possessed a holaspid thorax of only two segments, but the thorax is unknown in the majority of taxa.

Beyond this simple character the agnostid thorax is highly distinctive. Compared to most trilobites the axis is wide relative to the width of the segment as a whole and is divided into a median lobe and two lateral lobes by a pair of furrows. These characters are seemingly unique to agnostids. Additionally, the first thoracic segment is narrow (trans.) compared to the second, and the pleural tips of the first segment are angled backward whilst those of the second segment are angled forwards. These characters are clearly of importance during enrollment and are shared with at least some eodiscinids, such as *Cheledisus acifer* (see Rushton 1966: text-fig. 6). The angle of the pleural tips is shared more widely, including, e.g., *Costadiscus minutus* (Eodiscidae; see Babcock 1994: fig. 29.3-4) and *Tsunyidiscus niutitangensis* (see Zhang et al. 1980: pl. 5, fig. 3). These taxa have thoracic segments of approximately equal width (trans.), but the pleural tips of the third segment are pointed forwards.

Öpik's (1979) suggestion that the widening and division of the occipital ring into basal lobes and the similar modification of the thorax in agnostids are linked evolutionarily is not supported by examination of weymouthiids. The division of the occipital ring in *Cheledisus* is not accompanied by division of the thoracic axial lobes.

3.10 Cephalothoracic articulation

The agnostids share a special type of articulation between the cephalon and the thorax that is unknown in other trilobites. The anterior thoracic segment of agnostoids and condylopygoids lacks an articulating half ring and the anterior margin of the segment is modified with the result that, on enrollment, a small gap, the 'cephalothoracic aperture', is left between the axial lobe of the thorax and the occipital band of the cephalic axis (which, as argued above, represents the median part of the occipital ring). This was first recognised in agnostoids by Robison (1964: p. 515; 1984a: fig. 31) and in condylopygoids by Rushton (1979: p. 45). No similar structure has been described from any eodiscinid, although the thorax is unknown in a majority of species.

3.11 Segmentation of the pygidial axis

The form of the agnostid pygidial axis has probably been more widely regarded as highly distinctive than any other feature of the agnostids. Rasetti (1948) suggested that "[...] the agnostid pygidium is a very different structure from the usual pygidium of the other trilobites, including the eodiscids." Many authors have considered the agnostid pattern of axial segmentation impossible to homologise with that of other trilobites. This has led those authors who accepted a relationship between agnostids and eodiscinids to propose that the agnostid segmentation was secondarily derived from a primitively unfurrowed axis (Henningsmoen 1951: p. 181; Palmer 1955; Rushton 1966: p. 10). Rushton further suggested that this was retained in Early Cambrian taxa such as *Condylopyge amitina* Rushton (see Rushton 1966: p. 29, pl. 4, figs. 1-12; Fig. 1A) and *Peronopsis roddyi* (Resser & Howell 1938) (see Blaker & Peel 1997: p. 26, figs. 13-16, 25.4-5+7).

The agnostid pygidium is characterised by a usually well-defined axis that is variable in both outline and length and which only bears furrows anteriorly. The pygidial margin of most condylopygoids and many agnostoids is equipped with one pair (or occasionally more) of broad based, flattened marginal spines extending posteriorly in the plane of the border. In condylopygoids three pairs of pygidial ring furrows are defined, whereas in most agnostoids only two pairs of furrows are present. It is widely accepted (following Palmer 1955; Öpik 1963, 1967) that the agnostid posteroaxis consists of a number of segments which are not defined on the dorsal surface. These segments are sometimes indicated by small rounded pits or muscle insertion scars (termed notulae), and vary from four to nine in number (Öpik 1967: p. 67; Fig. 2J). In agnostoids a prominent node (hereafter, the axial node) is present on the second axial ring and a small node or nodes (posterior nodes) on the undivided posteroaxis. Robison (1984: p. 17, 1988: p. 42; Peng & Robison 2000: p. 11) has repeatedly argued against interpreting these nodes of agnostoids as phylogenetically significant. We agree that the presence of the terminal and other posterior nodes may be a somewhat unreliable character, since these nodes are generally very weak and their presence may be polymorphic within species. However, those that can be associated unambiguously with the terminus of the intranotular axis should be considered homologous (Pratt 1992: p. 31). Secondly, in a majority of taxa showing multiple posterior nodes (reviewed by Peng & Robison 2000: p. 11), one of these nodes is associated with a transverse sulcus (in the terminology of Robison 1988: p. 32; the 'rosette' of Öpik 1979: p. 19) which is likely to be phylogenetically significant and should not be dismissed as 'iteratively evolved' (Peng & Robison 2000: p. 10) without good evidence. The presence of the prominent axial node on the second segment, however, is constant in the group. In a number of taxa this is produced into a long spine (reviewed by Öpik 1979: table 5). In condylopygoids spines or nodes may be present on all three of the anterior axial segments, but that on the second segment is generally the most prominent (e.g., *Pleuroctenium granulatum*, see Rushton 1979) and can be regarded as homologous with the anterior axial node of the agnostoid pygidium.

The eodiscinid axis is generally fully segmented, although the ring furrows are effaced in a number of taxa. The number of segments in the eodiscinid pygidium is highly variable, particularly in weymouthiids, where it ranges from 6 (e.g., *Chelediscus acifer* Rushton, 1966) to at least 11 (*Bolboparia elongata* Rasetti, 1966: p. 20, pl. 5, figs. 12-13), a similar range to that found in agnostids. Two patterns of axial nodes and spines can be distinguished amongst eodiscinids. Firstly many taxa show segmental spines which, when not present on all axial rings, are generally effaced posteriorly (e.g., Fig. 2K). Secondly, many eodiscinids have a long, broad-based spine on a single axial segment (sometimes along with segmental nodes or spines on other segments). The best known of these are the terminal or sub-terminal spines of many Eodiscidae (e.g., Jell 1975). In weymouthiids such spines are generally on the anterior part of the axis and often on the second segment (e.g., *Acimetopus bilobatus* Rasetti, 1966; *Bolboparia elongata* Rasetti, 1966). As well as being on the homologous segment these spines resemble the anterior axial node of agnostids in distorting the ring furrows, which medially bend anteriorly and posteriorly to accommodate the broad base of the spine. This is very similar to the situation described, for example, by Rushton (1966: p. 29) for ptychagnostids. Possession of a prominent node or spine on the second pygidial segment may therefore be a good synapomorphy uniting agnostids and some weymouthiids. The homology of these spines across all Agnostida, however, is problematic because their position seems to be extremely plastic (see, for example, Blaker & Peel 1997).

The effacement of the posterior axial furrows remains a good synapomorphy for the Agnostida. In eodiscinids with effaced ring furrows, all the furrows are more-or-less evenly effaced. In trilobites, furrows on the dorsal surface form ridges projecting ventrally which are generally considered to provide attachment sites for muscles acting on the appendages (e.g., Fortey & Owens 1999). The loss of the posterior pygidial furrows in agnostids may therefore be explained by Müller & Walossek's (1987) observation that there were only three pairs of appendages under the pygidium of *Agnostus pisiformis*. This is supported by the presence of three pairs of notulae on the anteroaxis of the agnostid pygidium (see, e.g., *Rhaptagnostus cyclopygeformis* (Sun, 1924), as illustrated by Shergold et al. 1990: fig. 16.3b; Shergold & Laurie 1997: fig. 233.3a; *Lejopyge calva* Robison, 1964, illustrated by Robison 1984: fig. 24). Reduction of segmentation of the pygidial axis in Agnostida is likely to have been associated with reduction and loss of the posterior pygidial appendages.

No obvious homologues of the paired marginal spines can be identified amongst the eodiscinids, despite the considerable range of spinose margins known in the group (see, e.g., Rasetti 1966; Jell 1997). Pygidial spines in the Agnostida as a whole are probably modified from the segmental, laterally directed, marginal spines of polymeroid pygidia. The paired marginal spines have such a wide distribution (including most condylopygoids) that they are likely to be an agnostid synapomorphy, and their loss in some agnostoids a reversal.

The final feature of the agnostid pygidium that requires comment is the absence of pleural furrows, which is shared with all weymouthiids (e.g., Fig. 1A-B, D) except *Stigmadiscus stenometopus* Rasseti, 1967 (see pl. 5, figs. 1-4) and a few other eodiscinid taxa. Since the presence of pygidial pleural furrows is a feature of most eodiscinid and polymeroid trilobites it seems likely that it is plesiomorphic for eodiscinids. The loss of pleural furrows is therefore probably a synapomorphy of a weymouthiid-agnostid clade.

4 CLADISTIC ANALYSIS

The classification of eodiscinids has been remarkably unstable, even compared to that of other Cambrian trilobite groups, and the few attempts to resolve the phylogeny of the group have been largely unsuccessful. The phylogeny of the Eodiscina is therefore of considerable interest in itself, as well as because of its significance for the understanding of the origin of the Agnostina.

Eodiscinids are in many ways ideal subjects for phylogenetic analysis amongst trilobites. They are highly complex morphologically, with a number of unusual character complexes that are likely to be of phylogenetic importance. Most importantly, and in contrast to many groups of Cambrian trilobites, they have generally been well described (e.g., Bengtson et al. 1990; Blaker & Peel 1997; Jell 1975; Öpik 1975; Rasetti 1952, 1966, 1967). The group has recently been extensively reviewed by Jell (1997) and by S. Zhang (in Zhang et al. 1980).

The outline hypothesis of eodiscinid phylogeny identified above on the basis of comparative morphology was tested by cladistic analysis of a matrix of 79 eodiscinid taxa and three agnostids coded for 123 characters. Further details of this analysis, including discussion of previous studies of eodiscinid phylogeny, investigation of the impact of alternative assumptions and a full discussion of its systematic implications, will be presented elsewhere.

4.1 *Taxonomic sampling*

All taxa included in the analysis are listed in Table 1 (Appendix 2), along with authorship and important subsequent references (including descriptions of other species or genera where these were used during coding). These taxa represent all but one of the eodiscinid genera recognised as valid in the most recent review of the group (Jell 1997). Most genera were represented by the type species. In the cases of *Delgadella*, *Serrodiscus* and *Tannudiscus*, all material assigned to the type species is poorly preserved, the type species was therefore not included and these genera represented by better known species. In addition, 16 eodiscinid taxa were included to represent the range of morphology in polytypic genera, and hence test Jell's (1997) hypotheses of generic synonymy and/or genus monophyly. The completeness of material and availability of specimens or English language descriptions were also used as criteria for selecting taxa. Jell's (1997) taxonomy is followed throughout.

Agnostids are represented by two taxa from the Early Cambrian, the condylopygoid *Condylopyge amitina* Rushton 1966 and the agnostoid *Peronopsis roddyi* (Resser & Howell, 1938). These species are the earliest occurring and putatively phylogenetically basal (Rushton 1966; Blaker & Peel 1997) members of the two agnostid superfamilies. The ptychagnostid *Ptychagnostus gibbus* (Linnarsson 1869), was also included in the analysis.

Ptychagnostidae is a diverse and morphologically divergent Cambrian agnostid family and it has been suggested (although this has not subsequently been supported, as far as we are aware) that it may have a separate origin from other agnostids (Jell 1975: text-fig. 6). *Ptychagnostus gibbus* is one of the most widely distributed and thoroughly described Middle Cambrian ptychagnostids and was chosen to represent the family. The position of *P. gibbus* was considered in a cladistic analysis of the Ptychagnostidae by Westrop et al. (1996).

4.2 *Characters and coding*

The selected taxa were coded for 123 exoskeletal characters with a total of 299 character states. In total, the database includes 10086 observations (including missing data and inapplicable characters). Characters were based on the hypothesised comparative morphology of eodiscinids and agnostids presented above and on previous published comparisons of eodiscinid morphology. All characters used in previous studies of eodiscinid phylogeny (including Öpik 1975, Jell 1975, and Babcock 1994) are represented in this study in some form. All characters and character states are described, and some discussed in more detail, in Appendix 1.

The set of characters employed is intended to cover as much as possible of the known morphological variation within the eodiscinids. Variation in the density of character sampling across parts of the trilobite body or life-history stages therefore reflects differences in the level of known variability between the taxa under consideration, as opposed to investigative or descriptive bias. The 123 characters include 81 cephalic characters, 8 characters of the thorax and 32 characters of the pygidium (2 further characters are concerned with sculpture of the exoskeleton as a whole). This split of characters across organ systems is consistent with the relative paucity of articulated specimens in eodiscinids as a whole and the lack of assigned pygidia for many of the taxa included. The major gap in character construction is the exclusion of characters concerning ontogeny. Growth series are known for very few eodiscinid taxa (Chatterton & Speyer 1997; Rushton 1966), and the inclusion of ontogenetic characters would add little to the analysis.

Character state distributions for all species considered are shown in the matrix in Table 2 (Appendix 2). Character state assignments were determined primarily on the basis of published descriptions and illustrations. Major references used for the coding of each taxon are listed in Table 1 (Appendix 2). This reliance on literature was necessary given the very broad taxonomic and geographic scope of this study. Coding from the literature was supplemented by examination of specimens in the collections of the Natural History Museum (London), the Sedgwick Museum (University of Cambridge), the National Museum of Natural History (Smithsonian Institution, Washington DC) and the large collection of casts, including taxa from China and Russia, of the Institute for Cambrian Studies (Boulder, Colorado).

Eight autapomorphic characters (characters 28, 38, 85, 88, 90, 115, 121 and 122) were included in the character list and character distribution matrix for completeness - to provide a comprehensive database of eodiscinid morphology as a basis for further work. These cladistically uninformative characters were excluded from all analyses and are not included in the calculation of any tree statistics.

The broad taxonomic scope of this study, and the approach taken in selecting terminals, resulted in the inclusion of many taxa that are incompletely known, poorly preserved and little studied. Combined with the conservative approach taken to coding, this resulted in a total of 10.7% of the total observations being either missing data or uncertain.

4.3 *Methods*

These data were subjected to cladistic analysis using the software package PAUP* version 4 (Swofford 1999), beta test version 8 for Windows or version 6 for MacOS. All searches used a heuristic search algorithm with starting trees constructed by a random stepwise addition sequence. All characters were treated as unordered and of equal weight, and 'not applicable' characters treated as equivalent to missing data. Details of other analyses carried out to investigate the effect of these assumptions will be presented elsewhere. The treatment of continuous and 'not applicable' characters was discussed in more detail by Cotton (2001). Bremer support and bootstrap values were calculated for all. Bootstrap values were based on 200 bootstrap replicates, each consisting of a heuristic search with 5 addition sequence replicates. Bremer support values were calculated on the basis of heuristic searches with 20 addition sequence replicates.

Whilst it is generally accepted that eodiscinids evolved from polymerid trilobites by paedomorphosis (Stubblefield 1936; Jell 1975; Fortey 1990; Shergold 1991) the sister-group of the Agnostida within the Trilobita is unclear. Outgroup rooting was therefore not considered appropriate for this analysis. Instead, all analyses were unrooted. The resulting trees were rooted by treating the Tsunyidiscidae (consisting only of the genus *Tsunyidiscus*, following the recent revision of Jell 1997) as a monophyletic sister-group to all other taxa. *Tsunyidiscus* is the earliest known eodiscinid and shows a number of features (Jell 1997: p. 384) that are likely to be primitive for eodiscinids based on comparison with juvenile redlichioids (Fortey 1990; Jell 2003). These include the narrow glabella with well defined dorsal furrows, well defined eye ridges and palpebral lobes, and the furrowed pygidial pleurae. Whether the tsuniyidiscids are paraphyletic with respect to other Agnostida (Jell 1997: fig. 241) or a monophyletic sister-group to them has no impact on their use for rooting the trees produced here provided that the remaining members of the Agnostida form a monophyletic group.

4.4 *Results*

Seventy-two equally parsimonious trees 1119 steps long (consistency index = 0.423; rescaled consistency index = 0.257, retention index - 0.609; see Kitching et al. 1998, for definitions) were found after 30 addition sequence replicates. The strict consensus of these trees, along with selected bootstrap and bremer support indices, is shown in Figure 3. Clade numbers and letters referring to paraphyletic assemblages used below are shown on this figure. The same set of MPTs was obtained in 3 of the 30 addition sequence replicates, so it is unlikely that shorter trees exist.

In the strict consensus tree, the majority of species fall into two major sister clades, clades 1 and 2 of Figure 3, respectively containing 40 and 34 of the 85 taxa analysed. Other ingroup taxa formed a paraphyletic group (group *a* of Fig. 3) basal to these large clades, consisting of *Calodiscus* and *Korobovia*, *Tchernyshevioides*, a monophyletic *Sinodiscus*, and *Lenadiscus unicus* as successive sister-taxa to remaining ingroup Agnostida.

The larger of the two major clades (clade 1 of Fig. 3) includes a subclade (clade 3) containing all taxa assigned by Jell (1997) to the family Weymouthiidae with the exception of *Abakolia*, along with *Chelediscus* and the agnostids. Within the broader clade (clade 1), *Hebediscus attleborensis*, a monophyletic *Delgadella* and *Neocobboldia dentata* (group *b* of Fig. 3) formed successive outgroups to the weymouthiid, *Chelediscus* and agnostid clade (clade 3). This latter clade comprised three groups: A *Cephalopyge*, *Weymouthia* and *Runcinodiscus* clade, a monophyletic *Cobboldites* and a *Bathydiscus* and *Oodiscus* clade formed a basal paraphyletic group (group *c*) to a clade (clade 4) made up of two large subclades. The first of these (clade 5) contains the weymouthiid genera *Analox*, *Ninadiscus*, *Meniscuchus*, *Acimetopus*, *Acidiscus*, *Bolboparia*, *Stigmadiscus*, *Semadiscus*, *Leptochilodiscus* and *Litometopus*, and species of *Serrodiscus* arranged polyphyletically. The second clade (clade 6) consists of the weymouthiids *Tannudiscus*, *Mallagnostus* and *Jinghediscus*, *Chelediscus* and the agnostids. Within clade 6, *Mallagnostus* formed a basal paraphyletic assemblage to an unresolved trichotomy including *Tannudiscus altus*, a *Jinghediscus* and *Chelediscus* clade and a *Tannudiscus balanus* and agnostid clade. Within the agnostids, the condylopygoid *Condylopyge amitina* was found to be the sister-group to the agnostoids *Peronopsis* and *Ptychagnostus*.

The smaller of the main clades identified (clade 2 of Fig. 3) contains all the taxa assigned by Jell (1997) to the Eodiscidae and Yukoniidae, with the exception of *Lenadiscus unicus* (placed in Yukoniidae by Jell), alongside *Abakolia* (Weymouthiidae), *Pseudocobboldia* (Calodiscidae), *Dicerodiscus*, *Natalina*, *Neopagetina* and *Luvsanodiscus* (Hebediscidae). *Pseudocobboldia* and a *Dicerodiscus* and *Luvsanodiscus* clade (collectively group *d* of Fig. 3) formed successive outgroups to all other taxa (clade 7). The Yukoniidae formed a paraphyletic basal group (group *e*) to a clade consisting largely of Eodiscidae (clade 9) within clade 7. Within this yukoniid assemblage are four successive outgroup clades, a clade of *Hebediscina sardoa* and *H. yuqingensis*, a clade consisting of *H. blagonravovi*, Yukonides and a monophyletic *Egyngolia* (clade 8), an *Ekwipagetia* clade and an *Alaskadiscus* and *Yukonia* clade formed four successive outgroups to clade 9. Within clade 9, *Kiskinella*, *Sinopagetia* and a paraphyletic *Natalina* constituted the sister-group (clade 11) to all other taxa (clade 10). Clade 10 consisted of a clade (clade 12) containing *Abakolia*, *Dawsonia* and *Eodiscus* in opposition to a clade (clade 13) containing *Pagetia*, *Opsidiscus*, *Helepagetia*, *Pagetides*, *Neopagetina* and *Macannaia*.

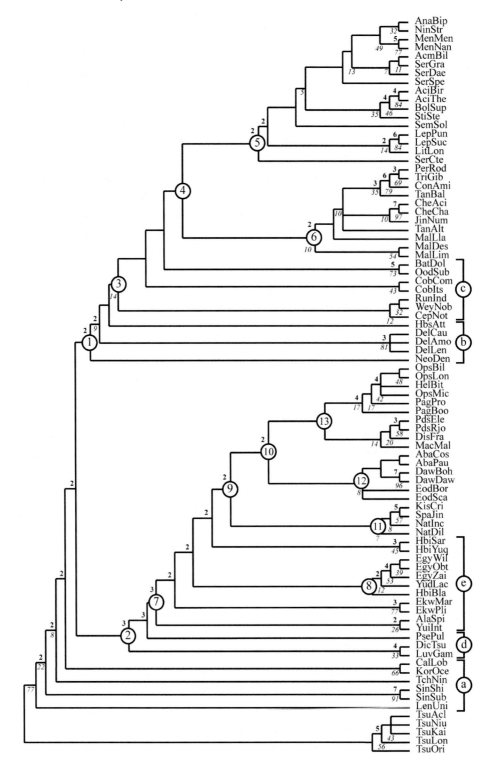

5. DISCUSSION

The assumption that *Tsunyidiscus*, which was used to root the trees, is the sister-group to other Agnostida, did not introduce significant bias. Other taxa that have been considered basal within the Eodiscina are closely related to (but do not form a clade with) *Tsunyidiscus*. Fortey (1990: p. 556), for example, suggested that *Sinodiscus changyanensis* S. Zhang in Zhang et al. 1980 could be the most primitive eodiscinid. *S. changyanensis* was not included in this analysis. However, rooting using either or both the species of *Sinodiscus* that were included would have produced similar results to rooting using *Tsunyidiscus*. Similarly the genera *Hebediscus* and *Neocobboldia*, suggested as basal members of two of the three eodiscinid lineages identified by Jell (1975: text-fig. 6), were found to be basal in this analysis. Hence, rooting using either of these genera would have had little impact on the topology of the trees presented here.

The results presented here strongly confirm that the agnostids are trilobites closely related to eodiscinids (e.g., Jell 1975; Fortey 1990b; Fortey & Theron 1994) and constitute strong evidence against the view that agnostids are more closely related to crustaceans than to eodiscinids and other trilobites (e.g., Walossek & Müller 1990; Shergold 1991; Bergström 1992). The paraphyly of the eodiscinids with respect to the agnostids, and division of more derived Agnostida into two large clades (recognised as an emended Agnostina and Eodiscina) are also strongly supported.

The considerable number of detailed similarities identified above between agnostids and some eodiscinids represents a convincing argument that the distinctiveness of the Agnostida has been considerably overstated, and strongly supports previous arguments that the agnostids are trilobites. In a cladistic context, a number of the characters previously thought to be synapomorphies of the Agnostida are instead likely to be synapomorphies of various nested clades uniting agnostids with some eodiscinids. In addition to synapomorphies uniting agnostids and trilobites as a whole, a much longer list of potential synapomorphies uniting agnostids and eodiscinids would need to be explained as a result of convergence if agnostids were excluded from the Trilobita (see, e.g., Fortey & Theron 1994: text-fig. 8).

The cladistic analysis supports the view that agnostids are closely related specifically to weymouthiid eodiscinids. The agnostids are found to be monophyletic and nested within a clade including the eodiscinid genera *Mallagnostus*, *Chelediscus*, *Tannudiscus* and *Jinghediscus*. This group formed part of a wider weymouthiid clade. The shortest trees in which the three agnostid taxa did not form a clade with the derived weymouthiids listed above were seven steps longer than the MPTs.

Figure 3. Strict consensus of 72 trees of 1119 steps resulting from cladistic analysis of the matrix shown in Table 2 (Appendix 2). Taxa are referred to by the six letter codes shown in Table 2. Clades and paraphyletic assemblages referred to in the text are indicated by encircled numbers over nodes and lower case letters, respectively. Bootstrap percentages based on 200 bootstrap replicates each of 5 addition sequence replicates are shown in italics below each node where the value was greater than 5%. Bremer support values are shown above nodes in bold type, for all nodes with a support value greater than 1.

The analysis confirms much of Jell's (1997: fig. 241) phylogeny. Like Jell, this study suggests that the Yukoniidae is paraphyletic with respect to Eodiscidae and the Weymouthiidae with respect to Agnostina. However, Jell's suggestions that the Calodiscidae forms a third major monophyletic lineage and that the Hebediscidae is paraphyletic with respect to the weymouthiid + agnostid clade are not supported. Instead, taxa assigned by Jell to Hebediscidae and Calodiscidae were intermingled in three paraphyletic groups - at the base of all Agnostida except Tsunyidiscidae, and at the base of the Eodiscidae + Yukoniidae clade and Weymouthiidae + Agnostina clade.

A range of authors have explicitly supported the view that there are no intermediate forms between agnostids and other trilobites (e.g., Kobayashi 1939: p. 73; Fortey 1997: p. 295). The analysis presented above suggests that the widely held view that "a suite of species connecting [agnostids] with some other taxon is not known" (Fortey 1997: p. 295) is incorrect and that the Weymouthiidae constitute just such a suite. In fact, of the characters discussed above (in comparative morphology), more support a clade of weymouthiids and agnostids than support the monophyly of the agnostids.

ACKNOWLEDGEMENTS

We would like to thank Adrian Rushton, Peter Jell, Derek Briggs, James Cotton and Helen McColm for comments on earlier drafts of this work. James Cotton also assisted with the phylogenetic analysis. TJC would like to thank Pete and Pat Palmer at the Institute for Cambrian Studies, Boulder, Colorado, and Ralph Chapman, Ali Olcott, Steve Schellenberg and Jen Young at the National Museum of Natural History, Washington, D.C., for their hospitality during the course of this research. This research was supported by the Geological Society of London Curry Research Studentship, the Department of Paleontology, The Natural History Museum, London, and the Department of Earth Sciences, University of Bristol.

INSTITUTIONAL ABBREVIATIONS

CPC: Commonwealth Palaeontological Collection, Geoscience Australia, Canberra, Australia
GSC: Geological Survey of Canada, Ottawa, Canada.
NHM: Department of Palaeontology, The Natural History Museum, London, U.K.
SBNM: National Museum, Prague, Czech Republic.
SM: Sedgwick Museum, University of Cambridge, U.K.

APPENDIX 1

Characters and character states used in phylogenetic analysis of Agnostida. Character state distributions are shown in Table 2 (Appendix 2). The conditions for characters to be coded as 'not applicable' (see text) are shown in square brackets after the character description. For example, character 2 could only be coded for taxa with a backwardly convex anterior border, coded as state 0 of character 1 (1:0), and is coded as 'not applicable' for taxa with other states of character 1. Autapomorphic characters are indicated in square brackets after the character description. Characters are numbered in approximate order of their position on the eodiscinid body from anterior to posterior, with the exception of characters that are variable only within the agnostid taxa considered (numbers 118 to 123), which are described after other characters.

1. Spines on the posterior cephalic border. 0: Present; 1: Absent.
2. Position and angle of posterior cephalic border spines [1:0]. 0: At genal angles, directed posterolaterally at approximately 45 degrees, 1: Adaxial to genal angles at geniculation, directed posteriorly subparallel to axis.
3. Length of posterior border spine [1:0]. 0: Short (approximately half or less the distance from axial furrows to genal angles), 1: Long (equal to or greater than distance from axial furrows to genal angles).
4. Anterior lateral cephalic border spines. 0: Absent, 1: Present.
5. Angle of anterior lateral cephalic border spines [4:1]. 0: Approximately perpendicular to cephalic axis, 1: Directed anterolaterally at approximately 45 degrees to axis.
6. Posterior lateral cephalic border spines (second pair of spines on the lateral cephalic border). 0: Absent in meraspids and holaspids, 1: Present in meraspids, lost in holaspids, 2: Present in holaspids.
7. Anterior cephalic margin bicuspate in anterior view. 0: Absent, 1: Present.
    This character was described in *Leptochilodiscus succinctus* by Bassett et al. (1976). It can also be made out in the type material of *L. punctulatus* Rasetti, 1966 (USNM146009 and USNM 146010), despite dorsoventral flattening.
8. Anterolateral cephalic border scrobiculate. 0: Absent, 1: Present.
    Despite Öpik's (1975: p. 33) assertion to the contrary, radial furrows can clearly be seen amongst the tubercles on the anterior cephalic border of at least two specimens of *Discomesite fragum* (Öpik 1975: pl. 5, figs. 1, 3). The series of pits on the anterior cephalic border in *Jinghediscus nummularius* Xiang and Zhang, 1985, and *Mallagnostus llarenai* (Richter & Richter, 1941) are here treated as homologous with the more fully developed scrobiculae found in other taxa.
9. Depth of cephalic border scrobiculae [8:1]. 0: Very shallow, 1: Moderate, 2: Deep.
10. Length of cephalic border scrobiculae [8:1]: 0: Short (small elongate pits in border furrow), 1: Moderately long (approximately half border width), 2: Long (considerably more than half border width).
11. Epiborder furrow. 0: Absent, 1: Present.
12. Cephalic border nodes. 0: Absent, 1: Present.
    The term nodes is used here (following Öpik 1975) for the row of large rounded hollow swellings found on the anterolateral cephalic borders of a range of eodiscinid genera, to distinguish them from the tubercular prosopon of other taxa. Tubercular prosopon of the anterior border can most easily be distinguished from the border nodes because it is matched by the sculpture of the genae and glabella (see, e.g., *Acimetopus bilobatus* Rasetti, 1966: pl. 4, figs. 1, 6, 7). The border nodes are likely to have accommodated the ventrally directed spines on the

pygidial border during enrollment (Jell in Bengtson et al. 1990: p. 259). In most taxa the cephalic border nodes form a row of pairs along much of the anterolateral cephalic border. Morphologically similar nodes in *Leptochilodiscus, Bolboparia* and *Ninadiscus* with rather different distributions are treated as homologous, following Jell (*op. cit.*).

13. Number of pairs of border nodes [12:1]. 0: 1-3, 1: 4-6, 2: 7-9, 3: 10+.
This is coded assuming that in species of *Tsunyidiscus* with border nodes the nodes continued onto the free cheeks. The total number of node pairs was therefore estimated from the number on the anterior border (3 pairs in *Tsunyidiscus niutitangensis*, 4 or 5 in *T. aclis*) and increased in proportion to the length of the free cheek margin. The estimates are 3 or 4 more pairs in *T. niutitangensis*, and 4 or 5 more in *T. aclis*.

14. Density of cephalic border nodes (minimum distance between adjacent nodes) [12:1]. 0: Separated by less than node diameter, 1: Separated by approximately node diameter, 2: Separated by more than node diameter.

15. Definition of nodes [12:1]. 0: Weakly defined, 1: Strongly defined.

16. Tubercles more strongly defined posteriorly than anteriorly [12:1]. 0: Absent, 1: Present, 2: Entirely absent or effaced anteriorly.

17. Border nodes more strongly defined anteriorly or absent posteriorly [12:1]. 0: Absent, 1: Present.

18. Unpaired sagittal anterior border node. 0: Absent, 1: Present.

19. Nodes inside posterolateral border furrow. 0: Absent, 1: Present.
A single pairs of low nodes are located inside the posterolateral angle of the cephalic border in *Meniscuchus menetus* and *M. nanus*. These differ from the nodes recognised by Character 12 above by being separated from the border by a clear furrow. The homology of these two types of nodes cannot be demonstrated (although I consider it likely) and they are here treated as distinct pending further investigation. Öpik (1975: p. 30) compared the nodes of *Meniscuchus* type to those found in *Bolboparia*, which are here regarded as homologous with the border nodes of other taxa. However, contrary to Öpik's assertion, the nodes of *Bolboparia* are clearly located on the cephalic border (see, e.g., Rasetti 1966: pl. 5, figs. 1, 4, 6, 13) and not inside it.

20. Cephalic border furrow effaced. 0: Absent, 1: Present.

21. Width of anterior border (sag.) in dorsal view as a proportion of cephalic length (sag.). 0: < 0.10, 1: = 0.1-0.15, 2: = 0.15-0.2, 3: = 0.2-0.25, 4: => 0.25.

22. Border expanded sagittally so that a line perpendicular to the axis may cut the border furrow in four places. 0: Absent, 1: Present.

23. Degree of sagittal border expansion (width of border in dorsal view, sag., compared to width of border, exsag.) [22:1]. 0: Very small (less than 1.2 time exsag. width), 1: Small (1.2 to 1.5 times exsag. width), 2: Moderate (between 1.5 and 2 times exsag. width), 3: Large (greater than twice exsag. width).

24. Anterolateral cephalic border crescentic in dorsal view. 0: Absent (border slightly, i.e., <= 1.2 times, wider sagittally than at 45 degrees, border as wide sagittally as at 45 degrees, or narrower sagittally than at 45 degrees), 1: Present (border at least 1.2 times wider sagittally than at 45 degrees).

25. Degree of expansion of crescentic anterior cephalic border (width of border in dorsal view, sag., as a proportion of width of border at 45° between axis and posterior cephalic margin) [24:1]. 0: 1.2-1.5, 1: 1.51-1.9, 2: > 1.9.

26. Cephalic border narrower (exsag.) anteriorly than posteriorly in dorsal view. 0: Absent, 1: Present.

27. Anterolateral cephalic border downsloping in lateral view. 0: Absent, 1: Present.

28. Pair of pores in anterolateral cephalic border. 0: Absent, 1: Present. [Autapomorphic for *Leptochilodiscus punctulatus*]

29. Shape of cephalon (sagittal length, excluding glabellar or occipital spines, as a proportion of maximum width in dorsal view). 0: <= 0.649, 1: 0.65-0.749, 2: 0.75-0.849, 3: >= 0.85.

30. Anterior margin of cephalon truncated in dorsal view. 0: Absent, 1: Present.
31. Outline of cephalon two-phased (as described by Öpik 1975: p. 30). 0: Absent, 1: Present.
32. Shape of cephalic outline in dorsal view. 0: Semicircular (maximum width in posterior 0.25 of length), 1: Rounded (maximum width approximately at cephalic mid-length).
33. Width of cephalic border furrow (sag., in dorsal view). 0: Narrow (half or less than width of cephalic border), 1: Moderate (approximately equal in width to cephalic border), 2: Wide (over 1.5 times width of cephalic border).
34. Pre-glabellar field. 0: Absent (glabella reaches anterior border furrow), 1: Present.
35. Sagittal pre-glabellar furrow [32:1]. 0: Absent, 1: Present.
36. Width of sagittal pre-glabellar furrow [32:1]. 0: Wide (greater than or equal to half the width of the anterior glabellar lobe), 1: Moderate (approximately half the width of the anterior glabellar lobe), 2: Narrow (less than half the width of the anterior glabellar lobe).
37. Anterior genae crossed by transverse furrow at the position of the anterior termination of the glabella. 0: Absent, 1: Present.

    A faint furrow is present in *Mallagnostus desideratus* (USNM 18327), that is similar to the stronger furrows seen in *Ladadiscus limbatus* (see Rushton 1966) and *Jinghediscus nummularius* (see Xiang and Zhang 1985: pl. 1, fig. 1).

38. Anterior genae with prominent caecal network. 0: Absent, 1: Present. [Autapomorphic for *Lenadiscus*]
39. Genae independently convex in anterior view, i.e., genae reach maximum height (dorsoventrally) adaxial to glabellar furrows rather than smoothly slope down to border furrow. 0: Absent, 1: Present.
40. Genae overhanging border anterolaterally in dorsal view. 0: Absent, 1: Present.

    Whilst the border furrow is effaced in *Analox*, the course of the furrow around the posterolateral corner of the cephalon shows that in this genus too the border is probably covered dorsally by the genae. The non-effaced posterolateral portion of the border furrow runs for a short distance under the convexity of the lateral margin of the cephalon (see Rasetti 1966: pl. 6, figs. 3, 7).

41. Facial sutures. 0: Present, 1: Absent.
42. Palpebral lobes. 0: Present, 1: Absent.
43. Palpebral lobes strongly elevated above genae [42:0]. 0: Present, 1: Absent.
44. Palpebral furrows [42:0]. 0: Present, 1: Absent.
45. Width of palpebral furrows [44:0]. 0: Narrow, 1: Wide (subequal in width to palpebral lobes in dorsal view).
46. Length of palpebral lobes [42:0]. 0: Short (max. length of lobes less than 0.2 of the sag. length of cephalon, in dorsal view), 1: Moderate (0.2-0.29 of length of cephalon), 2: Long (0.3 or more of length of cephalon).
47. Shape of palpebral lobes (length of lobe as a proportion of width of lobe, in dorsal view) [42:0]. 0: 1-1.9, 1: 2-2.9, 2: 3-3.9, 3: 4-4.9, 4: 5-5.9, 5: 6+.
48. Anteroposterior position of palpebral lobes (distance from base of cephalon to midpoint of lobes as a proportion of length of cephalon, in dorsal view) [42:0]. 0: < 0.3, 1: 0.3-0.39, 2: 0.4-0.49, 3: >= 0.5.
49. Lateral position of palpebral lobes [42:0]. 0: At cephalic border (external margin of palpebral lobe in contact with cephalic border furrow, in dorsal view), 1: Inside cephalic border.
50. Angle of palpebral lobes to sagittal line [42:0]. 0: Parallel or subparallel, 1: Angled inwards (> 15 degrees).
51. Eye ridges. 0: Absent, 1: Very weak, 2: Well defined.
52. Width of eye ridges [51:1 or 2]. 0: Wide (maximum width greater than half maximum width of palpebral lobes), 1: Narrow (less than half maximum width of palpebral lobes).
53. Length of glabella (sag. distance from base of cephalon to anterior termination of the glabella) as a proportion of sag. cephalic length (excluding anterior border), in dorsal view. 0: < 0.65, 1: 0.651-0.75, 2: 0.751-0.85, 3: 0.851-0.95, 4: > 0.95.

54. Maximum width of glabella as a proportion of maximum width of cephalon (trans., in dorsal view). 0: < 0.2, 1: 0.2-0.29, 2: 0.3-0.39, 3: => 0.4.
55. Cephalic axial furrows effaced on external surface. 0: Absent, 1: Present.
56. Form of occipital ring. 0: Vertical, 1: Angled posteriorly.
57. Posterior of glabella inflated. 0: Absent, 1: Dorsally, 2: Posterodorsally.
58. Form of SO. 0: Approximately even width across entire axis, 1: Weaker medially than laterally, 2: Divided.
59. Length of divided SO furrows [58:2]. 0: Long, 1: Short.
60. SO furrows bifid. 0: Absent, 1: Present.
61. Weak furrow crosses posterodorsal glabellar expansion dorsally [57:2]. 0: Absent, 1: Present.
62. SO evenly effaced on external surface. 0: Absent, 1: Moderately effaced, 2: Entirely effaced.
63. SO consisting of pits isolated from axial furrows connected by narrow (sag.) furrow. 0: Absent, 1: Present.
64. Occipital spine or node. 0: Absent, 1: Present.
65. Cranidial spine (following Jell 1975: p. 4): 0: Absent, 1: Present.
66. Form of cranidial spine [65:1]. 0: Long and strongly directed posteriorly, 1: Short and posterodorsally directed.
67. Occipital ring with strongly convex posterior margin compared to anterior margin. 0: Absent, 1: Present.
68. Basal lobes divided from median band of occipital ring. 0: Absent, 1: Present.
69. Pre-occipital glabellar spine, node or ridge (in holaspid). 0: Absent, 1: Present.
70. Long glabellar spine angled posterodorsally. 0: Absent, 1: Present.
71. Lateral furrows indicated by rounded pits isolated from axial furrows. 0: Absent, 1: Present.
72. S2 furrows entirely effaced. 0: Absent; 1: Present.
73. S2 furrows trans-glabellar. 0: Absent, 1: Weakly, 2: Strongly.
74. S3 furrows entirely effaced. 0: Absent, 1: Present.
75. S3 furrows trans-glabellar. 0: Absent, 1: Weakly, 2: Strongly.
76. Trans-glabellar S2 and S3 furrows merged medially [73: 1 or 2 and 75: 1 or 2]. 0: Absent, 1: Present.
77. Anterior glabella lobe expanded laterally (maximum width of lobe greater than posterior width of lobe). 0: Present, 1: Absent.
78. Anterior termination of glabella pointed. 0: Absent, 1: Present.
79. Anterior termination of glabella truncated: 0: Absent, 1: Present.
80. Angle of posterior axial furrows. 0: Strongly convergent anteriorly, 1; Weakly convergent or parallel, 2: Divergent anteriorly.
81. Number of thoracic segments. 0: 3, 1: 2.
    Coding of all thoracic characters for *Opsidiscus microspinus* and *Macannaia maladensis* is based on the similar taxa *Opsidiscus brevicaudatus* (Jell 1975: p. 78, pl. 26, figs. 1-2; pl. 28, figs. 4-8), and *Macannaia stenorhachis* (Jell 1975: p. 73, pl. 25, figs. 1-14), respectively.
82. Long axial spine on posterior thoracic segment. 0: Absent, 1: Present.
83. Thoracic axial spine geniculate [82:1]. 0: Absent, 1: Present.
84. Axial tubercles on thoracic segments. 0: Absent, 1: Present.
85. Anterior thoracic segment with long pleural spines. 0: Absent, 1: Present. [Autapomorphic for *Yukonia*]
86. Shape of pygidium (maximum width of pygidium as proportion of sag. length of pygidium, in dorsal view). 0: Long (<= 1.4), 1: Intermediate (1.41-1.7), 2: Wide (>= 1.71)
87. Width of pygidial border and furrow as a proportion of length of pygidium (sag., in dorsal view). 0: Very narrow (less than 0.05), 1: Narrow (0.05-0.09), 2: Wide (0.1-0.14), 3: Very wide (=> 0.15).
88. Pygidial border furrow effaced. 0: Absent, 1: Present. [Autapomorphic for *Cephalopyge*]
89. Width of pygidial border increases posteriorly. 0: Absent, 1: Present.

90. Pygidial border with pair of dorsally directed marginal spines. 0: Absent, 1: Present. [Autapomorphic for *Lenadiscus*]
91. Pygidial border with segmental spines or serrations. 0: Absent, 1: Present.
    The ventrally directed spines of *Serrodiscus*-like pygidia (see Jell in Bengtson et al. 1990; Rasetti 1966: p. 9) and the laterally directed border spines of other taxa (e.g., *Hebediscina sardoa* Rasetti, 1972: pl. 7, figs. 5-20), which have a wide distribution amongst polymerids, are treated as homologous since both are likely to reflect the primitive segmentation of the pygidium.
92. Pygidial border spines all visible in dorsal view [91:1]. 0: Absent, 1: Present. [Not applicable to taxa with 91:0]
    The situation in *Leptochilodiscus*, where the posteriormost spine pair is visible in dorsal view but other spines are directed ventrally (see Rasetti 1967; Bassett et al. 1976), is not considered to be homologous to that in other taxa. In *Leptochilodiscus*, the posterior spines are visible in dorsal view due not to their orientation with respect to the border but due to an upturning of the posterior part of the pygidial border.
93. Pygidial border doublure expanded and directed ventrally. 0: Absent, 1: Present.
    This recognises the potential homology between the vertically directed spines of *Serrodiscus* and similar genera, and the 'cuff' of *Meniscuchus*, *Analox* and *Bathydiscus* (see Öpik 1975: p. 22). This is supported by the somewhat intermediate situation in *Litometopus*.
94. Doublure expanded around entire margin, with smooth or denticulate edge. 0: Absent, 1: Present.
95. Pygidial border lowered postaxially. 0: Absent, 1: Present.
96. Pygidial border turned upwards postaxially. 0: Absent, 1: Present.
    These two distinctive modifications of the pygidial border were recognised by Jell (1975: p. 86) in his phenetic analysis.
97. Pygidial axial furrows effaced externally. 0: Absent, 1: Present.
98. Width of pygidial axis as a proportion of maximum width of pygidum. 0: < 0.25, 1: 0.25-0.29, 2: 0.3-0.34, 3: 0.35-0.39, 4: >= 0.4.
99. Pygidial axis short (less than 0.9 of the length of pygidium excluding border): 0: Absent, 1: Present.
100. Pygidial axis reaches border furrow. 0: Absent, 1: Present.
101. Pygidial axis overhangs border or border furrow posteriorly. 0: Absent, 1: Present.
102. Number of segments in pygidial axis, excluding terminal part. 0: 3, 1: 4, 2: 5, 3: 6, 4: 7, 5: 8, 6: 9, 7: 10, 8: 11, 9: > 11.
    In *Delgadella lenaicus*, the number of rings can be coded, despite effacement of the furrows, because the segments are indicated by clear muscle scars. The number of segments in *Dicerodiscus* follows Jell's (1997: p. 389) observation of 3 pairs of pleural furrows, which appear to cover the length of the pygidium.
103. Pygidial ring furrows entirely effaced. 0: Absent, 1: Present.
104. Pygidial ring furrows effaced medially. 0: Absent, 1: Present.
105. Lateral margins of pygidial axis convex in dorsal view. 0: Absent, 1: Present.
106. Broad-based spine on more than one segment of pygidial axis, or encroaching on other segments. 0: Absent, 1: Present.
107. Angle of broad-based spine on pyigidial axis [106:1]. 0: Approximately vertical (greater than 70 degrees from horizontal), 1: Posterodorsally directed (30 to 60 degrees from horizontal), 2: Posteriorly directed (less than 30 degrees).
108. Position of broad-based spine on pyigidial axis [106:1]. 0: Anterior (less than 1/3 of axis length from anterior margin of pygidium), 1: Median (1/3-2/3 axis length), 2: Terminal.
109. Broad-based terminal spine incorporates terminal piece of pygidial axis [108:2]. 0: Absent, 1: Present.
110. Segmental spines or nodes on pygidial axis. 0: Absent, 1: Present.

111. Pygidial pleural furrows. 0: Present, 1: Absent.
112. Strength of pygidial pleural furrows [111:0]. 0: Strongly incised (deep furrows), 1: Weakly incised (relatively wide but shallow), 2: Faint (very narrow and shallow markings which do not interrupt pleural convexity).
113. Pygidial interpleural furrows. 0: Present, 1: Absent.
114. Pygidial pleurae overhang border and border furrow posterolaterally in dorsal view. 0, Absent, 1, Present.
115. Pygidial pleurae crossed by raised ridges that converge abaxially. 0: Absent, 1: Present. [Autapomorphic for *Lenadiscus*]
116. Punctate sculpture. 0: Absent, 1: Present.
117. Prominent tuberculate sculpture. 0: Absent, 1; Present.
118. Thoracic axis divided into lateral and median lobes. 0: Absent, 1: Present.
119. Articulating half-ring of anterior thoracic segment. 0: Present, 1: Absent.
120. Posterior part of pygidial axis unsegmented. 0: Absent, 1: Present.
121. Posterior margin of pygidium with pair of broad-based posteriorly directed spines. 0: Absent, 1: Present. [Autapomorphic for *Ptychagnostus*]
122. Anterior glabella effaced compared to posterior glabella. 0: Absent, 1: Present. [Autapomorphic for *Peronopsis*]
123. Thoracic axis more than half the width of the thoracic segment. 0: Absent, 1: Present.

APPENDIX 2

Table 1. Taxa, authorship and important references for taxa included in the cladistic analysis. Type species of genera recognised as valid by Jell (1997) are indicated by an asterisk (*), type species of genera regarded as junior synonyms (according to Jell) are indicated by the name and authorship of the synonymous genus in square brackets.

Agnostina

Condylopyge amitina Rushton, 1966.

Peronopsis roddyi (Resser and Howell, 1938). Blaker and Peel 1997.

Ptychagnostus gibbus [*Triplagnostus* Howell, 1935] (Linnarsson, 1869). Öpik 1979; Robison 1982; Peng & Robison 2000.

Eodiscina

Abakolia minutus [*Costadiscus* Babcock, 1994] (Babcock, 1994).

Abakolia pauca* Bognibova in Chernysheva, 1971. Korobov 1980; Jell 1997.

Acidiscus birdi* Rasetti, 1966.

Acidiscus theristes Rushton, 1966.

Acimetopus bilobatus* Rasetti, 1966.

Alaskadiscus spinosus* (Palmer, 1968). Zhang et al. 1980.

Analox bipunctata* Rasetti, 1966.

Bathydiscus dolichometopus* Rasetti, 1966.

Bolboparia superba* Rasetti, 1966.

Calodiscus lobatus* (Hall, 1847). Lochman 1956; Rasetti 1967; Geyer 1988; Blaker & Peel 1997.

Cephalopyge notabilis* Geyer, 1988.

Chelediscus acifer* Rushton 1966.

Chelediscus chathamensis Rasetti, 1967.

Cobboldites comleyensis* (Cobbold, 1910). Fletcher 1972; Jell 1997.

Cobboldites itsariensis Geyer, 1988.

Dawsonia bohemicus [*Aculeodiscus* Šnajdr, 1950] (Šnajdr, 1950). Šnajdr 1958.

Dawsonia dawsoni* (Hartt in Dawson, 1868). Rasetti 1952.

Delgadella caudatus [*Delgadoia* Vogdes, 1917] (Delgado, 1904). Sdzuy 1961, 1962; Geyer 1988; Jell 1997.

Delgadella lenaicus [*Pagetiellus* Lermontova, 1940] (von Toll, 1899). Geyer 1988; Jell 1997.

Delgadella amouslekensis [*Pentagonalia* Geyer, 1988] (Geyer, 1988).

Dicerodiscus tsunyiensis* Zhang, 1964. Zhang et al. 1980.

Egyngolia willochra Jell in Bengtson et al., 1990.

Egyngolia obtusa* Korobov, 1980. Jell in Bengtson et al. 1990; Jell 1997.

Egyngolia zaicevi [*Mongolodiscus* Korobov, 1980] (Korobov, 1980). Jell in Bengtson et al. 1990.

Ekwipagetia marginata (Rasetti ,1967). Blaker and Peel 1997.

Ekwipagetia plicofimbria* Fritz, 1973.

Eodiscus borealis Westergård, 1946. Rushton 1966.

Table 1 continued.

*Eodiscus scanicus** (Linnarsson, 1883). Westergård 1946; Rasetti 1952; Hutchinson 1962; Babcock 1994.

*Hebediscina sardoa** Rasetti, 1972.

Hebediscina blagonravovi (Korobov, 1980). Jell in Bengtson et al. 1990.

Hebediscina yuqingensis (Zhang, 1980). Jell in Bengtson et al. 1990.

*Hebediscus attleborensis** (Shaler & Foerste, 1888). Shaw 1950; Hutchinson 1962.

*Helepagetis bitruncula** Jell, 1975.

*Jinghediscus nummularius** Xiang & Zhang, 1985; Jell 1997.

*Kiskinella cristata** Romanenko & Romanenko 1967. Jell 1997.

*Korobovia ocellata** Jell in Bengtson et al., 1990.

*Lenadiscus unicus** Repina in Khomentovskii & Repina, 1965. Korobov 1980.

*Leptochilodiscus punculatus** Rasetti, 1966. Rasetti 1967.

Leptochilodiscus succinctus [*Kerberodiscus* Bassett et al., 1976] (Bassett et al., 1976). Jell 1997.

*Litometopus longispinus** Rasetti, 1966.

*Luvsanodiscus gammatus** Korobov, 1980. Jell 1997.

*Macannaia maladensis** (Resser, 1939). Rasetti 1966; Jell 1975; Palmer & Halley 1979.

*Mallagnostus desideratus** (Walcott, 1890). Howell 1935; Jell 1997.

Mallagnostus limbatus [*Ladadiscus* Pokrovskaya, 1959] (Pokrovskaya, 1959). Rushton 1966, Jell 1997.

Mallagnostus llarenai (Rushton, 1966). Jell 1997.

*Meniscuchus menetus** Öpik, 1975.

Meniscuchus nanus (Palmer, 1968). Öpik 1975.

*Natalina incita** Romanenko in Repina & Romanenko, 1978. Jell 1997.

Natalina dilata [*Limbadiscus* Korobov, 1980] (Korobov, 1980). Jell 1997.

*Neocobboldia dentata** (Lermontova, 1940). Repina 1972; Jell 1997.

*Neopagetina rjonsnitzkii** (Lermontova, 1940). Jell 1997; Blaker & Peel 1997.

*Ninadiscus strobulatus** Korobov, 1980. Jell 1997.

*Oodiscus subgranulatus** Rasetti, 1966.

*Opsidiscus bilobatus** (Westergård, 1946). Jell 1975; Jell 1997.

Opsidiscus microspinus Jell, 1975.

Opsidiscus longispinus Babcock, 1994a.

*Pagetia bootes** Walcott, 1916. Rasetti 1951.

Pagetia prolata Jell, 1975.

*Pagetides elegans** Rasetti, 1945. Blaker & Peel 1997.

Pagetides fragum [*Discomesites* Öpik, 1975] (Öpik, 1975). Jell 1997.

*Pseudocobboldia pulchra** (Hupé, 1953). Geyer 1988.

*Runcinodiscus index** Rushton in Bassett et al., 1976. Jell 1997.

Table 1 continued.

Semadiscus sollennis* Romanenko in Repina & Romanenko, 1978. Jell 1997.

Serrodiscus speciosus [*Paradiscus* Kobayashi, 1943] (Ford, 1873). Rasetti 1952; Lochman 1956; Theokritoff 1964; Blaker & Peel 1997.

Serrodiscus gravestocki Jell in Bengtson et al., 1990.

Serrodiscus daedalus Öpik, 1975.

Serrodiscus ctenoa Rushton, 1966.

Sinodiscus shipaiensis* Zhang in Lu et al., 1974. Zhang in Zhang et al. 1980.

Sinodiscus subquadratus [*Tologoja* Korobov, 1980] (Korobov, 1980). Jell 1997.

Sinopagetia jinnanensis* Lin & Wu in Zhang et al., 1980. Zhang in Zhang et al. 1995; Jell 1997.

Stigmadiscus stenometopus* Rasetti, 1966. Rasetti 1967.

Tannudiscus altus Repina in Repina et al., 1964. Rushton 1966.

Tannudiscus balanus Rushton, 1966.

Tchernyshevioides ninae* Hajrullina in Repina et al., 1975. Jell 1997.

Tsunyidiscus niutitangensis* Zhang 1964. Zhang et al. 1980; Jell 1997.

Tsunyidiscus kaiyangensis [*Guizhoudiscus* Zhang in Zhang et al., 1980] Zhang in Zhang et al., 1980.

Tsunyidiscus aclis [*Mianxiandiscus* Zhang in Zhang et al., 1980] Zhou, 1975. Zhang in Zhang et al. 1980.

Tsunyidiscus orientalis [*Hupeidiscus* Zhang in Lu et al., 1974] (Zhang, 1953). Zhang et al. 1980.

Tsunyidiscus longquanensis [*Shizhudiscus* Zhang & Zhu in Zhang et al., 1980] Zhang & Zhu in Zhang et al., 1980. Zhang & Clarkson 1993.

Weymouthia nobilis* (Ford, 1872). Shaw 1950; Rasetti 1952.

Yukonia intermedia* Palmer, 1968.

Yukonides lacrinus* Fritz, 1972. Fritz 1973.

Table 2. Data matrix used in phylogenetic analyses of Agnostida. The full names of taxa are used in the first part of the matrix. In the second part, taxa are referred to by six-letter codes, and character numbers are shown at the top of the table; characters and states are described in Appendix 1. Missing data are indicated by a question mark, 'N' refers to non-applicable characters. Other capital letters indicate multistate uncertainty coding, as follows: A = {01} (126 instances), B = {12} (88 instances), C = {23} (54 instances), D = {34} (20), E = {45} (1), F = {56} (1), G = {67} (3), H = {78} (2), I = {89} (2), J = {012} (1), K = {123} (2), L = {234} (1), M = {678} (1).

Taxon	Code	Character numbers 0000000001111111111222222222233333333334444444445 1234567890123456789012345678901234567890123456789 0
Condylopyge amitina	ConAmi	0000N000NN00NNNNN00010N0N0103101010N000011NNNNNNNN
Peronopsis roddyi	PerRod	1NN0NA00NN00NNNNN00000N0N1103101A10N000011NNNNNNNN
Ptychagnostus gibbus	PtyGib	0010NA00NN00NNNNN00000N0N0103101011200001 1NNNNNNNN
Abakolia minutus	AbaMin	1NN0NA010200NNNNN000B0N110001000011100101 1NNNNNNNN
Abakolia pauca	AbaPau	1NN0NA00NN00NNNNN000C1110000A0000110000011NNNNNNNN
Acidiscus birdi	AciBir	00011200NN011200000010N110003000010N000011NNNNNNNN
Acidiscus theristes	AciThe	00111200NN011100000020N100003000010N000011NNNNNNNN
Acimetopus bilobatus	AcmBil	0000N200NN00NNNNN000B0N100003000010N000011NNNNNNNN
Alaskadiscus spinosus.	AlaSpi	1NN0NA00NN00NNNNN00000N0N000100020NN0000000002C200
Analox bipunctata	AnaBip	1NN0NA00NN00NNNNN001413???00300?00NN001111NNNNNNNN
Bathydiscus dolichometopus	BatDol	0100NA00NN00NNNNN0001110N010310000NN000011NNNNNNNN
Bolboparia superba	BolSup	0000N200NN0100020?00212???0020000110001111NNNNNNNN
Calodiscus lobatus	CalLob	1NN0N00NN00NNNNN00010N0N000B000B0NN00001????00301
Cephalopyge notabilis	CepNot	1NN0NA00NN00NNNNN001??????10B000?10N000?11NNNNNNNN
Chelediscus acifer	CheAci	0000N2011000NNNNN00000N0N010C001011200001 1NNNNNNNN
Chelediscus chathamensis	CheCha	00?0NA011100NNNNN00010N0N010C001011200001 1NNNNNNNN
Cobboldites comleyensis	CobCom	1NN0NA00NN00NNNNN00010N1000010000 0NN000011NNNNNNNN
Cobboldites itsariensis	CobIts	1NN0NA00NN00NNNNN000A0N10000B00000NN000011NNNNNNNN
Dawsonia bohemicus	DawBoh	1NN0NA012200NNNNN00020N1000010000 0NN001011NNNNNNNN
Dawsonia dawsoni	DawDaw	1NN0NA012200NNNNN00030N10000B0000 0NN001011NNNNNNNN
Delgadella caudatus	DelCau	1NN0NA00NN00NNNNN000A0N120003000110N00000011N23B00
Delgadella lenaicus	DelLen	1NN0NA00NN00NNNNN00000N100002000110N00000011N1?100
Delgadella amouslekensis	DelAmo	1NN0NA00NN00NNNNN00000N0N0001000110N00000011N1?100
Dicerodiscus tsunyiensis	DicTsu	1NN10A00NN00NNNNN00030N11000010020NN00001010010101
Egyngolia willochra	EgyWil	1NN0NA00NN00NNNNN00021111000100011110010000001B0?
Egyngolia obtusa	EgyObt	1NN0NA00NN00NNNNN00000N10000J000B11100100000000201
Egyngolia zaicevi	EgyZai	1NN0NA00NN00NNNNN00000N0N0002000211100?011NNNNNNNN
Ekwipagetia marginata	EkwMar	1NN0NA00NN10NNNNN00020N10000200010NN0010001001CA00
Ekwipagetia plicofimbria	EkwPli	1NN0NA00NN10NNNNN00020N0N100C00010NN001000100B2000
Eodiscus borealis	EodBor	1NN0NA012B00NNNNN00010N100002000111200101 1NNNNNNN0
Eodiscus scanicus	EodSca	A000NA010100NNNNN00000N0N000B0001112001011NNNNNNN0
Hebediscina sardoa	HbiSar	1NN0NA00NN00NNNNN00021010000C00020NN00000000021100
Hebediscina blagonravovi	HbiBla	1NN0NA00NN00NNNNN000A0N0N?00B000211?0010001?0B?C00
Hebediscina yuqingensis	HbiYuq	1NN0NA0???00NNNNN000C10110002000111000 1000100AB100
Hebediscus attleborensis	HbsAtt	1NN0NA00NN00NNNNN00020N120003???10NN00000001N112?1

Table 2 continued.

		Character numbers
Taxon	**Code**	000000000111111111122222222223333333333444444444 5 123456789012345678901234567890123456789012345 67890
Kiskinella cristata	KisCri	1NN0NA012200NNNNN00040N120001000011000100010111100
Korobovia ocellata	KorOce	1NN0NA00NN00NNNNN00010N0N000B00020NN00100000001301
Lenadiscus unicus	LenUni	0000NA00NN00NNNNN00010N10000C000010N0100000?010011
Leptochilodiscus punculatus	LepPun	1NN0NA10NN00NNNNN00000N0N101C100010N000011NNNNNNNN
Leptochilodiscus succinctus	LepSuc	1NN0NA10NN010?02000000N0N100C10000NN000111NNNNNNNN
Litometopus longispinus	LitLon	0010N200NN00NNNNN00020N11010100000NN000011NNNNNNNN
Luvsanodiscus gammatus	LuvGam	1NN0NA00NN00NNNNN00010N11000200020NN00000011N22100
Macannaia maladensis	MacMal	1NN0NA010200NNNNN000B0N11000C000111200100011N1D100
Mallagnostus desideratus	MalDes	1NN0NA00NN00NNNNN00010N???003001010N100011NNNNNNNN
Mallagnostus limbatus	MalLim	1NN0NA00NN00NNNNN00010N0N100C001110N100011NNNNNNNN
Mallagnostus llarenai	MalLla	1NN0NA00NN0112020000B0N0N000C001010N000011NNNNNNNN
Meniscuchus menetus	MenMen	1NN0NA00NN00NNNNN010C0N11000201000NN001111NNNNNNNN
Meniscuchus nanus	MenNan	1NN0NA00NN00NNNNN01021112000301000NN001111NNNNNNNN
Natalina incita	NatInc	1NN0NA00NN00NNNNN000412100003?0?11100010001001?100
Natalina dilata	NatDil	1NN0NA00NN00NNNNN00021110000C000111000100011N1?100
Neocobboldia dentata	NeoDen	1NN0NA00NN00NNNNN00000N12000200020NN00?000?0011100
Neopagetina rjonsnitzkii	NepRjo	1NN0NA010200NNNNN000B12110001000111100100011001D200
Ninadiscus strobulatus	NinStr	1NN0NA00NN0101101100D1112000B00000NN001111NNNNNNNN
Oodiscus subgranulatus	OodSub	01010200NN00NNNNN00000N0N0103100010N000011NNNNNNNN
Opsidiscus bilobatus	OpsBil	1NN0NA011100NNNNN000A0N0N1001100011A001010?1012B10
Opsidiscus microspinus	OpsMic	1NN0NA011B00NNNNN00010N110001100011B0010101102?110
Opsidiscus longispinus	OpsLon	1NN0NA00NN00NNNNN00000N0N1000100011000101001010B10
Pagetia bootes	PagBoo	1NN0NA011200NNNNN00020N10000200001120010001101?100
Pagetia prolata	PagPro	1NN0N0011210NNNNN000B0N110001000011100100011013100
Pagetides elegans	PdsEle	1NN0N0010200NNNNN000C1111000C00011110010001001DA00
Pagetides fragum	PdsFra	1NN0NA010B00NNNNN000211110002000A11000100011N11B00
Pseudocobboldia pulchra	PsePul	1NN0NA00NN00NNNNN00020N11000100010NN00001000000200
Runcinodiscus index	RunInd	1NN0NA00NN011200000000N100003000110N000011NNNNNNNN
Semadiscus sollennis	SemSol	1NN0NA00NN00NNNNN00020N120002000110N000011NNNNNNNN
Serrodiscus speciosus	SerSpe	1NN0NA00NN013B01000020N120002000A10N000011NNNNNNNN
Serrodiscus gravestocki	SerGra	1NN0NA00NN012111000020N11000300010NN000011NNNNNNNN
Serrodiscus daedalus	SerDae	1NN0NA00NN0120100100C0N110002001010N000011NNNNNNNN
Serrodiscus ctenoa	SerCte	0000N100NN011200100010N0N000200000NN000011NNNNNNNN
Sinodiscus shipaiensis	SinShi	1NN0NA00NN00NNNNN00000N0N000100010NN00000000011301
Sinodiscus subquadratus	SinSub	1NN0NA00NN00NNNNN00000N0N000100010NN00000000022301
Sinopagetia jinnanensis	SpaJin	1NN0NA011200NNNNN00030N120002000011000100011 02?200
Stigmadiscus stenometopus	StiSte	0100NA00NN00NNNNN00010N100001000010N000011NNNNNNNN
Tannudiscus altus	TanAlt	1NN0NA00NN00NNNNN000A0N0N010C001010N000011NNNNNNNN
Tannudiscus balanus	TanBal	0100N100NN00NNNNN0001110N110310100NN000011NNNNNNNN
Tchernyshevioides ninae	TchNin	0010NA00NN00NNNNN0002100N000000000NN00000000011301

Table 2 continued.

Taxon	Code	Character numbers 0000000001111111111222222222233333333334444444445 1234567890123456789012345678901234567890
Tsunyidiscus aclis	TsuAcl	0010NA00NN11C010010020N0N0002000A10N00000011N25200
Tsunyidiscus orientalis	TsuOri	0000NA00NN00NNNNN000A0N100003000110N00000010025000
Tsunyidiscus longquanensis	TsuLon	1NN0N000NN00NNNNN000B100N0001000010N00000010025200
Weymouthia nobilis	WeyNob	1NN0NA00NN0112?2000010N100003000A?0N00011NNNNNNNN
Yukonia intermedia	YuiInt	1NN0NA00NN00NNNNN00000N0N000000010NN00101000100301
Yukonides lacrinus	YudLac	1NN0NA011000NNNNN000A0N0N0002000B0NN00001000100200

Code	Character numbers 00000000000000000000000000000000000000000001111111111111111111111 5555555556666666666777777777788888888889999999999000000000011111111112222 1234567890123456789012345678901234567890123456789012345678901234567890123
ConAmi	0NCC0120N000010N011001002N000110N10030100N000004100?1111?0N01N10000111001
PerRod	0NB20120N000000N011000002N100110N00020000N000004100?0111?0N01N10000111011
PtyGib	0N2B0120N000000N011000002N110111010020000N000003100?001110N01N10000111101
AbaMin	0NB1012201?0001000?000001N110101000110000N0000000050000NNN10010000000000
AbaPau	11C10122?100001000?000001N100A?????220000N00000A000D0000NNN10010000??000?
AciBir	0N22001AN0N0110N001000010N0001?????0A000101000020008000NNN11N10000??000?
AciThe	0NB20000N0N0110N001000010N00010??100B0001010000B000H0001?0N11N10000000000
AcmBil	0N320120N000100N00110020210001?????000000N00000300150011 00N01N10001??000?
AlaSpi	11220120010001000000010?0N100110N00210001100002100C11?110N00210000000000
AnaBip	0N3B0120N001100N0011001011 0000?????000000N11000C00050010NNN01N11010??000?
BatDol	0N4C010200N0000N000001010N1002?????0B0000N111004100?1010NNN01N10000??000?
BolSup	0NB20000N0N0100N001000000N1100?????000001010000200090001 00N01N11001??000?
CalLob	11CB0000N0N0000N100000A0A0?0010????1C00011000003000B0000NNN00010000000000
CepNot	0NB31100N0N2000N000001010N100000N00001000N000014010?1010NNN01N10010000000
CheAci	0N120120N000000N010001002N110010N00000000N110004000C0010NNN11N10000000000
CheCha	0NB20120N0000?0N01?001001N1100??????0?00N11000300020010NNN11N10000??000?
CobCom	0N430?0??N2000N?00001010N10000????100000N00000D01041110NNN01N11010000000
CobIts	0NDC0100N0NB000N000001010N1001?????000000N000003010?10?0NNN01N11000??000?
DawBoh	0N2101220?0000110000000?0N110A1????210000N00000300120000NNN00010001000000
DawDaw	0N2101220?0000110000000?0N110110N00110000N000001001 30000NNN00010001000000
DelCau	0N?31?0???N2?00N?00001010N?11A00N00010000N000011000?1000NNN01N10000000000
DelLen	0N331000N0N2000N100001010N011000N00010000N00001?00071000NNN01N11000000000
DelAmo	0N331?0???N2?00N?00001010N0110?????010000N00001C000?1000NNN01N10000??000?
DicTsu	10220000N0NB010N10?0010?0N1110???????0000N00000??000???0NNN?00?0000??000?
EgyWil	111B012211 10001 10000 10001N1011?????2B000110000010003 0000NNN10200000??000?
EgyObt	11AB01221110001?00?010001N101100N?0220000N00000A00030000NNN1000?000000000
EgyZai	11A10122111000??00?010001N?011???????????????????????????????00???00?
EkwMar	11B201220010001000000000N1?01?????B10001100000C000201011 2000200000??000?
EkwPli	111B012200100010000000000N1001?????110001100000C0011001 12000200000??000?

Table 2 continued.

Code	Character numbers
	00000000000000000000000000000000000000000000000011111111111111111111
	55555555566666666667777777777888888888899999999990000000000111111112222
	12345678901234567890123456789012345678901234567890123456789012345567890123
EodBor	0N1101220??0001000?000000N110100N10A10000N00000A00030000NNN10010000000000
EodSca	0NA101220100001000000000AN11010??00AA0000N00000A000M0000NNN11N10010000000
HbiSar	20B1012201000010000000000N1001?????0B0001100000C00020000NNN10100000??000?
HbiBla	101101220010001?00?00000?N1001?????130001100000A00020000NNN1?2?0000??000?
HbiYuq	11B10122?100001100000000N1001?????02000110000020004000NNN10100000??000?
HbsAtt	B0230000N0N2010N000001010N1011??????20000N00001?010?10?0NNN01N10000??000?
HelBit	0N2B012200000010000000001N1101?????B10000N000002001000012210110001??000?
JinNum	0NA20120N001000N00?00?002N0000?????030100N000004100I1010NNN01N10000??000?
KisCri	1131012200?000??000001010N1000?????010000N00001000130000NNN00010000??000?
KorOce	11210000N0N0000N100001002N1001?????130001100000D00010000NNN10010100??000?
LenUni	2000012??0?0001000000????N0001?????1C0010N00000C10010100NNN01N10000??000?
LepPun	0N320100N0N0000N100001010N0101?????000001010010D000G1110NNN01N11010??000?
LepSuc	0N4C1100N0N0110N000001010N110110N00000001010011D000I1010NNN1N11010000000
LitLon	0N420101N0N0000N000001010N0001?????1A00010110003000?1010NNN01N10010??000?
LuvGam	102C0000N0N1010N100000000N101100N00A20001100000200040100NNN00101000000000
MacMal	11B101221?100011000000000N110111100120000N00001001301011201020000000000
MalDes	0N020?0??N2?00N000001010N1001?????0??????????????????????????????00??0?
MalLim	0N110100N0N0000N00?001010N10010????030100N00000C10070010NNN01N10000??000?
MalLla	0N120100N0N0000N001000000N1001?????030100N00000C?00H0010NNN11N10000000000
MenMen	A1320122000000N001002020000?????010000N11100D01140000NNN11N10000??000?
MenNan	0N320122000000N00?00010100000?????000000N???00401040000NNN11N10000??000?
NatInc	??2B000A?0NB?00N100001010N1001?????030000N000001010D0000NNN?0B10000??000?
NatDil	101B0??2?????00N?00001010N1001?????130000N00002000300000NNN??2?0000??000?
NeoDen	0N120000N0N0000N100001010N101100N?02300011000003000A0000NNN01N10000000000
NepRjo	11210122110000100000001N1?01?????110000N000000040000NNN10000000??000?
NinStr	0N3B0120N000000N00?00010201000?????00000?0??000B00170000NNN11N11000??000?
OodSub	0N33012BN000000N000001010N1002?????020000N000003100?1010NNN01N10000??000?
OpsBil	111101220??0001000?000002N1101?????AB0000N000001000B0000NNN10110000??000?
OpsMic	11B10122000000100000000001N110111100B10000N000001000010000NNN10110000000000
OpsLon	111001220000001000001002N1002???????????????????????????????????00???0?
PagBoo	111201220000001000000001N110111100110000N000001001C000122101N10000000000
PagPro	111101220000001000000001N110111100110000N00000A00?B000122110200000000000
PdsEle	112B012211000011000000001N1101?1100A10000N000001000400000NNN10200000000000
PdsFra	112B012210000110000000000N1101?????110000N00000100030000NNN10000000??000?
PsePul	103B0000N0N0000N000001010N1000?????B20001100000C?0000100NNN002?0000??000?
RunInd	0N331100N0N2000N000001010N0000?????000001010014000G1010NNN01N11000??000?
SemSol	0N320100N0N0010N000001010N0001?????????????????????????????????00???0?
SerSpe	0N320100N0N1000N0000001?A0000000N00010001100003000G0000NNN11N10000000000
SerGra	0N32012AN000000N001100202000??????000001110000D000F00?100N?1N1?000??000?

Table 2 continued.

Code	Character numbers
	0000000000000000000000000000000000000000000000001111111111111111111111
	5555555556666666666777777777788888888889999999999000000000011111111112222
	1234567890123456789012345678901234567890123456789012345678901234567890123
SerDae	0NCB01220000000N00110020200000?????020100N00000?00070000NNN?1N10000??000?
SerCte	0ND20100N0NB000N000001010N000100N00000001110000D00050110NNN01N11000000000
SinShi	20310000N0N0000N000000202000000NN00120000N00000D10010000NNN00000001000000
SinSub	20CB0000N0N0000N00000020200001?????130000N00000D10010000NNN00000000??000?
SpaJin	1?2201220000000N000001010N1000?????010000N00000100020000NNN01N10000??000?
StiSte	0N2B0111N0N0110N001010000N0001????????0?011100003???90011?1N10010000??000?
TanAlt	0NCC0120N000000N00?000002N0001?????030100N00000C000?1000NNN01N10000??000?
TanBal	0NC30120N000000N001001002N000100N10030100N00000D000E0110NNN01N10000000000
TchNin	20320000N0N0010N100000000N1000?????2B0000N00000A00030000NNN00010000??000?
TsuNiu	1?200120N010000N001100101000000N00220000N000000100D1000NNN00010000000000
TsuKai	1?C1012??01B000N001100A010000000N00B10000N000000100C1000NNN10210000000000
TsuAcl	1?20012??011000N001100A010001?????00000110000000040100NNN?0210000000000
TsuOri	1?210120N010000N001100101000000N00110000N000000100K0000NNN11N10000000000
TsuLon	0NC1012AN011000N001100101000000N00210001100000010040100NNN00010000000000
WeyNob	0N??1?0???N2??0N?00001010N??0?00N?002000?????1?????10?0NNN01N10000000000
YuiInt	11B1012200000010000001010N110100N012C0001100000B100K0101?0N00B0000000000
YudLac	111101221100001?000000000N1001?????A20001100000B100L0000NNN10200000??000?

REFERENCES

Babcock, L.E. 1994. Systematics and phylogenetics of polymeroid trilobites from the Henson Gletscher and Kap Stanton formations (Middle Cambrian), North Greenland. *Bull. Grønlands Geolog. Undersøg.* 169: 79-127.

Bassett, M.G., Owens, R.M. & Rushton, A.W.A. 1976. Lower Cambrian fossils from the Hell's Mouth Grits, St. Tudwal's Peninsula, north Wales. *J. Geolog. Soc. London* 132: 623-644.

Bengtson, S., Conway Morris, S., Cooper, B.J., Jell, P.A. & Runnegar, B.R. 1990. Early Cambrian fossils from South Australia. *Assoc. Austral. Palaeontol., Mem.* 9: 364 pp.

Bergström, J. 1992. The oldest arthropods and the origin of the Crustacea. *Acta Zoologica* 73: 287-291.

Blaker, M.R. & Peel, J.S. 1997. Lower Cambrian trilobites from North Greenland. *Meddel. Grønland* 35: 1-145.

Briggs, D.E.G., Fortey, R.A. & Wills, M.A. 1992. Morphological disparity in the Cambrian. *Science* 256: 1670-1673.

Briggs, D.E.G., Fortey, R.A. & Wills, M.A. 1993. How big was the Cambrian explosion? A taxonomic and morphologic comparison of Cambrian and Recent arthropods. In: Lees, D.R. & Edwards, D. (eds.), *Evolutionary Patterns and Processes*: 33-44. London: Linn. Soc. London, Linn. Soc. Symp. Ser

Chatterton, B.D.E. & Speyer, S.E. 1997. Ontogeny. In: Kaesler, R.L. (ed.), *Treatise on Invertebrate Paleontology, Part O, Arthropoda 1* (revised): 173-247. Boulder; Lawrence: Geol. Soc. Amer. and Univ. Kansas Press.

Chernysheva, N.E. (ed.) 1971. Amginskii iarus Altae-Saianskoi oblasti [The Amgan Stage in the Altay-Sayan region]. *Trudy Sibirsk. Nauchno-Issledov. Inst. Geol. Geof. Mineral. Syr.* 111: 1-267 [in Russian].

Cobbold, E.S. 1910. On some small trilobites from the Cambrian rocks of Comley (Shropshire). *Quart. J. Geol. Soc. London* 46: 19-51.

Cotton, T.J. 2001. The phylogeny and systematics of blind Cambrian ptychoparioid trilobites. *Palaeontology* 44: 167-207.

Cotton, T.J. & Braddy, S.J. 2004. The phylogeny of arachnomorph arthropods and the origin of the Chelicerata. *Trans. R. Soc. Edinburgh, Earth Sci.* 94: 169-193.

Dawson, J.W. 1868. *Acadian Geology; The Geological Structure, Organic Remains and Mineral Resources of Nova Scotia, New Brunswick and Prince Edward Island, etc.; 2nd edition*: pp. i-xxvi + 1-694. London.

Delgado, J.F.N. 1904. Faune Cambrienne du Haut-Alemtejo (Portugal). *Comm. Serv. Geol. Portugal* 5: 307-374.

Edgecombe, G.D. & Ramsköld, L. 1999. Relationships of Cambrian Arachnata and the systematic position of Trilobita. *J. Paleontol.* 73: 263-287.

Fletcher, T. 1972. *Geology and Lower to Middle Cambrian Trilobite Faunas of the SW Avalon, Newfoundland.* University of Cambridge, UK: unpublished Ph.D. dissertation.

Ford, S.W. 1872. Description of some new species of primordial fossils. *Amer. J. Sci., Ser. 3*, 3: 419-422.

Ford, S.W. 1873. Remarks on the distribution of the fossils in the Lower Potsdam rocks at Troy, N.Y., with descriptions of a few new species. *Amer. J. Sci., Ser. 3*, 6: 134-140.

Fortey, R.A. 1985. Pelagic trilobites as an example of deducing the life habits of extinct arthropods. *Trans. R. Soc. Edinburgh, Earth Sci.* 76: 219-230.

Fortey, R.A. 1990. Ontogeny, hypostome attachment and trilobite classification. *Palaeontology* 33: 529-576.

Fortey, R.A. 1997a. Classification. In: Kaesler, R.L. (ed.), *Treatise on Invertebrate Paleontology, Part O, Arthropoda 1* (revised): 289-302. Boulder; Lawrence: Geol. Soc. Amer. and Univ. Kansas Press.

Fortey, R.A. 1997b. Late Ordovician trilobites from Southern Thailand. *Palaeontology* 40: 397-449.

Fortey, R.A. & Owens, R.M. 1999. The trilobite exoskeleton. In: Savazzi, E. (ed.), *Functional Morphology of the Invertebrate Skeleton*: 537-562. New York: John Wiley & Sons, Ltd.

Fortey, R.A. & Theron, J.N. 1994. A new Ordovician arthropod *Soomaspis*, and the agnostid problem. *Palaeontology* 37: 841-861.

Fritz, W.H. 1972. Lower Cambrian trilobites from the Sekwi Formation type section, Mackenzie Mountains, northwestern Canada. *Bull. Geol. Surv. Canada* 212: 1-90.

Fritz, W.H. 1973. Medial Lower Cambrian trilobites from the Mackenzie Mountains northwestern Canada. *Paper Geol. Surv. Canada* 73-24: 1-43.

Geyer, G. 1988. Agnostida aus dem höheren Unterkambrium und der Mittelkambrium von Marokko; Teil 2: Eodiscina. *Neues Jahrb. Geol. Palaeontol., Abhandl.*, 177: 93-133.

Gould, S.J. 1989. *Wonderful life. The Burgess Shale and the Nature of History.* New York: Norton.

Gürich, G. 1907. Versuch einer Neueinteillung Trilobiten. *Zentralbl. Mineral. Geol. Palaeontol., Stuttgart*, 1907: 129-133.

Hall, J. 1847. Palaeontology of New York, vol. 1. In: *Natural History of New York, Part 4*. Albnay, New York: Carroll & Cook.

Henningsmoen, G. 1951. Remarks on the classification of trilobites. *Norsk Geol. Tiddskr.* 29: 174-217.

Hou Xianguang & Bersgtröm, J. 1997. Arthropods of the Lower Cambrian Chengjiang fauna, southwest China. *Fossils & Strata* 45: 1-116.

Howell, B.F. 1935. Some New Brunswick Cambrian agnostians. *Bull. Wagner Free Inst. Sci.* 10: 13-17.

Hunt, A.S. 1967. Growth, variation and instar development of an agnostid trilobite. *J. Paleontol.* 41: 203-208.

Hupé, P. 1953. Sur les zones de Trilobites du Cambrien inférieur marocain. *Compt. Rend. Acad. Sci., Paris,* 235: 480-481.

Hutchinson, R.D. 1962. Cambrian stratigraphy and trilobite faunas of southeastern Newfoundland. *Bull. Geol. Surv. Canada* 88: 1-156.

Jaekel, O. 1909. Über die Agnostiden. *Zeitschr. Deutsch. Geol. Gesellsch.* 61: 380-401.

Jell, P.A. 1970. *Pagetia ocellata*, a new Cambrian trilobite from northwestern Queensland. *Mem. Queensland Mus.* 15: 303-313.

Jell, P.A. 1975. Australian Middle Cambrian eodiscoids with a review of the superfamily. *Palaeontographica, Abt. A,* 150: 1-97.

Jell, P.A. 1997. Suborder Eodiscina. In: Kaesler, R.L. (ed.), *Treatise on Invertebrate Paleontology, Part O, Arthropoda 1* (revised): 383-404. Boulder; Lawrence: Geol. Soc. Amer. and Univ. Kansas Press.

Jell, P.A. 2003. Phylogeny of Early Cambrian trilobites. In: Lane, P.D., Siveter, D.J. & Fortey, R.A. (eds.), *Trilobites and their Relatives. Contrib. 3rd Int. Conf., Oxford 2001*. Spec. Pap. Palaeontol. 70: 45-57.

Kaesler, R.L. (ed.) 1997. *Treatise on Invertebrate Paleontology, Part O, Arthropoda 1* (revised). Boulder; Lawrence: Geol. Soc. Amer. and Univ. Kansas Press.

Khomentovskii, V.V. & Repina, L.N. 1965. *Nezhnii Kembrii Stratotipicheskogo razreza Sibiri* [*The Lower Cambrian Stratotype Section of Siberia*]. Sibirsk. Otdelenie, Inst. Geol. Geof., Akad. Nauk SSSR [in Russian].

Kitching, I.J., Forey, P.L., Humphries, C.J. & Williams, D.M. 1998. *Cladistics: The Theory and Practice of Parsimony Analysis, 2nd edition*. Oxford: Oxford Sci. Publ., Syst. Assoc. Publ. 11.

Kobayashi, T. 1939. On the agnostids (part 1). *J. Fac. Sci., Imperial Univ. Tokyo, Sec. II,* 4: 369-522.

Kobayashi, T. 1943. Brief notes on the eodiscids 2, phylogeny of the Dawsonidea. *Proc. Imperial Acad. Tokyo* 19: 43-47.

Kobayashi, T. 1944. On the eodiscids. *J. Fac. Sci., Imperial Univ. Tokyo, Sec. II,* 7: 1-74.

Korobov, M.N. 1980. Biostratigrafiia i miomernye trilobity nizhnego kembriia Mongolii [Biostratigraphy and miomeroid trilobites from the Lower Cambrian of Mongolia]. *Joint Sov.-Mongol. Sci.-Res. Geol. Exp., Trans.* 26: 5-108 [in Russian].

Lermontova, E.V. 1940. Klass Trilobity [Class Trilobita]. In: Vologdin, A.G. (ed.), *Atlas rukovodyashchikh form iskopaemykh fauna SSSR. 1. Kembriy* [*Atlas of the Leading Forms of the Fossil Faunas of the USSR; 1. Cambrian*]: 112-162. Moscow and Leningrad: St. Edit. Office Geol. Lit. [in Russian].

Linnarsson, J.G.O. 1869. Om Vestergötlands Cambriska och Silurska aflagringar. *Kongl. Svenska Vetensk.-Akad. Handl.* 34: 1-86.

Linnarsson, J.G.O. 1883. De undre Paradoxideslagren vid Andrarum. *Sveriges Geol. Undersök., Ser. C,* 54: 1-48.

Lochman, C. 1956. Stratigraphy, paleontology and paleogeography of the *Elliptocephala asaphoides* strata in Cambridge and Hoosic Quadrangles, New York. *Bull. Geol. Soc. Amer.* 67: 1331-1396.

Lu Yenhao, Zhang Wentang, Qian Yiyuan, Zhu Zhaoling, Lin Huanling, Zhou Zhiyi, Zhang Sengui and Yuan Jinliang. 1974. [Cambrian trilobites]. In: Nanjing Insitute of Geology and Paleontology, [*A Handbook of the Stratigraphy and Palaeontology of Southwest China*]: 82-107. Beijing: Science Press [in Chinese].

Müller, K.J. & Walossek, D. 1987. Morphology, ontogeny and life habit of *Agnostus pisiformis* from the Upper Cambrian of Sweden. *Fossils & Strata* 19: 1-124.

Olesen, J. & Walossek, D. 2000. Limb ontogeny and trunk segmentation in *Nebalia* species (Crustacea, Malacostraca, Leptostraca). *Zoomorphology* 120: 47-64.

Öpik, A.A. 1961. Alimentary caeca of agnostids and other trilobites. *Palaeontology* 3: 410-438.

Öpik, A.A. 1963. Early Upper Cambrian fossils from Queensland. *Austr. Bur. Mineral Resour., Geol. Geoph., Bull.* 64: 1-133.

Öpik, A.A. 1967. The Mindyallan fauna of north-western Queensland. *Austr. Bur. Mineral Resour., Geol. Geoph., Bull.* 74: 1-404.

Öpik, A.A. 1975. Cymbric Vale Fauna of New South Wales and Early Cambrian Biostratigraphy. *Bur.Mineral Resour., Geol. and Geophysics, Bull.* 159: 1-78.

Öpik, A.A. 1979. Middle Cambrian agnostids: systematics and biostratigraphy. *Austr. Bur. Mineral Resour., Geol. Geoph., Bull.* 172: 1-188.

Palmer, A.R. 1955. Upper Cambrian Agnostidae of the Eureka District, Nevada. *J. Paleontol.* 29: 86-101.

Palmer, A.R. 1968. Cambrian trilobites of East-Central Alaska *Prof. Paper US Geol. Surv.* 559-B: 1-86.

Palmer, A.R. & Halley, R.B. 1979. Physical stratigraphy and trilobite biostratigraphy of the Carrara formation (Lower and Middle Cambrian) in the southern Great Basin. *Prof. Paper US Geol. Surv.* 1047: 1-131.

Peng Shanchi & Robison, R.A. 2000. Agnostid biostratigraphy across the Middle-Upper Cambrian boundary in Hunan, China. *Paleontol. Soc. Mem.* 53: 1-104.

Pokrovskaya, N.V. 1959. Trilobitovaya fauna i Stratigrafiya Kembrijskihk otlochenij Tuvy [Trilobite fauna and stratigraphy of Cambrian deposits of Tuva]. *Trudy Geol. Inst. Akad. Nauk SSSR* 27: 1-199 [in Russian].

Pratt, B.R. 1992. Trilobites of the Marjuman and Steptoean stages (Upper Cambrian), Rabitkettle Formation, southern Mackenzie Mountains, northwest Canada. *Palaeontogr. Canadiana* 9: 1-170.

Ramsköld, L. & Edgecombe, G.D. 1991. Trilobite monophyly revisited. *Hist. Biol.* 4: 267-283.

Rasetti, F. 1945. Fossiliferous horizons in the "Sillery Formation" near Lévis, Quebec. *Amer. J. Sci.* 243: 305-319.

Rasetti, F. 1948. Lower Cambrian trilobites from the conglomerates of Quebec (exclusive of the Ptychopariidea). *J. Paleontol.* 22: 1-24.

Rasetti, F. 1951. Middle Cambrian stratigraphy and faunas of the Canadian Rocky Mountains. *Smiths. Misc. Coll.* 116 (5): 1-270.

Rasetti, F. 1952. Revision of the North American trilobites of the family Eodiscidae. *J. Paleontol.* 26: 434-451.

Rasetti, F. 1966. New Lower Cambrian trilobite faunule from the Taconic sequence of New York. *Smiths. Misc. Coll.* 148: 1-52.

Rasetti, F. 1967. Lower and Middle Cambrian trilobite faunas from the taconic sequence of New York. *Smiths. Misc. Coll.* 152: 1-111.

Rasetti, F. 1972. Cambrian trilobite faunas of Sardinia. *Att. Accad. Naz. Lincei, Mem., Ser. 8*, 11: 1-100

Rasetti, F. & Theokritoff, G. 1967. Lower Cambrian agnostid trilobites of North America. *J. Paleontol.* 41: 189-196.

Raymond, P.E. 1913. On the genera of the Eodiscidae. *The Ottawa Naturalist* 27: 101-106.

Repina, L.N. 1972. Trilobity Tarynskogo - Gorizonta razrezov nizhnego Kembriya r. Sukharikhi (Igarskii raion) [Trilobites of the Taryn - Gorizont in the Lower Cambrian section of Sukharikhi River, Igarskii Region]. In: Zhurvaleva, I.T. (ed.), *Problemy biostratigrafii i paleontologii nizhnego Kembriya Sibiri* [*Problems in Biostratigraphy and Paleontology of the Lower Cambrian of Siberia*]. Moscow: Inst. Geol. Geof., Sibirsk. Otdelenie, Akad. Nauk SSSR [in Russian].

Repina, L.N., Khomentovskii, V.V., Zhuravleva, I.T. & Rozanov, A.Y. 1964. *Biostratigrafiia nizhnego kembriia Sayano-Altaiskoi skladchatoi oblastii* [*Biostratigraphy of the Lower Cambrian of the Sayan-Altay Fold Region*]. Moscow: Inst. Geol. Geof., Sibirsk. Otdelenie, Akad. Nauk SSSR [in Russian].

Repina, L.N., Petrunina, Z.E. & Hajrullina, T.I. 1975. Trilobity [Trilobites]. In: Repina, L.N., Yaskovich, B.V., Aksarina, N.A., Petrunina, Z.E., Poniklenko, I.A., Rubanov, D.A., Bolgova, G.V., Golikov, A.N., Hajrullina, T.I. & Posokhova, M.M. (eds.), *Stratigrafiia i fauna nizhnego paleozoya severnykh predgorii Turkestanskogo i Altaiskogo khrebtov (yuzhnyi Tyan'-Shan)*. [*Stratigraphy and fauna of the Lower Paleozoic of the northern submontane belt of the Turkestan and Altai ridges (southern Tyan-shan)*]. Trudy Inst. Geol. i Geofizikii, Sibirsk. Otdelenie, Akad. Nauk SSSR, 278: 100-248 [in Russian].

Repina, L.N. & Romanenko, E.V. 1978. Trilobity i stratigrafiya nizhnego kembriya Altaya [Trilobites and stratigraphy of the Lower Cambrian of Altai]. *Trudy Inst. Geol. Geof., Sibirsk. Otdelenie, Akad. Nauk SSSR*, 382: 1-304 [in Russian].

Resser, C.E.1938. Cambrian system (restricted) of the Southern Appalachians. *Geol. Surv. Amer., Spec. Paper* 15: 1-140.

Resser, C.E. 1939. The *Ptarmigania* strata of the northern Wasatch Mountains. *Smiths. Misc. Coll.* 98 (24): 1-72.

Resser, C.E. & Howell, B.F. 1938. Lower Cambrian *Olenellus* Zone of the Appalachians. *Geol. Soc. Amer., Bull.* 49: 195-248.

Robison, R.A. 1964. Late Middle Cambrian faunas from western Utah. *J. Paleontol.* 38: 510-566.

Robison, R.A. 1972. Hypostoma of agnostid trilobites. *Lethaia* 5: 239-248.

Robison, R.A. 1982. Some Middle Cambrian agnostoid trilobites from Western North America. *J. Paleontol.* 56: 132-160.

Robison, R.A. 1984. Cambrian Agnostida of North America and Greenland, Part 1, Ptychagnostidae. *Univ. Kansas Paleontol. Contrib., Paper* 109: 1-59.

Robison, R.A. 1988. Trilobites of the Holm Dal Formation (late Middle Cambrian), central North Greenland. *Meddel. Grønland, Geosci.* 20: 23-103.

Robison, R.A. 1994. Agnostoid trilobites from the Henson Gletscher and Kap Stanton formations (Middle Cambrian), North Greenland. *Bull. Grønlands Geol. Undersøg.* 169: 54-78.

Robison, R.A. & Campbell, D.P. 1974. A Cambrian corynexochoid trilobite with only two thoracic segments. *Lethaia* 7: 273-282.

Romanenko, E.V. & Romanenko, M.F. 1967. Nekotorye voprosy paleogeografii i trilobity kembriia Gornogo Altaia [On some questions of the paleogeography and Cambrian trilobites of Gorny Altay]. *Izvestiya Altayskogo Otdela Geograf. Obshchestva Soyuza SSR* 8: 62-96 [in Russian].

Rozova, A.V. 1964. *Biostratigrafiya i opisanie trilobitov srednego i verkhnego kembriya severo-zapada Sibirskoy platformy* [*Biostratigraphy and Description of Trilobites of the Middle and Upper Cambrian of the Northwest Siberian Platform*]. Akad. Nauk SSSR, Sibirsk. Otdelenie, Inst. Geol. Geof. [in Russian].

Rushton, A.W.A. 1966. The Cambrian trilobites from the Purley Shales of Warwickshire. *Palaeontogr. Soc. Monogr.* 120: 1-55.

Rushton, A.W.A. 1979. A review of the Middle Cambrian Agnostida from the Abbey Shales, England. *Alcheringa* 3: 43-61.

Salter, J.W. 1864. A monograph of the British trilobites from the Cambrian, Silurian, and Devonian formations. *Monogr. Palaeontogr. Soc.* 1864: 1-80, pls. 1-6.

Schram, F.R. & Koenemann, S. 2001. Developmental genetics and arthropod evolution: Part I, on legs. *Evolution & Development* 3: 343-354.

Sdzuy, K. 1961. Das Kambrium Spaniens; Teil II: Trilobiten; 1. Abschnitt. *Akad. Wissensch. Lit. Mainz, Abhandl. Mathem.-Naturwissensch. Klasse* 1961: 499-594.

Sdzuy, K. 1962. Trilobiten aus dem Unter-Kambrium der Sierra Morena (S.-Spanien). *Senckenbergiana Lethaea* 43: 181-228.

Shaler, N.S. & Foerste, A.F. 1888. Preliminary description of North Attleborough fossils. *Bull. Mus. Comp. Zool. Harvard Coll., Ser.2,* 16: 27-41.

Shaw, A.B. 1950. A revision of several Early Cambrian trilobites from eastern Massachusetts. *J. Paleontol.* 24: 577-590.

Shergold, J.H. 1991. Protaspis and early meraspis growth stages of the eodiscoid trilobite *Pagetia ocellata* Jell, and their implications for classification. *Alcheringa* 15: 65-86.

Shergold, J.H. & Laurie, J.R. 1997. Suborder Agnostina. In: Kaesler, R.L. (ed.), *Treatise on Invertebrate Paleontology, Part O, Arthropoda 1* (revised): 331-383. Boulder; Lawrence: Geol. Soc. Amer. and Univ. Kansas Press.

Shergold, J.H., Laurie, J.R. & Sun Xiaowen. 1990. Classification and review of the trilobite order Agnostida Salter, 1864: An Australian perspective. *Austr. Bur. Mineral Resour., Geol. Geoph., Rep.* 296: 1-92.

Šnajdr, M. 1950. *Aculeodiscus* nov. gen. Ze středočeského středního kambria (Trilobitae). *Sborn. Stat. Geol. Úst. Českoslov. Repub., Paleontol., Praha,* 17: 201-212.

Šnajdr, M. 1958. Trilobiti českého středního kambria. *Rozpr. Ústřed. Úst. Geol.* 24: 1-280.

Speyer, S.E. & Chatterton, B.D.E. 1989. Trilobite larvae and larval ecology. *Hist. Biol.* 3: 27-60.

Stubblefield, C.J. 1936. Cephalic sutures and their bearing on current classification of trilobites. *Biol. Rev.* 11: 407-440.

Sun Yunzhu. 1924. Contribution to the Cambrian faunas of North China. *Palaeontol. Sinica, Ser.B,* 1 (4): 1-109.

Swofford, D.L. 1999. *PAUP*: Phylogenetic Analysis Using Parsimony (*and other methods)*. Sunderland, MA: Sinauer Assoc.

Theokritoff, G. 1964. Taconic stratigraphy of northern Washington County, New York. *Bull. Geol. Soc. Amer.* 75: 171-190.

Vogdes, A.W. 1917. Palaeozoic crustacea: The publications and notes on the genera and species during the past twenty years, 1895-1917. *Trans. San Diego Soc. Nat. Hist.* 3 (1): 1-141.

von Toll, E. 1899. Beiträge zur Kenntniss des Sibirschen Cambrium. *Imp. Akad. Nauk, Leningrad, Zap.,* 8 (10): 1-57.

Walcott, C.D. 1890. The fauna of the Lower Cambrian or *Olenellus* zone. *Ann. Rep. US Geol. Surv.* 10: 598-774.

Walcott, C.D. 1916. Cambrian trilobites. *Smiths. Misc. Coll.* 64: 303-456.

Walossek, D. & Müller, K.J. 1990. Upper Cambrian stem-lineage crustaceans and their bearing upon the monophyletic origin of Crustacea and the position of *Agnostus*. *Lethaia* 23: 409-427.

Westergård, A.H. 1946. Agnostidae of the Middle Cambrian of Sweden. *Sveriges Geol. Undersök., Ser. C, no. 477, Årsbok* 40 (1): 1-140.

Westrop, S.R., Ludvigsen, R. & Kindle, C.H. 1996. Marjuman (Cambrian) agnostoid trilobites of the Cow Head group, western Newfoundland. *J. Paleontol.* 70: 804-829.

Whittington, H.B. 1965. Trilobites of the Ordovician Table Head Formation, Western Newfoundland. *Bull. Mus. Comp. Zool., Harvard Univ.,* 132: 281-442.

Whittington, H.B. 1997a. Morphology of the Exoskeleton. In: Kaesler, R.L. (ed.), *Treatise on Invertebrate Paleontology, Part O, Arthropoda 1* (revised): 1-85. Boulder; Lawrence: Geol. Soc. Amer. and Univ. Kansas Press.

Whittington, H.B. 1997b. The Trilobite Body. In: Kaesler, R.L. (ed.), *Treatise on Invertebrate Paleontology, Part O, Arthropoda 1* (revised): 87-135. Boulder; Lawrence: Geol. Soc. Amer. and Univ. Kansas Press.

Whittington, H.B. & Kelly, S.R.A. 1997. Morphological terms applied to Trilobita. In: Kaesler, R.L. (ed.), *Treatise on Invertebrate Paleontology, Part O, Arthropoda 1* (revised): 313-329. Boulder; Lawrence: Geol. Soc. Amer. and Univ. Kansas Press.

Wills, M.A., Briggs, D.E.G. & Fortey, R.A. 1994. Disparity as an evolutionary index: a comparison of Cambrian and Recent arthropods. *Paleobiology* 20: 93-130.

Xiang Liwen & Zhang Tairong. 1985. Chapter 3, Systematic description of trilobites. In: *Stratigraphy and Trilobite Faunas of the Cambrian in the Western Part of Northern Tianshan, Xinjiang*. Minist. Geol. Mineral Resour., Geol. Mem , Ser.2, 4: 1-243 [in Chinese with English summary].

Zhang Wentang. 1953. Some Lower Cambrian trilobites from Western Hupei. *Acta Palaeontol. Sinica* 1 (3): 121-149.

Zhang Wentang (ed.) 1964. *Atlas of Palaeozoic Fossils of Northern Guizhou*. Nanjing: Nanjing Insit. Geol. Palaeontol., Acad. Sinica [in Chinese].

Zhang Wentang. 1980. On the Miomera and Polymera (Trilobita). *Scientia Sinica* 23: 223-234.

Zhang Wentang, Lu Yenhao, Zhu Zaoling, Qian Yiyuan, Lin Hunling, Zhou Zhiyi, Zhang Sengui & Yuan Jinliang. 1980. [Cambrian trilobite faunas of southwestern China.] *Palaeontol. Sinica, N.S. B,* (16) 159 :1-497, 134 pl [in Chinese].

Zhang Xiguang & Clarkson, E.N.K. 1993. Ontogeny of the eodiscid trilobite *Shizhudiscus longquanensis* from the Lower Cambrian of China. *Palaeontology* 36: 785-806.

Zhang Xiguang & Pratt, B.R. 1999. Early Cambrian trilobite larvae and ontogeny of *Ichangia ichangensis* Chang, 1957 (Protolenidae) from Henan, China. *J. Paleontol.* 73: 117-128.

Zhou Zhiqiang. 1975. Trilobita. In: Li Yaoxi, Song Lisheng, Zhou Zhiqiang & Yang Jingyao. [*Stratigraphical Gazetteer of the Lower Palaeozoic,Western Dabashan Mountains*]: 372 pp., 70 pls. Beijing: Geol. Publ. House [in Chinese].

III DEVELOPMENT AND EVOLUTION

Heads, Hox and the phylogenetic position of trilobites

GERHARD SCHOLTZ[1] & GREGORY D. EDGECOMBE[2]

[1] *Institut für Biologie/Vergleichende Zoologie, Humboldt-Universität zu Berlin, Berlin, Germany*
[2] *Australian Museum, Sydney, Australia*

ABSTRACT

The Arachnomorpha or Arachnata concepts have resolved Trilobita as most closely related to Chelicerata amongst extant Arthropoda. An alternative position of trilobites in the stem lineage of Mandibulata is suggested by their pattern of head tagmosis. The antennae of trilobites and Mandibulata are considered non-homologous with the antennae of Onychophora and stem lineage Euarthropoda: they represent 'secondary' and 'primary antennae', respectively. In extant taxa, 'secondary antennae' are deutocerebral, post-ocular, and are connected to deutocerebral olfactory neuropils, whereas 'primary antennae' are pre-ocular and connected to protocerebral olfactory neuropils. In fossils, an insertion at the antero-lateral margin of the hypostome rather than more anteriorly on the head allows 'secondary antennae' to be identified. A deutocerebral mouthpart, of which the onychophoran jaw and the chelicera are examples, is regarded as plesiomorphic for Arthropoda. A loss of 'primary antennae' and modification of the deutocerebral mouthpart into a sensory antenna defines the Mandibulata. Trilobites share a 'secondary antenna' and a clearly-delimited head tagma with mandibulates. Given the extensive homoplasy forced by the Arachnata concept (reversals in pycnogonids and arachnids), a trilobite/mandibulate alliance may be better supported.

Dedicated to Fred Schram on the occasion of his retirement. Our article challenges canonical views about arthropods, and we were forced to question ideas that we have long considered the best explanation of facts. In doing so, we venture into a territory from which Fred Schram has never shied away. Fred's synthesis of data from living and fossil arthropods, his efforts to integrate classical morphological and evo-devo perspectives, and his willingness to explore dangerous ideas have inspired our reappraisal of the trilobite problem.

1 INTRODUCTION

The last decade has seen dramatic changes of our views on arthropod development, morphology, palaeontology, phylogeny, and evolution (see, for example, the books edited by Fortey & Thomas 1997; Edgecombe 1998; Deuve 2001; Scholtz 2004). The comparative molecular approach to embryology, cell lineage studies, new microscopic techniques with a high morphological resolution, and phylogenetic analyses based on molecular and refined morphological data sets led to an increased interest in the body organisation, development and evolution of arthropods and to new and controversial hypotheses about arthropod relationships. New and sometimes surprising solutions emerged from molecular and morphological approaches to long standing and highly controversial issues such as head segmentation, trunk tagmosis, and limb homologies in arthropods. Here we show how the current views about arthropod tagmosis patterns, which are mainly based on molecular developmental genetics, influence our interpretations of fossils. This does not imply principally untestable inferences about developmental patterns and processes in fossil groups but it leads to a framework based on data from Recent arthropods which allows new interpretations of fossil structures and relationships.

We reappraise the phylogenetic placement of trilobites by examining current data on head and trunk segmentation of extant arthropods, including brain anatomy, developmental, and gene expression evidence. These data lead to a radically altered thinking about the basic alignment of the head segments in the major arthropod groups. The new view of heads invites a reconsideration of trilobites and other arthropod fossils and their affinities within the Arthropoda.

2 WHO ARE THE TRILOBITA?

Trilobita is the most species-rich extinct clade within the Arthropoda. Trilobites are known from more than 10,000 species that, in total, span some 275 million years from the Early Cambrian to the end of the Permian. Though trilobites are among the most familiar fossil organisms (Fig. 1), their systematic position within the Arthropoda remains contested (Westheide 1996). Because they are a well characterized fossil group in terms of their appendage structure, tagmosis, ontogeny, and exoskeletal form (Fig. 1), the precise phylogenetic position of Trilobita has important consequences for broader issues in arthropod phylogeny.

Following the discovery of the antennae and biramous appendages of trilobites in the late 19th century, most workers regarded trilobites as most closely related to crustaceans (Beecher 1893; Raymond 1920), and trilobite-crustacean affinities (Hu 1971) or a trilobitomorph ancestry for Crustacea have been maintained by some later investigators (Sanders 1957; Hessler & Newman 1975). The Arachnomorpha concepts of Heider (1913) and later Størmer (1944) provided a major shift in thinking about trilobite relationships because they related Trilobita with Chelicerata. According to Størmer, Arachnomorpha was conceived as a group that encompassed Trilobita, a variety of extinct taxa in the Trilobitomorpha, and the Chelicerata. The idea that chelicerates are the closest living relatives of trilobites has been maintained by most recent workers. Bergström (1979, 1980) placed particular emphasis on the structure of the lamellar setae on the appendages (exopods or book gills) as an indicator of a trilobite-chelicerate relationship. Subsequently, a laterally splayed stance of the limbs and the dorsal penetration of the eyes were cited as additional apomorphic characters for Arachnomorpha (Bergström 1992). Parsimony analyses that have included a variety of fossil taxa have also resolved trilobites within an arachnomorph clade that includes the chelicerates as its only extant member (Wills et al. 1995, 1998; Cotton & Braddy 2004). An exception to the prevailing idea of trilobite-chelicerate affinities was the Gnathomorpha concept of Boudreaux (1979), in which trilobites were instead resolved as stem lineage Mandibulata, based principally on a shared head/trunk tagmosis pattern. Unlike Boudreaux, most palaeontologists have considered trilobites and trilobitomorphs to provide evidence in favour of a group that unites crustaceans with chelicerates rather than with other mandibulates. The trilobite-chelicerate-crustacean group (TCC of Cisne 1974, 1975) corresponds to the so-called Schizoramia sensu Bergström (1979, 1980). Neontological data, both morphological and molecular, conflict with the Schizoramia or TCC concepts, which attests to the central role of extinct taxa in formulating this grouping.

The Arachnata concept of Lauterbach (1973, 1980b, 1983) was developed in the context of Trilobita, in its traditional sense, being paraphyletic with respect to Chelicerata. Lauterbach proposed three characters in support of the Olenellinae (Early Cambrian taxa usually regarded as trilobites) being the sister group to Chelicerata, with the remaining trilobites then being the sister group to that assemblage. Lauterbach's characters were subsequently rejected (Ramsköld & Edgecombe 1991), and the idea of trilobite paraphyly was countered by a greater amount of evidence in favour of trilobite monophyly (Fortey & Whittington 1989). In the present study, we consider Trilobita to be a monophyletic group, as indicated by such apomorphic characters as a low magnesian calcite cuticle, uniquely mineralised eyes, and circumocular ecdysial sutures. A distinctive mode of segment shedding in ontogeny, with thoracic segments released from the anterior margin of a transitory pygidium and a pygidium as a tagma of unreleased segments, defines Trilobita or a slightly more inclusive trilobitomorph clade (Edgecombe & Ramsköld 1999).

Figure 1. Two representatives of Trilobita showing the characteristic body organisation with a head bearing compound eyes and a trunk with a thoracic region and a posterior pygidium with fused segments. (A) *Paradoxides gracilis* from the Cambrian of Bohemia, Czech Republic, length 11 cm (Zoologische Lehrsammlung, Humboldt-Universität zu Berlin). (B) *Ptychopyge excavato-zonata* from the Ordovician of southern Sweden, length 4.5 cm (private collection GS).

3 THE PHYLOGENETIC FRAMEWORK OF RECENT ARTHROPODS

Arthropod phylogeny is a hotly debated issue. There is an almost general agreement upon the monophyly of arthropods including tardigrades, onychophorans and euarthropods based on morphological and molecular grounds (see contributions in Fortey & Thomas 1997; Weygoldt 1986; Ax 1999; Nielsen 2001; Giribet et al. 2001; Kusche et al. 2003; Mallatt et al 2004). However, whether tardigrades or onychophorans or both together are the sister group of euarthropods is not clear (see Dewel et al. 1999). Within the Euarthropoda, which comprises the Crustacea, Hexapoda, Myriapoda and Chelicerata, all combinations are favoured by different authors and backed up by different kinds of evidence. Even the monophyly of each large euarthropod group has been contested (Myriapoda: Dohle 1980; Kraus 2001; Negrisolo et al. 2004, Hexapoda: Nardi et al. 2001; Crustacea: Wilson et al. 2000; Sinakevitch et al. 2003; Schram & Koenemann 2004; Chelicerata *s. lat.*, including Pycnogonida: Giribet et al. 2001). Here we consider euarthropods as monophyletic and discuss only characters of onychophorans, which we treat as sister group to euarthropods because the relevant data for tardigrades are either lacking or are difficult to interpret (Dewel et al. 1999).

Despite molecular analyses in favour of a chelicerate/myriapod sister group relationship (Paradoxopoda of Mallatt et al. 2004; Myriochelata of Pisani et al. 2004), we think there is still ample evidence for mandibulate monophyly (Wägele 1993; Giribet et al. 2001; Kusche et al. 2003). This concerns the head with its differentiated appendages: antennae, mandibles and maxillae in corresponding segmental register. In particular, the presence of a mandible with similar substructures has to be mentioned, together with the expression of the appendage genes *Distal-less* and *dachshund* in the mandibles, the expression patterns of the *Hox* genes in the head, the organisation of brain neuropils, the pattern of serotonin-immunoreactive neurons, and the structure of the ommatidia of the compound eyes with four crystalline cone cells, primary pigment cells, as well as interommatidial pigment cells (Scholtz et al, 1998; Bitsch 2001; Prpic et al. 2001, Scholtz 2001; Hughes & Kaufman 2002b; Loesel et al. 2002; Richter 2002; Edgecombe et al 2003; Müller et al. 2003; Prpic & Tautz 2003; Fanenbruck et al. 2004; Harzsch 2004a; Loesel 2004). Accordingly, we interpret Chelicerata and Mandibulata as sister groups, forming Euarthropoda. We do not enter the debate of infra-mandibulate affinities (Atelocerata versus Tetraconata, see Dohle 2001, Richter 2002) as this is not relevant for our argumentation.

4 CONFLICTS IN THE ARACHNATA CONCEPT

None of the mentioned synapomorphies for Arachnata or Trilobita + Chelicerata (summarised by Kraus 1976; Lauterbach 1980b; Ax 1984; Weygoldt 1986) is shared by all members of the group. Few are shared by Pycnogonida, such that the prevalent idea that Pycnogonida is more closely related to Euchelicerata than is Trilobita (see Dunlop & Arango 2004) forces the supposed 'synapomorphies' to be reversed/lost in pycnogonids. More problematic (given the almost universal acceptance of monophyly of Euchelicerata) is that most of the proposed trilobite/chelicerate synapomorphies are also lacking in arachnids. These characters are for the most part shared only by trilobites/trilobitomorphs and 'merostomes' or, within the latter grade, only Xiphosura. This scattered systematic distribution invites a re-interpretation of the characters as convergent in trilobites and xiphosurans, rather than homologies that are forced to reverse in pycnogonids and arachnids. Similarity of corresponding characters occurring in crustaceans, myriapods, and hexapods further weakens these characters as indications of trilobite/chelicerate affinities.

Trilobation. The trilobed tergum was considered by Størmer (1944) as a character supporting trilobite and chelicerate relationship (Arachnomorpha). Weygoldt (1986) treated trilobation as an autapomorphy of Arachnata, and Wills et al. (1998, character 9) as an autapomorphy of Arachnomorpha except for *Burgessia*. The validity of trilobation has already been critically discussed by Hessler & Newman (1975). Along with Sanders (1957), these authors convincingly showed that the shape of cephalocarid crustaceans is not far removed from the conditions found in trilobites. This is also true for other crustaceans (Isopoda, Brachyura), myriapods (Arthropleurida (Kraus & Brauckmann 2003), Polydesmidae), and hexapods (Zygentoma) that have paratergal lobes. Moreover, absence of trilobation in most arachnids (notable exceptions are the Ricinulei and the extinct Trigonotarbida (e.g., Dunlop 1996)) and pycnogonids forces multiple losses.

Widened head shield with genal spines. The present character is a more precise expression of "widening and broadening of the front end of the body", cited by Weygoldt (1986) as a synapomorphy for Trilobita + Chelicerata. This widening is, however, restricted to trilobitomorphs and Xiphosura, being absent in arachnids, eurypterids, and pycnogonids.

Exoskeleton hard and strong on the dorsal side, soft on the ventral side. This character is as described by Weygoldt (1986). It accurately describes the situation in trilobites, in which the tergum is calcified but the sternum is unmineralised, indeed such that the shape of the sternites has been observed in only a single trilobite species (Whittington, 1993). A comparison with Xiphosura is more or less reasonable but neither arachnids, eurypterids, nor pycnogonids are accurately described by this character.

Lateral eyes penetrating dorsal surface of head shield. Bergström (1992) distinguished Arachnata from Crustacea and its stem lineage by the incorporation of the eyes into the dorsal head shield in the former, versus fundamentally anteroventral eyes in the latter. Some trilobite-allied taxa such as the xandarellid *Cindarella* have stalked, anteroventral eyes (Ramsköld et al. 1997). The putative apomorphy pertains to Trilobita and Euchelicerata, but is absent (or, more accurately, inapplicable) in pycnogonids, which lack lateral facetted eyes. Bergström subsequently rejected the homology of dorsal eyes in trilobites and chelicerates, mapping them on their arthropod cladogram as convergently evolved (Bergström & Hou 2003). Lateral eyes incorporated into the dorsal head shield are also found among crustacean representatives such as Notostraca and Isopoda. Furthermore, the lateral eyes of hexapods and myriapods are included into the head capsule. This evidence suggests that considerable homoplasy plagues this character.

Laterally splayed appendages. Bergström (1992) and Hou & Bergström (1997) considered the orientation of the limbs to be a character of fundamental significance in arthropod systematics. They distinguished Arachnata from a crustacean clade based on the former having "laterally splayed appendages" and the latter having "pendant" appendages. The interpretation of this information in fossils was critiqued by Edgecombe & Ramsköld (1999).

Lamellar setae on exopods. The imbricated lamellar setae on the cephalic and trunk exopods of trilobites and other trilobitomorphs have been homologised with the respiratory lamellae of chelicerates, i.e., the lamellate book gills of Xiphosura and Eurypterida (Bergström 1979). The uncertainty in identifying the homologue of an exopod shaft in xiphosurans and eurypterids is a problem for establishing this homology and, as noted by Cotton & Braddy (2004: 181), "there are certainly major morphological differences between book-gills and trilobite-type exopods." Book gills are traditionally considered to be homologous with book lungs in scorpions and tetrapulmonate arachnids. Whether arachnid book lungs have a single or multiple origins is debatable (Shultz 1990, character 51). Lamellar setae (and indeed exopods) are lacking in pycnogonids, forcing a reversal/loss under the traditional Arachnata hypothesis.

5 HEAD SEGMENTATION - ONCE MORE

5.1 *The persisting problems*

The number and nature of segments and other elements involved in head formation of arthropods has been an issue of constant controversial debates for more than a century (Goodrich 1897; Weber 1952; Siewing 1963; Rempel 1975; Scholtz 1995, 1997, 2001; Rogers & Kaufman 1997; Queinnec 2001). Modern methods of molecular developmental biology such as the comparative analysis of gene expression patterns raised the hope of an end of this "endless dispute" (Rempel 1975). However, although some former hypotheses concerning head segmentation can be clearly ruled out by the outcome of the new methods (Scholtz 2001) we still face the same old problems: What kind of morphological or genetic evidence is enough to indicate the presence of a segment, or limb? How do we prove serial homology of the different parts involved in head formation? Is ontogenetic transformation indicative of evolutionary change? Accordingly, there is a still ongoing debate about the presence or absence of an anterior non-segmental part, the so-called acron (Scholtz 2001, Budd 2002), about the origin of the tritocerebrum (Page 2004), and about the nature of the labrum as either a simple outgrowth or a highly modified pair of limbs, and if the limb nature of the labrum is considered, whether it represents a pre-antennal limb pair (Budd 2002; Urbach & Technau 2003) or the limbs of the tritocerebrum (Haas et al. 2001; Boyan et al. 2002). It is beyond the scope of this article to discuss all these problematic issues of head segmentation. Accordingly, in the following we only mention those characters that are relevant to the points we want to raise[1].

---

[1] We consider the labrum being the fused pair of appendages (or the basal parts thereof) of the intercalary (tritocerebral) segment (Haas et al. 2001; Boyan et al. 2002) as very unlikely. In crustaceans we find both together, appendages (second antennae with endites) in the tritocerebral segment and a labrum. Furthermore, the labral musculature stems from very anterior mesoderm parts, and this is true for hexapods as well as crustaceans (Siewing 1963). The origin of the tritocerebrum from the mandibular neuromere (Page 2004) seems unlikely as well because the existence of a complete neuromere in the corresponding segment clearly predates the evolutionary origin of hexapods or mandibulates (see also Harzsch 2004b).

5.2 Mandibulata

Among Recent arthropods, only the Crustacea, Hexapoda and Myriapoda, together forming the Mandibulata, show a clear head tagma or cephalon that is characterised by its sensory and feeding functions and morphologically by a head shield or head capsule and a posterior limit which is clearly separated from trunk segments. This posterior boundary of the head of Recent Mandibulata is situated posterior to the second maxillary segment (Fig. 2) but there is evidence from cephalocarid crustaceans (Lauterbach 1980a), from fossils (Walossek & Müller 1990) and from development (Scholtz 1997) that the posterior head boundary was originally one segment anterior, i.e., posterior to the first maxillary segment (Fig. 3). In addition to the second maxillae, one or more trunk segments can undergo 'cephalisation', i.e., they become fused or otherwise transformed to support head function (maxillipeds).

The expression of the segment-polarity gene *engrailed* in the head shows a distinct pattern that is almost identical in myriapods, crustaceans, and hexapods (Fig. 2). As in the trunk, *engrailed* is expressed in transverse stripes at the posterior margin of each head segment (e.g., Patel et al. 1989; Fleig 1994; Scholtz 1995; Manzanares et al. 1996; Rogers & Kaufman 1997; Abzhanov & Kaufman 1999; Hughes & Kaufman 2002a; Kettle et al. 2003; Janssen et al. 2004). The anteriormost stripe marks the ocular-protocerebral region (Fig. 2). However, *engrailed* is also expressed in the labrum of some hexapod species (e.g., Fleig 1994, Rogers & Kaufman 1997) but neither in the labrum of crustaceans nor in that of myriapods (e.g., Scholtz 1995; Manzanares et al. 1996; Hughes & Kaufman 2002a; Kettle et al. 2003; Janssen et al. 2004).

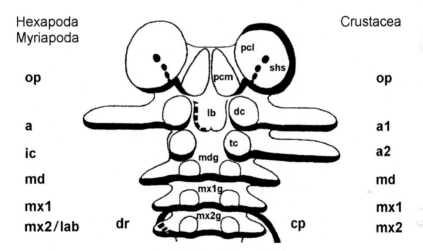

Figure 2. Schematic representation of *engrailed* expression in the heads of mandibulates (bold areas and stripes, anterior is up). The left side shows the situation in myriapods and hexapods, the right side that of crustaceans. The stippled line indicates *engrailed* expression in the labrum (lb) of some hexapod species. Dorsal ridge expression (dr) and secondary head spots (shs) have not been described for myriapods. Abbreviations: a = antenna, cp = carapace, dc = deutocerebrum, ic = intercalary segment, lab = labium, md = mandible, mdg = mandibular ganglion, mx = maxilla, mxg = maxillary ganglion, op = ocular-protocerebral region, pcl = lateral protocerebrum, pcm = median protocerebrum, tc = tritocerebrum (modified after Scholtz 2001).

146 *Scholtz & Edgecombe*

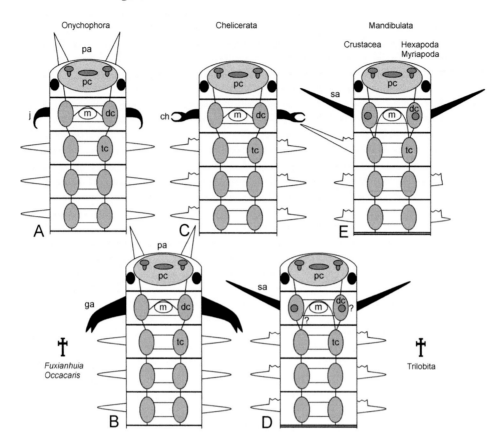

Figure 3. Alignment of head regions and segments of extant (upper row) and extinct (lower row) arthropods showing 'primary antennae' (pa) and 'secondary antennae' (sa) (anterior is up). (A) Onychophora. (B) Stem lineage euarthropod (e.g., *Fuxianhuia, Occacaris*). (C) Chelicerata. (D) Trilobita. (E) Mandibulata (left Crustacea, right Myriapoda/Hexapoda). The appendages (j = jaw, ch = chelicera, ga = frontal/great appendage, and sa = 'secondary antenna') of the deutocerebrum (dc) are shown in black. Elements of the nervous system are shown in grey (light grey: ganglia, dark grey: specialized neuropil areas like the central body, the paired mushroom bodies in the protocerebrum (pc), and the olfactory neuropil in the deutocerebrum). Eyes are represented by lateral black dots in the protocerebrum. Each pair of segmental ganglia is schematically connected by an anterior and a posterior commissure. The mouth (m) lies in the deutocerebral segment between the anterior and the posterior commissure. The stomatogastric nervous system is symbolized by a loop anterior to the mouth. Plesiomorphically it is connected to the deutocerebrum, in mandibulates it is mainly connected to the tritocerebrum (tc). The elements of the nervous system are only inferred in the fossil taxa, and it is not clear whether a deutocerebral olfactory neuropil and a tritocerebral connection of the stomatogastric nervous system were present in Trilobita (question marks). In animals with 'great appendage' or 'frontal appendage' (ga) we find different extensions of head shields covering a varying number of segments (not shown). The posterior border of the head of trilobites and mandibulates (ground pattern, ending posterior to the first maxilla segment) is marked by a double line.

The expression of *Hox* genes in the heads of myriapods, crustaceans and hexapods is similar in terms of anterior boundaries of the expression of *labial, proboscipedia, Deformed,* and *Sex combs reduced*. Furthermore, these genes show only a restricted overlap in their expression domains and in most cases a similar posterior expression boundary. This reflects the morphological differentiation and diversification of the head appendages. The head/trunk boundary is characterised by the anterior border of *Antennapedia* expression (for review see Hughes & Kaufman 2002b).

5.2.1 Ocular-protocerebral region

The ocular-protocerebral region is the anteriormost head part (Figs. 2, 3). Often it is referred to as the acron which conceptually means that it is the asegmental anterior end of the body (see Scholtz 2001). It bears the eyes and the anteriormost brain part, the protocerebrum. The main neuropil areas which are relevant for our discussion are the central body and the mushroom bodies (Fig. 3). The commissures are all pre-oral (Hanström 1928; Bullock & Horridge 1965).

5.2.2 Antennal-deutocerebral segment

The second part of the head is considered by many authors as the first true segment (see Scholtz 2001). Externally it is recognisable by a pair of antennae which are the main chemosensory and often tactile organs of the head of myriapods, crustaceans and hexapods. The corresponding brain part, the deutocerebrum, bears the olfactory neuropils which are connected with the antennae and serve for olfactory sensory processing (Strausfeld et al 1995; Fanenbruck et al. 2004) (Fig. 3). The deutocerebral commissure seems to run pre- and post-stomodaeally. This has at least been shown for a hexapod representative (Boyan et al. 2003).

5.2.3 Second antennal/intercalary segment

The following segment bears the tritocerebrum. It has a post-stomodaeal commissure and neuropils that are connected to the second antennae in crustaceans. In myriapods and hexapods this segment (intercalary segment) is somewhat reduced because it lacks appendages (Figs. 2, 3). However, it is clearly recognisable in hexapod embryos by its neuroblasts (Urbach & Technau 2003) and in myriapods and hexapods by its *engrailed* expression (Rogers & Kaufman 1997; Kettle et al. 2003; Janssen et al. 2004) and as the anteriormost segment in which *Hox* genes (*labial, proboscipedia*) are expressed (Hughes & Kaufman 2002b). The tritocerebrum in all mandibulates is the main connection to the stomatogastric nervous system (Hanström 1928, Bullock & Horridge 1965).

5.2.4 Gnathal segments and the head/trunk boundary

In all three mandibulate groups the gnathal segments show similar structures, at least the mandibles and the first maxillae. The mandibles are morphologically and genetically similar (Scholtz et al. 1998; Edgecombe et al. 2003; Prpic & Tautz 2003). The second maxillae as head appendages and the second maxillary segment in general might not be part of the ground pattern of Mandibulata or of each of its subgroups (see above). In extant mandibulates, the gnathal region is characterised by the expression domains of *Deformed* and *Sex comb reduced* (Hughes & Kaufman 2002b).

5.3 *Chelicerata*

Chelicerata do not possess a head *sensu stricto* because there is no posterior boundary separating it morphologically and functionally from the trunk (Fig. 3). Only pycnogonids and some arachnids such as schizomids, solifuges, palpigrades, and mites show a 4-segmented proterosoma (cephalosoma, propeltidium) (Kraus 1976; Moritz 1993; Vilpoux & Waloszek 2003; Dunlop & Arango 2004) - a condition that is most likely to be apomorphic for these taxa given the current views on their placement within the Chelicerata and the apparent differences between these structures (Weygoldt & Paulus 1979; Shultz 1990; Moritz 1993; Ax 1999; Wheeler & Hayashi 1998; Giribet et al. 2002). Accordingly, all we see is an anterior tagma, the prosoma, which serves the functions of feeding, sensory orientation and walking. Other than in some arachnids in which the pedipalp has raptorial modifications, the only true head appendage is the chelicera, which is the anteriormost limb and which is mainly used for feeding. Plesiomorphically the limb bases of the walking limbs also played a role in food collecting and processing, as can be seen in Xiphosura and Eurypterida. The prosoma is followed by the opisthosoma, which mainly has the functions of respiration and reproduction. The alignment of the anterior segments of chelicerates with those of the mandibulates is problematic because there is only a limited amount of morphological correspondence. Traditionally, the cheliceral segment was homologized with the second antennal/intercalary segment of crustaceans, hexapods and myriapods. Accordingly, the absence of a deutocerebrum has been interpreted as a secondary loss (reviewed by Scholtz 2001). However, recently the comparative analysis of gene expression patterns has changed our view.

As in the Mandibulata, the anteriormost *engrailed* expression in Chelicerata is found in the ocular region (absent in the eyeless mite *Archegozetes longisetosus* Telford & Thomas 1998) followed by regular transverse stripes in the posterior region of all prosomal and opisthosomal segments (Damen 2002). An *engrailed* expression in the labrum has not been found (Telford & Thomas 1998; Damen 2002).

Hox gene expression in chelicerates is characterised by large overlapping domains. The expression areas of the genes *labial, proboscipedia*, and *Hox3* span basically the entire length of the prosoma reflecting the small degree of differentiation in this tagma and the absence of a proper head (Telford & Thomas 1998; Damen et al. 1998). The alignment of the anterior *engrailed* expression (Damen 2002) and, in particular, the anterior boundaries of the expression of *Hox* genes of hexapods, myriapods, crustaceans and chelicerates revealed that the cheliceral segment bears the deutocerebrum and is homologous to the (first) antennal segment of mandibulates (Damen et al. 1998; Telford & Thomas 1998; Hughes & Kaufman 2002b). This interpretation is supported by an analysis of brain morphogenesis in the horseshoe crab (Mittmann & Scholtz 2003). Accordingly, the 'head' of chelicerates is composed as follows.

5.3.1 *Ocular-protocerebral region*
The first body part bears the compound eyes, the anteriormost brain part which contains the optic neuropils, the central body and the mushroom bodies (Fig. 3). The latter are well developed. The protocerebral commissures lie in front of the stomodaeum.

5.3.2 Cheliceral-deutocerebral segment

The segment of the chelicerae, which follows next, bears the deutocerebrum. The deutocerebrum shows distinct neuropil regions but it is devoid of an olfactory neuropil (olfactory lobe). As in mandibulates, the deutocerebral commissures run anterior and posterior to the stomodaeum. Different from the other euarthropods, the main connection to the stomatogastric nervous system is found in the deutocerebrum of chelicerates (Mittmann & Scholtz 2003) (Fig. 3). In the embryo, this brain part lies at the midlateral margin of the circumesophageal nerve ring, in a comparable position to the deutocerebrum of mandibulates (Mittmann & Scholtz 2003).

5.3.3 Pedipalpal-tritocerebral segment

The next segment externally bears the first walking legs, which are often transformed into pedipalps that can be involved in mating, defence and feeding. As in all arthropods, the tritocerebral commissure runs posterior to the stomodaeum (Fig. 3). The tritocerebral segment is, as in Mandibulata, the anteriormost region in which *Hox* genes (*labial, proboscipedia*) are expressed (Hughes & Kaufman 2002b).

5.3.4 The other prosomal segments and the prosoma/opisthosoma boundary

Posterior to the pedipalpal-tritocerebral segment follow four more prosomal segments bearing walking limbs in Recent Xiphosura and Arachnida. In contrast, in pycnogonids the demarcation of the prosoma/opisthosoma is not clear since there are often more walking limbs present (Vilpoux & Waloszek 2003). Moreover, the xiphosuran stem lineage representative *Weinbergina* possesses an additional pair of walking limbs (Stürmer & Bergström 1981). This leg occupies the position of the chilaria of Recent xiphosurans. Interestingly, the chilaria segment is interpreted as the first opisthosomal segment. All this demonstrates that tagmatization in Chelicerata might not be as stable as commonly suggested. This variation is also reflected at the level of *Hox* gene expression where *Ultrabithorax/abdominal-A* shows variation with respect to its anterior boundary (Popadić & Nagy 2001).

5.4 Onychophora

Onychophora do not show a proper head with a distinct external demarcation of the trunk. Instead, some anterior segments and their limbs are modified as sense organs, feeding structures, and means of defence. The number of onychophoran anterior (head) segments and their relationships to those of euarthropods has always been a controversial issue (see Holmgren 1916; Hanström 1928; Pflugfelder 1948; Schürmann 1995; Scholtz 1997; Eriksson et al. 2003). Recent investigations of the ontogeny and anatomy of the onychophoran head and brain using new techniques such as fluorescent dyes and confocal-laser-scan-microscopy with high morphological resolution clarified some of the contentious issues (Eriksson & Budd 2000; Eriksson et al. 2003). In adult onychophorans, segmentation is partly concealed by a characteristic annulation of the cuticle and by not-so evident ganglia (Schürmann 1995). In contrast, the metameres and neuromeres are clearly visible during embryogenesis. The expression pattern of the segment-polarity gene *engrailed* is correlated with this embryonic segmentation in a pattern comparable to that of euarthropods (Wedeen et al. 1997). Accordingly, *engrailed* expression has been detected at the posterior margin of each segment beginning with the jaw segment. It is not clear whether *engrailed* is also expressed in the ocular-protocerebral area of onychophorans (Wedeen et al. 1997).

Apart from results on the expression of *Ultrabithorax/abdominal-A*, which is restricted to the last lobopods and the terminus (Grenier et al. 1997), we have no further data on *Hox* gene expression patterns in onychophorans.

5.4.1 Ocular-protocerebral region

This anteriormost region is characterized externally by the antennae and the eyes. Eriksson et al. (2003) have clearly shown that the antennae are formed anterior to the eyes (Fig. 3). Internally we find the protocerebrum with the central body, the optic ganglia, the mushroom bodies and olfactory neuropils (Holmgren 1916; Hanström 1928; Schürmann 1995; Strausfeld et al. 1995; Strausfeld et al. 1998). In the embryo, the commissures connecting the lateral protocerebral halves occupy a pre-oral position (Eriksson et al. 2003).

5.4.2 Jaw-deutocerebral segment

This segment bears the sickle shaped jaws, which represent highly modified limbs serially homologous to the walking limbs (Fig. 3). As in chelicerates, this brain part shows no olfactory neuropil and it is connected to the stomatogastric nervous system via head nerves 10 and 11 according to Eriksson & Budd (2000). In the embryo the anlagen of the deutocerebrum are in a comparable position to those of euarthropods, about half way on the circumesophageal nerve ring (Eriksson et al. 2003). In the adult this brain part lies dorsally and is fused to the protocerebrum with apparent pre- and postesophageal commissural elements (see figs. 9, 10, 15 in Eriksson & Budd 2000).

5.4.3 Slime papilla-'tritocerebral' segment

The neuromere of the segment of the slime papillae is the first strictly postesophageal element of the onychophoran brain. The slime papillae are modified limbs which are used for catching prey and for defence by extruding a sticky secretion. They are clearly apomorphic for the terrestrial crown-group onychophorans. The neuromere lies ventrally and its commissures run posterior to the stomodaeum. In the embryo it occupies a position at the posterior margin of the circumesophageal nerve ring and in the adult a narrowing of the central nervous system marks the posterior boundary of the 'head' (Hanström 1928; Eriksson & Budd 2000; Eriksson et al. 2003).

5.4.4 The trunk segments

There is no clear cut external boundary between the head and the homonomous trunk. As mentioned above, the neuromeres of the trunk of the Onychophora are not clearly recognizable in adults because the pericarya are only slightly concentrated in segmental ganglia and the number of commissures is very high and not strictly segmental (Schürmann 1995).

6 TRANSFORMATION OF HEAD STRUCTURES: THE CONCEPT OF 'PRIMARY' AND 'SECONDARY ANTENNAE'

6.1 *Antennae of onychophorans are not homologous to antennae of euarthropods*

The different positions of the antennae of onychophorans (pre-ocular, protocerebral) and of mandibulates (post-ocular, deutocerebral) indicate that the onychophoran antennae are not homologous to the (first) antennae of myriapods, crustaceans, and hexapods (see also Budd 2002; Eriksson et al. 2003) (Fig. 3). Chelicerata lack antennae but their cheliceral-deutocerebral segment corresponds to the (first) antennal/deutocerebral segment of mandibulates as is evident from segmental gene and *Hox* gene expressions, from its position, and from brain anatomy (Damen et al. 1998; Telford & Thomas 1998; Mittmann & Scholtz 2003; Simonnet et al. 2004) (Fig. 3). However, the (first) antennae of mandibulates and the chelicerae display additional intrinsic similarities. Notably, both limb types are uniramous and lack any traces of gnathobases in adults and throughout development. This is true for the developing (first) antennae of crustaceans, myriapods and hexapods (Lauterbach 1980a) but also for the chelicerae of *Limulus*, which are the only prosomal limbs showing neither an indication of an outer branch nor a gnathobasic primordium as is indicated by *Distal-less* expression patterns (Mittmann & Scholtz 2001). Recent genetic data on proximal-distal pattern formation show more similarities between chelicerae and antennae in so far as, in contrast to other limbs, no intermediate genetic region is formed (Prpic & Damen, 2004). Given this segmental alignment of mandibulates and chelicerates, one can conclude that the original protocerebral antennae of onychophorans were lost either in the stem lineage of euarthropods or they underwent independent losses in the chelicerate and mandibulate lineages. However, even if we consider homology between the chelicerae and antennae we have to address the question of what came first, a mouthpart or a sense organ?

6.2 *A post-ocular segment with mouthparts and the absence of a proper head is plesiomorphic for Euarthropoda*

Once we accept the above mentioned alignment of head segments between onychophorans and euarthropods one could conclude that the jaws of the onychophoran deutocerebral segment correspond with the chelicerae of the Chelicerata (Fig. 3). This could mean that the ancestral condition for the euarthropods is that the first post-ocular head appendage was a feeding limb rather than a sensory limb. Interestingly, the jaws of Onychophora and the chelicerae of Chelicerata are the only mouthparts of Recent arthropods that process food with the appendage tip. In contrast, in mandibles and maxillae the gnathobasic parts are used for biting, chewing, and grinding food particles. That does not necessarily mean that the specialised 3-segmented chelicera is a euarthropod ground pattern character, but rather that a related structure used for food processing existed. However, if the results of the cladistic analysis of Giribet et al. (2001) that the pycnogonids are the sister groups of all other euarthropods should be further corroborated, we would have even more reason for the assumption of a jaw/chelicera-like first post-ocular head appendage in the euarthropod stem species.

Accordingly, the (first) antenna of the Mandibulata must be regarded as an evolutionary novelty which appeared in the mandibulate stem lineage. This means that a former feeding appendage of jaw/cheliceral structure became transformed into a mainly sensory appendage (Fig. 3). The transformation and change of function of the appendages of the deutocerebral segment, from feeding to sensorial, correlates with a shift of the connection of the stomatogastric nervous system, which in onychophorans and chelicerates is mainly innervated by the deutocerebrum, whereas in mandibulates the main connection lies in the tritocerebrum (Fig. 3). In addition to these changes we find the differentiation of a true head tagma comprising the ocular-protocerebral region and the segments of the antennae and three following segments. All this is covered by the head shield. Hence, we conclude that a proper head might not have been present in the euarthropod stem species and the condition of an ill-defined head/trunk boundary in Chelicerata must be regarded as plesiomorphic.

6.3 *Plasticity of olfactory organs*

The onychophoran protocerebral antennae are connected with the mushroom bodies which also lie in the protocerebrum (Schürmann 1995; Strausfeld et al. 1995; Eriksson et al. 2003)[2]. The deutocerebral (first) antennae of mandibulates are connected to the protocerebral mushroom bodies via the newly developed olfactory neuropils (glomeruli) in the deutocerebrum (Strausfeld et al. 1995, 1998). The role of the mushroom bodies in Chelicerata is very interesting in this regard. Since chelicerates lack antennae, the chemosensory receptors are distributed to different degrees in a taxon-specific manner on the prosomal limbs (Strausfeld et al. 1998). For instance, the whip spiders (Amblypygi) possess first walking limbs modified as large antenna-like sensory organs. Accordingly we see olfactory glomeruli in the corresponding ganglion that are connected to elaborated mushroom bodies in the protocerebrum (Strausfeld et al. 1998). This demonstrates the somewhat intermediate position of chelicerates: the 'primary antennae' as found in onychophorans are lost and the 'secondary antennae' of Mandibulata are not present.

This scenario shows that the original brain part of arthropods which was involved in olfaction, or in more general terms, chemosensory processing was the anteriormost one (protocerebrum) as is typical for other bilaterians such as vertebrates (telencephalon) and annelids (supraesophageal ganglion). There is good reason to assume that this is a character which was already present in the stem species of Bilateria. With the evolution of an anteroposterior body axis it became very useful to possess sense organs and nerve complexes for processing chemical cues in the anteriormost body region because that is the region which is the first to encounter useful or dangerous chemicals in the environment.

6.4 *'Primary antennae' and 'secondary antennae'*

To summarize: we propose the discrimination between a 'primary antenna' and a 'secon-

---

[2] The onychophoran situation is very similar to that in the annelid *Nereis* where the mushroom bodies in the anteriormost brain part in the prostomium are connected to the chemosensory palps (Strausfeld et al. 1995). In the light of the Articulata hypothesis (see Scholtz 2002) one is tempted to interpret onychophoran antennae as transformed palp-like structures and the head region of the protocerebrum as an acron which accordingly is a transformed prostomium.

dary antenna'. Among Recent arthropods, the 'primary antenna' is present in the onychophorans (Fig. 3). This 'primary antenna' is situated in front of the eyes and it is connected to the protocerebrum, and in particular to the mushroom bodies. Posterior to the 'primary antenna' and the eye region is the deutocerebral segment bearing the main mouthpart. In onychophorans this is represented by the jaws, in chelicerates by the chelicerae (Fig. 3). The 'primary antenna' is lost in the euarthropods, either once in the stem linage of Euarthropoda or independently in the lineages leading to Chelicerata and Mandibulata[3]. Chelicerata use different specializations of their prosomal limbs and the corresponding ganglia for olfaction, these limbs being connected to the protocerebral mushroom bodies. The Mandibulata evolved a 'secondary antenna' which lies posterior to the eyes and which is derived from the mouthpart of the deutocerebral segment (Fig. 3). This antenna is again connected to the protocerebral mushroom bodies via olfactory glomeruli in the deutocerebrum.

How plausible is the loss of a sensory antenna, and what are the circumstances under which this can happen? There are some examples among extant arthropods, in particular in several crustacean lineages and in hexapods, for a reduction or loss of sensory antennae. For instance, notostracan crustaceans have only tiny first antennae, and the second antennae are absent. Their functions are shifted towards the first thoracic appendages. These are distinctly transformed compared to the other trunk legs bearing elongated processes (endites) which are used as tactile and chemical sense organs. A similar situation can be found in the proturan hexapods. They lack antennae and the first thoracic limbs replace their function leading to a unique 4-legged locomotion. Notostraca and Protura live in very different environments (aquatic, terrestrial) and the reasons for the loss or reduction of sensory antennae is not clear. However, these examples show that antennae can get lost and that their function can be taken over by more posterior appendages.

The lifestyle and behaviour of Recent Xiphosura might be a clue for a hypothetical scenario of the loss of 'primary antennae' in Cambrian arthropod/euarthropod stem lineage representatives. *Limulus* is well armed and protected and it digs itself through the sediment to collect food that is not very agile such as molluscs, annelids or other dead animal remnants. Accordingly, *Limulus* is equipped with chemical and tactile sensory organs at the margins of its body and on the limbs which work on short distances or direct contact (see Mittmann & Scholtz 2001). In this environment, long sensory antennae are easily broken off and are not necessary for distance perception of chemical and tactile information to detect food or predators.

7 THE CONCEPT OF 'PRIMARY' AND 'SECONDARY ANTENNAE' RELATED TO FOSSILS

How does the concept of 'primary' and 'secondary antennae' relate to the fossil record? A number of Cambrian arthropod fossils show antenniform appendages at their heads. Such fossils include forms like *Fuxianhuia, Occacaris,* and *Marrella* but also Trilobita, Phos-

---

[3] One could speculate that the frontal processes of Remipedia or Cirripedia larvae (Schram 1986) might be vestigial 'primary antennae'; the same might be true for the small frontal protuberances that have been described in a fossil pycnogonid larva (Waloszek & Dunlop 2002). Data about the innervation of these structures e.g., in Remipedia might contribute to this issue.

phatocopina and *Rehbachiella* and many others (e.g., Walossek 1993; Hou & Bergström 1997; Hou 1999; Maas et al. 2003)[4] (Figs. 3, 4). Some of these fossils are considered as stem lineage arthropods, and others as stem lineage representatives of euarthropods and their subgroups, respectively. Is there any possibility to distinguish between a 'primary' or 'secondary antenna'? This discrimination is quite important since it allows us to ally a fossil with different hierarchical levels of Recent Arthropoda. Given that the plesiomorphic condition would be the existence of a (protocerebral) antenna and a (deutocerebral) jaw-like mouthpart we would suggest that the combination of the two is indicative of a 'primary antenna'. This makes the so-called 'great appendage' of some Cambrian fossils such as *Fuxianhuia, Occacaris* or *Branchiocaris* a homologue of the plesiomorphic jaws of Onychophora and the chelicerae of Chelicerata (Figs. 3, 4). On the other hand, an antenna in combination with no specialised whole limb mouthpart would indicate the presence of a 'secondary antenna' and would suggest that the fossil taxon in question is a stem lineage representative of the Mandibulata (Figs. 3, 4).

Some intrinsic evidence allows us to more directly discern between 'primary' and 'secondary antennae'. Antennae that occur together with great appendages insert somewhere at the anterior margin of the head and the great appendage is attached to the anterolateral region of the hypostome/labrum, which shows a slight narrowing (notch) in the attachment area, e.g., in *Fuxianhuia* (Hou & Bergström 1997) (Fig. 4). The latter is exactly the attachment site of the antennae of stem lineage crustaceans like Phosphatocopina but also of the trilobites and their relatives (Fig. 4). Even in Recent crustaceans a corresponding position of the first antenna can be seen, e.g., in anostracan larvae (Olesen 2004) or in Cephalocarida (Sanders 1963). This correspondence fits with the above assumed transformation of the original post-ocular jaw to a 'secondary antenna' in the mandibulate lineage. Accordingly, an antenna attached to the anterolateral region of the hypostome/labrum is considered as a 'secondary antenna'.

The concept of the 'primary' and 'secondary antennae' contradicts the views presented by Budd (2002), Chen et al. (2004), and Cotton & Braddy (2004). Budd (2002) suggested the homology between the various great appendages found in stem lineage arthropods and the antennae of Recent Onychophora. Moreover, he homologised the antennae of Cambrian arthropods with those of Recent Mandibulata. Hence, he suggested that the great appendage/onychophoran antenna is reduced in euarthropods and most likely transformed into the labrum (see also Eriksson et al. 2003). However, this hypothesis faces the problem that the antennae in Cambrian arthropods such as *Fuxianhuia, Occacaris*, and *Branchiocaris* (Briggs 1976; Chen et al. 1995; Hou & Bergström 1997; Hou 1999) are topologically anterior to the supposed frontal appendage ('great appendage'). Budd (2002) tried to solve this problem by evoking a ventral rotation of the great appendage to follow the mouth, resulting in an antero-dorsal position of the antennae. In our opinion the dorsal position of antennae combined with a ventral mouth in *Fuxianhuia* and onychophorans makes this explanation not very convincing. Furthermore, if antennae are considered as sense organs, an original position posterior to the mouthpart seems unlikely. Recent examples of posterior sense organs such as in some chelicerates (e.g., Amblypygi) and in Protura among hexapods are obviously of secondary nature (see above). The hypothesis favoured here accounts for the

---

[4] We do not consider a supposed pair of antennae in *Fortiforceps* (Hou & Bergström 1997) to provide evidence for 'primary antennae' because the structures in the fossils are of unknown identity and are omitted in new interpretations of this taxon (Bergström & Hou 2003: fig. 5B).

observed order of the antenna and raptorial mouthpart in *Fuxianhuia* without needing to posit an ad hoc rotation. Likewise, our hypothesis positions the antenniform appendage and 'great appendage' of *Occacaris* as they were originally described (Hou 1999). The antenniform appendage that attaches near the anterior margin of the head is a 'primary antenna', and the 'great appendage' is posterior to it (Hou & Bergström 1997) rather than anterior (Budd 2002). Accordingly, our view of head evolution contradicts the hypothesis that the euarthropod labrum is derived from the frontal/great appendage (Budd 2002).

We agree with Chen et al. (2004) and Cotton & Braddy (2004) who argue for homology between chelicerae and 'great appendages' in Cambrian arthropods. Chen et al. (2004) suggest, however, that the mouth-part like 'great appendages' are derived from older antenniform sensory structures; accordingly they interpret the 'great appendage' as an apomorphy which evolved in the lineage of the Chelicerata. In contrast, Cotton & Braddy (2004) adopt the more traditional view of the alignment of arthropod head segments. They suggest that the (1st) antennae were lost in the chelicerate stem lineage and that the 'great appendages' and the chelicerae are homologous to the 2nd antennae of crustaceans.

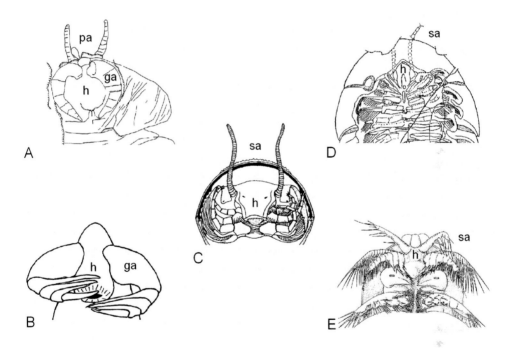

Figure 4. Anterior region of selected fossil taxa (anterior is up). (A) *Fuxianhuia* (modified after Hou and Bergström 1997). (B) *Parapeytoia* (modified after Hou et al. 1995). (C) *Phacops* (modified after Bruton & Haas 2003). (D) *Xandarella* (modified after Bergström & Hou 1998). (E) *Rehbachiella* (modified after Walossek 1993). 'Primary antennae' (pa) are situated at the frontal margin of the 'head' anterior to the 'frontal/great appendages' (ga) as can be seen in *Fuxianhuia*. The latter is attached to the hypostomal/labral region (h). The 'secondary antennae' (sa) occupy an anterolateral position at the hypostomal/labral region corresponding to that of the great appendage. According to this view, *Parapeytoia* must have lost the 'primary antennae' whereas the trilobitomorphs *Phacops* and *Xandarella* show 'secondary antennae' like the mandibulate crustacean *Rehbachiella*.

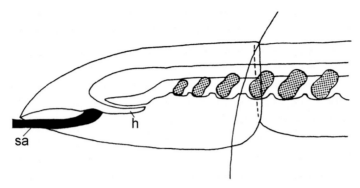

Figure 5. The head of trilobites and their allies as represented by a schematic lateral view of the head in a naraoiid. Anterior is to the left. The head shield covers the segment of the 'secondary antenna' (sa) and three and a half postantennal segments represented by attachment areas of appendages (dotted regions). The curved line indicates the hinge/flexure of the head, h = hypostome (modified after Edgecombe & Ramsköld 1999).

8 THE HEAD OF TRILOBITES

8.1 *Head appendages*

Appendages are known for 20 species of trilobites (see Hughes 2003b, table 1 for a list). In all trilobites for which antennae are preserved, a single flagelliform antenna is present, composed of annulated articles. The antenna is attached against the hypostome, being accommodated by a groove along the lateral side of the hypostome, called the antennal notch (Figs. 4, 5). The antenna is the sole pre-oral cephalic appendage.

Three pairs of biramous postoral appendages are present anterior to the cephalo-thoracic articulation (Cisne 1975, 1981; Whittington 1975; Bruton & Haas 1999, 2003). A debate over whether or not certain trilobites had four pairs of biramous cephalic appendages (Bergström & Brassel 1984; Hou & Bergström 1997) stems from the fourth postoral appendage pair being positioned at the cephalo-thoracic articulation (see Edgecombe & Ramsköld 1999 for discussion) (Fig. 5). The fourth pair is functionally part of the thorax. Because the phylogenetic spread of species observed to have an antenna and three biramous cephalic appendages spans the cladogram for Trilobita, this arrangement can be regarded as part of the trilobite ground pattern. The conservative number of glabellar furrows (generally an occipital furrow and three furrows separating the glabellar lateral lobes) and apodemes across the Trilobita is consistent with a fixed number of cephalic appendages.

The structure of the biramous appendages on the cephalon closely resembles and grades into those of the trunk (thorax and pygidium). Some reconstructions have shown a marked increase in size of the endopods and exopods between the posteriormost cephalic appendage and the first thoracic appendage (*Triarthrus*: Cisne 1975, fig. 2) but the basis for these claims has been convincingly refuted. The cephalic appendages of *Triarthrus* grade evenly in size into those on the thorax (Whittington & Almond 1987). In the best preserved specimens of *Phacops* (e.g., Bruton & Haas 1999, text-fig. 16), the posterior two pairs of cephalic appendages have endopods at least as large as those on the anterior thoracic segments. The only reasonable evidence that the trilobite biramous cephalic appendages are differentiated from those of the thorax comes from the shape of the coxopodite and setation of the endopod. The coxopodites of the cephalic appendages of *Triarthrus* appear to be deeper (dorsoventrally) and less rectangular in outline than those of the thorax, and cephalic limbs lack strong setae on the endopods (Whittington & Almond 1987). Cisne (1981) considered it likely that the endites on the cephalic coxopodites were more ventrally directed than those of the thorax. Fortey & Owens (1999) considered these differences to indicate a differentiation of cephalic and thoracic functions in *Triarthrus* which they related to particle feeding habits. In any case, these difference in setation, shape and possibly orientation of the coxopodites are the only evidence for differentiation of the cephalic appendages as mouthparts in Trilobita.

In summary, trilobites have a head/trunk tagmosis pattern, with antennae and three post-antennal appendages covered by a head shield, but the post-antennal head appendages are only subtly differentiated from trunk appendages.

8.2 Ontogeny of head and trunk

The development of trilobites is well known from the earliest calcified stages, called protaspides. We refer the reader to Hughes' (2003a,b) summary of trilobite ontogeny in the context of tagmosis. The most significant point for the present discussion is the early fixation of head segmentation in trilobite ontogeny. Head segments are specified by the earliest calcified stages; whether they are all precisely synchronous is unknown in the absence of data from the embryo. By the time of calcification, the glabellar lobes and furrows of protaspides indicate a complete complement of cephalic segments, presumed to correspond to the antenna and three post-antennal appendages of later growth stages.

In contrast to the early fixation of the head, the segments of the trunk are added sequentially. Typically one thoracic segment is shed from the generative zone in the so-called transitory pygidium at each moult (Chatterton & Speyer 1997; Hughes 2003a,b; Minelli et al. 2003). Trilobite ontogeny thus indicates the basic distinction between the head and trunk in this group.

9 TRILOBITA AS STEM LINEAGE REPRESENTATIVES OF MANDIBULATA

With the concept of 'primary' and 'secondary antennae' in mind and the degree of homoplasy forced by the Arachnomorpha or Arachnata concepts, the phylogenetic position of the Trilobita may be reconsidered. From what is mentioned above it is evident that trilobites possess a 'secondary antenna'. This is based on the fact that no specialised great appendage follows the antenna and on the position of the antennal attachments against the anterolateral region of the hypostome, accommodated by the antennal notch. In addition to this 'secondary antenna', trilobites exhibit a head comprising three post-antennal segments. This head is covered by a head shield and is formed as a distinct tagma, clearly differentiated from the trunk, early in ontogeny. These characters together suggest a position of trilobites in the stem lineage of the Mandibulata (Fig. 6).

Several authors claim that a head comprising four segments is part of the euarthropod ground pattern (e.g., Walossek & Müller 1990; Scholtz 1997). This hypothesis is in large part influenced by an assumed close relationship between trilobites and chelicerates. On the other hand, Chelicerata do not show any indication of a proper head tagma (see above). Recently, Chen et al. (2004) (re)analysed the heads of some Cambrian arthropods such as *Fortiforceps*, *Yohoia*, *Alacomenaeus*, *Leanchoilia*, and *Haikoucaris* which they (as well as Cotton & Braddy 2004) interpret as stem lineage Chelicerata. These authors conclude that a head shield covering four segments was part of the ground pattern of chelicerates or euarthropods. Apart from the fact that interpretations of the fossil specimens are sometimes somewhat ambiguous, Chen et al. (2004: 15) themselves concede that fewer head segments are also likely and that the 4-segmented condition in chelicerates might be the result of a parallel evolution to that in mandibulates.

The morphological arguments made herein for Trilobita apply to a few other groups of trilobitomorphs known from soft-part preservation, mostly from the Cambrian. The monophyly of a group that unites trilobites with naraoiids, helmetiids, telopeltids and xandarellids is indicated by shared details of exopod structure (Edgecombe & Ramsköld 1999; Bergström & Hou 2003; Cotton & Braddy 2004), notably a division of the exopod into an inner lobe that bears the imbricated lamellar setae and is hinged along the length of the coxopodite, and an outer lobe that bears a fringe of short setae. All of these taxa resemble trilobites in having a head shield that covers a pair of pleural, flagelliform antennae that attach against a hypostome and three or more pairs of postoral biramous appendages. In the case of naraoiids, helmetiids and tegopeltids the head segmentation (antenna + three pairs of biramous appendages anterior to the head/trunk articulation) matches that of Trilobita (Fig. 5). Head/trunk tagmosis further corresponds to that of trilobites in that the biramous appendages of the head are not significantly morphologically differentiated from those in the trunk. Accordingly, these taxa form the core of a trilobitomorph clade and a phylogenetic repositioning of Trilobita on the mandibulate stem lineage accommodates them as well.

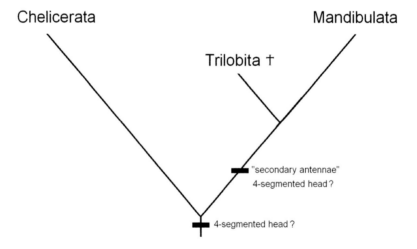

Figure 6. Cladogram of the Euarthropoda with the herein suggested phylogenetic position of Trilobita as a stem lineage taxon of the Mandibulata.

ACKNOWLEDGEMENTS

We thank Jason Dunlop for critical comments and many fruitful discussions, Fabian Scholtz for the photographs in Figure 1, and Hans-Hartmut Krueger for the determination of *Ptychopyge*. GS' studies are supported by the Deutsche Forschungsgemeinschaft.

REFERENCES

Abzhanov, A. & Kaufman, T.C. 1999. Homeotic genes and the arthropod head: expression patterns of the *labial, proboscipedia*, and *Deformed* genes in crustaceans and insects. *Proc. Nat. Acad. Sci. USA* 96: 10224-10229.

Ax, P. 1984. *Das phylogenetische System*. Stuttgart: Fischer.

Ax, P. 1999. *Das System der Metazoa II*. Stuttgart: Fischer.

Beecher, C.E. 1893. On the thoracic legs of *Triarthrus*. *Am. J. Sci.* 66: 467-470.

Bergström, J. 1979. Morphology of fossil arthropods as a guide to phylogenetic relationships. In: Gupta, A. (ed.), *Arthropod Phylogeny*: 3-56. New York: Van Nostrand Reinhold Co.

Bergström, J. 1980. Morphology and systematics of early arthropods. *Abh. Naturwiss. Ver. Hamburg, N.F.*, 23: 7-42.

Bergström, J. 1992. The oldest arthropods and the origin of Crustacea. *Acta Zool.* 73: 287-291.

Bergström, J. & Brassel, G. 1984. Legs in the trilobite *Rhenops* from the Lower Devonian Hunsrück Slate. *Lethaia* 17: 67-72.

Bergström, J. & Hou, X. 1998. Chengjiang arthropods and their bearing on early arthropod evolution. In: Edgecombe, G.D. (ed.), *Arthropod Fossils and Phylogeny*: 151-184. New York: Columbia Univ. Press.

Bergström, J. & Hou, X. 2003. Arthropod origins. *Bull. Geosci., Czech Geol. Surv.*, 78: 323-334.

Bitsch, J. 2001. The arthropod mandible: morphology and evolution. Phylogenetic implications. *Ann. Soc. Entomol. Fr., N.S.*, 37: 305-321.

Boudreaux, H.B. 1979. *Arthropod Phylogeny - with Special Reference to Insects*. New York: Wiley.

Boyan, G.S., Williams, J.L.D., Posser, S. & Bräunig, P. 2002. Morphological and molecular data argue for the labrum being non-apical, articulated, and the appendage of the intercalary segment in the locust. *Arthrop. Struct. Dev.* 31: 65-76.

Boyan, G., Reichert, H., & Hirth, F. 2003. Commissure formation in the embryonic insect brain. *Arthrop. Struct. Dev.* 32: 61-77.

Briggs, D.E.G. 1976. The arthropod *Branchiocaris* n. gen., Middle Cambrian, Burgess Shale, British Columbia. *Geol. Surv. Canada, Bull.* 264: 1-29.

Bruton, D.L. & Haas, W. 1999. The anatomy and functional morphology of *Phacops* (Trilobita) from the Hunsrück Slate (Devonian). *Palaeontographica, Abt. A,* 253: 29-75, pls. 1-15.

Bruton, D.L. & Haas, W. 2003. Making *Phacops* come alive. *Spec. Pap. Palaeontol.* 70: 331-347.

Budd, G.E. 2002. A palaeontological solution to the arthropod head problem. *Nature* 417: 271-275.

Bullock, T.H. & Horridge, G.A. 1965. *Structure and Function in the Nervous System of Invertebrates*. San Francisco: Freeman.

Chatterton, B.D.E. & Speyer, S.E. 1997. Ontogeny. In: Whittington, H.B. (ed.), *Treatise on Invertebrate Paleontology, Part O, Arthropoda I, Trilobita* (revised): 173-247. Boulder; Lawrence: Geol. Soc. Amer. and Univ. Kansas Press.

Chen, J.-Y., Edgecombe, G.D., Ramsköld, L. & Zhou, G.-Q. 1995. Head segmentation in Early Cambrian *Fuxianhuia*: implications for arthropod evolution. *Science* 268: 1339-1343.

Chen, J.-Y., Waloszek, D. & Maas, A. 2004. A new "great appendage" arthropod from the Lower Cambrian of China and the phylogeny of the Chelicerata. *Lethaia* 37: 3-20.

Cisne, J.L. 1974. Trilobites and the origin of arthropods. *Science* 186: 13-18.

Cisne, J.L. 1975. Anatomy of *Triarthrus* and the relationships of the Trilobita. *Fossils & Strata* 4: 45-63.

Cisne, J.L. 1981. *Triarthrus eatoni* (Trilobita): anatomy of its exoskeletal, skeletomusculature, and digestive systems. *Palaeontogr. Amer.* 9: 99-142.

Cotton, T.J. & Braddy, S.J. 2004. The phylogeny of arachnomorph arthropods and the origin of the Chelicerata. *Trans. R. Soc. Edinburgh, Earth Sci.,* 94: 169-193.

Damen, W.G.M. 2002. Parasegmental organization of spider embryo implies that the parasegment is an evolutionary conserved entity in arthropod embryogenesis. *Development* 129: 1239-1250.

Damen, W.G.M., Hausdorf, M., Seyfarth, E.-A. & Tautz, D. 1998. A conserved mode of head segmentation in arthropods revealed by the expression pattern of Hox genes in a spider. *Proc. Nat. Acad. Sci. USA* 95: 10665-10670.

Deuve, T. (ed.) 2001. *Origin of the Hexapoda. Ann. Soc. Entomol. Fr., N.S.,* 37.

Dewel, R.A., Budd, G.E., Castano, D.F., & Dewel, W.C. 1999. The organization of the subesophageal nervous system in tardigrades: insights into the evolution of the arthropod hypostome and tritocerebrum. *Zool. Anz.* 238: 191-203.

Dohle, W. 1980. Sind die Myriapoden eine monophyletische Gruppe? *Abh. Naturwiss. Ver. Hamburg, N.F.,* 23: 45-104.

Dohle, W. 2001. Are the insects terrestrial crustaceans? A discussion of some new facts and arguments and the proposal of the proper name "Tetraconata" for the monophyletic unit Crustacea + Hexapoda. *Ann. Soc. Entomol. Fr., N.S.,* 37: 85-103.

Dunlop, J.A. 1996. Systematics of the fossil arachnids. *Rev. Suisse Zool., Vol. Hors Série*: 173-184.

Dunlop, J.A. & Arango, C.P. 2004. Pycnogonid affinities: a review. *J. Zool. Syst. Evol. Res.* (in press).

Edgecombe, G.D. (ed.) 1998. *Arthropod Fossils and Phylogeny*. New York: Columbia Univ. Press.

Edgecombe, G.D., & Ramsköld, L. 1999. Relationships of Cambrian Arachnata and the systematic position of Trilobita. *J. Paleontol.* 73: 263-287.

Edgecombe, G.D., Richter, S. & Wilson, G.D.F. 2003. The mandibular gnathal edges: homologous structures throughout Mandibulata? *African Invertebr.* 44: 115-135.

Eriksson, B.J. & Budd, G.E. 2000. Onychophoran cephalic nerves and their bearing on our understanding of head segmentation and stem-group evolution of Arthropoda. *Arthrop. Struct. Dev.* 29: 197-209.

Eriksson, B.J., Tait, N.N. & Budd, G.E. 2003. Head development in the onychophoran *Euperipatoides kanangrensis* with particular reference to the central nervous system. *J. Morphol.* 255: 1-23.

Fanenbruck, M., Harzsch, S. & Wägele, J.-W. 2004. The brain of the Remipedia (Crustacea) and an alternative hypothesis on their phylogenetic relationships. *Proc. Nat. Acad. Sci. USA* 101: 3868-3873.

Fleig, R. 1994. Head segmentation in the embryo of the Colorado beetle *Leptinotarsa decemlineata* as seen with anti-en immunostaining. *Roux's Arch. Dev. Biol.* 203: 227-229.

Fortey, R.A. & Owens, R.M. 1999. Feeding habits in trilobites. *Palaeontology* 42: 429-465.

Fortey, R.A. & Thomas, R.H. 1997. (eds.), *Arthropod Relationships*. London: Chapman & Hall.

Fortey, R.A. & Whittington, H.B. 1989. The Trilobita as a natural group. *Hist. Biol.* 2: 125-138.

Giribet, G., Edgecombe, G.D. & Wheeler, W.C. 2001. Arthropod phylogeny based on eight molecular loci and morphology. *Nature* 413: 157-161.

Giribet, G., Edgecombe, G.D., Wheeler, W.C. & Babitt, C. 2002. Phylogeny and systematic position of Opiliones: a combined analysis of chelicerate relationships using morphological and molecular data. *Cladistics* 18: 5-70.

Goodrich, E.S. 1897. On the relation of the arthropod head to the annelid prostomium. *Quart. J. Micr. Sci.* 40: 247-268.

Grenier, J.K., Garber, T.L., Warren, R., Whitington, P.M. & Carroll, S. 1997. Evolution of the entire arthropod *Hox* gene set predated the origin and radiation of the onychophoran/arthropod clade. *Curr. Biol.* 7: 547-553.

Haas, M.S., Brown, S.J. & Beeman, R.W. 2001. Pondering the procephalon: the segmental origin of the labrum. *Dev. Genes Evol.* 211: 89-95.

Hanström, B. 1928. *Vergleichende Anatomie des Nervensystems der wirbellosen Tiere*. Berlin: Springer.

Harzsch, S. 2004a. Phylogenetic comparison of serotonin-immunoreactive neurons in representatives of the Chilopoda, Diplopoda, and Chelicerata: implications for arthropod relationships. *J. Morphol.* 259: 198-213.

Harzsch, S. 2004b. The tritocerebrum of Euarthropoda: a "non-*drosophilo*centric" perspective. *Evol. Dev.* 6: 303-309.

Heider, K. 1913. Entwicklungsgeschichte und Morphologie der Wirbellosen. In: Hinneberg, P. (ed.), *Die Kultur der Gegenwart, Teil 3, Abt. 4, Bd. 2*: 176-332. Leipzig: Teubner.

Hessler, R.R. & Newman, W.A. 1975. A trilobitomorph origin for the Crustacea. *Fossils & Strata* 4: 437-459.

Holmgren, N. 1916. Zur vergleichenden Anatomie des Gehirns von Polychaeten, Onychophoren, Xiphosuren, Arachniden, Crustaceen, Myriapoden und Insekten. *Vet. Akad. Handl. Stockholm* 56: 1-303.

Hou, X. 1999. New rare bivalved arthropods from the Lower Cambrian Chengjiang fauna, Yunnan, China. *J. Paleontol.* 73: 102-116.

Hou, X, Bergström, J. 1997. Arthropods of the Lower Cambrian Chengjiang fauna, southwest China. *Fossils & Strata* 45: 1-116.

Hou, X., Bergström, J., & Ahlberg, P. 1995. *Anomalocaris* and other large animals in the Lower Cambrian Chengjiang fauna of southwest China. *Geol. För. Stockholm Förhandl.* 117: 163-183.

Hu, C.-H. 1971. Ontogeny and sexual dimorphism of Lower Paleozoic Trilobita. *Palaeontogr. Americ.* 7: 27-155.

Hughes, C.L. & Kaufman, T.C. 2002a. Exploring myriapod segmentation: the expression patterns of *even-skipped*, *engrailed*, and *wingless* in a centipede. *Dev. Biol.* 247: 47-61.

Hughes, C.L. & Kaufman, T.C. 2002b. Hox genes and the evolution of the arthropod body plan. *Evol. Dev.* 4: 459-499.

Hughes, N.C. 2003a. Trilobite body patterning and the evolution of arthropod tagmosis. *BioEssays* 25: 386-395.

Hughes, N.C. 2003b. Trilobite tagmosis and body patterning from morphological and developmental perspectives. *Integr. Comp. Biol.* 43: 185-206.

Janssen, R. Prpic, N.-M., & Damen, W.G.M. 2004. Gene expression suggests decoupled dorsal and ventral segmentation in the millipede *Glomeris marginata* (Myriapoda: Diplopoda). *Dev. Biol.* 268: 89-104.

Kettle, C., Johnstone, J., Jowett, T., Arthur, H. & Arthur, W. 2003. The pattern of segment formation, as revealed by *engrailed* expression, in a centipede with a variable number of segments. *Evol. Dev.* 5: 198-207.

Kraus, O. 1976. Zur phylogenetischen Stellung und Evolution der Chelicerata. *Ent. Germ.* 3: 1-12.

Kraus, O. 2001. "Myriapoda" and the ancestry of the Hexapoda. *Ann. Soc. Entomol. Fr., N.S.,* 37: 105-127.

Kraus, O. & Brauckmann, C. 2003. Fossil giants and surviving dwarfs. Arthropleurida and Pselaphognatha (Atelocerata, Diplopoda): characters, phylogenetic relationships and construction. *Verh. Naturwiss. Ver. Hamburg, N.F.,* 40: 5-50.

Kusche, K., Hembach, A., Hagner-Holler, S., Genauer, W. & Burmester, T. 2003. Complete subunit sequences, structure and evolution of the 6 x 6-mer hemocyanin from the common house centipede, *Scutigera coleoptrata. Eur. J. Biochem.* 270: 2860-2868.

Lauterbach, K.-E. 1973. Schlüsselereignisse in der Evolution der Stammgruppe der Euarthropoda. *Zool. Beitr., N.F.,* 19: 251-299.

Lauterbach, K.-E. 1980a. Schlüsselereignisse in der Evolution des Grundplans der Mandibulata (Arthropoda). *Abh. Naturwiss. Ver. Hamburg, N.F.,* 23: 105-161.

Lauterbach, K.-E. 1980b. Schlüsselereignisse in der Evolution des Grundplans der Arachnata (Arthropoda). *Abh. Naturwiss. Ver. Hamburg, N.F.,* 23: 163-327.

Lauterbach, K.-E. 1983. Synapomorphien zwischen Trilobiten- und Chelicerantenzweig der Arachnata. *Zool. Anz.* 210: 213-238.

Loesel, R. 2004. Comparative morphology of central neuropils in the brain of arthropods and its evolutionary and functional implications. *Acta Biol. Hung.* 55: 39-51.

Loesel, R., Nässel, D.R. & Strausfeld, N.J. 2002. Common design in a unique midline neuropil in the brains of arthropods. *Arthrop. Struct. Dev.* 31: 77-91.

Maas, A., Waloszek, D. & Müller, K.J. 2003. Morphology, ontogeny and phylogeny of the Phosphatocopina (Crustacea) from the upper Cambrian "Orsten" of Sweden. *Fossils & Strata* 49: 1-238.

Mallatt, J.M., Garey, J.R. & Shultz, J.W. 2004. Ecdysozoan phylogeny and Bayesian inference: first use of nearly complete 28S and 18S rRNA gene sequences to classify the arthropods and their kin. *Mol. Phylogenet. Evol.* 31: 178-191.

Manzanares, M., Williams, T.A., Marco, R. & Garesse, R. 1996. Segmentation in the crustacean *Artemia*: engrailed staining studied with an antibody raised against the *Artemia* protein. *Roux's Arch. Dev. Biol.* 205: 424-431.

Minelli, S., Fusco, G. & Hughes, N.C. 2003. Tagmata and segment specification in trilobites. *Spec. Pap. Palaeontol.* 70: 31-43.

Mittmann, B. & Scholtz, G. 2001. *Distal-less* expression in embryos of *Limulus polyphemus* (Chelicerata, Xiphosura) and *Lepisma saccharina* (Insecta, Zygentoma) suggests a role in the development of mechanoreceptors, chemoreceptors, and the CNS. *Dev. Genes Evol.* 211: 232-243.

Mittmann, B. & Scholtz, G. 2003. Development of the nervous system in the "head" of *Limulus polyphemus* (Chelicerata: Xiphosura): morphological evidence for a correspondence between the segments of the chelicerae and of the (first) antennae of Mandibulata. *Dev. Genes Evol.* 213: 9-17.

Moritz, M. 1993. Unterstamm Arachnata. In: Gruner, H.-E. (ed.), *Lehrbuch der Speziellen Zoologie (Begründet von A. Kästner), Band I, 4.Teil*: 64-442. Jena: Gustav Fischer Verlag.

Müller, C.H.G., Rosenberg, J., Richter, S. & Meyer-Rochow, V.B. 2003. The compound eye of *Scutigera coleoptrata* (Linnaeus, 1758) (Chilopoda: Notostigmophora): an ultrastructural reinvestigation that adds support to the Mandibulata concept. *Zoomorphology* 122: 191-209.

Nardi, F., Spinsanti, G., Boore, J.L., Carapelli, A., Dallai, R. & Frati, F. 2003. Hexapod origins: monophyletic or paraphyletic? *Science* 299: 1887-1889.

Negrisolo, E., Minelli, A., & Valle, G. 2004. The mitochondrial genome of the house centipede *Scutigera* and the monophyly versus paraphyly of myriapods. *Mol. Biol. Evol.* 21: 770-780.

Nielsen, C. 2001. *Animal Evolution, 2nd edition.* Oxford: Oxford Univ. Press.

Olesen, J. 2004. On the ontogeny of the Branchiopoda (Crustacea): contribution of development to phylogeny and classification. In: Scholtz, G. (ed.), *Crustacean Issues 15, Evolutionary Developmental Biology of Crustacea*: 217-269. Lisse: Balkema.

Page, D.T. 2004. A mode of arthropod brain evolution suggested by *Drosophila* commissure development. *Evol. Dev.* 6: 25-31.

Patel, N.H., Kornberg, T.B. & Goodman, C.S. 1989. Expression of *engrailed* during segmentation in grasshopper and crayfish. *Development* 107: 201-212.

Pflugfelder, O. 1948. Entwicklung von *Paraperipatus anboinensis* n. sp. *Zool. Jahrb. Anat.* 69: 443-492.

Pisani, D., Poling, L.L., Lyons-Weiler, M. & Hedges, S.B. 2004. The colonization of land by animals: molecular phylogeny and divergence times among arthropods. *BMC Biology* 2.

Popadić, A. & Nagy, L. 2001. Conservation and variation in *Ubx* expression among chelicerates. *Evol. Dev.* 3: 391-396.

Prpic, N.-M. & Damen, W. 2004. Expression patterns of leg genes in the mouthparts of the spider *Cupiennius salei* (Chelicerata: Arachnida). *Dev. Genes Evol.* 214: 296-302.

Prpic, N.-M. & Tautz, D. 2003. The expression of the proximodistal axis patterning genes *Distal-less* and *dachshund* in the appendages of *Glomeris marginata* (Myriapoda: Diplopoda) suggests a special role of these genes in patterning the head appendages. *Dev. Biol.* 260: 97-102.

Prpic, N.-M., Wigand, B., Damen, W.G.M. & Klingler, M. 2001. Expression of *dachshund* in wild-type and *Distal-less* mutant *Tribolium* corroborates serial homologies in insect appendages. *Dev. Genes Evol.* 211: 467-477.

Quéinnec, E. 2001. Insights into arthropod head evolution. Two heads in one: the end of the "endless dispute"? *Ann. Soc. Entomol. Fr., N.S.,* 37: 51-69.

Ramsköld, L. & Edgecombe, G.D. 1991. Trilobite monophyly revisited. *Hist. Biol.* 4: 267-283.

Ramsköld, L., Chen, J.-Y., Edgecombe, G.D. & Zhou, G.-Q. 1997. *Cindarella* and the arachnate clade Xandarellida (Arthropoda, Early Cambrian) from China. *Trans. R. Soc. Edinburgh, Earth Sci.,* 88: 19-38.

Raymond, P.E. 1920. The appendages, anatomy and relationships of trilobites. *Mem. Conn. Acad. Arts Sci.* 7: 1-169.

Rempel, J.G. 1975. The evolution of the insect head: the endless dispute. *Quaest. Entomol.* 11: 7-25.

Richter, S. 2002. The Tetraconata concept: hexapod-crustaceans relationships and the phylogeny of Crustacea. *Org. Divers. Evol.* 2: 217-237.

Rogers B.T. & Kaufman, T.C. 1997. Structure of the insect head in ontogeny and phylogeny: a view from *Drosophila*. *Int. Rev. Cytol.* 174: 1-84.

Sanders, H.L. 1957. Cephalocarida and crustacean phylogeny. *Syst. Zool.* 6: 112-128.

Sanders, H.L. 1963. The Cephalocarida. Functional morphology, larval development, comparative external anatomy. *Mem. Conn. Acad. Arts Sci.* 15: 1-80.

Scholtz, G. 1995. Head segmentation in Crustacea - an immunocytochemical study. *Zoology* 98: 104-114.

Scholtz, G. 1997. Cleavage, germ band formation and head segmentation: the ground pattern of the Euarthropoda. In: Fortey, R.A. & Thomas, R.H. (eds.), *Arthropod Relationships*: 317-332. London: Chapman & Hall.

Scholtz, G. 2001. Evolution of developmental patterns in arthropods - the analysis of gene expression and its bearing on morphology and phylogenetics. *Zoology* 103: 99-111.

Scholtz, G. 2002. The Articulata hypothesis - or what is a segment? *Org. Divers. Evol.* 2: 197-215.

Scholtz, G. (ed.) 2004. *Evolutionary Developmental Biology of Crustacea*. Lisse: Balkema.

Scholtz, G., Mittman, B. & Gerberding, M. 1998. The pattern of *Distal-less* expression in the mouthparts of crustaceans, myriapods and insects: new evidence for a gnathobasic mandible and the common origin of Mandibulata. *Int. J. Dev. Biol.* 42: 801-810.

Schram, F.R. 1986. *Crustacea*. Oxford: Oxford Univ. Press.

Schram, F.R. & Koenemann, S. 2004. Developmental genetics and arthropod evolution: on body regions of Crustacea. In: Scholtz, G. (ed.), *Crustacean Issues 15, Evolutionary Developmental Biology of Crustacea*: 75-92. Lisse: Balkema.

Schürmann, F.W. 1995. Common and special features of the nervous system of Onychophora: A comparison with Arthropoda, Annelida and some other invertebrates. In: Breidbach, O. & Kutsch, W. (eds.), *The Nervous Systems of Invertebrates: An Evolutionary and Comparative Approach*: 139-158. Basel: Birkhäuser.

Shultz, J.W. 1990. Evolutionary morphology and phylogeny of Arachnida. *Cladistics* 6: 1-38.

Siewing, R. 1963. Das Problem der Arthropodenkopfsegmentierung. *Zool. Anz.* 170: 429-468.

Simmonet, F., Deutsch, J. & Quéinnec, E. 2004. *hedgehog* is a segment polarity gene in a crustacean and a chelicerate. *Dev. Genes Evol.* (in press).

Sinakevitch, I., Douglass, J.K., Scholtz, G., Loesel, R. & Strausfeld, N.J. 2003. Conserved and convergent organization in the optic lobes of insects and isopods, with reference to other crustacean taxa. *J. Comp. Neurol.* 467: 150-172.

Størmer, L. 1944. On the relationships and phylogeny of fossil and recent Arachnomorpha. *Skrift. Utgitt Norske Vidensk.-Akad. Oslo. I. Math.-Naturvitensk. Klasse* 5: 1-158.

Strausfeld, N.J., Buschbeck, E.K. & Gomez, R.S. 1995. The arthropod mushroom body: Its functional roles, evolutionary enigmas and mistaken identities. In: Breidbach, O. & Kutsch, W. (eds.), *The Nervous Systems of Invertebrates: An Evolutionary and Comparative Approach*: 349-381. Basel: Birkhäuser.

Strausfeld, N.J., Hansen, L., Li, Y., Gomez, R.S. & Ito, K. 1998. Evolution, discovery, and interpretations of arthropod mushroom bodies. *Learning & Memory* 5: 11-37.

Stürmer, W. & Bergström, J. 1981. *Weinbergina*, a xiphosuran arthropod from the Devonian Hunsrück Slate. *Paläontol. Z.* 55: 237-255.

Telford, M.J. & Thomas, R.H. 1998. Expression of homeobox genes shows chelicerate arthropods retain their deutocerebral segment. *Proc. Nat. Acad. Sci. USA* 95: 10671-10675.

Urbach, R. & Technau, G.M. 2003. Early steps in building the insect brain: neuroblast formation and segmental patterning in the developing brain of different insect species. *Arthrop. Struct. Dev.* 32: 103-123.

Vilpoux, K. & Waloszek, D. 2003. Larval development and morphogenesis of the sea spider *Pycnogonum litorale* (Ström, 1762) and the tagmosis of the body of Pantopoda. *Arthrop. Struct. Dev.* 32: 349-383.

Wägele, J.W. 1993. Rejection of the 'Uniramia' hypothesis and implications on the mandibulate concept. *Zool. Jahrb. Syst.* 120: 253-288.

Walossek D. 1993. The Upper Cambrian *Rehbachiella* and the phylogeny of Branchiopoda and Crustacea. *Fossils & Strata* 32: 3-202.

Walossek, D. & Müller, K.J. 1990. Upper Cambrian stem-lineage crustaceans and their bearing upon the monophyly of Crustacea and the position of *Agnostus*. *Lethaia* 23: 409-427.

Waloszek, D. & Dunlop, J.A. 2002. A larval sea spider (Arthropoda: Pycnogonida) from the Upper Cambrian 'Orsten' of Sweden, and the phylogenetic position of pycnogonids. *Palaeontology* 45: 421-446.

Weber, H. 1952. Morphologie, Histologie und Entwicklungsgeschichte der Articulaten II. Die Kopfsegmentierung und die Morphologie des Kopfes überhaupt. *Fortschr. Zool.* 9: 18-231.

Wedeen, C.J., Kostriken, R.G., Leach, D. & Whitington, P. 1997. Segmentally iterated expression of an *engrailed*-class gene in the embryo of an Australian onychophoran. *Dev. Genes Evol.* 270: 282-286.

Westheide, W. 1996. Trilobita. In: Westheide, W. & Rieger, R. (eds.), *Spezielle Zoologie, Teil 1: Einzeller und Wirbellose Tiere*: 445-448. Stuttgart: Gustav Fischer.

Weygoldt, P. 1986. Arthropod interrelationships - the phylogenetic-systematic approach. *Z. Zool. Syst. Evol.-forsch.* 24: 19-35.

Weygoldt, P. & Paulus, H.F. 1979. Untersuchungen zur Morphologie, Taxonomie und Phylogenie der Chelicerata. *Z. Zool. Syst. Evol.-forsch.* 17: 85-116, 177-200.

Wheeler, W.C. & Hayashi, C.Y. 1998. The phylogeny of extant chelicerate orders. *Cladistics* 14: 173-192.

Whittington, H.B. 1975. Trilobites with appendages from the Middle Cambrian Burgess Shale, British Columbia. *Fossils & Strata* 4: 97-136.

Whittington, H.B. 1993. Anatomy of the Ordovician trilobite *Placoparia*. *Phil. Trans. R. Soc. London, B,* 339: 109-118.

Whittington, H.B. and Almond, J.E. 1987. Appendages and habits of the Upper Ordovician trilobite *Triarthrus eatoni*. *Phil. Trans. R. Soc. London, B,* 317: 1-46.

Wills, M.A., Briggs, D.E.G. & Fortey, R.A. & Wilkinson, M. 1995. The significance of fossils in understanding arthropod evolution. *Verh. Deutsch. Zool. Ges.* 88 (2): 203-215.

Wills, M.A., Briggs, D.E.G., Fortey, R.A., Wilkinson, M & Sneath, P.H.A. 1998. An arthropod phylogeny based on fossil and recent taxa. In: Edgecombe, G.D. (ed.), *Arthropod Fossils and Phylogeny*: 33-105. New York: Columbia Univ. Press.

Wilson, K., Cahill, V., Ballment, E. & Benzie, J. 2000. The complete sequence of the mitochondrial genome of the crustacean *Penaeus monodon*: are malacostracan crustaceans more closely related to insects than to branchiopods? *Mol. Biol. Evol.* 17: 863-874.

Resolving arthropod relationships: Present and future insights from evo-devo studies

STEVEN HRYCAJ & ALEKSANDAR POPADIĆ

Department Of Biological Sciences, Wayne State University, Detroit, Michigan, U.S.A.

ABSTRACT

In the past two centuries, the field of arthropod phylogeny has been the subject of intense discussion. Traditionally, relationships based on morphology and fossil evidence of the four major arthropod lineages have suggested a closer relationship between myriapods and insects to the exclusion of crustaceans. It was also generally recognized that the chelicerates branched off as a basal group. However, recent molecular studies analyzing sequence data strongly contradict these groupings, and instead suggest the following relationships: (i) [insects + crustaceans], and (ii) [chelicerates + myriapods]. As is evident from this lack of congruence, future resolution of arthropod relationships must rely upon a re-evaluation of traditionally assigned morphological homologies. The field of evolutionary developmental biology (evo-devo) has the potential to accomplish this by emphasizing the developmental mechanisms governing formation of particular morphological features. For example, previous studies of gene expression patterns have revealed that all arthropod mandibles are gnathobasic. As a result of these analyses, this feature (mandibular composition) can no longer be used to group myriapods and insects. Future investigations should shift the focus to delineating the genetic mechanisms of structural development down to their most specific events, encompassing regulatory mechanisms to the level of individual target genes. These detailed genetic networks can then be used to establish true homologies of complex morphological traits such as tracheal systems and Malpighian tubules.

1 INTRODUCTION

Arthropods represent the most diverse animal phylum and the origins of this vast diversity have fascinated scientists for decades. Such high levels of morphological variation have also enabled arthropods to successfully inhabit nearly every ecological niche on earth. Although all arthropods share unifying features such as segmented bodies, jointed appendages, and a hard exoskeleton, they also exhibit distinct differences in the organization of their body plans. Based on these differences, we can recognize four major extant arthropod lineages: chelicerates, myriapods, crustaceans and insects. Even when these members of groups exhibit a shared feature such as a defined head region, selection has acted to modify the structures rendering them unique to each species in their respective subphyla. A full understanding of the origins of such complexity has the potential to enhance our understanding of the relationships among these main lineages.

For the past two centuries, arthropod phylogeny has been subject to vigorous discussion. Traditionally, relationships between the different arthropod subphyla have been primarily determined by analysis of fossil evidence and by comparison of adult structures. Although all combinations of relationships among the four groups have been proposed at one time or another, only the mandibulate theory and the 'TCC' view have been seriously considered (Kukalova-Peck 1992; Telford & Thomas 1995). The mandibulate theory unites the insects, myriapods, and crustaceans into the Mandibulata, with trilobites and chelicerates branching off as basal lineages (Fig. 1A). In contrast, the 'TCC' view unites the trilobites, crustaceans, and chelicerates, thereby separating them from the insects and myriapods. Although these two hypotheses differ substantially, they share a close relationship between the myriapods and insects. The grouping of the insects and myriapods as monophyletic is based on the presence of five shared derived characteristics: a tracheal system, Malpighian tubules, absence of appendages homologous to the second antennae of crustaceans, unbranched legs, and a mandible composed of a whole limb (Manton 1964). Based on these morphological similarities, insects and myriapods were thought by (Sharov 1966) to be sister groups and in the united into Atelocerata.

In the past decade, this close relationship between insects and myriapods has been challenged by molecularly based phylogenies (Boore et al. 1995; Cook et al. 2001; Friedrich & Tautz 1995; Gonzalez-Crespo & Morata 1996; Kusche & Burmester 2001; Regier & Shultz 1997). In 1995, a pair of studies comparing nuclear ribosomal gene sequences and gene rearrangements within mitochondrial genomes concluded that insects may be a sister group to crustaceans (Fig. 1B), not myriapods (Boore et al. 1995; Friedrich & Tautz 1995). This finding attracted a lot of attention, because it was the first in recent times to question the existence of Atelocerata.

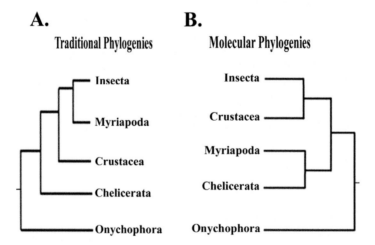

Figure. 1. Two models of arthropod relationships. (A) Classical textbook model with Insecta and Myriapoda forming 'Atelocerata' (Brusca & Brusca 1990). (B) An emerging view based on recent molecular data (Boore et al. 1998; Hwang et al. 2001; Kusche & Burmester 2001).

Subsequent analyses of mitochondrial gene rearrangements, and additional sequence comparisons between both nuclear elongation factors and homeotic (*Hox*) genes, all strongly supported a close insect/crustacean relationship (Boore et. al. 1998; Cook et. al. 2001, Regier & Schultz 1997). These analyses also suggested that myriapods and chelicerates may be sister groups (Fig. 1B). This second conclusion has been further corroborated by additional studies of a variety of gene sequences (Cook et al. 2001; Friedrich & Tautz 1995; Kusche & Burmester 2001).

Despite intense research on resolving arthropod relationships, a universally accepted phylogeny remained elusive. As is evident from the above summary, the emerging view from molecular based phylogenies is in direct conflict with the previously favored groupings, including the Atelocerata. This underscores the need to revisit the various morphological traits/structures used to argue for traditionally assigned homologies. The emerging field of evolutionary developmental biology (evo-devo) can offer unique insight into the possible resolution of arthropod relationships by focusing on developmental processes governing the formation of structure(s) under study. Comparing these genetic mechanisms in a wide array of arthropod taxa can therefore provide a significant contribution to the field of arthropod systematics.

2 EVO-DEVO SUCCESS STORY: THE ORIGIN OF ARTHROPOD MANDIBLES

Arthropod mandibles are feeding appendages functioning as 'jaws' and are used to bite and chew food (Manton 1964). The specific part of this appendage that directly manipulates food (i.e., limb base versus limb tip) has been used to support a close phylogenetic relationship among myriapods, crustaceans, and insects. As shown in Figure 2A, mandibles can be composed of either a whole limb or the basal portion only (Manton 1964). The jaws of insects and myriapods have been considered to be whole limbs (composed of both coxopodite and telopodite), manipulating food with the tip of the appendage (Manton 1964; Telford & Thomas 1995). In contrast, crustacean mandibles were thought to be gnathobasic (composed of coxopodite only), the result of either a reduction or absence of the distal portion of the appendage (Manton 1964; Telford & Thomas 1995). Thus, the structure of mandibles has been recognized as a key morphological feature uniting insects and myriapods into Atelocerata (Sharov 1966).

Because current molecular phylogenies do not support the Atelocerata concept, there is an increasing need for a critical re-examination of traditionally assigned character states (Boore et al. 1995; Cook et al. 2001; Friedrich & Tautz 1995; Kusche & Burmester 2001; Regier & Shultz 1997). In the case of mandibles, this means to unambiguously determine which arthropods have a mandibular telopodite and which do not. The principal limitation of the traditional approach is that it lacks an in depth understanding of the developmental processes and genetic mechanisms responsible for these differences. This may restrict our ability to fully elucidate the evolution of a particular trait in the arthropod under study. The recent move to integrate the fields of evolutionary and developmental biology holds promise for overcoming this problem. By using a set of novel molecular markers (cross-reacting antibodies to gene products that regulate various developmental processes) and comparing their expression patterns in developing animals, it is now becoming possible to better understand the molecular basis, and hence, the origins of particular morphological features.

In *Drosophila melanogaster* embryos and larval imaginal disks, both expression and functional analyses have revealed that the gene *Distal-less* (*Dll*) is essential for formation and development of the distal portion of all appendages (Gonzalez-Crespo & Morata 1996; Gorfinkiel et al. 1997). Moreover, subsequent comparative RNA inhibition experiments have indicated that the function of *Dll* is conserved across arthropods, from chelicerates to insects (Prpic et al. 2001; Schoppmeier & Damen 2001). Of equal importance, a cross-specific antibody that recognizes the DISTAL-LESS protein has been used successfully to study the expression of this gene in a variety of protostomes and deuterostomes (Panganiban et al. 1997; Popadić et al. 1998a). Both its specificity to distal portions of the appendages and its broad cross-reactivity make the *Dll* antibody an ideal molecular marker for delineating the true composition of arthropod mandibles. Coupled with its conserved function, an analysis of *Dll* expression provides a straightforward, unambiguous way of determining whether the distal portion of an appendage is present or not. This is critically important if one considers how difficult it is to infer the mandibular origins based solely on morphology. As an illustration, the dissected adult insect and crustacean mandibles are shown in Figures 2B and C. Even to a trained eye, these appendages appear very similar, and yet, in the former, the mandible is thought to consist of a whole limb, whereas in the latter it encompasses a limb-base only.

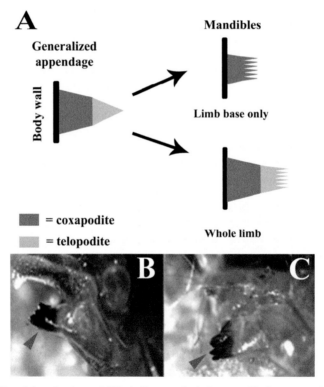

Figure 2. [See Plate 1 in color insert.] (A) A diagram depicting mandibular composition (whole limb versus limb base only). (B) Dissected adult mandibles of the pillbug, *Armadillidium vulgare*. (C) Dissected adult mandibles of the firebrat, *Thermobia domestica*. In B and C, arrowheads point to dissected mandibular appendages.

The developmental origins of arthropod mandibles have been the focus of several recent studies (Grenier et al. 1997; Popadić 1996; Popadić et al. 1998b; Scholtz et al. 1998), and a summary of *Dll* expression patterns in representative embryos of major arthropod groups is depicted in Figure 3. Among arthropods, chelicerates have a unique body plan that lacks a defined 'head' region (Brusca & Brusca 1990). Instead, their bodies consist of an anterior prosoma that bears all walking and feeding appendages, and a limbless posterior opisthosoma. It is now generally accepted that in other arthropod lineages, some of the appendages corresponding to chelicerate walking legs have become incorporated into the head and became transformed into mouthparts.

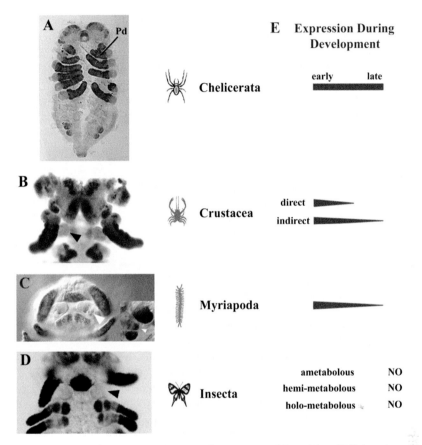

Figure 3. [See Plate 2 in color insert.] Expression patterns of *Distal-less* (*Dll*) in arthropod embryos. (A) Spider *Achaeranea tepidariorum*; note that although chelicerates do not have mandibles, it is thought that the pedipalpal segment (Pd) corresponds to it in other arthropods (Weygoldt 1979). (B) Pillbug, *Armadillidium vulgare*, a terrestrial isopod. (C) Millipede *Oxidus gracilis*, inset shows an earlier stage of development. (D) Firebrat, *Thermobia domestica*, a representative of a basal insect lineage. Arrowheads point to mandibles in myriapod, crustacean and insect embryos. (E) Diagram of *Dll* expression in the mandibular appendages in representative embryos of major classes of arthropods. Filled rectangles depict high, continuous expression of *Dll* from early to late development, whereas filled arrowheads indicate reduction of *Dll* expression in later developmental stages. There are no larval stages in chelicerates and myriapods. Modified from (Popadić et al. 1998b).

As shown for spider embryos, all of the limbs of the prosoma express *Dll* (Fig. 3A). This pattern persists throughout embryogenesis, and continues even after dorsal closure, to the stage just before hatching, when embryos resemble miniature adults (Popadić et al. 1998b). Thus, based on *Dll* expression, all six pairs of prosomal appendages are whole-limbs, consisting of both a basal coxopodite and a distal telopodite. This finding suggests that the ancestral state of all arthropod appendages (both walking and feeding) is of the whole limb type.

Among crustaceans, the vast majority of species undergo indirect development, which consists of a succession of larval stages, usually beginning with a nauplius (Schram 1986). These larvae are free living, feeding organisms that can be morphologically quite different from adults. Thus, if one is interested in adult morphology, there is always the question of whether and how much a particular larval feature contributes to adult form. Directly developing crustaceans, on the other hand, hatch from the egg as miniature adults. The nauplius and other larval stages are suppressed or occur sequentially within the egg. For this reason, studying directly developing terrestrial species such as isopods provides a way to circumvent the potential problem of larval/adult differences. *Dll* expression analysis in early embryos of several directly developing crustaceans has shown a greatly reduced level of mandibular expression when compared to that of the antennal and maxillary segments (Popadić et al. 1998b; Scholtz et al. 1998). This is followed by a complete absence of mandibular *Dll* expression (Fig. 3B) in later stages of development as the limb buds begin to extend (Popadić et al. 1998b). Late nauplius larvae in indirectly developing species also lack mandibular *Dll* expression (Popadić et al. 1998b). Together, these observations provide strong evidence for the gnathobasic nature of mandibles in all crustaceans.

As previously mentioned, it has long been recognized that both myriapods and insects exhibit a mandible composed of both proximal and distal portions (Manton 1964). If this is true, then *Dll* antibody staining in myriapod and insect embryos should reveal *Dll* expression in the distal portion of the mandibular appendage. In two myriapod taxa studied, millipedes and centipedes, *Dll* expression can be observed in mandibular primordia during early embryogenesis (Fig. 3C inset) (Grenier et al. 1997; Popadić et al. 1998a; Popadić et al. 1998b; Scholtz et al. 1998). However, as development progresses, the *Dll* signal gradually decreases (Popadić et al. 1998b). By late embryogenesis, *Dll* cannot be detected in the tips of the mandibles, and with the exception of a few cells in the middle, is essentially absent from the entire limb (Fig. 3C). The other appendages, however, continue to express *Dll* throughout embryogenesis. This difference in expression pattern suggests that millipede mandibles lack, or have only a vestigial telopodite (distal tip).

Studies of *Dll* expression in insect embryos were equally revealing (Popadić et al. 1998b; Scholtz et al. 1998). From those of basal lineages such as *Thermobia domestica* (firebrats) to those of highly derived flies, there is a complete absence of *Dll* expression in the mandibles (Fig. 3D). Furthermore, this lack of *Dll* expression is observed from very early to late developmental stages, showing that *Dll* is never turned on in these appendages. Coupled with genetic analysis in *Tribolium castaneum* embryos (Prpic et al. 2001), these data show that insect mandibles are missing the distal part of the appendage. Therefore, the analysis of *Dll* patterns in embryos of representative ateloceratan taxa have provided direct evidence that insect and myriapod mandibles are lacking the telopodite and are gnathobasic.

Thus, in direct contrast with the traditional view, these comparative studies show that adult mandibles of insects, crustaceans, and myriapods are gnathobasic (Popadić 1996; Popadić et al. 1998b; Prpic & Tautz 2003; Prpic et al. 2001). Also, the study of expression of another appendage gene, *dachshund* (*dac*), recently provided additional strong evidence for this new view. *dac* is required to establish the medial portion of arthropod appendages (Prpic & Tautz 2003), and consequently, is expressed in the middle of whole-limb appendages such as legs. However, in mandibles *dac* is localized to the tip, exactly where expected if one assumes that the telopodite is absent. Thus, as a direct result of expression analyses of developmental genes such as *Dll* and *dac*, the structure of the mandibles can no longer be used to unite insects and myriapods in Atelocerata to the exclusion of crustaceans.

3 ARTHROPOD APPENDAGES: EMERGING VIEW

Due to individual adaptations to a variety of functions, arthropods exhibit the greatest amount of appendage diversity within the animal kingdom in terms of size, shape, and leg anatomy (Beklemishev 1964; Brusca & Brusca 1990). Traditionally, the existence of shared derived anatomical features has been used to suggest closer relationships within the arthropods. For example, one of the characteristics unifying insects and myriapods into Atelocerata is the presence of unbranched (uniramous) legs, which are anatomically distinct from the branched (multiramous) legs observed in a few chelicerates and most crustaceans (Manton 1964; Telford & Thomas 1995). What are the genetic underpinnings of this diversity? Results of recent molecular studies provide a significant insight into the genetic basis of leg evolution in arthropods.

In the past two decades, through the efforts of numerous research groups, the developmental mechanisms governing leg patterning in *Drosophila* imaginal disks have been determined to a fairly detailed level (Abu-Shaar & Mann 1998; Cohen et al. 1989; Gonzalez-Crespo & Morata 1996; Lecuit & Cohen 1997; Mardon et al. 1994). Secretion of the signal molecules *decapentaplegic* (*dpp*) and *wingless* (*wg*) acting in a combinatorial mode, activate various downstream genes across the proximo-distal (PD) axis of the developing leg disc (Fig. 4A). High levels of *dpp* and *wg* expression in the distal portion of the appendage activate *Distal-less* (*Dll*), thereby causing formation of the distal portion of the tibia, tarsus, and pre-tarsus (Cohen et al. 1989; Gonzalez-Crespo & Morata 1996; Lecuit & Cohen 1997). Moderate levels of *dpp* and *wg* in the medial portion of the leg activate *dachshund* (*dac*), a gene responsible for formation of the femur, and proximal tibia (Lecuit & Cohen 1997; Mardon et al. 1994). The proximal portion of the developing appendage (coxa and trochanter), is regulated through the effects of two other leg patterning genes, *homothorax* (*hth*) and *extradenticle* (*exd*) (Lecuit & Cohen 1997). A comparison between the genes governing appendage development in embryos of the more basal species such as cricket (*Acheta domesticus*) and the highly derived flies (*D. melanogaster*) has revealed expression patterns and dynamics of these genes to be highly similar (Abzhanov & Kaufman 2000). In addition to this general conservation, subtle differences have been observed in embryos of the flour beetle *Tribolium castaneum* in which the *dac/Dll* region of overlap is much smaller than in *Drosophila* (Prpic et al. 2001). However, the overall basic expression patterns of the genes determining the proximal, medial and distal portions of developing appendages (*hth/exd*, *dac* and *Dll*, respectively) are a conserved feature throughout the insects. This finding further indicates that the *Drosophila* PD axis patterning mechanism as a whole (Fig. 4A) is conserved in most insects. Recent studies focused on determining the expression of leg patterning genes in embryos of other arthropods (myriapods, crustaceans, and chelicerates) indicate that PD axis patterning is generally conserved, although some individual species-specific differences occur (Abzhanov & Kaufman 2000; Prpic et al. 2003; Prpic & Tautz 2003; Prpic et al. 2001). It is now recognized that PD axis patterning predates the evolution of arthropods (Prpic et al. 2003).

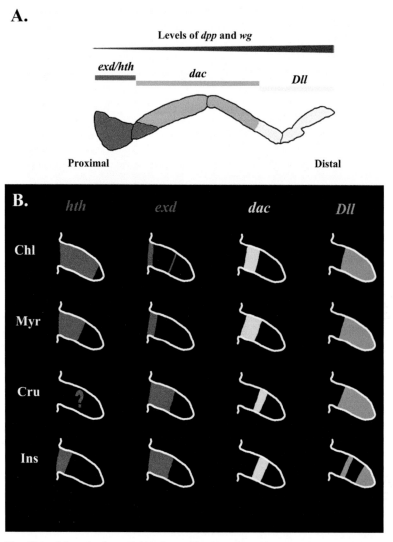

Figure 4. [See Plate 3 in color insert.] (A) Generalized leg PD axis patterning in *Drosophila melanogaster*. (B) Expression patterns of the four leg patterning genes in representatives of major arthropod lineages. The generated summary of expression patterns is based on the data from the following studies: (Abzhanov & Kaufman 2000; Prpic et al. 2003; Prpic & Tautz 2003; Prpic et al. 2001). The question mark denotes absence of data regarding *hth* expression in crustaceans. Abbreviations: Chl = chelicerates; Myr = myriapods; Cru = crustaceans; Ins = insects; *exd* = *extradenticle*; *hth* = *homothorax*; *dac* = *dachshund*; *Dll* = *Distal-less*.

When the expression domains of the major leg patterning genes in the embryos and larvae of various arthropods are compared, a number of similarities and differences are detected. This is evident in Figure 4B, which depicts the spatial expression of the four leg patterning genes in representatives of the major arthropod groups during early leg embryogenesis (approximately 30%). A closer examination of *Dll* expression reveals that insects exhibit a unique 'ring and sock' pattern (Abzhanov & Kaufman 2000; Popadić et al. 1998b). More specifically, the proximal domain of *Dll* expression is confined to the proximal portion of the femur (i.e., ring), while the distal domain encompasses distal half of the tibia and the entire tarsus with the exception of the pre-tarsus (i.e., sock). Consequently, there is a gap in expression between the tibia and femur. *Dll* expression in myriapod embryos begins proximal to that of insects, and encompasses the entire medial and distal portion of the appendage (Prpic & Tautz 2003). Likewise, in chelicerate and crustacean embryos, *Dll* expression is present in all podomeres except the most proximal (coxa in chelicerates; coxa and trochanter in crustaceans) (Abzhanov & Kaufman 2000; Grenier et al. 1997; Popadić et al. 1998b). Collectively, these data suggest that *Dll* patterning is conserved throughout the arthropods with a derived pattern in insects.

As previously mentioned, the *dac* gene specifies development of the medial portion of appendages. As is evident in Figure 4B, comparison of the *dac* expression among representative embryos of the major arthropod groups yields two findings: (i) similarity between insects and crustaceans, and (ii) between myriapods and chelicerates (Abzhanov & Kaufman 2000; Prpic et al. 2003; Prpic & Tautz 2003; Prpic et al. 2001). In insect embryos, *dac* is generally expressed in the femur, tibia and first tarsal segment (Prpic et al. 2001). In crustaceans, *dac* expression is confined to a single medial segment known as the merus (Abzhanov & Kaufman 2000). However, in chelicerate and myriapod embryos *dac* expression is restricted to trochanter and femur (Abzhanov & Kaufman 2000). Thus, insects and crustaceans exhibit medial *dac* expression, while myriapods and chelicerates have more proximal patterns. This indicates that *dac* may serve as a good candidate gene for comparative studies of arthropod development and has a potential to resolve further relationships between the major groups. Future comparison of regulatory mechanisms controlling *dac* expression in arthropod embryos could yield significant insights into the evolution of arthropod leg patterning.

Throughout the investigated arthropod embryos, *exd* is one of two genes that are responsible for the formation of the proximal portion of developing appendages. Comparison of *exd* expression patterns among arthropods can be used to contribute further to understanding of phylogenetic relationships. As is evident in Figure 4B, the expression boundaries of *exd* in crustaceans and insects are similar to each other, as are those in chelicerates and myriapods. In insects and terrestrial crustaceans (isopods), *exd* expression starts proximally. Later in development, *exd* expression extends significantly more distally than in chelicerate and myriapod embryos (Abzhanov & Kaufman 2000; Prpic et al. 2003). Chelicerate *exd* expression is observed as a single stripe in the medial portion of the developing appendage with a second, more proximal domain (Prpic et al. 2003). Myriapod expression is similar, but lacks the band of medial expression (Prpic & Tautz 2003). However, it is important to recognize that the above data is based on expression at approximately 30% of development, when specific leg segments are still unrecognizable in most arthropod embryos.

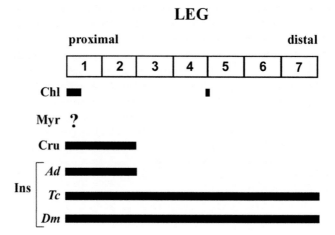

Figure 5. Expression of *extradenticle* (*exd*) in specific leg podomeres during later stages of development. Based on the data from (Abzhanov & Kaufman 2000; Prpic et al. 2003; Rauskolb et al. 1995). The question mark denotes the absence of currently available information regarding *exd* expression in distinct leg segments in myriapods. Abbreviations: Chl = chelicerates; Myr = myriapods; Cru = crustaceans; Ins = insects; *Ad* = *Acheta domestica*; *Tc* = *Tribolium castaneum*; *Dm* = *Drosophila melanogaster*.

More precise data on *exd* expression in later developmental stages in which leg segmentation is distinguishable has recently become available for several species (Fig. 5). Within the insects, two patterns have been detected: (i) in embryos of the more derived, holometabolous insects such are *Tribolium* and *Drosophila*, *exd* expression is ubiquitous, encompassing all podomeres, and; (ii) in those more basal, hemimetabolous insects such as *Acheta domesticus*, *exd* expression is restricted to the coxa and trochanter (Abzhanov & Kaufman 2000). In crustacean embryos, *exd* expression is also restricted to the coxa and basis (Abzhanov & Kaufman 2000). Contrastingly, in chelicerate embryos, the medial pattern of *exd* is found in a single stripe around the joint separating the tibia and patella, with proximal expression confined to the proximal portion of the coxa (Prpic et al. 2003). The specific podomeres in which *exd* is expressed in myriapod embryos has yet to be determined. Although incomplete, available data of *exd* expression has revealed a shared expression pattern in both basal insects and crustaceans (Fig. 5). However, before any conclusions can be reached, it is important that equally detailed *exd* expression studies be performed on additional species throughout each subphylum, especially in myriapods.

Homothorax (*hth*), the last of the four major leg patterning genes to be discussed, has been analyzed in all major arthropod lineages with the exception of the crustaceans (Abzhanov & Kaufman 2000; Prpic et al. 2003; Prpic & Tautz 2003). Within the chelicerates (Fig. 4B), *hth* expression is present throughout the appendage except in the apex (Prpic et al. 2003). In contrast, insect *hth* expression is localized to the proximal-most portion of each developing leg (Prpic et al. 2003). Myriapod expression is intermediate to the other two patterns (Fig. 4B), extending more proximally than in insects, but also retracting more distally than in chelicerates (Prpic & Tautz 2003). As illustrated in Figure 4B, comparison of the mechanisms of spatial *hth* regulation between derived insect and the basal chelicerate embryos may provide a novel insight into the evolution of PD axis patterning. Disparities observed between the two groups could serve as meaningful reference points, with each difference representing two extreme ends of a continuum. This can then be used to determine if crustacean *hth* expression is similar to that of embryos of any other group or whether it is unique. Based on data to date, it seems as if *hth* expression retracts in a proximal direction in the embryos of the more recent arthropod taxa. This would suggest that the crustacean *hth* pattern may be more similar to the situation observed in insects. However, as previously stated, future detailed crustacean *hth* expression analyses will be essential in order to make such an inference.

The emerging view from the above comparative analyses indicates that studies of appendage development have a potential to provide a significant, in-depth understanding of appendage evolution in arthropods. Whereas the genes responsible for PD axis specification are generally conserved, there is also a fair amount of variation in their expression domains, and consequently, their regulation. Some of this variation, as is the case with the 'ring and sock' pattern of *Dll* in insects (fig. 4B), is unique to a particular lineage and would have no phylogenetic relevance. At the same time, the evo-devo studies have revealed a presence of shared variation in the observed expression patterns. For example, *exd* is localized in the first two leg podomeres in both basal insects and crustaceans (Fig. 5). *dac* expression patterns are also highly similar between insects and crustaceans, and quite different from that observed in myriapods (Fig. 4B). While highly indicative, these results also require a much better understanding of the origins of variation in gene regulation that exist in nature. To address this question, future evo-devo studies should shift focus from the gene expression analyses to the elucidation of the actual genetic mechanisms that govern the observed changes. Such research endeavors would not be trivial, but if they reveal a common regulatory mechanism, it would add a significant, independent support for a close relationship between insects and crustaceans.

4 NEXT CHALLENGE: ORIGINS OF COMPLEX FEATURES SUCH AS TRACHEAL SYSTEMS

Tracheal systems in most arthropods comprise a branching network of tubules that facilitate gas exchange in a terrestrial environment. The presence of a tracheal system also serves as an important phylogenetic character, as reflected in the name Tracheata, which is also used to describe the assembly of insects and myriapods (Brusca & Brusca 1990; Kraus & Kraus 1994). As previously mentioned, recent molecular evidence argues against this concept, suggesting instead that insects are more closely related to crustaceans (Boore et al. 1995; Friedrich & Tautz 1995; Hwang et al. 2001). The former implies the homology of tracheae in insects and myriapods, and thus a single origin in a common ancestor, whereas the latter suggests that these systems evolved independently. Distinguishing between these two hypotheses can be greatly facilitated by the understanding of the developmental basis and genetic architecture of arthropod tracheogenesis.

The respiratory system in most insects consists of a network of branched epithelial tubes ramifying throughout the body. The branching structure is organized in three levels: primary, secondary and terminal (Ohshiro et al. 2002; Sutherland et al. 1996). The terminal branches end close to or within tissues, directly delivering oxygen to and removing carbon dioxide from them (Brusca & Brusca 1990). The system is bilaterally symmetrical, although each individual tube may be of different length and/or diameter. The cellular mechanisms of tube formation have been shown to differ in each of the three stages of branching (Ohshiro et al. 2002; Sutherland et al. 1996). The myriapod tracheal system is morphologically similar to that of insects, although some differences exist. Among the main myriapod lineages, most of Chilopoda (centipedes) have branched trachea, whereas most of Diplopoda (millipedes) have segmental clusters of unbranched respiratory tubules (Hilken 1997; Snodgrass 1935). Insight of tracheal systems in chelicerates is primarily based on studies of spiders. Only advanced spiders have one or two pairs of tubular tracheae which can branch throughout the body (Foelix 1996). Furthermore, some species possess only very short tubes, while others have highly branched tubes that pervade the prosoma and even the extremities. Most crustaceans are marine and use gills for gas exchange. However, terrestrial forms are characterized by the presence of thin-walled, blind ending sacs (pseudotracheae) in which the diffusion of gasses transpires. In summary, the presence of some form of tracheae is not restricted to insects and myriapods, and such structures are found in other terrestrial arthropods (e.g., arachnids) and even onychophorans. This fact highlights the need to obtain the detailed understanding of the molecular aspects of tracheogenesis in order to discuss the origins of tracheal systems in various arthropods.

The *Drosophila* tracheal system is the only one within the arthropods with its developmental genetic basis thoroughly studied. In the past decade, these studies have identified key genes essential for tracheogenesis in this insect. As illustrated in Figure 6, tracheogenesis is initiated when protein kinase B (PKB) interacts with *tracheales* (*trh*), a gene encoding a bHLH-PAS transcription factor (Jin et al. 2001). Embryos in which *trh* has been knocked out never experience the initial invagination event, resulting in the ectodermal cells of the tracheal placode remaining at the surface and with no formation of a tracheal system (Jin et al. 2001). *Ventral veinless/drifter* (*vvl/dfr*), another gene which encodes the POU-domain transcription factor CF1a, is also required for tracheogenesis (Bradley & Andrew 2001). The transcription factors encoded by both *trh* and *vvl/dfr* then directly regulate the activation of *branchless* (*bnl*), an insect fibroblast growth factor (FGF) homolog (Bradley & Andrew 2001; Sutherland et al. 1996). *Branchless* then directly activates the gene *breathless* (*btl*) whose effects are essential for primary branch formation (Bradley & Andrew 2001; Ohshiro et al. 2002; Sutherland et al. 1996). Mutations of *btl* in embryos result in a fully internalized but undifferentiated sac of ectodermal cells (Ohshiro et al. 2002). Recently, it has been found that the *decapentaplegic* (*dpp*), *epidermal growth factor receptor* (*EGFR*) and *wingless* (*WG/WNT*) signaling pathways also affect primary branching by aiding the specific migration of primary branches (Bradley & Andrew 2001). While expression of additional genes are required for proper tracheogenesis, it is the effects of *trh* and *vvl/dfr* that drive the cascade of events (Fig. 6). Due to the close similarity of tracheal structure and function in the vast majority of other insects, the *Drosophila* respiratory system can likely serve as a model representing tracheal system development in all insects. It is this level of understanding that needs to be achieved for tracheogenesis in other arthropod embryos.

The research on tracheal structure and function represents a well studied aspect of arthropod biology which has been treated in great detail by a number of authors (Dohle 1985; Fahlander 1938; Hilken 1997). For the purpose of this article, it is sufficient to say that there is a diversity of views with regard to tracheal origins. These views range from strongly supportive of the single origin in insects and myriapods (Wägele & Stanjek 1995) to ones postulating that tracheae originated independently at least four (Kraus 1998), and perhaps as much as six times (Dohle 1988). We may begin to resolve these conflicting views by assessing the complexity of the molecular regulation used to develop a system of invaginated ectodermal tubes with respiratory function. By this we mean characterizing the genes and their interactions in the pathway, and assessing their presence or absence in diverse taxa. Are all the genes present in all taxa, and hence ancestral? Are parts of the pathway used in other developmental contexts or are they novel? It must be recognized that it is now generally accepted that morphological novelties arise by the tinkering of already existing developmental machinery, and not by generating developmental networks *de nuovo* (Wilkins 2002). For example, development of both the *Drosophila* tracheal system and mammalian lungs is partly regulated through the same signaling pathway suggesting a common scheme for patterning branching morphogenesis (Metzger & Krasnow 1999). Thus, it is likely that some similarities will be observed when comparing details of insect and myriapod tracheogenesis. Nevertheless, we should still be able to detect an appreciable degree of regulatory differences if these groups are not sister taxa. The opposing view, favoring Atelocerata, could also be further tested by studying genetic mechanisms of tracheogenesis in chelicerates such as spiders. This is because advanced spiders possess a tracheal system that is evolutionarily distinct from the ones present in insects and myriapods. Thus, similarities observed between the chelicerate and insect/myriapod systems can be attributed to developmental canalization. Any remaining similarities between insects and myriapods would then likely reflect common ancestry. Future evo-devo studies of this kind have a potential to elucidate the origins of arthropod tracheal systems.

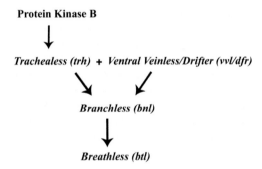

Figure 6. Genetic regulation of tracheogenesis in *Drosophila melanogaster*, based on data from (Bradley & Andrew 2001; Jin et al. 2001; Ohshiro et al. 2002; Sutherland et al. 1996).

5 CONCLUSION

In the past decade, primarily by comparing gene expression patterns, the field of evolution of development has offered a new insight into the evolution of arthropod body plans. With regard to arthropod systematics, several highly focused investigations have now revealed that all adult arthropod mandibles are gnathobasic in nature (Grenier et al. 1997; Popadić et al. 1998b; Scholtz et al. 1998). As a consequence, this character (mandibular composition) can no longer be used to group insects and myriapods together to the exclusion of crustaceans. Another recent study comparing brain morphologies has provided new insight into the phylogenetic position of another arthropod, that of remipede crustaceans. Since their discovery in 1979, remipedes were considered a basal, proto-crustacean lineage (Schram 1986). However, it has now been revealed that the remipede brain is highly organized and well differentiated at a level of complexity matched only by the brain of 'higher' crustaceans (Malacostraca) and Hexapods (Fanenbruck et al. 2004). This surprising result therefore argues in favor of a remipede-malacostracan-hexapod clade. Collectively, these studies highlight the potential of utilizing novel approaches to help further clarify arthropod relationships.

Our motivation for writing this article was to provide a brief, 'evo-devo centric' perspective on how to resolve the issue of homologies of morphological traits that have been traditionally used in arthropod systematics. As more data emerges, the complexity of the developmental mechanisms governing the formation of various arthropod features (such as tracheae and Malpighian tubules) has become much more apparent. As exemplified by appendage development, the genetic cascade of events responsible for structural formation is known to include many target genes and numerous regulatory events. It is therefore necessary that future studies of evo-devo reach beyond the comparison of expressional domains of just key genes and instead focus on delineating these developmental pathways in their entirety. Only by determining the genetic mechanisms of structural development down to their lowest levels (encompassing specific regulatory events, target genes, etc.) in key taxa, will we be able to reach a true understanding of the origins of the key morphological features in arthropods.

ACKNOWLEDGEMENTS

The authors would like to thank E.M. Golenberg for his thoughtful comments. We also thank Alessandro Minelli and an anonymous reviewer for the meticulous reading and commenting of the manuscript. This work was in part supported by a grant from the National Institutes of Health to A.P.

REFERENCES

Abu-Shaar, M. & Mann, R.S. 1998. Generation of multiple antagonistic domains along the proximodistal axis during *Drosophila* leg development. *Development* 125: 3821-3830.

Abzhanov, A. & Kaufman, T.C. 2000. Homologs of *Drosophila* appendage genes in the patterning of arthropod limbs. *Dev. Biol.* 227: 673-689.

Beklemishev, W. 1964. *Principles of the Comparative Anatomy of Invertebrates*. Chicago: Univ. Chicago Press.

Boore, J.L., Collins, T.M., Stanton, D., Daehler, L.L. & Brown, W.M. 1995. Deducing the pattern of arthropod phylogeny from mitochondrial DNA rearrangements. *Nature* 376: 163-165.

Boore, J.L., Lavrov, D.V. & Brown, W.M. 1998. Gene translocation links insects and crustaceans. *Nature* 392: 667-668.

Bradley, P.L. & Andrew, D.J. 2001. *ribbon* encodes a novel BTB/POZ protein required for directed cell migration in *Drosophila melanogaster*. *Development* 128: 3001-3015.
Brusca, R. & Brusca, G. 1990. *Invertebrates*. Sunderland: Sinauer Assoc.
Cohen, S.M., Bronner, G., Kuttner, F., Jurgens, G. & Jackle, H. 1989. *Distal-less* encodes a homoeodomain protein required for limb development in *Drosophila*. *Nature* 338: 432-434.
Cook, C.E., Smith, M.L., Telford, M.J., Bastianello, A. & Akam, M. 2001. Hox genes and the phylogeny of the arthropods. *Curr. Biol.* 11: 759-763.
Dohle, W. 1985. Phylogenetic pathways in the Chilopoda. *Bijdr. Dierk.* 55: 55-66.
Dohle, W. 1988. *Myriapoda and the Ancestry of Insects*. Manchester, UK: The Manchester Polytechnic.
Fahlander, K. 1938. Beiträge zur Anatomie und systematischen Einteilung der Chilopoden. *Zool. Bidr. Uppsala* 17: 1-148.
Fanenbruck, M., Harzsch, S. & Wägele, J.W. 2004. The brain of the Remipedia (Crustacea) and an alternative hypothesis on their phylogenetic relationships. *Proc. Nat. Acad. Sci. USA* 101: 3868-3873.
Foelix, R. 1996. *Biology of Spiders*. New York: Oxford Univ. Press.
Friedrich, M. & Tautz, D. 1995. Ribosomal DNA phylogeny of the major extant arthropod classes and the evolution of myriapods. *Nature* 376: 165-167.
Gonzalez-Crespo, S. & Morata, G. 1996. Genetic evidence for the subdivision of the arthropod limb into coxopodite and telopodite. *Development* 122: 3921-3928.
Gorfinkiel, N., Morata, G. & Guerrero, I. 1997. The homeobox gene *Distal-less* induces ventral appendage development in *Drosophila*. *Genes. Dev.* 11: 2259-2271.
Grenier, J.K., Garber, T.L., Warren, R., Whitington, P.M. & Carroll, S. 1997. Evolution of the entire arthropod Hox gene set predated the origin and radiation of the onychophoran/arthropod clade. *Curr. Biol.* 7: 547-553.
Hilken, G. 1997. Vergleich von Tracheensystemen unter phylogenetischem Aspekt. *Verhandl. Naturwiss. Ver. Hamburg, N.F.*, 37.
Hwang, U.W., Friedrich, M., Tautz, D., Park, C.J. & Kim, W. 2001. Mitochondrial protein phylogeny joins myriapods with chelicerates. *Nature* 413: 154-157.
Jin, J., Anthopoulos, N., Wetsch, B., Binari, R.C., Isaac, D.D., Andrew, D.J., Woodgett, J.R. & Manoukian, A.S. 2001. Regulation of *Drosophila* tracheal system development by protein kinase B. *Dev. Cell* 1: 817-827.
Kraus, O. & Kraus, M. 1994. Phylogenetic system of the Tracheata (Mandibulata): on 'Myriapoda'-Insecta interrelationships, phylogenetic age and primary ecological niches. *Verhandl. Naturwiss. Ver. Hamburg* 34: 5-31.
Kukalova-Peck, J. 1992. The "Uniramia" do not exist: The ground plan of the Pterygota as revealed by Permian Diaphanopterodea from Russia. *Can. J. Zool.* 70: 236-255.
Kusche, K. & Burmester, T. 2001. Diplopod hemocyanin sequence and the phylogenetic position of the Myriapoda. *Mol. Biol. Evol.* 18: 1566-1573.
Lecuit, T. & Cohen, S.M. 1997. Proximal-distal axis formation in the *Drosophila* leg. *Nature* 388: 139-145.
Manton, S. 1964. Mandibular mechanisms and the evolution of arthropods. *Phil. Trans. R. Soc. London, Ser. B*, 247: 1-183.
Mardon, G., Solomon, N.M. & Rubin, G.M. 1994. *dachshund* encodes a nuclear protein required for normal eye and leg development in *Drosophila*. *Development* 120: 3473-3486.

Metzger, R.J. & Krasnow, M.A. 1999. Genetic control of branching morphogenesis. *Science* 284: 1635-1639.

Ohshiro, T., Emori, Y. & Saigo, K. 2002. Ligand-dependent activation of *breathless* FGF receptor gene in *Drosophila* developing trachea. *Mech. Dev.* 114: 3-11.

Panganiban, G., Irvine, S.M., Lowe, C., Roehl, H., Corley, L.S., Sherbon, B., Grenier, J.K., Fallon, J.F., Kimble, J., Walker, M., Wray, G.A., Swalla, B.J., Martindale, M.Q. & Carroll, S.B. 1997. The origin and evolution of animal appendages. *Proc. Nat. Acad. Sci. USA* 94: 5162-5166.

Pisani, D., Poling, L.L., Lyons-Weiler, M. & Hedges, S.B. 2004. The colonization of land by animals: molecular phylogeny and divergence times among arthropods. *BMC Biol.* 2: 1.

Popadić, A. 1996. Origin of the arthropod mandible. *Nature* 380: 395.

Popadić, A., Abzhanov, A., Rusch, D. & Kaufman, T.C. 1998a. Understanding the genetic basis of morphological evolution: the role of homeotic genes in the diversification of the arthropod bauplan. *Int. J. Dev. Biol.* 42: 453-461.

Popadić, A., Panganiban, G., Rusch, D., Shear, W.A. & Kaufman, T.C. 1998b. Molecular evidence for the gnathobasic derivation of arthropod mandibles and for the appendicular origin of the labrum and other structures. *Dev. Genes Evol.* 208: 142-150.

Prpic, N.M., Janssen, R., Wigand, B., Klingler, M. & Damen, W.G. 2003. Gene expression in spider appendages reveals reversal of *exd/hth* spatial specificity, altered leg gap gene dynamics, and suggests divergent distal morphogen signaling. *Dev. Biol.* 264: 119-140.

Prpic, N.M. & Tautz, D. 2003. The expression of the proximodistal axis patterning genes *Distal-less* and *dachshund* in the appendages of *Glomeris marginata* (Myriapoda: Diplopoda) suggests a special role of these genes in patterning the head appendages. *Dev. Biol.* 260: 97-112.

Prpic, N.M., Wigand, B., Damen, W.G. & Klingler, M. 2001. Expression of *dachshund* in wild-type and *Distal-less* mutant *Tribolium* corroborates serial homologies in insect appendages. *Dev. Genes Evol.* 211: 467-477.

Rauskolb, C., Smith, K.M., Peifer, M. & Wieschaus, E. 1995. *extradenticle* determines segmental identities throughout *Drosophila* development. *Development* 121: 3663-3673.

Regier, J.C. & Shultz, J.W. 1997. Molecular phylogeny of the major arthropod groups indicates polyphyly of crustaceans and a new hypothesis for the origin of hexapods. *Mol. Biol. Evol.* 14: 902-913.

Scholtz, G., Mittmann, B. & Gerberding, M. 1998. The pattern of *Distal-less* expression in the mouthparts of crustaceans, myriapods and insects: new evidence for a gnathobasic mandible and the common origin of Mandibulata. *Int. J. Dev. Biol.* 42: 801-810.

Schoppmeier, M. & Damen, W.G. 2001. Double-stranded RNA interference in the spider *Cupiennius salei*: the role of *Distal-less* is evolutionarily conserved in arthropod appendage formation. *Dev. Genes Evol.* 211: 76-82.

Schram, F.R. 1986. *Crustacea*. New York: Oxford Univ. Press.

Sharov, A.G. 1966. *Basic Arthropodan Stock*. New York: Pergamon.

Snodgrass, R.E. 1935. *Principles of Insect Morphology*. New York: McGraw-Hill Book Comp., Inc.

Sutherland, D., Samakovlis, C. & Krasnow, M.A. 1996. *branchless* encodes a *Drosophila* FGF homolog that controls tracheal cell migration and the pattern of branching. *Cell* 87: 1091-1101.

Telford, M. & Thomas, R. 1995. Demise of Atelocerata? *Nature* 376: 123-124.

Wägele, J.W. & Stanjek, G. 1995. Arthropod phylogeny inferred from 12S rRNA revisited: monophyly of Tracheata depends on the sequence alignment. *J. Zool. Syst. Evol. Res.* 33: 75-80.

Wilkins, A.S. 2002. *The Evolution of Developmental Pathways*. Sunderland: Sinauer Assoc.

Color insert

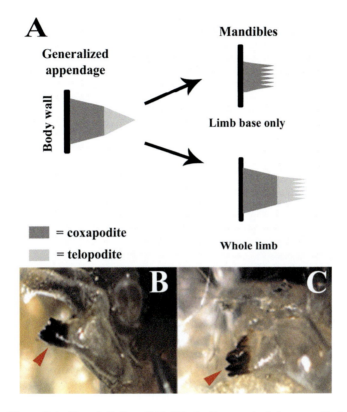

Plate 1. [See Figure 2 in Hrycaj & Popadić.] (A) A diagram depicting mandibular composition (whole limb versus limb base only). (B) Dissected adult mandibles of the pillbug, *Armadillidium vulgare*. (C) Dissected adult mandibles of the firebrat, *Thermobia domestica*. In B and C, arrowheads point to dissected mandibular appendages.

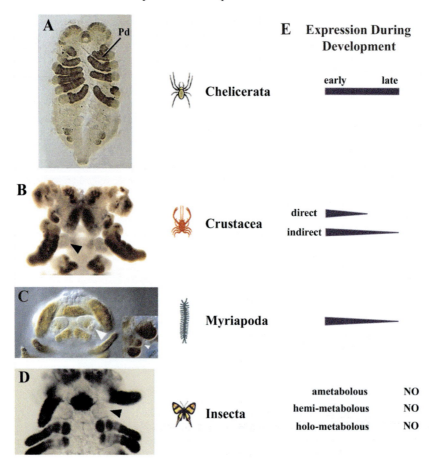

Plate 2. [See Figure 3 in Hrycaj & Popadić.] Expression patterns of *Distal-less* (*Dll*) in arthropod embryos. (A) Spider *Achaeranea tepidariorum*; note that although chelicerates do not have mandibles, it is thought that the pedipalpal segment (Pd) corresponds to it in other arthropods (Weygoldt 1979). (B) Pillbug, *Armadillidium vulgare*, a terrestrial isopod. (C) Millipede *Oxidus gracilis*, inset shows an earlier stage of development. (D) Firebrat, *Thermobia domestica*, a representative of a basal insect lineage. Arrowheads point to mandibles in myriapod, crustacean and insect embryos. (E) Diagram of *Dll* expression in the mandibular appendages in representative embryos of major classes of arthropods. Filled rectangles depict high, continuous expression of *Dll* from early to late development, whereas filled arrowheads indicate reduction of *Dll* expression in later developmental stages. There are no larval stages in chelicerates and myriapods. Modified from (Popadić et al. 1998b).

Color insert

A.

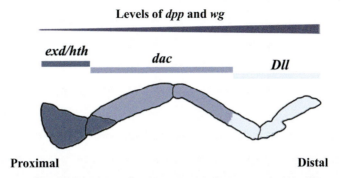

B.

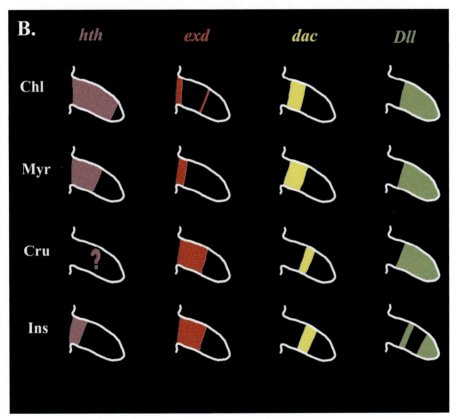

Plate 3. [See Figure 4 in Hrycaj & Popadić.] (A) Generalized leg PD axis patterning in *Drosophila melanogaster*. (B) Expression patterns of the four leg patterning genes in representatives of major arthropod lineages. The generated summary of expression patterns is based on the data from the following studies: (Abzhanov & Kaufman 2000; Prpic et al. 2003; Prpic & Tautz 2003; Prpic et al. 2001). The question mark denotes absence of data regarding *hth* expression in crustaceans. Abbreviations: Chl = chelicerates; Myr = myriapods; Cru = crustaceans; Ins = insects; *exd* = *extradenticle*; *hth* = *homothorax*; *dac* = *dachshund*; *Dll* = *Distal-less*.

IV COMPARATIVE MORPHOLOGY

Evolution of eye structure and arthropod phylogeny

COLETTE BITSCH & JACQUES BITSCH

Evolution & Diversité Biologique, Université Paul Sabatier, Toulouse, France

ABSTRACT

A review of the main cytological types of arthropodan eyes leads us to suggest their possible evolutionary changes across the major groups of arthropods. Simple eyes of the ocellus type, typically with many photoreceptor cells beneath a single corneal lens, display plesiomorphic features which have probably been inherited from some non-arthropodan phylum. Ocellus-type eyes are present in many chelicerate lineages, both in the median and the lateral position on the prosoma, while only median ocellus-type eyes have been retained in most mandibulates. Compound eyes consisting of numerous contiguous highly organized ommatidia are found in many arthropod lineages. They most probably derived from simple ocellus-like eyes. The ommatidia of Mandibulata appear to be more evolved than those of Xiphosura, having acquired a more elaborate dioptric apparatus including a crystalline cone, and retinulae typically composed of eight photoreceptor cells. In different mandibulate taxa, a regressive evolution of the compound eyes has led to variously modified visual organs, such as 'dispersed eyes' and more highly modified stemmata which are found in particular in many holometabolous insect larvae. The evolution of the lateral eyes in some hexapods may even have led to ocellus-like lateral eyes, for instance in *Cicindela* larvae and in some adult insects exhibiting unicorneal eyes. The latter cases thus appear to result from a reversal to a plesiomorphic state. The lateral simple eyes present in several myriapod groups appear to be either similar to isolated and simplified ommatidia, or not very different from more complex stemmata such as those found in certain insect larvae.

The data and suggestions reported are used in a discussion about the phylogenetic relationships among the major arthropod groups. Compound eyes were acquired very early by several extinct arthropodan lineages, such as Trilobita and Eurypterida, and it can be assumed that compound eyes not very different from those of living Xiphosura were already present in the hypothetical euarthropodan ancestor. In this view, the ocellus-type lateral eyes of arachnids would have derived from compound eyes of the xiphosuran type. In an alternative hypothesis, the simple lateral eyes of arachnids might be regarded as conserved ancient structures, but in this case the compound eyes of Xiphosura and of some extinct lineages would have evolved independently from the compound eyes of Mandibulata. The great similarities between the compound eyes of Crustacea and Hexapoda have been used as an argument to support the assemblage of the two groups in the presumed clade Pancrustacea (= Tetraconata), excluding Myriapoda. However, the compound eyes of the scutigeromorph centipedes can be regarded as derived from the ground pattern of mandibulate eyes, and the simple eyes of other chilopods and of diplopods present an organization similar to the stemmata of various insect larvae. Thus, the structure of the eyes in Myriapoda supports the placement of this group within the Mandibulata and is compatible with the clade Atelocerata (= Tracheata) which is found in most cladistic analyses based on morphological data and in some combined analyses using morphological and molecular data.

This paper is dedicated to Prof. Frederick R. Schram on occasion of his retirement and in homage to his fundamental contribution to crustacean biology and arthropod phylogeny.

1 INTRODUCTION

The phylogenetic relationships between the major arthropod groups have been discussed for a long time and remain intensely debated. Among the data upon which the phylogenetic analyses were based, numerous morphological features were used, such as gross morphology, ultrastructural observations and developmental studies. The visual systems, including the different types of eyes and the elaborate optic centers of the brain, have been intensively studied by many workers to infer phylogenetic relationships amongst the arthropods. Various investigations of the late 19th century and early 20th century, using histological observations, concerned the comparative structure of the eyes in many arthropod taxa. They have been summarized in several textbooks, in particular in those of Snodgrass (1935, 1952), Weber (1933) and Bullock & Horridge (1965) and they have shown the great structural diversity of the visual organs in arthropods, depending on the taxonomic groups and also in relation to the lighting of the animal's environment, the behavior of each species and the types of vision. Since 1960, many electron microscopic investigations have provided new cytological details, especially about the photoreceptor cells. Several of these publications will be cited throughout the present paper, but it is appropriate to mention here the substantial contributions and reviews presented by Paulus (1979, 1986, 2000). Many recent investigations have more specially concerned the development of the visual organs and their underlying genetic mechanisms.

The present review is focused on the comparative cyto-morphology of arthropod eyes and does not consider the optic centers of the protocerebrum. A brief survey of the different types of eyes suggests new interpretations of the evolutionary changes having affected arthropod eyes. Then, these data and conclusions are compared with the different phylogenetic hypotheses on arthropod relationships that have been presented during the last decades.

2 THE DIFFERENT STRUCTURAL TYPES OF ARTHROPOD EYES

Arthropod visual organs present a considerable diversity in number, position and structure. They have been the subject of numerous morphological and functional investigations. Considering only their structural organization, the visual organs of arthropods can be divided into a few types that should be briefly redefined due to divergent meanings that have been given to the names used for their description. The traditional division between simple and compound eyes is clear: simple eyes have a single dioptric apparatus for a variable number of receptor cells, whereas compound eyes have many dioptric apparatuses, each corresponding to a visual unit, or ommatidium.

Compound eyes, well developed in most arthropod groups, are typically composed of many contiguous similar ommatidia, each ommatidium appearing as a highly differentiated visual unit (Fig. 1). However, compound eyes may be reduced to various extents in different taxa, especially in soil and cave-inhabiting species, or in parasitic species. Some taxonomic groups present *dispersed lateral eyes* in which the ommatidia, in a restricted number, are distinctly separated from each other.

Within the *simple eyes*, two major types can be distinguished by their cytological characteristics. *Ocelli* have a dioptric apparatus made of a single cuticular lens which covers a retina composed of numerous receptor cells whose distal processes have differentiated arrays of microvilli forming the rhabdomeres. The receptor cells of an ocellus may be contiguous, their connected rhabdomeres forming a continuous, netlike rhabdom (Fig. 3A), or the receptor cells are arranged into clusters designated as retinulae, each retinula containing a cylindrical or star-like rhabdom in an axial position. In the latter case, the different rhabdoms are separated from each other (Fig. 3B). Ocellus-type eyes may occupy a median or a lateral position on the head.

Stemmata are most often made up of a single retinula whose structure is little different from that found in the ommatidia of compound eyes. In fact, stemmata look like separate and variously modified ommatidia. The dioptric apparatus typically consists of a biconvex corneal lens and a vitreous or crystalline body built up of four Semper cells. The retina is most often composed of a variable number of photoreceptor cells arranged around an axial rhabdom (Fig. 2B-D). Stemmata are always found laterally on the head. They were first described in holometabolous insect larvae, but stemma-like eyes, as they are defined here, also occur in some other taxa, for instance in myriapods (Fig. 2E, F).

The usual distinction between median and lateral eyes depends not only on their position on the cephalic region, but also on their connection with the brain. The photoreceptor cells of the median eyes project their axons into a median neuropil of the protocerebrum, the median eye center, whereas the lateral eyes send their axons to the optic centers located laterally on each side of the protocerebrum.

2.1 Compound Eyes

The compound eyes consist of numerous structural and functional units, the ommatidia, on each side of the head. They are present in living xiphosuran Chelicerata, in most Crustacea and Hexapoda, and in scutigeromorph Myriapoda. They were also present in ancient groups, such as the Trilobita, Eurypterida and some diplopodan Myriapoda. Their comparative structure is abundantly documented (for reviews, see in particular Paulus 1979, 2000; Meinertzhagen 1991; Caveney 1998). Thus, only a few general comments are made here, special attention is being given to some particular cases for phylogenetic purposes.

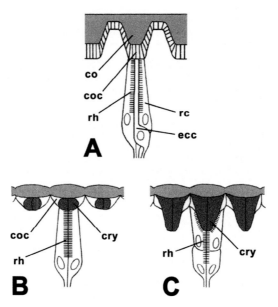

Figure 1. Schematic representation of ommatidia in compound eyes of various arthropods. (A) *Limulus polyphemus* (Xiphosura), simplified after Fahrenbach 1975. (B) Basic type of ommatidium of Crustacea and Insecta, after Paulus 2000. (C) *Scutigera coleoptrata* (Chilopoda), after Müller et al. 2003. For clarity, the pigment cells have not been represented, and only the nuclei of the retinula cells have been drawn. Abbreviations: co = cornea or corneal lens (in light grey); coc = corneagenous cells; cry = crystalline cone (in dark grey); ecc = eccentric and arhabdomeric cell; rc = retinula cells; rh = rhabdom.

2.1.1 *Compound eyes of* Limulus

Among recent Chelicerata, only the Xiphosura (*Limulus* and related genera) possess a pair of well-developed lateral compound eyes in addition to a pair of median eyes. The *Limulus* compound eyes have been described in particular by Demoll (1914) and in a series of papers by Fahrenbach (1968, 1969, 1975, 1999). The cuticle covering each compound eye has a smooth external surface, but internally, it forms honeycomb-like protrusions (called the 'cuticular cones') evoking the corneal lenses of facetted eyes (Fig. 1A).

Numerous translucent corneagenous cells lie beneath each 'lens', but no crystalline cone is present. The photoreceptor units corresponding to the 'lenses' are isolated from each other by pigment cells. Each visual unit, or retinula, is composed of about 10-13 rhabdomeric cells which are radially arranged around the axial and arhabdomeric dendritic process of a neuron. This neuron has been designated as the 'eccentric cell' owing to the lateral position of its cellular body in the retinula; it is considered as autapomorphic for Xiphosura. Fahrenbach (1999) states that "double eccentric cells in one ommatidium are not uncommon".

2.1.2 *Compound eyes of Crustacea and Hexapoda*

Most Crustacea and Insecta possess lateral compound eyes which display many common patterns. Typically, each ommatidium includes a dioptric apparatus which consists of a biconvex cuticular lens secreted by two corneagenous cells (called primary pigment cells in insects) and a second non-cuticular lens produced by four underlying cells (tetrad of Semper cells) composing a quadripartite crystalline cone (Fig. 1B). Several accessory pigment cells, the distal and proximal ones, isolate each ommatidium from the neighboring photoreceptor units. Each retinula typically includes eight receptor cells which form a common rhabdom aligned along the optical axis of the ommatidium. The rhabdom is generally of the fused type, or sometimes of the open type.

The developmental sequences leading to the formation of compound eyes have been studied in great detail in the fruit fly *Drosophila melanogaster* (Tomlinson 1988; Dickson & Hafen 1993; Wolf & Ready 1993; Freeman 1997; Matsuo et al. 1997; Frankfort & Mardon 2002; Voas & Rebay 2004). Some other studies have compared the eye development of the hemimetabolous *Schistocerca* and of the holometabolous *Tribolium* with that of *Drosophila* (Friedrich et al. 1996; Friedrich & Benzer 2000; Friedrich 2003; Liu & Friedrich 2004). Each ommatidium develops as a cluster of eight prospective retinula cells (R-cells) assembled by a process of recruitment involving intercellular contacts. A central cell, noted R8, is first specified and behaves as a founder cell for the ommatidium by initiating the sequential recruitment of the remaining retinula cells; the cells R2 and R5, R3 and R4, R1 and R6 are added in pairs, then R7 is recruited. Afterwards, the four cone cell precursors and the pigment cells become incorporated into the ommatidial unit. Lastly, the retinula cells project axons into the optic neuropil.

In the branchiopod crustaceans *Triops* and *Lepidurus*, the ommatidial pattern formation corresponds "in an astonishing manner to that of insects" (Melzer et al. 1997, 2000). Also in the eye of the decapods *Procambarus* and *Homarus*, the pattern formation and the ommatidial differentiation present remarkable similarities with those of *Drosophila* and other insects (Hafner & Tokarski 1998, 2001).

Thus, the compound eyes of crustaceans and insects share many structural similarities, especially the basic arrangement of the cells and the highly specialized components of the ommatidia appear to be strongly conserved in both groups. Developmental studies suggest that the highly specific patterning processes of eyes are likely to be conserved in both groups.

2.1.3 *Some variations in the ommatidia of Crustacea and Hexapoda*
The cuticle, which obligatorily covers all arthropod eyes, usually forms the hexagonally edged and biconvex facets of the compound eyes. However, in the eyes of different crustacean taxa, as in Cumacea and Amphipoda (Hallberg 1977; Hallberg et al. 1980) and in Branchiura (Meyer-Rochow et al. 2001), the cuticular cornea is not thickened into a lens.

The underlying crystalline cone is typically of the 'eucone' type, formed by four cone cells arranged in radial quadrates around the optical axis of the ommatidium. However, the crystalline cones of Ascothoracida and Cirripedia are built by only three cells (Hallberg et al. 1985), while five cone cells have been reported in cladocerans and a number varying from 3 to 5 in anostracans (Wolken & Gallik 1965; Hallberg 1977). In various malacostracan groups, such as Anaspidacea, Euphausiacea, Lophogastrida and Mysida, Richter (1999) noted the presence of bipartite crystalline cones produced by two cone cells, while two accessory cone cells are not involved, or to a minor extent, in the formation of the cone. In insects, authors traditionally distinguished several types of ommatidia depending on the structure of the crystalline cone. A true cone is absent from the so-called 'pseudocone eyes'; in the 'acone eyes' the Semper cells are optically clear. A variable number of cone cells, ranging from 3 to 6, has been reported for instance in the ommatidia of *Ephestia kuehniella* (Fischer & Horstmann 1971).

The common ground pattern of the ommatidium of Crustacea and Hexapoda presumably includes eight retinula cells (Melzer et al. 1997; Diersch et al. 1999). Eight retinula cells build up a fused rhabdom in *Triops* and *Lepidurus*. In these genera, the cells referred to as R1-R6 are symmetrically aligned along the optical axis of the ommatidium and hence can be seen as three pairs of cells exhibiting bilateral symmetry. In contrast, the two other cells, R7 and R8, are asymmetrically arranged. The R7-cell forms its rhabdomere only in the distal region adjacent to the cone, and its pericaryon is also distally located. In fact the number of retinula cells per ommatidium varies from four to eleven depending on the crustacean taxa. In many malacostracans, for example in the isopod *Ligia*, the ommatidium consists of seven regular retinula cells and an eighth eccentric cell (Keskinen et al. 2002). In the eye of most mysids, each ommatidium is made up of seven 'regular' retinular cells and an eighth 'aberrant' cell forming a distal rhabdom; however, this R8 cell is absent in some genera (Hallberg 1977).

The insect ommatidium usually consists of eight retinula cells forming a fused rhabdom (Melzer et al. 1997; Caveney 1998). However, the number of retinula cells may vary according to the taxa: only 6 or 7 in some species, 9 in the bee (Eisen & Youssef 1980), 10 to 13 in *Ephestia kuehniella* (Fischer & Horstmann 1971); the number of retinula cells per ommatidium may vary from 7 to 21 in the passalid beetles (Gokan & Meyer-Rochow 2000).

The arrangement of the rhabdomeres shows a high diversity according to varied visual adaptations. A closed rhabdom, in which the nearby rhabdomeres meet along the ommatidial axis, is regarded as an ancestral structure (Fischer et al. 2000). A banded rhabdom organization, in which layers of microvilli are arranged perpendicular to each other, is known in decapod crustaceans and in few insect species, such as the archaeognathan *Allomachilis*, the coleopteran *Creophilus* and the lepidopteran *Pieris* (Meyer-Rochow 1971). The occurrence of a banded rhabdom in decapods and in some hexapod species may result from a convergence in relation to some special function in light reception. An open rhabdom, in which most of the rhabdomeres are spaced away from the optic axis, is regarded as an apomorphic character. Open rhabdoms have been found in various insect groups, e.g. Dermaptera, Coleoptera and brachyceran Diptera (Caveney 1998). In Heteroptera, the open rhabdom is produced by six peripheral retinula cells and by two central cells (Fischer et al. 2000).

Thus, the typical pattern of the mandibulate ommatidium may present different variations which concern the number and arrangement of the cone cells and the receptor cells, and the organization of the rhabdom. In addition, many mandibulate taxa have atypical eyes which will be mentioned below in the section dealing with the dispersed eyes and stemmata.

2.1.4 *Compound eyes in Scutigeromorpha*

Among extant Myriapoda, only the Scutigeromorpha possess compound eyes that have been described in particular by Hanström (1934) and by Paulus (1979). The recent investigations of Müller et al. (2003) give further information and make corrections to some previous observations. The eye of *Scutigera coleoptrata* is composed of about 150 ommatidia, and that of *Thereuopoda clunifera* of about 600 visual units. In *Scutigera*, each ommatidium has a biconvex corneal lens and eight to ten pigment cells that are in direct contact with the corneal lens. These corneagenous cells, despite their enhanced number, are presumed to be homologous to the hexapod primary pigment cells. In most ommatidia four cone cells have built up a large crystalline cone (Fig. 1C) that is subdivided into eight compartments, but at the periphery of the eye, some ommatidia have an additional cone cell and the crystalline cone may present up to ten compartments. The nuclei of the cone cells are located in cellular processes which pass between the retinular cells and extend up to the basal lamina. The presence of such inter-retinular processes has also been reported in various crustacean and hexapodan ommatidia (Melzer et al. 1997). In *Scutigera*, a distal layer of 9-12 retinula cells forms an open rhabdom, whereas a proximal layer of 4 cells forms a closed rhabdom. In addition, an average of 15 interommatidial pigment cells separate the ommatidia from each other. Thus, despite their higher number of photoreceptor cells, the ommatidia of the scutigeromorphs present a common pattern with those of hexapods and crustaceans, including in particular four cone cells and an axial rhabdom.

2.2 Dispersed lateral eyes and stemmata

In this section we gather different cases of modified eyes ranging from visual organs similar to compound eyes but with dispersed ommatidia, to simple stemmata-type eyes, with transitional examples that may be observed in a single taxonomic group.

2.2.1 Reduced and dispersed eyes in some adult Crustacea and Insecta

A reduction in size and structure of compound eyes is frequent in soil-inhabiting, cave, deep-sea or parasitic forms. Among Crustacea, the terrestrial Isopoda have small sessile compound eyes composed of a restricted number of widely spaced ommatidia (Nemanic 1975; Nilsson 1978). Each ommatidium of *Porcellio scaber*, for instance, has a dioptric apparatus consisting of a crystalline cone apparently made up of only two cone cells; the retinula consists of seven radially arranged sensory cells and a central process that, according to Nemanic (1975), is an extension of an eighth cell whose cellular body is located in an apical position, near the crystalline cone. All eight retinula cells contribute to the formation of the fused rhabdom. Several isopod families comprise cave or parasitic species that present variously reduced eyes. In the Trichoniscidae, which include many cave taxa, the different species can be distributed into several groups according to their eye structure: in some species the eyes are made of three separate ommatidia, in others of only one modified ommatidium resembling a stemma of some holometabolous insect larvae, while in other species the eye is replaced by a clump of 6-8 little differentiated cells (Lattin 1939).

A reduction in size of the compound eyes, resulting from a decrease in the number of ommatidia, is also found in adults of various insect species. Among the carabid beetles, for instance, different degrees in the regression of the eyes are known, from large normal eyes to complete disappearance of the eyes (Bernard 1932).

2.2.2 Lateral eyes of Collembola and Zygentoma

The species of these two groups of apterygote hexapods (when they are not blind) have lateral eyes which appear as separate ommatidia. Already present in immature stages, the eyes grow during development by the formation of additional ommatidia. Among the remaining apterygote hexapods, Protura and Diplura are eyeless, whereas Archaeognatha possess large facetted eyes.

A comparative study of the eye structure in several genera of Collembola has been put forward by Paulus (1972, 1977, 1979) using ultrastructural observations. There are at most eight simple eyes on each side of the head, but in many genera their number is reduced or the eyes are absent. Each simple eye has a structure similar to that of a pterygote ommatidium. The dome-shaped cornea is only slightly thickened. Four Semper cells form a crystalline body that can be non-segmented or quadripartite. Two corneagenous cells are probably homologous with the primary pigment cells usually present in insect ommatidia. The retinula usually consists of eight photoreceptor cells, arranged either in only one layer (Fig. 2A), or in two layers. The rhabdom is composed of eight rhabdomeres, or less, depending on the genera; the rhabdomeres often show a complicated arrangement and do not present a strict radial symmetry. In *Entomobrya muscorum*, for example, a distal layer of five retinular cells forms an open rhabdom, at the center of which lies the rhabdomeric process of a sixth, eccentric cell; the two remaining cells, numbered 7 and 8, occupy a central and more proximal position (Paulus 1977). The evolution of the eyes within the different families of Collembola has involved a series of reductions having successively affected the number of ommatidia, the dioptric apparatus, the rhabdom and finally the optic center (Barra 1973). An extreme reduction of the dioptric apparatus, observed in the eyes of *Anurida maritima*, has led to the loss of the crystalline cone (Paulus 1979).

The lateral eyes of the Lepismatidae consist of 12 separate ommatidia, while in the presumed primitive family Lepidothrichidae they contain between 40 and 50 ommatidia with slightly separate lenses (Elofsson 1970; Paulus 1974). In the lepismatids, each ommatidium includes a biconvex cornea and a crystalline cone made up of four Semper cells. Each retinula consists of seven photoreceptor cells arranged in two layers. A distal layer of four rhabdomeric cells forms a square rhabdom around a central area which, in *Lepisma saccharina*, is occupied by the apical process of a poorly-defined sensory cell (an 8th sensory cell or an extension of one of the proximal receptor cells?). According to Brandenburg (1960) the central process of *Lepisma* differentiates a rhabdomere on one of its sides, while in *Ctenolepisma* rhabdomeres are present on all sides of the central process (Paulus 1974). The proximal layer of the retinula consists of three receptory cells forming a tri-branched rhabdom[1].

---

[1] The ommatidium depicted in figure 6.13 of the paper of Paulus (1979) has been erroneously assigned to *Thermobia domestica*, but it does not correspond to the description given in that paper, nor to figure 18 given by Paulus (1974) for *Lepisma saccharina*. The ommatidium depicted on figure 6.13 is more probably a stemma of some holometabolous larva, maybe of a coleopteran larva.

Evolution of eye structure and arthropod phylogeny 193

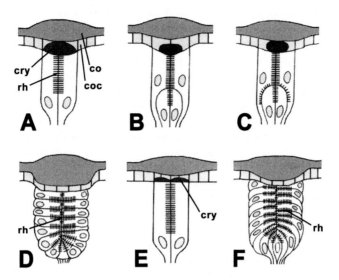

Figure 2. Schematic representation of stemmata and stemma-like simple lateral eyes in some hexapods and in two myriapods. (A) *Dicyrtoma ornata* (Collembola), simplified after Paulus 1974. (B) Larva of *Mamestra brassicae* (Lepidoptera), after Gilbert 1994. (C) Larva of *Lesteva* sp. (Coleoptera), after Paulus 2000. (D) Larva of *Gyrinus natator* (Coleoptera), after Weber 1933. (E) Lateral simple eye of *Polyxenus lagurus* (Diplopoda), after Spies 1981. (F) Lateral simple eye of *Polybothrus fasciatus* (Chilopoda), after Bedini 1968. Same abbreviations as in Figure 1.

2.2.3 *Stemmata in holometabolous insect larvae*

Most holometabolous insect larvae possess between one and seven lateral simple eyes on each side of the head. These eyes, which disappear at the time of metamorphosis while compound eyes develop from imaginal discs, are usually designated as stemmata (reviews in Paulus 1986, 2000, and in Gilbert 1994) or more rarely as lateral ocelli (Toh & Tateda 1991). In fact, two main types of lateral simple eyes should be distinguished within holometabolous insect larvae on the basis of their cytological features. Simple eyes consisting of a single retinula with an axial rhabdom, are regarded here as stemmata *s. str.*, whereas other simple eyes having a receptor epithelium, which presents either a netlike rhabdom or several separate rhabdoms, are regarded as ocelli. This second type of simple eye will be considered in a following section.

Stemmata are present in the larvae of many holometabolous orders, such as Mecoptera, Neuropteroidea, Coleoptera, Hymenoptera, Trichoptera, Lepidoptera and Diptera. Only a few examples are given here. The stemmata of mecopteran larvae still present a structure little different from that of the ommatidia of compound eyes. In the larvae of *Panorpa*, for instance, each stemma has a cuticular lens, a crystalline cone formed by four Semper cells surrounded by two pigment cells, and a retinula made up of eight receptor cells arranged in two layers of four cells forming a two-tiered central rhabdom (Gilbert 1994). Similar stemmata have been described in detail by Melzer et al. (1994) in the larvae of the mecopteran genus *Nannochorista*. In caterpillars (lepidopteran larvae), each stemma (Fig. 2B) presents an almost spherical corneal lens, a crystalline cone formed by only three Semper cells, and a retinula including 6-8 photoreceptor cells arranged in two layers around an axial rhabdom (Lin et al. 2002). In coleopteran larvae, several subtypes of stemmata can be found, sometimes within a single family, genus or species. A crystalline cone is still present in some taxa, but lacking in others. The retinula can have a single layer of receptor cells, or it is most often two- or multilayered, then composed of an increased number of receptor cells. In *Lesteva* (Staphylinidae) larvae, each retinula is made of two layers of receptor cells and of a bilayered rhabdom whose proximal part appears branched (Fig. 2C). In *Cantharis* larvae, each stemma is composed of two retinulae beneath a single biconvex corneal lens, each retinula consists of 7-8 receptor cells arranged in two layers. In *Drilus* larvae, the stemmata are composed of three retinulae, each consisting of more than 30 receptor cells. An example of multilayered retinula is to be found in the larvae of *Gyrinus* (Fig. 2D).

2.2.4 Simple lateral eyes of Myriapoda

Several chilopod lineages possess simple lateral eyes whose structure is similar to that of certain stemmata of insect larvae, not of ocelli, according to the definitions used in the present paper. Their number on each side of the head varies depending on the family and the species. Fifteen visual units are present in the centipede *Polybothrus fasciatus* (Bedini 1968) and 40 in *Lithobius forficatus* (Bähr 1974), whereas there is only one pair of simple eyes in Henicopidae and Craterostigmomorpha. The Cryptopidae and Geophilomorpha are eyeless. Each stemma of *Lithobius* presents a biconvex cornea, two corneagenous cells, and a deeply sunken retina. The numerous photoreceptor cells (up to 110) are arranged in several superposed layers. The distal receptor cells form a common axial rhabdom that appears as massive and cylindrical as represented by Bähr (1974), but star-like in cross section as represented by Paulus (2000). The proximal receptor cells have differentiated microvilli on all sides of their distal process. The structure of the simple eyes of *Polybothrus*, as described by Bedini (1968), presents a remarkable similarity with the stemmata of *Gyrinus* larvae (compare figs. 2F and 2D).

The eyes of Diplopoda appear as simple lenses arranged in several rows. The number of lenses can reach 25 on each side, but the Polydesmoidea lack eyes. The visual units may be well separated, or closely contiguous, but all present a common general structure which has been studied in particular, using electron microscopy, by Bedini (1970) in *Glomeris* and by Spies (1981) in several species of different families. In the Chilognatha, numerous receptor cells, from 25 to about 500 depending on the species, have built up an axial, cylindrical or stellate, and multilayered rhabdom. The case of *Polyxenus lagurus* is especially interesting because the number of retinular cells is only seven, four distal and three proximal, forming a fused, bilayered rhabdom (Fig. 2E). Moreover, Spies (1981) reports the presence in the *Polyxenus* eye of two modified epidermal cells located between the corneagenous cells and the distal border of the rhabdom. Each of these two cells contains "lentiform, translucent enclosures of a vitreous nature". Spies stresses the similarity of the *Polyxenus* visual units with the simple eyes of Poduromorpha (Collembola) and Zygentoma, and with the stemmata of Lepidoptera and Trichoptera larvae. In further observations of the eye of *Polyxenus*, Paulus (2000) found up to four vitreous cells instead of two, and he noted the presence of "cone cell roots which [are] normally a continuation of Semper cells in a mandibulate ommatidium". These additional observations reinforce the homology of these modified cells with Semper cells and hence support the view that the visual units of Diplopoda are derived from a typical mandibulate ommatidium.

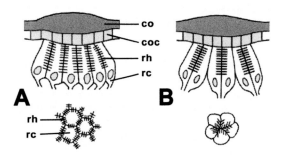

Figure 3. Schematic representation of simple eyes of the ocellus type. (A) Ocellus with a net-like rhabdom. (B) Ocellus consisting of retinulae, each with an axial rhabdom. Above, longitudinal section of the ocellus; below, transverse section at the level of retinula cells. Same abbreviations as in Figure 1.

2.3 *The ocelli*

Different kinds of ocelli are found in arthropods. Most Chelicerata have both median and lateral eyes of the ocellus type. According to the arrangement of the photoreceptor cells in the retina, a netlike rhabdom may be present (Fig. 3A), as in the eyes of various spiders (Homann 1971; Uehara et al. 1977; Paulus 1979), or separate rhabdoms occupy the axis of the numerous retinulae (Fig. 3B), as in the lateral eyes of jumping spiders (Eakin & Brandenburger 1971) and in the lateral eyes of most scorpions (Spreitzer & Melzer 2003). The retina may be of everse (direct) type, the rhabdomeric processes being directed toward the light, or inverse (indirect) type, the rhabdoms occupying a proximal position in the retina, the nuclei being located more distally. In the mite *Neocarus texanus* and in the scorpion *Parabuthus transvaalicus*, Kaiser & Alberti (1991) and Spreitzer & Melzer (2003) have successively described the presence of arhabdomeric cells that project between the retinula cells, as do the so-called eccentric cells of *Limulus* compound eyes.

The nauplius eyes, present in larvae of most groups of Crustacea, and in some adults, correspond to other types of median ocelli. Their structure varies depending on the crustacean order (Elofsson 1963, 1965, 1966, 1992). The dioptric apparatus is reduced or absent. The number of receptor cells within an eye cup varies from only a few to several hundred. The rhabdom can be netlike, as in *Artemia salina* (Rasmussen 1971), or irregular. In the nauplius of the copepod *Doropygus seclusus*, for example, the median eye is composed of three contiguous ocelli, each presenting only three functional retinula cells and six 'primordial' cells, but one of the functional cells degenerates at the copepodit stage, so that eight retinula cells are present in each ocellus of the adult (Dudley 1969). A number of 9 retinula cells in each dorsal ocellus and 5 to 10 in the ventral ocellus have been reported in other copepod species (Vaissière 1961; Fahrenbach 1964).

The frontal ocelli of Hexapods (Insecta *s. lat.*) are present in the adults of most orders and in immatures of many non-holometabolous insects. They exhibit a great diversity in number and structure depending on the taxon (Goodman 1970, 1981; Yoon et al. 1996). The cornea may be thin or thickened and form a biconvex lens. A layer of corneagenous cells, often flat but sometimes elongated, lies beneath the corneal lens. A crystalline cone is usually lacking, although a cellular biconvex structure, called a 'crystalline body' has been reported for example in the frontal ocellus of *Cloeon* (Hesse 1901; Goodman 1970). However, determining the true nature of this structure would require further investigations. The number of receptor cells varies from a few cells to over 10,000. In some cases, the receptor cells produce a netlike rhabdom, but usually they are arranged into clusters of 2 to 7 cells forming isolated rhabdoms which appear Y-shaped, X-shaped, or star-shaped in cross section.

In Hexapoda, the term ocellus is generally reserved for the median simple eyes, whereas the term stemma is applied to simple lateral eyes (Gilbert 1994). However, this view does not take into account the cytological structure of the simple eyes. In some holometabolous insect species, the larval simple eyes, although usually called stemmata, display cytological features which characterize the ocelli according to the definition adopted in the present paper (Fig. 3). In these larval eyes, the numerous receptor cells are no longer arranged in a cup with a large axial rhabdom, as is the case of the stemmata of certain other insect larvae, but they are radially arranged and form a netlike rhabdom. The two types of larval eyes, the stemmata and ocellus-type eyes, may exist in different species within the same taxonomic group. This is the case of some coleopteran larvae, such as *Cicindela* (Friederichs 1931; Toh & Tateda 1991; Toh & Mizutani 1994), *Atheta* (Paulus 2000) and *Lytta* (Heming 1982), of the megalopteran *Protohermes* (Yamamoto & Toh 1975), of the Tenthredinidae larvae (Hesse 1901; Meyer-Rochow 1974; Paulus 1979) and of some larvae of nematoceran Diptera (Paulus 1979).

The lateral eyes of the adults of some insect orders or families, in the past depicted as 'unicorneal eyes', also have a structure similar to that of ocelli. Such atypical eyes are known in particular in some ectoparasitic taxa, as Siphonaptera, Mallophaga and Anoplura. The lateral eye of *Ceratophyllus gallinae* (Siphonaptera) comprises more than a hundred receptor cells which form an irregular, netlike rhabdom (Wachmann 1972a). The head of the Mallophaga bears two cuticular lenses on each side (Amblycera) or only one lens (Ischnocera). Each corneal lens covers one, or most often two retinulae, each consisting of 6-8 sensory cells forming an axial branching rhabdom; there is no trace of a crystalline cone (Wundrig 1936). Thus, it appears that the simple eyes of Mallophaga present transitional features between stemmata and true ocelli. In *Pediculus corporis* (Anoplura), the retina consists of about 50 adjoining visual cells forming a netlike rhabdom (Wundrig 1936). Ocellus-like eyes are also found in some Coccoidea (Homoptera). While the presumed primitive families, such as the Margarodidae and Ortheziidae, have facetted eyes, the males of the more evolved Coccidae only possess simple lateral eyes. For example, the big unicorneal eyes of the male of *Pseudococcus* have numerous receptor cells that present microvilli regularly arranged around their adjoining distal segment, so forming a netlike rhabdom (Paulus 1979).

In the endoparasitic Strepsiptera, the winged males have big eyes that are composed of large, round lenses separated by rows of microtrichia. The number of the lenses varies from about 10 to 150 depending on the genus (Kinzelbach 1967, 1971). Each convex corneal lens corresponds to a simple eye. A 'pseudocone' made of cuticle is well developed in *Stylops*, but is reduced in other genera, and absent in *Xenos*. The retina includes between 50 and 100 receptor cells forming a deep cup. In *Xenos*, the rhabdom has been described as "an irregular, roughly hexagonal continuous mesh" (Buschbeck et al. 2003), whereas in *Stylops* the rhabdom appears as a cylindrical structure surrounding a granular extracellular material (Wachmann 1972b). So, the dispersed eyes of the Strepsiptera are composed of visual units which appear to be highly modified ommatidia having acquired the cytological structure of ocelli.

Lastly, a very interesting example is that of the ants of the genus *Eciton*. The males have large compound eyes, whereas the queen and workers have only one small lateral eye on each side. Each worker's eye consists of a biconvex corneal lens, a corneagenous epithelium and a retina in which numerous clusters of usually ten receptor cells each form an axial rhabdom (Werringloer 1932). So, the small lateral eyes of female adult *Eciton* present features characteristic of ocelli with retinulae. A study of the development of this atypical eye, from the pupal stage, shows the initial formation of a number of rudimentary, but well-recognizable ommatidia, each with four transparent cells similar to Semper cells and two cells similar to corneagenous cells (Werringloer 1932). However, crystalline cones are never differentiated. Then, the further development of the ommatidia is stopped, the Semper cells are no longer discernible and a unique corneal lens is formed above the retina that acquires the characteristics of the adult simple eye. Thus, the transformation of a prospective compound eye into an ocellus-like organ takes place during the metamorphosis of female *Eciton*.

3 ORIGIN AND EVOLUTION OF ARTHROPOD EYES

The views developed in the following sections are summarized in Figure 4.

3.1 *Origin and evolution of compound eyes*

The compound eyes of euarthropods are complex and highly specialized visual organs present in most extant lineages. Compound eyes were already present in the first known Trilobita of the Lower Cambrian, about 520 million years ago (Gal et al. 2000). Most trilobites had holochroal eyes composed of numerous small closely packed lenses, whereas the suborder Phacopina exhibited schizochroal eyes composed of large separate lenses (Cronier & Clarkson 2001). Another group of Paleozoic arthropods, the Eurypterida, assigned to the Chelicerata, possessed a pair of compound eyes. Some Paleozoic aquatic scorpions might also have possessed compound eyes. Large facetted eyes were present in some diplopod Myriapoda from the Upper Carboniferous (Spies 1981; Shear 1998). Although the internal structure of the eyes of these Paleozoic taxa has not been preserved, it is clear that facetted eyes and also eyes with dispersed lenses have a great antiquity among arthropods. Compound eyes have also been described in some non-arthropodan phyla, as in some polychaete annelids and bivalve molluscs (Nilsson 1994), but their ommatidia present a cell composition different from that of euarthropod ommatidia. Compound eyes with cuticular lenses and organized retinulae can be regarded as an autapomorphy of the Euarthropoda.

The structural complexity and high differentiation of the arthropod compound eye indicate that this organ is most probably derived from another, less specialized visual organ. The following data suggest that the compound eye has evolved from an ocellus-type eye consisting of a thin cuticular cornea, a corneagenous epithelium and a simple layer of photoreceptor cells arranged in several neighboring retinulae.

Ocelli are more ancient structures than arthropod-type compound eyes. Some examples of ocelli exhibiting a simple structure are known, for instance, in the nauplius eyes of crustaceans. In the case of the female ant *Eciton*, mentioned in the previous section, the dedifferentiation of an incipient compound eye with rudimentary ommatidia, at the time of the metamorphosis, shows that the same cellular material can express either the pattern of a compound eye or that of an ocellus. It can be suggested that a change in the expression of certain developmental genes, possibly in response to some physiological modifications at the time of metamorphosis, may explain this morphological transformation. Whatever the actual situation may be, the fact that the transformation of a developing compound eye into an ocellus-like eye occurs in the female *Eciton*, indicates that a reciprocal transformation of ocelli into compound eyes may also have taken place.

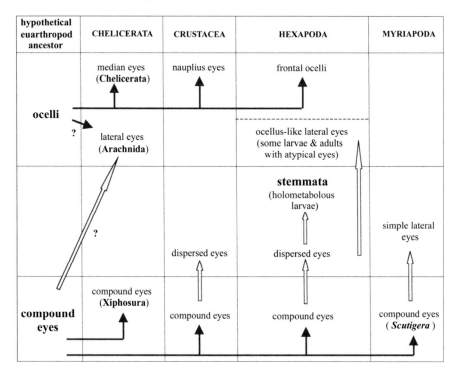

Figure 4. Possible evolution of the different types of visual organs in the major euarthropod groups from a hypothetical euarthropod ancestor. The major evolutionary pathways are indicated by black lines and arrows, while some regressive evolutions are indicated by white arrows (see the text for explanations).

The compound eyes of *Limulus* present several peculiarities, such as the cornea not divided into facets but with internally regular protrusions, the absence of crystalline cones, and the arrangement of the retinulae which include an eccentric arhabdomeric cell. However, the structure of the xyphosuran eyes is not basically different from that of mandibulate compound eyes; so, a common origin from some arthropod ancestor can be reasonably considered. Consequently, a homology of the compound eyes of Xiphosura and those of Mandibulata seems probable. Nevertheless, the ommatidia of mandibulate eyes display a more highly specialized pattern, probably resulting from a more advanced evolution: the cuticular corneal lenses have become individualized, a second lens (the crystalline cone) has been acquired, typically formed by four Semper cells, and the number of photoreceptor cells in each retinula has typically become stabilized to eight cells.

In fact, the different groups of mandibulates present a great diversity in the number and arrangement of the eyes components, to such an extend that a common origin of the compound eyes has been sometimes questioned. In the case of ostracod crustaceans, for example, most lineages lack compound eyes, with the exception of a subgroup, the Myodocopida. Phylogenetic analyses of Ostracoda, based on 18S and 28S rRNA genes and character reconstruction, indicate that compound eyes were probably absent in ostracod ancestors (Oakley & Cunningham 2002; Oakley 2003), and this suggests that the myodocopid eyes have evolved independently from those of other crustacean groups. In a more general way, it could be assumed that compound eyes in different arthropod lineages have been formed independently. But Oakley (2003) rather proposes the hypothesis of a switchback evolution according to which "genes that code for myodocopid compound eyes were maintained during ostracod evolution in the absence of the eyes themselves".

The eyes of *Scutigera* exhibit some particularities that explain that they have formerly been called 'pseudofacetted eyes' by Adensamer (1894), and that the view of their non-homology with the eyes of insects has been accepted by Paulus (1979, 2000). According to Paulus, the *Scutigera* eyes are probably derived from lateral stemmata similar to those observed in *Lithobius*. However, an alternative assumption seems to be more plausible, according to which the compound eye of the Scutigeromorpha corresponds to a plesiomorphic state for Myriapoda, while the stemmata of other myriapodan groups result in a regressive evolution from compound eyes. This conception is also supported by the reinvestigation of Müller et al. (2003) who have shown that *Scutigera* has indeed true compound eyes, probably homologous to the typical eyes of Crustacea and Insecta, even if some details are different, especially a large number of photoreceptor cells. The *Scutigera* eyes have most probably derived from some common mandibulate ancestor. Nevertheless, in order to confirm this view, further investigations on the development of *Scutigera* ommatidia, and a comparison with that of insect ommatidia, should be useful. The assumed plesiomorphic state of *Scutigera* eyes is also consistent with the most usual view according to which the Scutigeromorpha occupy a 'basal' position within the Chilopoda (Borucki 1996; Kraus 1998; Edgecombe 2004; Edgecombe & Giribet 2004).

It can be concluded that the euarthropods probably acquired compound eyes very early by transformation of simple, ocellus-type eyes which were already present in some ancestral non-arthropod lineage. The euarthropod ancestor might have possessed compound eyes little different in structure from those of recent Xiphosura and possibly from those of extinct taxa related to Chelicerata. A further evolutionary step has been achieved by the Mandibulata in which the compound eyes have acquired new specializations in relation with functional adaptations. The *Scutigera* eyes merely represent a variety of mandibulate compound eyes.

3.2 *Origin and evolution of dispersed eyes and stemmata*

Compound eyes reduced to a small number of separate visual units, whose structure can be similar to typical or to variously modified ommatidia, occur among many mandibulate lineages, most often in relation with a dark environment. The modifications of the eye structure, which may lead to totally eyeless forms, appear as physiological adaptations, and likely result from convergent evolution. When they are not eyeless, the species of two apterygote hexapod orders, the Collembola and Zygentoma, present different types of dispersed eyes that are clearly derived from the compound eyes that were presumably present in some hexapod ancestor. An interesting evolution of simple eyes within the Collembola has been suggested by Paulus (1979: fig. 6.12): from a typical hexapod ommatidium, a regressive evolution has successively affected the dioptric apparatus, the primary pigment cells and the retinula, leading to the various kinds of simple eyes observed in different genera. Reorganization of the retinula into two layers is also observed in the separate ommatidia of the lepismatids.

The different subtypes of stemmata found in holometabolous insect larvae can be regarded as being derived from typical ommatidia. Paulus (1986) has proposed some possible evolutionary pathways leading to the different types of larval stemmata within the Coleoptera (his fig. 8): the less modified stemmata still exhibit a crystalline cone and a reduced number of receptor cells, more modified stemmata lack a crystalline cone and have a single layer of retinula cells, lastly still more modified stemmata have a bilayered rhabdom built by two layers of receptor cells. In several coleopteran genera the distal receptor cells form a central rhabdom, while the less numerous proximal cells are arranged into two adjacent clusters, each forming a distinct rhabdom. Paulus has interpreted these special stemmata as resulting from the fusion of two or several modified ommatidia, which would account for the increased cell number and for the arrangement of the receptor cells in two or several layers.

In a detailed study on the embryonic development of the stemma-type larval eyes of *Tribolium*, Liu & Friedrich (2004) confirm the hypothesis that the stemmata of holometabolous larvae derive from prospective ommatidia. More precisely, the authors show that the two stemmata of each *Tribolium* lateral eye derive from developing photoreceptors lying at the posterior margin of the embryonic eye field, a position corresponding to that of the posterior-most ommatidia in the adult compound eye of non-holometabolous insects and probably of the insect ancestor. Taking into account the expression pattern of the gene *glass* used as a marker for the onset of photoreceptor cell differentiation, Liu & Friedrich observe the formation of five discrete cell clusters in the embryonic larval eye primordia. They claim that these primary clusters later assemble into two larger 'epiclusters' corresponding to the differentiated stemmata present on each side of the larval head. These observations seem to support the hypothesis of a fusion of several cell clusters to form a single stemma.

Except in the Scutigeromorpha which have compound eyes, the lateral eyes of Myriapoda, when present, are simple eyes of the stemma type. In each simple eye, the cup arrangement of the receptor cells forming an axial rhabdom, recalls the structure of the stemmata of certain holometabolous insect larvae, and likely results from convergent evolution. In fact, there is a great variety in the structure of the lateral eyes of myriapods. The presumably primitive eyes of *Polyxenus* concern a taxon which is usually regarded as the sister group to the remaining diplopods (Edgecombe 2004). In the *Polyxenus* eyes, each stemma is not very different from an ommatidium of a compound eye: it has a corneal lens, up to four reduced vitreous body cells and a retina of only seven cells forming a fused rhabdom. In the eyes of other diplopods and of chilopods, there is no longer any trace of vitreous body cells, and the photoreceptor cells whose number is often much greater, are arranged into two or several layers, but similar modifications are also found in the stemmata of different holometabolous insect larvae, and so the simple eyes of myriapods can be regarded as being derived from the ommatidia of compound eyes. The evolution of the lateral eyes, from mandibulate compound eyes to myriapodan simple eyes, is assumed to have involved a reduction in the number of ommatidia, the breaking up of the compound eye into separate ommatidia, the reduction of the crystalline cones, and the amplification and rearrangement of the retinula cells and of the rhabdom. The view according to which the myriapodan simple eyes are derived from mandibulate compound eyes was also accepted by Paulus (1979, 1986) and by Spies (1981).

In a more recent paper, Paulus (2000) takes into account new views on the phylogeny of arthropods, and he considers the possibility of an alternative interpretation, according to which the "Myriapoda eyes represent the ancestral/primitive state in the Mandibulata", whereas the compound eyes of Crustacea and Insecta are thought to be derived structures. However, in our opinion, the comparative data mentioned above rather support the Atelocerata hypothesis, according to which the myriapod eyes have derived from mandibulate compound eyes.

3.3 Origin and evolution of the ocelli

Simple eyes of the ocellus-type are widely distributed among most arthropod lineages, but they present a great diversity in number, position and structure, which has led to doubt about their homology in the different groups. Even within the Crustacea, Elofsson (1965, 1966) distinguished several types of nauplius eyes that may be not strictly homologous with each other. Median ocellus-type eyes which presumably occurred in the euarthropod ground pattern, have been most likely inherited from some other related phylum. Various types of pigment-cup ocelli are found in representatives of non-arthropodan phyla, such as Turbellaria, Annelida and Mollusca. Ocelli with a netlike rhabdom are present in particular in various polychaetes (Rhode 1991; Arendt et al. 2002) and in the pulmonate mollusc *Planorbarius* (Zhukov et al. 2002). A pair of lateral ocelli with a network-like rhabdom, also occurs in *Peripatus* (Eakin & Westfall 1965). In Pantopoda, the four median ocelli display a cuticular lens and an inverse retina, the sensory cells form a latticed rhabdom and each ocellus is surrounded by a reflecting tapetum and pigment layers (Hess et al. 1996).

During the evolution of the major arthropod groups, the median eyes have retained a relatively simple structure, forming the nauplius eyes of Crustacea and the frontal ocelli of Hexapoda, whereas the lateral visual organs usually present the structure of compound eyes.

The origin of the lateral ocellus-type eyes of arachnids remains an open question. The structure of these simple eyes is little different from that of the median eyes, and both are reminiscent of the ocelli found in some outgroups. If it is assumed that compound eyes were already present in the chelicerate ancestor, it should be accepted that the simple lateral eyes of arachnids, but not the median eyes, result from an evolutionary change involving a reversal to a plesiomorphic feature. This hypothesis has been proposed in particular by Paulus (1979, 2000) who suggested that the transformation of a facetted eye into a group of lateral ocelli involved the reduction of the number of ommatidia, the disintegration of the eye into several parts, and the fusion of several facets to form a common corneal lens for each ocellus. Yet, an alternative hypothesis can be put forward, according to which the arrangement of several ocellus-type simple eyes in a transverse line on the cephalic region, as found in various arachnids, corresponds to an ancestral condition for chelicerates and, maybe, for euarthropods. In this view, it should be assumed that the compound eyes of Xiphosura and several chelicerate-related fossil lineages result from an evolution independent from that having led to the compound eye of Mandibulata. However, this second hypothesis appears to be a less parsimonious solution.

In Hexapoda, the compound eyes have undergone different types of structural modifications giving rise to dispersed eyes, stemmata or even ocellus-type eyes. Ocellus-like simple eyes are found in particular in some holometabolous insect larvae, in the place of ommatidia-derived stemmata present in most holometabolous larvae. This is, for example, the case of the larva of *Lytta viridana* (Meloidae): its single lateral eye, described as a 'stemma', actually corresponds to a cup-shaped ocellus, the retina of which consists of a large number of photoreceptor cells (Heming 1982). This ocellus originates in the embryo as a field of pseudostratified epidermal cells in which a little number of mitoses can be observed; neither photoreceptor cell cluster formation nor fusion of cell clusters have been reported. Heming considers that all the lateral simple eyes of holometabolous larvae are homologous, but have evolved differently and independently from the compound eyes of some common progenitor species. In the case of the atypical eyes of some adult insects, such as in the parasitic orders Siphonaptera and Anoplura, the ocellus-type lateral eyes have taken the place of compound eyes which are usually present in insects, in particular in closely related orders, and therefore which surely occurred in common ancestors. Among Coccoidea, lateral ocellus-type unicorneal eyes are present in evolved families, such as Coccidae, while usual compound eyes occur in primitive families. The atypical eyes of male Strepsiptera are composed of separate visual units, each having the structure of an ocellus, but this is certainly a secondary condition as compared to that of presumably closely related orders (Coleoptera?) that possess typical compound eyes.

A very informative example is given by the development of the simple eyes of the workers and queens of the ant *Eciton*: the eye *Anlage*, in the pupa, begins to form rudimentary ommatidia, in particular with four Semper cells each, but their differentiation is soon stopped and the retinal organization of an ocellus takes place under a common corneal lens. In other words, lateral ocellus-like eyes of female *Eciton* have developed from a cellular material initially destined to form compound eyes. The presence of lateral ocelli in larvae and some adults of hexapods have probably involved reversal to a plesiomorphic state, which might be explained by a process of switch-back evolution.

4 PHYLOGENETIC IMPLICATIONS

The preceding views on the evolution of eye structure can be used in a discussion on the phylogenetic relationships between the major euarthropod groups. It is well known that, depending on the methodologies used and on the taxa included, many divergent assumptions concerning the monophyly of the traditional major classes (Chelicerata, Crustacea, Myriapoda and Hexapoda) and their assemblage in larger clades have been proposed by different authors, expressing divergent views and in particular a dramatic conflict between morphological and molecular phylogenies.

4.1 *Euarthropoda*

At the present time, there is a general agreement concerning the monophyly of Euarthropoda (Arthropoda except the Onychophora and Tardigrada). The presence of compound eyes made up of many juxtaposed visual units, or ommatidia, each including a cuticular lens and a cluster of radially arranged photoreceptor cells with differentiated microvilli bearing photopigments, is usually regarded as an autapomorphy of the Euarthropoda and so supports the clade (Paulus 1979, 2000). However, an alternative hypothesis which cannot be totally ruled out proposes that the simple eyes of the ocellus-type, already present in several non-arthropodan phyla including hypothetical ancestors of arthropods, occurred in a euarthropod ancestor and that compound eyes have been acquired independently in different arthropod lineages, at least twice, in the Trilobita, Eurypterida and Xiphosura on the one hand, in the Mandibulata on the other hand.

4.2 *Chelicerata*

The monophyly of the Chelicerata is generally accepted (Shultz 1990; Wheeler & Hayashi 1998; Edgecombe et al. 2000; Giribet et al. 2001) and, among extant chelicerates, the Xiphosura and Arachnida are usually regarded as monophyletic sister groups (Shultz 2001). In recent Chelicerata, compound eyes are only present in Xiphosura, but different fossil chelicerate taxa or allied clades, such as Trilobita, also possessed compound eyes, sometimes in the form of dispersed eyes (schizochroal type) as in the Phacopida. The special organization of the compound eyes of *Limulus* as compared with mandibulate eyes, and especially the occurrence of an arhabdomeric eccentric cell in each visual unit, can be considered as an autapomorphy of Xiphosura, unless it reveals structural characteristics that were already present in the eyes of a hypothetical euarthropod ancestor. If it is assumed that the chelicerate ancestor had compound eyes, it should be accepted that the lateral simple eyes of arachnids result from a reverse evolutionary pathway. Whatever it may be, the simple eyes of arachnids appear to have undergone different evolutionary changes depending to the taxonomic groups, leading in particular to highly differentiated eyes of the jumping spiders. These cytological changes, probably linked to special modes of vision, are not found in other arthropod groups, and so, may imply an evolution of the chelicerate eyes independent from that of mandibulates.

4.3 *Mandibulata*

The traditional subdivision of the Euarthropoda into two major lineages, the Chelicerata and the Mandibulata, is supported by many analyses based either on morphological data (Weygoldt 1986; Zrzavý et al. 1998; Waloszek 2003; Bitsch & Bitsch 2004) or on molecular and combined data (Wheeler 1998; Giribet & Ribera 1998; Edgecombe et al. 2000;

Giribet et al. 2001). In turn, the monophyletic Mandibulata are subdivided into two major branches, the Crustacea and Atelocerata (or Tracheata), the latter including Hexapoda and Myriapoda. Monophyletic Mandibulata are also found in various combined analyses (Wheeler 1998; Giribet & Ribera 1998; Edgecombe et al. 2000; Giribet et al. 2001; Edgecombe 2004). The comparative structure of the lateral eyes provides an additional argument in favor of monophyletic mandibulates. The striking similarities between the compound eyes of Crustacea and Hexapoda are easily explained if they are assumed to have both derived from a structure already present in some common ancestor and if the eyes of Myriapoda are thought to be also derived from compound eyes of the mandibulate type (see below).

Yet, as indicated below, other assemblages of the major arthropod lineages have been suggested leading to the rejection of the mandibulate clade.

4.4 The chelicerate-crustacean assemblage

Using especially paleontological data, several authors have suggested grouping the Arachnomorpha (Trilobita and Chelicerata) with the Crustacea to form the clade 'Schizoramia', so called owing to their branched appendages (Bergström 1992; Waggoner 1996; Bergström & Hou 1998; Wills et al. 1998). This clade is opposed to the 'Uniramia', with unbranched appendages, including Hexapoda and Myriapoda (this uniramian group corresponds to the Atelocerata or Tracheata of morphologists). However, the clade Schizoramia was not found in analyses based on morphological or molecular data. It is not supported by the comparative anatomy of eyes either: the compound eyes present in recent Xiphosura have a structure distinctly different from that of crustacean and other mandibulate compound eyes. Nothing is known concerning the internal structure of the eyes of extinct Arachnomorpha.

4.5 The chelicerate-myriapod assemblage

Several molecular-based studies favor the grouping Chelicerata + Myriapoda, and so reject the clade Mandibulata (Friedrich & Tautz 1995; Hwang et al. 2001; Negrisolo et al. 2004). Two recent studies, based on different nuclear, mitochondrial and ribosomal genes, have also found the assemblage Chelicerata + Myriapoda, and have designated it under the name of *Myriochelata* (Pisani et al. 2004) or *Paradoxopoda* (Mallatt et al. 2004). Comparative analysis of the neurogenesis in a diplopod myriapod is also consistent with this grouping (Dove & Stollewerk 2003). However, as noted by Mallatt et al., "the hypothesis that chelicerates and myriapods form a monophyletic group [...] was never supported by morphology-based studies". A similar remark was presented by Pisani et al. who wrote that "we are unable to identify any morphological trait diagnostic of this clade". Concerning the visual organs, the simple eyes of arachnids and myriapods present fundamentally different structures: the arachnid simple eyes are of the ocellus-type with a net-like rhabdom or with separate rhabdoms in the axis of the retinulae, whereas the simple lateral eyes of myriapods are stemma-like organs, apparently derived from ommatidia of compound eyes. In addition, although two groups among Chelicerata and Myriapoda possess compound eyes, those of

Xiphosura have a structure that is very different from that of Scutigeromorpha. These observations are not consistent with the presumed clade Chelicerata + Myriapoda.

4.6 The crustacean-hexapod assemblage

Several recent analyses have led to phylogenetic trees supporting a clade comprising Crustacea and Hexapoda, but excluding Myriapoda (Friedrich & Tautz 1995, 2001; Giribet & Ribera 1998; Shultz & Regier 2000; Giribet et al. 2001; Regier & Shultz 2001; Kusche et al. 2003; Negrisolo et al. 2004). This grouping was named *Pancrustacea* by Zrzavý & Štys (1997), then renamed *Tetraconata* by Dohle (2001) taking into account the quadripartite crystalline cone in the ommatidia of both Crustacea and Hexapoda (review in Richter 2002). The Tetraconata concept is also supported by different structural and developmental aspects of the central nervous system (review in Harzsch et al. in press).

Nevertheless, the arguments supporting the presumed clade Pancrustacea should not be regarded as definitive (Richter 2002; Harzsch 2003). In limiting the discussion to the structure of the visual organs, it appears that the presumed close relationship between Crustacea and Hexapoda is in good agreement with the great similarity of the cellular composition and neuronal circuitry of the compound eyes and optic lobes in the two groups (Osorio et al. 1997; Nilsson & Osorio 1998; Strausfeld 1998; Richter 2002; Harzsch et al. in press). There is no doubt as to the homology of the compound eyes in Crustacea and Hexapoda. However, this conclusion does not necessarily support the hypothetical taxon Pancrustacea. In our opinion, it can rather be taken as an argument in favor of the grouping Mandibulata including the Myriapoda.

4.7 The myriapod-hexapod assemblage

The monophyly of the assemblage Myriapoda + Hexapoda, forming the traditional *Atelocerata* (= *Tracheata*) has been much debated (reviews in Dohle 2001; Klass & Kristensen 2001). The clade Atelocerata is found in most studies based on morphological characters (Kraus & Kraus 1994; Kraus 1998, 2001; Bitsch & Bitsch 2004) and in some combined analyses using both morphological and molecular data, such as those of Wheeler (1998) and Edgecombe et al. (2000). However, in the latter analysis, the Atelocerata were not strongly supported.

Several data from eye structure appear to be in favor of the Atelocerata concept. The compound eyes present in the Scutigeromorpha consist of numerous ommatidia whose structure is not very different from that of insect ommatidia, in particular due to the presence of large crystalline cones built up by two to four cone cells. In this respect, the scutigeromorph myriapodans should be included in the so-called Tetraconata. Owing to the probable basal position of the Scutigeromorpha within the Chilopoda (Borucki 1996; Kraus 1998; Bitsch & Bitsch 2004; Edgecombe 2004; Edgecombe & Giribet 2004), it can be assumed that their eyes have been acquired from the compound eyes of some mandibulate ancestor. Moreover, the lateral simple eyes present in different myriapod taxa can be regarded as ommatidia that have been little to highly modified. They are reminiscent of, and in some cases similar to, the simple stemma-type eyes present in several hexapodan groups, in particular in holometabolous insect larvae.

4.8 Phylogenetic position of the Myriapoda

The position of the Myriapoda is of crucial importance in a discussion about the interrelationships among the major arthropod groups, but several divergent views on this topic have been presented by different authors.

The traditional grouping of Myriapoda with Hexapoda has been examined in the preceding section. Some authors, taking into account both molecular and morphological features (Giribet et al. 2001; Edgecombe 2004), have found the Myriapoda in basal position within the Mandibulata, as sister group to the assemblage Crustacea + Hexapoda, so rejecting the clade Atelocerata. If this conclusion is accepted, it means that the compound eyes of scutigeromorphs have been directly inherited from some mandibulate ancestor, not from an atelocerate ancestor, and that the lateral simple eyes of centipedes and millipedes have resulted from a regressive evolution of ommatidia, evolution which should have been parallel to that found in several hexapod groups.

The question of the monophyletic or paraphyletic status of Myriapoda has also been discussed. Some authors, such as Kraus & Kraus (1994) and Kraus (1998, 2001) have considered the Chilopoda (Opisthogoneata) as the sister group of the assemblage of the Progoneata (the remaining myriapodan groups) and Hexapoda. The recent analysis of Edgecombe (2004) finds the Myriapoda as a clade composed of Chilopoda and Progoneata, but the monophyly of Myriapoda is not strongly supported. If only the structure of eyes is taken into account, the Scutigeromorpha appear to be clearly separated from all other myriapod taxa by the presence of compound eyes, while the simple eyes of other Chilopoda and of Diplopoda present a general common pattern which does not favor a clear separation between the two groups, at least as far as greater structural details are not considered.

4.9 Hexapoda

Within the Hexapoda, it should be noted that the primarily wingless orders formerly grouped under the taxon 'Apterygota' present very different types of visual organs and also include many eyeless species. The Collembola have a limited number of simple, stemma-like lateral eyes that are probably derived from compound eyes of the mandibulate type. The Archaeognatha and the Zygentoma, formerly grouped in the class of Thysanura because they share many common features, have quite different eyes. Whereas the Archaeognatha (Machilidae) possess large and contiguous compound eyes, the Zygentoma present separate ommatidia, whose number is large in the primitive family of Lepidothrichidae, but reduced in the more advanced Lepismatidae, or absent in various taxa. Clearly, the grouping Apterygota is not supported by the characteristics of the eyes.

The numerous types of compound eyes found in the Pterygota, but not considered in the present paper, can be regarded as modifications from a common pattern. We have focused our attention only on different regressive evolutions that have affected the pterygote eyes in the larvae of many holometabolous families and in the adults of some orders, leading either to 'separated eyes', or to more modified stemmata, or even to lateral ocellus-type simple eyes. This evolutionary change of compound eyes into lateral ocelli appears to be a remarkable case of reversion; it may indicate that the genetic program which controls the formation of simple ocellus-type eyes has been conserved in the hexapods, and probably in all arthropods.

4.10 Conclusions

The evolution of the different types of eyes in the major arthropodan groups, as suggested here, is consistent with the results of phylogenetic analyses based on abundant morphological data. Complex and highly differentiated compound eyes were present in several extinct groups related to Chelicerata and are present in extant Xiphosura; they are probably derived from lateral visual organs already present in the euarthropod ground pattern. Compound eyes probably result from the transformation of simpler, ocellus-type visual organs similar to those found in several non-arthropod phyla, and so they probably represent an autapomorphy of euarthropods. Recent Arachnida only possess simple eyes of the ocellus type. The lateral ocelli of arachnids may have derived from compound eyes already present in the chelicerate ancestor, or they have been inherited from some non-arthropodan ancestor.

Several lineages of Mandibulata have retained simple ocellus-type organs, yet only in a median position. The lateral eyes of Mandibulata are usually of the compound type and present a more highly differentiated structure than those of Xiphosura, each ommatidium typically including a crystalline cone made up of four cone cells. Such a 'Tetraconata' condition is characteristic of the compound eyes of most Crustacea and Hexapoda. Nevertheless, the striking similarities between the compound eyes of these two groups are not regarded here as supporting the presumed clade Pancrustacea (= Tetraconata), but rather as being in favor of an ancestor common to all Mandibulata, including the Myriapoda. The comparative cytological features of the simple and compound eyes no longer support several hypothetical assemblages that have been suggested using paleontological or molecular data, such as the Schizoramia (Chelicerata + Crustacea) and the Myriochelata or Paradoxopoda (Chelicerata + Myriapoda).

Among Myriapoda, only the scutigeromorph centipedes possess compound eyes that can be regarded as derived from the mandibulate ground pattern. In several chilopod and diplopod groups, the visual organs are only represented by stemma-like simple eyes, some of which are little different from the stemmata of certain holometabolous insect larvae. Thus, the structure of the myriapodan eyes argues in favor of placing Myriapoda within the Mandibulata, and this is compatible with the clade Atelocerata, grouping Hexapoda and Myriapoda.

ACKNOWLEDGEMENTS

We would like to thank Gregory D. Edgecombe for thoughtful suggestions on a first draft of the manuscript. We are grateful to Dr. Carsten H.G. Müller for detailed comments and to an anonymous referee for general remarks on the manuscript.

REFERENCES

Adensamer, T. 1894 [1893]. Zur Kenntnis der Anatomie und Histologie von *Scutigera coleoptrata*. *Verh. kais.-k. zool.-bot. Ges. Wien* 43: 573-578.

Arendt, D., Tessmar, K, Compos-Baptista M.M-I. de, Dorresteijn, A. & Wittbrodt, J. 2002. Development of pigment-cup eyes in the polychaete *Platynereis dumerilii* and evolutionary conservation of larval eyes in Bilateria. *Development* 129: 1143-1154.

Bähr, R. 1974. Contribution to the morphology of chilopod eyes. *Symposia Zool. Soc. London* 32: 383-404.

Barra, J.-A. 1973. Structure et régression des photorécepteurs dans le groupe *Lepidocyrtus - Pseudosinella* (Insectes, Collemboles). *Ann. Spéléol.* 28: 167-175.

Bedini, C. 1968. The ultrastructure of the eye of a centipede *Polybothrus fasciatus* (Newport). *Monit. zool. Ital., N.S.*, 1: 41-63.

Bedini, C. 1970. The fine structure of the eye in *Glomeris* (Diplopoda). *Monit. zool. Ital., N.S.*, 4: 201-219.

Bernard, F. 1932. Comparaison de l'oeil normal et de l'oeil régressé chez quelques Carabiques. *Bull. biol. Fr. Belg.* 66: 111-147.

Bergström, J. 1992. The oldest arthropods and the origin of the Crustacea. *Acta Zool.* 73: 287-291.

Bergström, J. & Hou, X. 1998. Chenjiang arthropods and their bearing on early arthropod evolution. In: Edgecombe, G.D. (ed.), *Arthropod Fossils and Phylogeny*: 151-184. New York: Columbia Univ. Press.

Bitsch, C. & Bitsch, J. 2004. Phylogenetic relationships of basal hexapods among the mandibulate arthropods: a cladistic analysis based on comparative morphological characters. *Zool. Scr.* 33: 511-550.

Borucki, H. 1996. Evolution und phylogenetisches System der Chilopoda (Tracheata). *Verh. Naturwiss. Ver. Hambg., N.F.*, 35: 95-226.

Brandenburg, J. 1960. Die Feinstruktur des Seitenauges von *Lepisma saccharina* L. *Zool. Beitr., N.F.*, 5: 291-300.

Bullock, T.H. & Horridge, G.A. 1965. *Structure and Function in the Nervous Systems of Invertebrates, Vol. I and II*. San Francisco: W.H. Freeman & Co.

Buschbeck, E., Ehmer, B. & Hoy, R. 2003. The unusual visual system of the Strepsiptera: external eye and neuropils. *J. Comp. Physiol., A*, 189: 617-630.

Caveney, S. 1998. Compound eyes. In: Harrison, F.W. & Locke, M. (eds.), *Microscopic Anatomy of Invertebrates; Vol. 11B, Insecta*: 423-445. Wiley-Liss, Inc.

Cronier, C. & Clarkson, E.N.K. 2001. Variation of eye-lens distribution in a new late Devonian phacopid trilobite. *Trans. R. Soc. Edinburgh, Earth Sci.*, 92: 103-113.

Demoll, R. 1914. Die Augen von *Limulus*. *Zool. Jahrb., Abt. Anat.*, 38: 443-464.

Diersch, R., Melzer, R.R. & Smola, U. 1999. Morphology of the compound eyes of two ancestral phyllopods, *Triops cancriformis* and *Lepidurus apus* (Notostraca: Triopsidae). *J. Crust. Biol.* 19: 313-323.

Dikson, B. & Hafen, E. 1993. Genetic dissection of eye development in *Drosophila*. In: Bate, M. & Martinez-Arias, A. (eds.), *The Development of Drosophila melanogaster*: 1327-1362. New York: Cold Spring Harbor Laboratory Press.

Dohle, W. 2001. Are the insects terrestrial crustaceans? A discussion of some new facts and arguments and the proposal of the proper name 'Tetraconata' for the monophyletic unit Crustacea + Hexapoda. *Ann. Soc. entomol. Fr., N.S.*, 37: 85-103.

Dove, H. & Stollewerk, A. 2003. Comparative analysis of neurogenesis in the myriapod *Glomeris marginata* (Diplopoda) suggests more similarities to chelicerates than to insects. *Development* 130: 2161-2171.
Dudley, P. 1969. The fine structure and development of the nauplius eye of the copepod *Doropygus seclusus* Illg. *Cellule* 68: 7-35.
Eakin, R.M. & Brandenburger, J.L. 1971. Fine structure of the eyes of jumping spiders. *J. Ultrastruct. Res.* 37: 618-663.
Eakin, R.M. & Westfall, J.A. 1965. Fine structure in the eye of *Peripatus* (Onychophora). *Z. Zellforsch. mikrosk. Anat.* 68: 278-300.
Edgecombe, G.D. 2004. Morphological data, extant Myriapoda, and the myriapod stem-group. *Contrib. Zool.* 73: 207-252.
Edgecombe, G.D. & Giribet, G. 2004. Adding mitochondrial sequence data (16S rRNA and cytochrome *c* oxidase subunit I) to the phylogeny of centipedes (Myriapoda: Chilopoda): an analysis of morphology and four molecular loci. *J. Zool. Syst. Evol. Res.* 42: 89-134.
Edgecombe, G.D., Wilson, G.D.F., Colgan, D.J., Gray, M.R. & Cassis, G. 2000. Arthropod cladistics: combined analysis of histone H3 and U2 snRNA sequences and morphology. *Cladistics* 16: 155-203.
Eisen, J.S. & Youssef, N.N. 1980. Fine structural aspects of the developing compound eye of the honey bee, *Apis mellifera* L. *J. Ultrastruct. Res.* 71: 79-94.
Elofsson, R. 1963. The nauplius eye and frontal organs in Decapoda (Crustacea). *Sarsia* 12: 1-68.
Elofsson, R. 1965. The nauplius eye and frontal organs in Malacostraca (Crustacea). *Sarsia* 19: 1-54.
Elofsson, R. 1966. The nauplius eye and frontal organs in non-Malacostraca (Crustacea). *Sarsia* 25: 1-128.
Elofsson, R. 1970. Brain and eyes of Zygentoma (Thys., Lepismatidae). *Entomol. Scand.* 1: 1-20.
Elofsson, R. 1992. To the question of eyes in primitive crustaceans. *Acta Zool.* 73: 369-372.
Fahrenbach, W.H. 1964. The fine structure of a nauplius eye. *Z. Zellforsch.* 62: 182-197.
Fahrenbach, W.H. 1968. The morphology of the eyes of *Limulus*, I: Cornea and epidermis of the compound eye. *Z. Zellforsch. mikrosk. Anat.* 87: 279-291.
Fahrenbach, W.H. 1969. The morphology of the eyes of *Limulus*, II: Ommatidia of the compound eye. *Z. Zellforsch. mikrosk. Anat.* 93: 451-483.
Fahrenbach, W.H. 1975. The visual system of the horseshoe crab *Limulus polyphemus*. *Int. Rev. Cyt.* 41: 285-349.
Fahrenbach, W.H. 1999. Merostoma. In: Harrison, F.W. & Locke, M. (eds.), *Microscopic Anatomy of Invertebrates, 8A*: 21-115. New York: Wiley-Liss.
Fischer, A. & Horstmann, G. 1971. Der Feinbau des Auges der Mehlmotte, *Ephestia kuehniella* Zeller (Lepidoptera, Pyralididae). *Z. Zellforsch.* 116: 275-304.
Fischer, C., Mahner, M. & Wachmann, E. 2000. The rhabdom structure in the ommatidia of the Heteroptera (Insecta), and its phylogenetic significance. *Zoomorphology* 120: 1-13.
Frankfort, B.J. & Mardon, G. 2002. R8 development in the *Drosophila* eye: a paradigm for neural selection and differentiation. *Development* 129: 1295-1306.
Freeman, M. 1997. Cell determination strategies in the *Drosophila* eye. *Development* 124: 261-270.
Friederichs, H.F. 1931. Beiträge zur Morphologie und Physiologie der Sehorgane der Cicindelinen (Col.). *Z. Morphol. Oekol. Tiere* 21: 1-172.
Friedrich, M. 2003. Evolution of insect eye development: first insights from fruit fly, grasshopper and flour beetle. *Integr. Comp. Biol.* 43: 508-521.

Friedrich, M. & Benzer, S. 2000. Divergent decapentaplegic expression patterns in compound eye development and the evolution of insect metamorphosis. *J. Exp. Zool. (Mol. Dev. Evol.)* 288: 39-55.

Friedrich, M., Rambold, T. & Melzer, R.R. 1996. The early steps of ommatidial development in the flour beetle *Tribolium castaneum* (Coleoptera Tenebrionidae). *Dev. Genes Evol.* 206: 136-146.

Friedrich, M. & Tautz, D. 1995. Ribosomal DNA phylogeny of the major extant arthropod classes and the evolution of myriapods. *Nature* 376: 165-167.

Friedrich, M. & Tautz, D. 2001. Arthropod rDNA phylogeny revisited: a consistency analysis using Monte Carlo simulation. *Ann. Soc. entomol. Fr., N.S.,* 37: 21-40.

Gal, J., Gabor, H., Clarkson, E.N.K. & Haiman, O. 2000. Image formation by bifocal lenses in a trilobite eye? *Vis. Res.* 40: 843-853.

Gilbert, C. 1994. Form and function of stemmata in larvae of holometabolous insects. *Annu. Rev. Entomol.* 39: 323-349.

Giribet, G., Edgecombe, G.D. & Wheeler, W.C. 2001. Arthropod phylogeny based on eight molecular loci and morphology. *Nature* 413: 157-161.

Giribet, G. & Ribera, C. 1998. The position of arthropods in the animal kindgdom: A search for a reliable outgroup for internal arthropod phylogeny. *Mol. Phylogenet. Evol.* 9: 481-488.

Gokan, N. & Meyer-Rochow, V.B. 2000. Morphological comparisons of compound eyes in Scarabaeoidea (Coleoptera) related to the beetles' daily activity maxima and phylogenetic positions. *J. Agric. Sci.* [Tokyo] 45: 15-61.

Goodman, L.J. 1970. The structure and function of the insect dorsal ocellus. *Adv. Insect Physiol.* 7: 97-195. London: Academic Press.

Goodman, L.J. 1981. Organization and physiology of the insect dorsal ocellar system. In: Autrum, H. (ed.), *Comparative Physiology and Evolution of Vision in Invertebrates, Vol. VII 6C*: 201-286. Berlin: Springer-Verlag.

Hafner, G.S. & Tokarski, T.R. 1998. Morphogenesis and pattern formation in the retina of the crayfish *Procambarus clarkii*. *Cell Tissue Res.* 293: 535-550.

Hafner, G.S. & Tokarski, T.R. 2001. Retinal development in the lobster *Homarus americanus*. Comparison with compound eyes of insects and other crustaceans. *Cell Tissue Res.* 305: 147-158.

Hallberg, E. 1977. The fine of the compound eyes of mysids (Crustacea, Mysidacea). *Cell Tissue Res.* 184: 45-65.

Hallberg, E., Elofsson, R. & Grygier, M.J. 1985. An ascothoracid compound eye (Crustacea). *Sarsia* 70: 167-171.

Hallberg, E., Nilsson, H.L. & Elofsson, R. 1980. Classification of amphipod compound eyes - the fine structure of the ommatidial units (Crustacea, Amphipoda). *Zoomorphologie* 94: 279-306.

Hanström, B. 1934. Bemerkungen über das Komplexauge der Scutigeriden. *Lunds Univ. Aarsskr., N.F. Andra Avd.,* 2 Bd. 30 Nr. 6: 3-13.

Harzsch, S. 2003. Ontogeny of the ventral nerve cord in malacostracan crustaceans: a common plan for neuronal development in Crustacea, Hexapoda and other Arthropoda? *Arthrop. Struct. Dev.* 32: 17-37.

Harzsch, S., Sandeman, D. & Chaigneau, J. In press. Morphology and development of the central nervous system. In: Forest, J. & Vaupel Klein, J.C. von (eds.), *Treatise on Zoology - Crustacea*. Leiden: Koninkl. Brill Acad. Publ.

Heming, B.S. 1982. Structure and development of the larval visual system in embryos of *Lytta viridana* Leconte (Coleoptera, Meloidae). *J. Morphol.* 172: 23-43.

Hess, M., Melzer, R.R. & Smola, U. 1996. The eyes of a "nobody", *Anoplodactylus petiolatus* (Pantopoda; Anoplodactylidae). *Helgol. Meeresunters.* 50: 25-36.

Hesse, R. 1901. Untersuchungen über Organe der Lichtempfindung bei niederen Tieren, VII: Von den Arthropodenaugen. *Z. wiss. Zool.* 70: 347-473.

Homann, H. 1971. Die Augen der Araneae. Anatomie, Ontogenie und Bedeutung für die Systematik (Chelicerata, Arachnida). *Z. Morphol. Tiere* 69: 201-272.

Hwang, U.W., Friedrich, M., Tautz, D., Park, C.J. & Kim, W. 2001. Mitochondrial protein phylogeny joins myriapods with chelicerates. *Nature* 413: 154-157.

Kaiser, T. & Alberti, G. 1991. The fine structure of the lateral eyes of *Neocarus texanus* Chamberlin and Mulaik, 1942 (Opilioacarida, Acari, Arachnida, Chelicerata). *Protoplasma* 163: 19-33.

Keskinen, E., Takaku, Y., Meyer-Rochow, V.B. & Hariyama, T. 2002. Postembryonic eye growth in the seashore isopod *Ligia exotica* (Crustacea, Isopoda). *Biol. Bull.* 202: 223-231.

Kinzelbach, R. 1967. Zur Kopfmorphologie der Fächerflügler (Strepsiptera, Insecta). *Zool. Jahrb., Abt. Anat.,* 84: 559-684.

Kinzelbach, R. 1971. Morphologische Befunde an Fächerflüglern und ihre phylogenetische Bedeutung (Insecta: Strepsiptera), I: Vergleichende Morphologie. *Zoologica* 119: I-XIII, 1-128.

Klass, K.-D. & Kristensen, N.P. 2001. The ground plan and affinities of hexapods: recent progress and open problems. *Ann. Soc. entomol. Fr., N.S.,* 37: 265-298.

Kraus, O. 1998 [1997]. Phylogenetic relationships between higher taxa of tracheate arthropods. In: Fortey, R.A. & Thomas, R.H. (eds.), *Arthropod Relationships*: 295-303. London: Chapman & Hall.

Kraus, O. 2001. "Myriapoda" and the ancestry of the Hexapods. *Ann. Soc. entomol. Fr., N.S.,* 37: 105-127.

Kraus, O. & Kraus, M. 1994. Phylogenetic system of the Tracheata (Mandibulata): on "Myriapoda" - Insecta interrelationships, phylogenetic age and primary ecological niches. *Verh. Naturwiss. Ver. Hambg., N.F.,* 34: 5-31.

Kusche, K., Hembach, A., Hagner-Holler, S., Gebauer, W. & Burmester, T. 2003. Complete subunit sequences, structure and evolution of the 6x6-mer hemocyanin from the common house centipede, *Scutigera coleoptrata*. *Eur. J. Biochem.* 270: 2860-2868.

Lattin, G. de 1939. Untersuchungen an Isopodenaugen. *Zool. Jahrb., Abt. Anat.,* 65: 417-468.

Lin, J.T., Hwang, P.-C. & Tung, L.-C. 2002. Visual organization and spectral sensitivity of lateral eyes in the moth *Trabela vishnou* Lefebur (Lepidoptera: Lasiocampidae). *Zool. Stud.* 41: 366-375.

Liu, Z. & Friedrich, M. 2004. The *Tribolium* homologue of *glass* and the evolution of insect larval eyes. *Dev. Biol.* 269: 36-54.

Mallatt, J.M., Garey, J.R. & Shultz, J.W. 2004. Ecdysozoan phylogeny and Bayesian inference: first use of nearly complete 28S and 18S rRNA gene sequences to classify the arthropods and their kin. *Mol. Phylogenet. Evol.* 31: 178-191.

Matsuo, T., Takahashi, K., Kondo, S., Kaibuchi, K. & Yamamoto D. 1997. Regulation of cone cell formation by Cance and Ras in the developing *Drosophila* eye. *Development* 124: 2671-2680.

Meinertzhagen, I.A. 1991. Evolution of the cellular organization of the arthropod compound eye and optic lobe. In: Cronly-Dillon, J.R. & Gregory, R.L. (eds.), *Evolution of the Eye and Visual System*: 341-363. Houndmills: MacMillan Press.

Melzer, R.R., Diersch, R., Nicastro, D. & Smola, U. 1997. Compound eye evolution: highly conserved retinula and cone cell pattern indicate a common origin of the insect and crustacean ommatidium. *Naturwissenschaften* 84: 542-544.

Melzer, R.R., Michalke, C. & Smola, U. 2000. Walking on insects paths? Early ommatidial development in the compound eye of the ancestral crustacean, *Triops cancriformis*. *Naturwissenschaften* 87: 308-311.

Melzer, R.R., Paulus, H.F. & Kristensen, N.P. 1994. The larval eye of nannochoristid scorpionflies (Insecta, Mecoptera). *Acta Zool.* 75: 201-208.

Meyer-Rochow, V.B. 1971. A crustacean-like organization of insect-rhabdoms. *Cytobiologie.* 4: 241-249.

Meyer-Rochow, V.B. 1974. Structure and function of the larval eye of the sawfly, *Perga* (Hymenoptera). *J. Insect Physiol.* 20: 1565-1591.

Meyer-Rochow, V.B., Au, D. & Keskinen, E. 2001. Photoreception in fishlice (Branchiura): the eyes of *Argulus foliaceus* Linné, 1758 and *A. coregoni* Thorell, 1865. *Acta Parasit.* 46: 321-331.

Müller, C.H.G., Rosenberg, J., Richter, S. & Meyer-Rochow, V.B. 2003. The compound eye of *Scutigera coleoptrata* (Linnaeus, 1758) (Chilopoda: Notostigmophora): an ultrastructural reinvestigation that adds support to the Mandibulata concept. *Zoomorphology* 122: 191-209.

Negrisolo, E., Minelli, A. & Valle, G. 2004. The mitochondrial genome of the house centipede *Scutigera* and the monophyly versus paraphyly of myriapods. *Mol. Biol. Evol.* 21: 770-780.

Nemanic, P. 1975. Fine structure of the compound eye of *Porcellio scaber* in light and dark adaptation. *Tissue Cell* 7: 453-468.

Nilsson, D.-E. 1994. Eyes as optical alarm systems in fan worms and ark clams. *Philos. Trans. R. Soc. London, B,* 346: 195-212.

Nilsson, D.-E. & Osorio, D. 1998 [1997]. Homology and parallelism in arthropod sensory processing. In: Fortey, R.A. & Thomas, R.H. (eds.), *Arthropod Relationships*: 333-347. London: Chapman & Hall.

Nilsson, H.L. 1978. The fine structure of the compound eyes of shallow-water asellotes, *Jaera albifrons* Leach and *Asellus aquaticus* L. (Crustacea: Isopoda). *Acta Zool.* 59: 69-84.

Oakley, T.H. 2003. On homology of arthropod compound eyes. *Integr. Comp. Biol.* 43: 522-530.

Oakley, T.H. & Cunningham, C.W. 2002. Molecular phylogenetic evidence for the independent evolutionary origin of an arthropod compound eye. *Proc. Nat. Acad. Sci. USA* 99: 1426-1430.

Osorio, D., Bacon, J.P. & Whitington, P.M. 1997. The evolution of arthropod nervous systems. *Amer. Sci.* 85: 244-253.

Paulus, H.F. 1972. Zum Feinbau der Komplexaugen einiger Collembolen. Eine vergleichend-anatomische Untersuchung (Insecta, Apterygota). *Zool. Jahrb., Abt. Anat.,* 89: 1-116.

Paulus, H.F. 1974. The compound eyes of apterygote insects. In: Horridge, G.A. (ed.), *The Compound Eye and Vision of Insects*: 3-19. Oxford: Clarendon Press.

Paulus, H.F. 1977. Das Doppelauge von *Entomobrya muscorum* Nicolet (Insecta, Collembola). *Zoomorphologie* 87: 277-293.

Paulus, H.F. 1979. Eye structure and the monophyly of Arthropoda. In: Gupta, A.P. (ed.), *Arthropod Phylogeny*: 299-383. New York: Van Nostrand Reinhold Co.

Paulus, H.F. 1986. Evolutionswege zum Larvenauge der Insekten - ein Modell für die Entstehung und die Ableitung der ozellaren Lateralaugen der Myriapoda von Fazettenaugen. *Zool. Jahrb., Abt. Syst.,* 113: 353-371.

Paulus, H.F. 2000. Phylogeny of the Myriapoda-Crustacea-Insecta: a new attempt using photoreceptor structure. *J. Zool. Syst. Evol. Res.* 38: 189-208.

Pisani, D., Poling, L.L., Lyons-Weiler, M. & Hedges, S.B. 2004. The colonization of land by animals: molecular phylogeny and divergence times among arthropods. *BMC Biology* 2: 1-10.

Rasmussen, S. 1971. Die Feinstructur des Mittelauges und des ventralen Frontalorgans von *Artemia salina* L. (Crustacea: Anostraca). *Z. Zellforsch.* 117: 576-596.

Regier, J.C. & Shultz, J.W. 2001. Elongation factor-2: a useful gene for arthropod phylogenetics. *Mol. Phylogenet. Evol.* 20: 136-148.

Rhode, B. 1991. Ultrastructure of prostomial photoreceptors in four marine polychaete species (Annelida). *J. Morphol.* 209: 177-188.

Richter, S. 1999. The structure of the ommatidium of the Malacostraca (Crustacea) - a phylogenetic approach. *Verh. Naturwiss. Ver. Hambg.*, *N.F.*, 28: 161-204.

Richter, S. 2002. The Tetraconata concept: hexapod-crustacean relationships and the phylogeny of Crustacea. *Organ. Diversity Evol.* 2: 217-237.

Shear, W.-A. 1998 [1997]. The fossil record and evolution of the Myriapoda. In: Fortey, R.A. & Thomas, R.H. (eds.), *Arthropod Relationships*: 211-219. London: Chapman & Hall.

Shultz, J.W. 1990. Evolutionary morphology and phylogeny of Arachnida. *Cladistics* 6: 1-38.

Shultz, J.W. 2001. Gross muscular anatomy of *Limulus polyphemus* (Xiphosura, Chelicerata) and its bearing on evolution in the Arachnida. *J. Arachn.* 29: 183-303.

Shultz, J.W. & Regier, J.C. 2000. Phylogenetic analysis of arthropods using two nucelar protein-encoding genes supports a crustacean + hexapod clade. *Proc. R. Soc. London, Ser. B, Biol. Sci.*, 267 (1447): 1011-1019.

Snodgrass, R.E. 1935. *Principles of Insect Morphology*. New York: McGraw Hill.

Snodgrass, R.E. 1952. *A Textbook of Arthropod Anatomy*. Cornell Univ.

Spies, T. 1981. Structure and phylogenetic interpretation of diplopod eyes (Diplopoda). *Zoomorphology* 98: 241-260.

Spreitzer, A. & Melzer, P.R. 2003. The nymphal eyes of *Parabuthus transvaalicus* Purcell, 1899 (Buthidae): an accessory lateral eye in a scorpion. *Zool. Anz.* 242: 137-143.

Strausfeld, N.J. 1998. Crustacean-insect relationships: the use of brain characters to derive phylogeny amongst segmented invertebrates. *Brain Behav. Evol.* 52: 186-206.

Toh, Y. & Mizutani, A. 1994. Structure of the visual system of the larva of the tiger beetle (*Cicindela chinensis*). *Cell Tissue Res.* 278: 125-134.

Toh, Y. & Tateda, H. 1991. Structure and function of the ocellus. *Zool. Sci.* 8: 395-413.

Tomlinson, A. 1988. Cellular interactions in the developing *Drosophila* eye. *Development* 104: 183-193.

Uehara, A., Toh, Y. & Tateda, H. 1977. Fine structure of the eyes of orb-weavers, *Argiope amoena* L. Koch (Araneae: Argiopidae). 1. The anteromedial eyes. *Cell Tissue Res.* 182: 81-91.

Vaissière, R. 1961. Morphologie et histologie comparées des yeux des Crustacés Copépodes. *Arch. Zool. exp. gén.* 100: 1-25.

Voas, M.G. & Rebay, I. 2004. Signal integration during development: insights from the *Drosophila* eye. *Dev. Dyn.* 229: 162-175.

Wachmann, E. 1972a. Das Auge des Hühnerflohs *Ceratophyllus gallinae* (Schrank) (Insecta, Siphonaptera). *Z. Morphol. Oekol. Tiere* 73: 315-324.

Wachmann, E. 1972b. Zum Feinbau des Komplexauges von *Stylops spec*. (Insecta, Strepsiptera). *Z. Zellforsch. mikrosk. Anat.* 123: 411-424.

Waggoner, B.M. 1996. Phylogenetic hypotheses of the relationships of arthropods to Precambrian and Cambrian problematic fossil taxa. *Syst. Biol.* 45: 190-222.

Waloszek, D. 2003. The 'Orsten' window - a three-dimensionally preserved Upper Cambrian meiofauna and its contribution to our understanding of the evolution of Arthropoda. *Paleontol. Res.* 7: 71-88.

Weber, H. 1933. *Lehrbuch der Entomologie.* Stuttgart: Gustav Fischer Verlag.

Werringloer, A. 1932. Die Sehorgane und Sehzentren der Dorylinen nebst Untersuchungen über die Facettenaugen der Formiciden. *Z. wiss. Zool.* 141: 432-524.

Weygoldt, P. 1986. Arthropod interrelationships - the phylogenetic systematic approach. *Z. Zool. Syst. Evolut.-forsch.* 24: 19-35.

Wheeler, W.C. 1998 [1997]. Sampling, groundplans, total evidence and the systematics of arthropods. In: Fortey, R.A. & Thomas, R.H. (eds.), *Arthropod Relationships*: 87-96. London: Chapman & Hall.

Wheeler, W.C. & Hayashi, C.Y. 1998. The phylogeny of the extant chelicerate orders. *Cladistics* 14: 173-192.

Wills, M.A., Briggs, D.E.G., Fortey, R.A., Wilkinson, M. & Sneath, P.H.A. 1998. An arthropod phylogeny based on fossil and recent taxa. In: Edgecombe, G.E. (ed.), *Arthropod Fossils and Phylogeny*: 33-109. New York: Columbia Univ. Press.

Wolf, T. & Ready, D.F. 1993. Pattern formation in the *Drosophila* retina. In: Bate, M. & Martinez-Arias, A. (eds.), *The Development of* Drosophila melanogaster: 1277-1325. New York: Cold Spring Harbor Laboratory Press.

Wolken, J.J. & Gallik, G.J. 1965. The compound eye of a crustacean, *Leptodora kindtii*. *J. Cell Biol.* 26: 968-973.

Wundrig, A. 1936. Die Sehorgane der Mallophagen, nebst vergleichenden Untersuchungen an Liposceliden und Anopluren. *Zool. Jahrb., Abt. Anat.,* 62: 45-110.

Yamamoto, K. & Toh, Y. 1975. The fine structure of the lateral ocellus of the Dobsonfly larva. *J. Morphol.* 146: 415-430.

Yoon, C.S., Hirosawak, K. & Suzuki, E. 1996. Studies on the structure of ocellar photoreceptor cells of *Drosophila melanogaster* with special reference to subrhabdomeric cisternae. *Cell Tissue Res.* 284: 77-85.

Zhukov, V.V., Bobkova, M.B. & Vakolyuk, I.A. 2002. Eye structure and vision in the freshwater pulmonate mollusc *Planorbarius corneus*. *J. Evol. Biochem.Physiol.* 38: 419-430.

Zrzavý, J., Hypša, V. & Vlášková, M. 1998 [1997]. Arthropod phylogeny: taxonomic congruence, total evidence and conditional combination apporaches to morphological and molecular data sets. In: Fortey, R.A. & Thomas, R.H. (eds.), *Arthropod Relationships*: 97-107. London: Chapman & Hall.

Zrzavý, J. & Štys, P. 1997. The basic body plan of arthropods: insights from evolutionary morphology and developmental biology. *J. Evol. Biol.* 10: 353-367.

Appendage loss and regeneration in arthropods: A comparative view

DIEGO MARUZZO, LUCIO BONATO, CARLO BRENA, GIUSEPPE FUSCO & ALESSANDRO MINELLI

Department of Biology, University of Padova, Padova, Italy

ABSTRACT

Evidence for loss and regeneration of arthropod appendages is reviewed and discussed in terms of comparative developmental biology and arthropod phylogeny. The presence of a preferential breakage point is well documented for some, but not all, lineages within each of the four major groups - chelicerates, myriapods, crustaceans and hexapods. Undisputed evidence of true autotomy, however, is limited to isopods, decapods and some basal pterygotes, and claimed for other groups. Regeneration of lost appendages is widespread within arthropods, even if not present or documented in some groups. During regeneration, growth and differentiation of epidermis, nerves, muscles and tracheae are to some extent mutually independent, thus sometimes failing to reproduce their usual developmental interactions, with obvious consequences on the reconstruction of the lost part of the appendage. In the regeneration of appendages composed of 'true segments', all the segments the animal is able to regenerate are already present (with extremely rare exceptions) following the first post-operative molt, whereas the regeneration of flagellar structures is often accomplished in steps, e.g., the first regenerate may show a reduced number of flagellomeres. Lack of autotomy is likely to be the plesiomorphic condition in arthropods, a condition maintained in the Myriochelata (myriapods plus chelicerates). Autotomy evolved within the Pancrustacea, perhaps close to the origin of a Malacostraca-Hexapoda clade, and was subsequently lost by some lineages, e.g., the Hemipteroidea and the endopterygote insects. A diaphragm reducing the risk of hemorrhage at the preferred breakage point of the appendage is generally associated with autotomizing appendages, but this anatomical specialization has been lost in some groups, including one (the Dictyoptera) where autotomy is still present.

This article is dedicated to Fred Schram in deep appreciation of his lasting contribution to arthropod phylogeny and comparative morphology, with the warmest wishes of the authors.

1 INTRODUCTION

Studies on regeneration were fashionable one hundred years ago, when developmental biology was turning from a merely descriptive into an experimental science, but was technically and conceptually limited to mechanical approaches. This was the time of the *Ent-*

wicklungsmechanik (mechanics of development), with Needham's chemical embryology, not to mention molecular developmental genetics, still decades ahead. Today, new powerful means of investigation are progressively resolving developmental mechanisms in terms of gene expression patterns, transcription factors, and molecular dialogues between cells. As for arthropods, modern studies on regeneration have mostly focused on those species of decapod crustaceans in which leg autotomy, followed by regeneration, is a frequent, natural occurrence.

Most of the bulky descriptive literature produced in the past century on animal regeneration is now ignored. To be sure, many old studies on regeneration lacked clear experimental design and what was reported in print was often little more than a scientifically shallow narrative. There was, however, an important approach in that literature, one that later sank into oblivion and has come back into the focus of front line research only recently, with the advent of evolutionary developmental biology: the comparative approach. Przibram's (1909) monograph gives us an idea of the diversity of organisms brought by biologists into the lab to perform experiments on regeneration. It also gives an idea of the propensity of many researchers of that generation to collect and comparatively analyze both experimental evidence and accurate descriptions of museum specimens. This latter aspect, however, which was eventually so productive for other fields of biology - as in Bateson's (1894) masterpiece, *Materials for the study of variation* - is one of the weaknesses of the old literature on regeneration. Reports were very often based on observation of field-caught specimens rather than on experimentation. Incomplete regeneration was the default hypothesis to explain the origin of defective appendages, e.g., those with segments reduced in number or size. Alternative explanations - defective embryonic or post-embryonic development without any traumatic removal of the original appendage, or of part of it - were seldom advanced. With hindsight, and with the benefit of better knowledge of arthropod development, we are sometimes able to evaluate the likelihood of that putative evidence of regeneration, but the reliability of many records remains nevertheless uncertain. This is mainly due to the fact that our current awareness of developmental processes is primarily limited to a few model species, from which we cannot safely generalize for all arthropods, as this review will show.

In conjunction with the experimental work on the regeneration of arthropod appendages recently started in our lab, we felt the need to systematically explore the literature on this subject, in order to summarize a scattered, unequal but nevertheless precious trove of comparative information that has not been re-evaluated in the context of evolutionary developmental biology.

In our summary of data about regeneration in different groups, we primarily focus on the walking legs; data on the remaining appendages is given whenever available. Our choice is a consequence of the prevailing focus of regeneration research in most arthropod groups. It should not be construed as depending on a concept of the walking leg as the default arthropod appendage, from which all other kinds are derived, a concept one of us has recently refuted (Minelli 2003a).

1.1 *Nomenclature*

In the following review, we will often mention the presence, or absence, of a preferred breakage point (also called 'autotomy plane' in the literature), using the acronym PBP.

The term appendotomy will be used for any loss of a more or less extended distal section of an arthropod appendage at the level of a PBP. Many authors used the term autotomy as a synonym of appendotomy, but a more restricted use of the former is recommended (Bliss 1960; Roth & Roth 1984). Following Bliss (1960), we will speak of autotomy when appendotomy is produced "by means of a reflex that is usually unisegmental", autospasy "when the appendage is pulled by an outside agent against resistance provided by the animal's weight or its efforts to escape" and autotilly if it occurs "with the assistance of mouthparts, claws, or walking legs of the animal itself."

1.2 The scope of the present review

We have mainly limited our attention to the following points: presence (and, if any, relative position along the appendage) of a preferred breakage point (PBP); occurrence of true autotomy (see above, par. 1.1); natural occurrence of regeneration following appendotomy; occurrence of regeneration following experimental or accidental amputation of a distal part of the appendage; dependence of regeneration on the proximo-distal level of the breakage and, in particular, relative to the PBP; timing of regeneration process, in terms of number of molts required to get a regenerate, or to complete its growth and differentiation. As far as information is available, we have also paid attention to the sequence with which the segments of the regenerating appendage differentiate and to the possibly different process of regenerating 'true segments' in comparison with regenerating flagellar structures.

Several important features of arthropod appendage regeneration are deliberately omitted from the present review. In particular, we will not report on the detailed histological information available for some regenerating appendages. Other aspects we will not cover are hormonal or nervous control, regeneration following experimental grafting, and heteromorphic regeneration (for example, an antenna replacing the eye-stalk of a decapod crustacean, or a leg-like regenerate replacing the antenna of a stick insect).

As to the adaptive value of regeneration, we shall only mention the very high frequency of specimens with regenerated appendages sometimes found in natural populations (up to 40% in some brachyuran crustaceans). Besides its obvious importance as a means to restore accidentally damaged appendages resulting from intraspecific fighting or from a foe's offence, regeneration is sometimes a complement of an autotomy mechanism that provides an excellent way to escape from predators or to get rid of a riotous exuvium. On the other hand, it would also be interesting to analyze how much the different regenerative behavior of different appendages within one and the same animal (e.g., decapod chelipeds versus walking legs, or orthopteran forelegs versus hindlegs) can be explained by 'phylogenetic inertia', and how much instead by current functional constraints. We will only briefly discuss the 'assimilation' of autotomy followed by regeneration into the regular ontogenetic schedule of some male fiddler crabs, to the extent that these events are required to obtain functional heterochely. It would be worthwhile, but again outside the scope of the present review, to also discuss considerations of resource allocation during regeneration, and how they affect growth rate and frequency of later molts. Finally, we will leave out of the picture the developmental events that depend on growth and differentiation of imaginal discs; therefore, our brief consideration of holometabolous insects will be limited to the regenera-

2 A SURVEY BY MAJOR TAXA

2.1 *Pycnogonida and Chelicerata*

2.1.1 *Pycnogonida*
Appendotomy has been observed in the basal region of the legs of *Phoxichilus*. The gut branch extending into the broken leg is left undamaged and is eventually cut off by the new epidermis growing in from the leg stump. In *Nymphon*, PBP is between the first and the second segment of the leg. In both species, appendotomy is followed by regeneration (Dohrn 1881; Gaubert 1892).

In some pycnogonids, e.g. *Colossendeis*, the chelicerae are deciduous and fall off during the last pre-adult instar (Kaestner 1968; Bain 2003). This behavior could be described as appendotomy without regeneration.

2.1.2 *Xiphosura*
The appendages of *Limulus* do not show any trace of PBP (Wood & Wood 1932); however, telson and limbs can regenerate, at least during the larval stages (Clare et al. 1990 and references therein).

2.1.3 *Scorpiones*
Based on the experimental work of Wood (1926) on *Centrurus*, scorpion appendages do not show a PBP.

Scorpions seem to be able to regenerate only the pretarsus, which can arise from whatever level at which the leg has been cut. At first, the regenerate is often smaller than the original pretarsus, but can attain full size following more molts. The segment proximal to it can develop traits characteristic of the missing segments, such as the number of sensory setae and the presence of the spine-like setae usually restricted to basitarsus and tarsus (Rosin 1964). The only documented case of regeneration of anything more than just the pretarsus is Vachon's (1957) report on a specimen that, on the second molt after the amputation made in the first post-embryonic ('larval') stage, dissociated its regenerated pretarsus into tarsus and pretarsus. By contrast, Rosin (1964) did not report any further increase of segments in the regenerating limb during subsequent stages (amputation made at nymphal stages).

Following Rosin (1964), scorpions are also able to regenerate the tip of the sting and the distal part of the chelicerae. Both Vachon (1957), who worked with 'larvae' of *Euscorpius carpathicus*, and Rosin (1964), who worked with nymphs of different species, complained high mortality after removal of appendages.

2.1.4 *Opiliones*
A PBP has been found at the trochanter-femur articulation of the species of harvestmen

thus far investigated (Wood 1926; Roth & Roth 1984).

In *Leiobunum nigropalpi*, appendotomy does not damage any muscles and the resulting wound is very small and easily clotted. It takes just a slight tension to separate the leg at this point, but there is no evidence of any specialized structure (Wood 1926) to produce autotomy. A second PBP has been reported for *Sclerobunus* near the base of the femur, but apparently, it is seldom used (Roth & Roth 1984).

Even though appendotomy is frequent, it is traditionally thought that harvestmen are unable to regenerate limb segments, despite some old, indirect evidence, as summarized in Przibram (1909).

2.1.5 *Acari*

Most mites do not have a PBP (Rockett & Woodring 1972), but in *Opilioacarus* (Notostigmata) the legs are easily detached between coxa and trochanter (Vitzthum 1943).

The regenerative power of mites is very poor and only the ticks (Ixodida) (Rockett & Woodring 1972; Belozerov 2001) and *Opilioacarus* (Coineau & Legendre 1975) are known to regenerate a missing limb completely. Among the mites, in some species, the site of amputation becomes the definitive end of the appendage; in others, there can be an unpredictable degree of reduction of the segmentation in the remaining segments, which sometimes become smaller and distorted. In most of the operated specimens of *Scheloribates nudus*, the distorted distal segments ended in a long terminal seta and in *Fuscuropoda agitans* there was regeneration of one (very rarely two) small distorted segment(s) with modified setal patterns (Rockett & Woodring 1972).

Again, in mites other than ticks, mortality following leg amputation is fairly high, possibly dependent on slow coagulation of the hemolymph: in *Tetranychus neocaledonicus* and *Pimeliaphilus podapolypophagus*, for which the highest mortality has been reported (more than 90% and 100%, respectively), there seems to be almost no coagulation (Rockett & Woodring 1972).

Ticks are the only mites that can regenerate a full appendage (both legs and mouth parts; Belozerov 2001), but there are differences between hard ticks (Ixodidae) and soft ticks (Argasidae) (Rockett & Woodring 1972; Belozerov 2001). Hard ticks have lower mortality and faster regeneration since they are able to reproduce a complete limb after just one molt (actually, some differences remain in Haller's organ, in the number and topography of sensilla and some other details; Belozerov 2001). In soft ticks, on the other hand, a limb with the complete number of segments, but with reduced size and chetotaxy, emerges at the first molt after amputation; it will require two to four molts to eventually obtain full size (Belozerov 2001).

Rockett & Woodring (1972) stated that amputations made in hard ticks in the quiescent stage always resulted in death, but in later experiments (reviewed in Belozerov 2001) on both engorged larvae and nymphs, mortality was low and regeneration strongly dependent on the time of amputation within the instar. In soft ticks, amputation during apolysis results in the absence of regeneration (Belozerov 2001).

Neither the site of amputation, the number of amputated legs, nor the time of amputation within a given instar, excluding amputations on quiescent stages or during apolysis, seem to influence the regeneration process (Rockett & Woodring 1972; Belozerov 2001).

Interestingly, the only mites that exhibit mitoses during post-larval life are the ticks, but few species of mites have been studied in this respect and no data are available for *Opilio-*

acarus. This suggests a strong association between regeneration and the presence of post-larval mitoses (Rockett & Woodring 1972).

2.1.6 *Amblypygi*

There is a PBP at the patella-tibia articulation both in the forelegs (the whips, with flagellar tibia and tarsus) and in the walking legs. The anatomy of this joint has been studied in detail by Wood (1926) in *Tarantula*. This joint lacks strength of articulation and interlacing chitin fibers as usually present in the other joints and only one muscle is disturbed, but not injured, during breakage. When the leg is torn apart, this muscle, which arises in the patella, loses its distal attachment, but remains intact, and retracts within the patella, including the distal tip.

Autotomy is not known, whereas leg autospasy has been described by Weygoldt (1984).

Regeneration is possible only at the PBP: there is no regeneration following breakage at points proximal to it, whereas amputations distal to the PBP trigger appendotomy, followed by regeneration (Weygoldt 1984; Igelmund 1987). The walking legs take just one molt to gain full size and the complete number of segments. At the first molt, the regenerating whips are smaller, but with a higher number of segments than their undamaged counterparts. The following molts allow them to obtain full size, but they maintain their extra number of segments (Weygoldt 1984; Igelmund 1987). In *Heterophrynus elaphus*, the number of segments in the regenerated tibia increases by approximately 60%, in the tarsus about 30%. This number might be age-dependent, since regenerated whips from younger animals seem to have fewer segments than those from adults (Igelmund 1987).

2.1.7 *Araneae*

Depending on the species, spiders can have a PBP in the coxa-trochanter articulation, in the patella-tibia articulation or at mid-length on the patella, or no PBP at all (for a review and a list of species, see Roth & Roth 1984). Some species seem to have more than one PBP (Roth & Roth 1984), which would confirm Wood's (1926) concept of a PBP in arachnids that is a simple point of structural weakness.

The structure of the PBP has been studied in detail only for species that have it at the coxa-trochanter articulation; only one muscle is damaged after parting at this point and the opening left is small (Wood 1926).

A PBP also exists in the palps, at least in some species. In *Tidarren*, for example, the males usually self-remove (autotilly) one of their palps at the coxa-trochanter joint soon after their last molt (Roth & Roth 1984; Knoflach & van Harten 2000).

Whether spiders exhibit real autotomy has long been questioned and seems unlikely (Wood 1926; Roth & Roth 1984). Spiders regenerate chelicerae, palps, labium, legs and even the spinnerets (Bonnet 1930; Mikulska et al. 1975). A comprehensive list of early studies about regeneration of legs and palps in different spiders is found in Przibram (1909).

In *Dolomedes* and *Tegenaria*, the regenerate is, at first, shorter and has less than the full complement of sensory hairs; nevertheless, it already possesses all segments (Bonnet 1930; Mikulska et al. 1975).

Vachon (1956, 1967) amputated the legs of *Coelotes terrestris* at a very early stage, the 'larva', when appendages are still incompletely segmented. The first regenerate exhibited reduced segmentation and increased its segment number during subsequent molts, as it would have done in undisturbed development.

The relationship between PBP and regeneration is varied. Some spiders, but not all, regenerate from the PBP as well as from any point distal to it (Bonnet 1930; Mikulska et al. 1975). No data are available about regeneration from levels proximal to the PBP. *Latrodectus variolus* has a PBP at the coxa-trochanter articulation, but regenerates only following amputation in the middle of the femur or distal to that point. However, despite its good regeneration power, it shows appendotomy without regeneration (Randall 1981). Lack of regeneration from the PBP has also been recorded for several other spiders (Vollrath 1990).

There may be differences between regenerated and undamaged appendages in the number of teeth on the leg claws (Bonnet 1930; Mikulska et al. 1975), of sensory setae (Bonnet 1930; Mikulska et al. 1975) and lyriform organs, but not always (Bonnet 1930; Vollrath 1995).

2.2 *Myriapoda*

2.2.1 *Scutigeromorpha*

The long legs of *Scutigera* have a PBP between coxa and trochanter (Verhoeff 1902-1925); no muscle stretches through this articulation, just one nerve does and a diaphragm prevents excessive loss of blood (Herbst 1891).

In *Scutigera*, appendotomy occurs by autotilly or autospasy, whereas autotomy is unlikely (Cameron 1926). Regeneration always starts from the PBP since any cut distal to it results in appendotomy (Cameron 1926).

The regenerating leg is already complete as soon as it appears, after the first post-operative molt or a molt later, depending on the timing of the amputation within the intermolt (Verhoeff 1902-1925; Cameron 1926).

2.2.2 *Lithobiomorpha*

A PBP along the leg is found between coxa and trochanter (Verhoeff 1902-1925).

In *Lithobius*, Verhoeff (1902-1925) described the leg regenerating from the PBP as consisting at first of prefemur, femur, tibia and a tarsus of one article and completely lacking setae, epidermal glands, muscles (these are just 'sketched') and tendons. With another molt, the appendage becomes longer (about half the length of an undamaged leg) and possesses a trochanter, a second tarsal article, a claw and its tendon. The musculature is also developing, although it is still gracile. Many sensory setae and epidermal glands have appeared too, but most of the spines are still missing. A further molt leads to a complete appendage that is slightly smaller than an undamaged one. Regeneration of legs is also possible from any level distal to the PBP. The mechanism does not seem to be the same in all species and/or stadia. In *Bothropolys asperatus*, for instance, the regenerating legs have, at first, an incomplete number of segments if the damage occurred in a larva, but the full number of segments if it was suffered by a post-larval specimen (Murakami 1958).

The antennae can regenerate too; their segment number usually increases with subsequent molts. The number of segments shown after the first post-operative molt not only depends on the point of amputation, but also on the instar and the intermolt stage at the time of the operation (Verhoeff 1902-1925; Scheffel 1987, 1989; Weise 1991).

2.2.3 *Scolopendromorpha*

A PBP is present between coxa and trochanter, at least in the last pair of legs.

Autospasy of the last pair of legs was reported in *Rhysida* (Cloudsley-Thompson 1961), appendotomy in *Cryptops* and *Alipes* (Lawrence 1953).

Regeneration of the legs has never been documented through experiment, but indirect evidence of regeneration of the last pair of legs (reduced size and incomplete armature of spines) has long been noted (Newport 1844). No comparable data are available for the remaining legs.

As for the antennae, the species with a fixed number of antennomeres can only increase the length, but not the number of the segments left after amputation. True regeneration, with an increasing number of segments, seems only possible in those species that usually add a few antennomeres during post-embryonic development (Lewis 2000). In *Scolopendra*, Lewis (1968) observed antennae composed of a few proximal antennomeres of expected size followed by a variable number of very short ones and interpreted those distal articles as regenerated. In this group, the regeneration of antennomeres is far from accurate, sometimes leading to atypically high numbers (Lewis 2000).

2.2.4 *Geophilomorpha*

No evidence is available for the presence of a PBP and the occurrence of regeneration in geophilomorph legs and antennae (Lewis 2000; Minelli et al. 2000). A vague reference to appendotomy of the last pair of legs is found in Lawrence (1953).

2.2.5 *Diplopoda*

Appendotomy apparently does not exist in this group (Lawrence 1953).

In an early study, Newport (1844) reported on regeneration in two chilognathan genera he referred to as "*Julus*" and "*Spirostreptus*". Legs regenerated with all the articles at the first post-operative molt, but with reduced size; it is not known how much their size increased with subsequent molts. The antennae also regenerated, but, after the first post-operative molt, both number and size of their articles could be defective depending on where the amputation occurred. It is not known if the number of segments increased with subsequent molts.

2.3 *Crustacea*

2.3.1 *Branchiopoda*

In this group there is no evidence of any PBP.

Regeneration of appendages was observed in anostracans (*Branchipus*: Przibram 1909) and notostracans (*Lepidurus*), particularly in endites of the thoracic legs (Rogers 2001). By contrast, the antennae of cladocerans (*Daphnia*, *Simocephalus*, *Ceriodaphnia*) never regenerated their articulated axis if experimentally cut, but only reproduced their complex system of 'muscular setae' (Agar 1930). It is worthwhile to notice that the regenerated setae are not as morphologically and functionally diversified as the original setae. Regenerated setae, whose number and size increase at each postoperative molt, are variable among specimens in both number and features, and are not affected by factors such as age, food availability and number of repeated operations (Agar 1930).

Regeneration of furcal rami was also reported in notostracans (*Apus* = *Triops*) (Rabes 1907). The number of flagellar units increases in successive molts, and the size of the regenerating filament converges toward that of the undamaged one.

2.3.2 Ostracoda, Copepoda, Cirripedia, Branchiura

Based on the scarce information available, including Wood & Wood's (1932) experimental work on *Lepas*, no PBP has been documented in these crustacean groups.

Przibram (1909) summarized limited early evidence of regeneration of antennae and furcal rami in cyclopid and diaptomid copepods. As for cirripedes, regeneration was reported in *Balanus* for the cirri (Darwin 1854) and the penis (Klepal & Barnes 1974). No evidence of appendage regeneration is available for branchiurans but for the puzzling record of a specimen of *Argulus* that produced two extra pairs of functional, thoracic-like legs following the ablation of the abdomen (Kocian 1930).

2.3.3 Phyllocarida and Stomatopoda

No PBP has been documented in these groups but for Wood & Wood's (1932) weak experimental evidence of a possible PBP at the joint between ischium and merus in *Squilla*.

Regeneration of antennae and furcal rami was reported in *Nebalia* (Przibram 1909). It is worth noting that the heteromorphic regeneration of the eye-stalk as a functional antennule was reported in *Squilla* (see Paulian 1938).

2.3.4 Isopoda

No experimental evidence of a PBP was found by Wood & Wood (1932) in the isopods *Oniscus*, *Cylisticus*, *Lygia* and *Sphaeroma*, but the presence of a PBP is documented in the legs of *Asellus* (Needham 1947) and in the proximal part of the basis in the legs of *Porcellio* (Noulin 1962, 1984). This PBP is not crossed by muscles and possesses a diaphragm (Noulin 1962; Needham 1965). A PBP, apparently without diaphragm, was also observed in the antennae of *Asellus* (Wege 1911), between the third and the fourth article.

Autotomy of the legs has been documented in some isopods (*Asellus*: Needham 1947; *Porcellio*: Noulin 1962), but could not be observed in other taxa (Wood & Wood 1932).

Regeneration has been documented in different kinds of isopod appendages, but in different species (for brief summaries see Przibram 1909; Vernet & Charmantier-Daures 1994).

The mechanism of leg regeneration was studied by Needham (reviewed in 1965) in *Asellus*: after the formation of a scab, a folded, regenerating limb is produced inside a cuticular sac and the regenerate appears following the next molt.

In the regenerating antenna of *Asellus*, the number of segments increases through successive molts, converging with the number in the undamaged appendage; the peduncle and the most distal part of the antennal flagellum are formed first, followed by more flagellomeres in between (Wege 1911).

2.3.5 Amphipoda

A PBP was not found by Wood & Wood (1932) in the legs of *Gammarus*, *Caprella* and *Orchestia*, but a point of less resistance seems, nevertheless, to exist both in the caprellids (Calman 1909) and in *Orchestia* (between basis and ischium; Charniaux-Cotton 1957).

Regeneration was documented for legs (different species, see Przibram 1909), antennae

(*Gammarus*: Dixey 1938; Paulian 1938), and the second pair of gnathopods (*Orchestia*: Charniaux-Cotton 1957). In this latter case, gnathopods are sexually dimorphic and the regeneration of the derived male appendage proceeds through an intermediate stage similar to the condition in juveniles and in adult females (Charniaux-Cotton 1957). Timing of regeneration and completeness of the regenerate depend on the intermolt stage at the time of amputation (reviews in Bliss 1960; Vernet & Charmantier-Daures 1994).

2.3.6 *Decapoda: Brachyura*

The existence of a PBP along the leg was well documented in almost all of the investigated species of brachyurans (reviews in Wood & Wood 1932; Bliss 1960). The PBP is most often localized at the joint between basis and ischium, which is usually a non-functional articulation (Wood & Wood 1932; Bliss 1960). This PBP, however, is functionally weak or even not detectable at all in some fossorial species, such as *Ranina ranina*, probably as a derived condition (Wood & Wood 1932; Juanes & Smith 1995).

The anatomical structure of the PBP was accurately studied in some representative brachyurans (Wood & Wood 1932; Bliss 1960; Adiyodi 1972). Microcanals spanning the whole depth of the cuticle are more abundant here than in other regions of the appendage. No muscle develops through the PBP; instead, a specialized "autotomizer" muscle is inserted just proximal to it. The only nerve developing through the PBP is locally tapered and, thus, weakly resistant and the blood vessels crossing the PBP have a valve preventing bleeding.

Autotomy of the legs was well documented in most investigated brachyurans and the mechanism involved was studied in detail in some species (Wood & Wood 1932; review in Bliss 1960). Contraction of the autotomizer muscle pulls the anterior-basal part of the basis into the coxa; the mechanical resistance of the distal margin of the coxa against the anterior surface of the basi-ischium results in breaking of the integument between basis and ischium. The nerve and the blood vessels are cut. Autotomy is infrequent soon after ecdysis, when the exoskeleton is more flexible, possibly depending on the lack of mechanically suitable conditions (Wood & Wood 1932).

Full regeneration of legs from the PBP was documented in most brachyurans, both in nature and under experimental conditions (Bliss 1960; Vernet & Charmantier-Daures 1994). The papilla emerging after the detachment of the scar develops into an external cuticular sac; the limb bud grows and differentiates inside this sac, as a double-folded limb, during the intermolt period. An early phase of fast growth and articular differentiation of the regenerating limb, involving mitotic proliferation, is usually followed by a temporary developmental stasis, which is variable in duration in relation to the intermolt cycle. After a further phase of growth, mainly involving protein and water accumulation, at the first molt the regenerating limb emerges out of the cuticular sac. According to Adiyodi (1972), in *Paratelphusa*, an intermediate segment (the merus) is the first to develop during the differentiation of the regenerating leg, followed by the remaining segments.

An efficient and quite rapid regenerative process is known for the five pairs of pereiopods when these appendages are detached at the level of their PBP. Efficiency and rapidity, however, are often different for the different pairs of legs of the same specimen and were found to be influenced by both internal factors such as the developmental stage and the physiological status of the specimen (Paulian 1938; Spivak & Politis 1989; Juanes & Smith 1995), and external factors such as environmental temperature, water salinity and concentration of some chemicals (Bliss 1960); also population density was found to affect the regenerative process (Juanes & Smith 1995).

Apart from the regeneration associated with the PBP, regenerative processes are also known from points distal to the PBP or even proximal to it. Lost dactyli of the chelipeds, for instance, can regenerate through the formation of a swollen hard tip, within which a new dactylus grows until its emergence at the first molt (Bliss 1960). In all these cases, however, growth and differentiation are slower and less efficient than regeneration from the PBP and the regenerate is not always complete.

Brachyurans are able to regenerate more than one leg at the same time. Autotomy of more than one leg occurs frequently in free-living specimens, and in these cases, all appendages undergo regeneration. Sometimes, when more than one leg is affected, the frequency of molts is increased. The growth of regenerating appendage(s) can limit the growth of intact limbs, particularly when several appendages are regenerating at the same time (Hopkins 1985). When different appendages are detached at different times, hormone-controlled regulative processes have been observed to synchronize the growth schedules of the regenerating appendages (Skinner 1985; Mykles 2001).

Regenerated legs show the same properties as the original legs in terms of ability to (re)autotomize and regenerative potential. The PBP, in particular, is reproduced very early during the regenerative process and appendotomy can be demonstrated well before the regeneration of the appendage is completed (Hopkins 1993). Thus, a leg may regenerate more than once (e.g., McConaugha 1991).

The regenerate acquires full size usually within two or three molts after detachment, but sometimes requires only one molt (e.g., Ameer Hamsa 1982).

In heterochelous brachyurans, the asymmetric pair of chelipeds is composed of two morphologically and functionally different limbs. The regeneration of one of these appendages, e.g., left or right cheliped, can affect the asymmetric condition. In some species (e.g., some *Uca*), regeneration produces a cheliped that is subequal to the original limb and the asymmetric pattern is maintained (e.g., Morgan 1920, 1923; Yamaguchi 2001). However, in other species of *Uca* the regenerate may be either subequal to the original cheliped or of the opposite type, depending on the developmental stage when regeneration occurs and on the reciprocal regulation between the two chelipeds. In some species of *Uca*, indeed, appendotomy and subsequent regeneration of a cheliped of the first pair is a regular event during male development and seems to be required to release differentiation of two asymmetric chelipeds, something of vital importance to the adult male crab (Hartnoll 1988; Yamaguchi 2001).

2.3.7 *Decapoda other than Brachyura*

A PBP at the joint between basis and ischium of the legs was documented in some, but not all, of the investigated species of non-brachyuran decapods (reviews in Wood & Wood 1932; Bliss 1960). This PBP is functionally weak or even not detectable at all in some species with morphological and behavioral adaptations to a fossorial life, such as the anomurans *Hippa* and *Emerita talpoidea*, as a probably derived condition (Wood & Wood 1932; Weis 1982; Juanes & Smith 1995). Instead, a different PBP, one at the joint between coxa and basis, was documented in *Hippa* (Wood & Wood 1932) and possibly also in the palinuran *Willemoesia* (Calman 1909).

A PBP was also documented in the antennae, between third and fourth article, in some species at least (e.g., *Palinurus*; Wood & Wood 1932). Other kinds of appendages such as the uropods, however, seem to lack a PBP (e.g., Toyota et al. 2003).

Autotomy of the legs was well documented in most investigated anomurans, as well as in some other non-brachyuran decapods (e.g., Wood & Wood 1932). However, the autotomic properties are often different among different pairs of legs: in some Astacidea (e.g., *Homarus*, *Cambarus*) and Thalassinidea (*Gebia* = *Upogebia*), autotomy was documented for the first cheliped pair only, while only autospasy was found in the walking legs. In some pagurids, autotomy was documented for all the three anterior pairs of pereiopods, but only autospasy for the two posterior (reduced) pairs; in other decapods (e.g., *Palinurus* and *Galathea*), similar autotomic properties were reported for all five pairs of pereiopods (Wood & Wood 1932; review in Bliss 1960).

Full regeneration of legs from the PBP was well documented in many non-brachyuran decapods (Bliss 1960; Vernet & Charmantier-Daures 1994), as was the regeneration of the antennae (e.g., in *Procambarus*: Mellon & Tewari 2000). A comprehensive list of early studies on the regeneration of different appendages in a number of decapod species was presented by Przibram (1909).

In anomurans the process of regeneration is very similar to that observed in brachyurans (see under 2.3.6), the limb bud growing and differentiating inside an external cuticular sac and emerging only later. In other decapods, conversely, an external limb bud emerges early after the detachment of the scab formed at the PBP; growth and structural differentiation of the limb proceed gradually during the intermolt period; in particular, the chela develops from a longitudinal furrow at the tip and the articular joints develop from annular, transversal furrows along the limb. As to the temporal sequence during which the different joints appear, there is disagreement among reports that deal with different species (e.g., Nouvel-van Rysselberge 1937; Bliss 1960; Govind & Read 1994; Read & Govind 1997a, b). It is not clear how much this reflects actual interspecific differences rather than the different quality of the studies.

In the king crab *Paralithodes camtschatica*, the regenerate requires four to seven molts, and up to seven years, to recover full size (Niwa and Kurata 1964; Edwards 1972).

Regenerative processes have been documented also from points different from the PBP as well as in appendages lacking any PBP (e.g., Bliss 1960). Experimental work on the uropods of *Marsupenaeus* documented high variability in the shape of the outgrowths produced after removing these appendages (Toyota et al. 2003). Ablation of eye stalks, which is commonly practiced in industrially reared decapods, usually does not induce any regenerative process. However, sometimes it results in heteromorphic regenerates and occasionally a full-size and structurally complete eye stalk can be reproduced (*Penaeus*: Desai & Achuthankutty 2000).

The effect of appendotomy and regeneration of one of the two heterochelous chelipeds on the heterochely of the same pair of appendages has been investigated in some non-brachyurans decapods (Govind & Read 1994; Read & Govind 1997a, b; Mariappan et al. 2000). Research on *Alpheus* documented high plasticity in the regenerative program and the regulative role of the underlying asymmetric nervous ganglia. In particular, the original asymmetric pattern of the first pair of chelipeds may be completely reversed or even changed into a symmetric pattern, producing a pair of subequal chelipeds (Govind & Read 1994; Read & Govind 1997a, b).

2.4 Hexapoda

2.4.1 Collembola

No evidence is available about a possible PBP.

Little is known about regeneration. The antennae of *Orchesella*, *Tomocerus*, *Folsomia* and *Heteromurus* can regenerate, but never obtain full size and usually regenerate just one segment, sometimes two. The regenerated segment(s) can be longer than the original one(s) (Ernsting & Fokkema 1983).

2.4.2 Diplura

The cerci of *Campodea* seem to show a PBP and, thus, some kind of appendotomy (Condé 1955).

Regeneration occurs in the antennae (Condé 1955): in the regenerate, the distal-most segment is much longer than in the original appendage (Condé 1955). Regeneration is also possible in legs (Lawrence 1953).

2.4.3 Archaeognatha and Zygentoma

The maxillary palps and legs of machilids usually break at a PBP (Wygodzinsky 1941).

Regeneration of the antennal segments, of the three posterior appendages and at least tibia and tarsus of the legs has been recorded in *Machilis* (Przibram & Werber 1907) and in *Thermobia* and *Lepisma* (Sweetman 1934). The pattern of sensilla in the regenerated palp of *Thermobia* shows irregularities that are not corrected during subsequent molts (Larink 1983).

2.4.4 Ephemeroptera

In the leg of mayflies, there is a PBP between trochanter and femur (Nilsson 1986).

Regeneration occurs, but regenerated legs are small and malformed. If still undersized at the last nymphal instar, the legs fail to grow to full size in the adult (Nilsson 1986). Mayfly nymphs can also regenerate antennae, posterior appendages and lateral gills (Przibram 1909).

2.4.5 Odonata

In zygopterans, the leg has a PBP between trochanter and femur. No muscle crosses this articulation and there is no evidence that it is functionally jointed. A fibrous diaphragm, with a small central gap, separates trochanter and femur. The presence of a large muscle in the trochanter suggests the existence of autotomy (Child & Young 1903).

Zygopterans have high regenerative capabilities. Regeneration (Child & Young 1903) can occur after cuts at any level along the leg. Cuts at different levels along the three-segmented tarsus produce different results. A cut at the base of the most distal tarsomere results in an incompletely regenerated tarsus. Amputation at the base of the second tarsomere results in disintegration of the remaining proximal tarsomere, and the subsequent regeneration from the tibio-tarsal articulation produces a complete tarsus within five or six molts. After the first molt, the tarsus is composed of only one segment and the claws; after the second molt, it has two segments, and after three to four additional molts its full complement of three segments. Amputations in the distal part of the tibia can result in regeneration from the cutting plane, while more proximal cuts, as well as amputations along the femur, produce appendotomy. Regeneration from the PBP is rapid, but incomplete regarding number of segments and function. After the first molt, the leg has an unsegmented tarsus with unarticulated claws of reduced size and unusual shape. Following the next molt, the tarsus almost always has two segments, but additional molts fail to produce a three-segmented tarsus with functional claws. Regeneration from levels more proximal than the PBP is also possible and follows the same path just described.

Child & Young (1903) observed that, during regeneration, the development of joints is closely correlated with the development of muscle insertions. Their interpretation is that muscles can apparently not keep pace with the growth of the exoskeleton during regeneration, and thus, influence the segmentation process. By contrast, during undisturbed development, the tegumentary structures grow more slowly and muscles can adequately attach to them.

Leg regeneration has been also described in anisopterans (see Przibram 1909). Zygopterans can regenerate the tracheal gills (Child & Young 1903) and in the anisopterans *Anax imperator* and *Aeshna cyanea*, Degrange & Seassau (1974) reported the regeneration of the mask.

2.4.6 *Blattodea*

In the leg of cockroaches, a PBP exists between trochanter and femur, but there is no diaphragm and two muscles cross this articulation (Bordage 1905).

Autotomy is documented in this group (e.g., Penzlin 1963).

Cockroaches have well-developed regenerative capabilities. Regeneration can occur after cuts at any level along the leg. After the first post-operative molt, the regenerating leg is always composed of the final number of segments. However, in *Blaberus craniifer*, amputations along the coxa are followed by a 'two-phase' regeneration, during which an unsegmented bud appears at first. It takes one more molt to produce a segmented leg. In *Blaberus craniifer* and in *Blattella germanica,* experimental amputations distal to the midlength of the third tarsomere produce a regenerate with the full number of five tarsomeres, while more proximal amputations produce a regenerate with four tarsomeres only (Bullière & Bullière 1985; Tanaka et al. 1992). By contrast, in *Periplaneta americana* and *Panchlora maderae* (= *Rhyparobia maderae*), every amputation made along the tarsus results in the loss of the remaining tarsal segments (apparently, a kind of appendotomy), and the subsequent regeneration begins at the tibia-tarsus articulation. In these species, regeneration invariably produces a four-segmented tarsus (Bordage 1905; Penzlin 1963). Amputation along the femur induces appendotomy in *Blaberus craniifer* (Bullière & Bullière 1985).

The length of the regenerate is correlated with the time of amputation within the intermolt period (Penzlin 1963). In *Periplaneta americana*, Kaars et al. (1984) observed that in the undamaged femur, nerve and trachea are closely associated and branch together at regular intervals, while in regenerate femurs, nerve and trachea are not closely associated and have a different branching pattern. Their interpretation is that tissue-level interactions during regeneration differ from those during embryogenesis. In *Periplaneta*, wound healing following appendotomy involves cell movement and cell division only in the distal half of the trochanter, while the formation of the blastema involves cell movement and cell division in the temporarily and reversibly fused trochanter and coxa (Truby 1985).

The antennae can also regenerate (Penzlin 1963; Urvoy 1963; Schafer 1973). At the first post-operative molt, the regenerate is composed of at least the first three articles

(Urvoy 1963), and the number of flagellar segments increases with subsequent molts (Schafer 1973).

Palps and cerci also regenerate (Penzlin 1963; Urvoy & Les Bris 1968). Following the first post-operative molt, the new appendage is often merely a bud or has reduced segmentation. The full number of segments is obtained after three to five molts, depending on the level of amputation and the time it is performed within the intermolt period (Urvoy & Les Bris 1968).

2.4.7 *Isoptera*
There is only indirect evidence of a PBP between trochanter and femur (Myles 1986).

It seems that legs can regenerate and attain full size, probably in approximately three molts (Myles 1986).

2.4.8 *Mantodea*
In this group (Bordage 1905), there is a PBP at the joint, marked externally by a furrow, between trochanter and femur, which are largely fused together.

Studies on autotomy and regeneration were carried out by Bordage (1905) on several species including *Mantis religiosa* and *M. prasina* (= *Paramantis prasina*). At least in these species, only meso- and metathoracic legs autotomize. Autotomy is produced by the strong extensor muscle that crosses the PBP. By contracting, it retracts in part into the trochanter. At the level of the PBP, there is no internal diaphragm. As a result, after autotomy, hemorrhage is only partly avoided by obstructing muscle fibers.

Regeneration following autotomy is usually fast and the legs are fully functional after the first molt. The regenerated leg, always with tetramerous tarsus, is usually smaller and slightly lighter than the undamaged leg (with pentamerous tarsus), but apparently does not differ from the former in ornamentation. Malformations are very rare, besides incomplete subdivision of the tarsus. Amputation distal to two thirds of the femur triggers appendotomy. Cuts in the last two tarsomeres do not produce any regenerate.

2.4.9 *Orthoptera and Dermaptera*
Legs have a PBP between trochanter and femur (Bordage 1905; Brousse-Gaury 1958). It is not crossed by muscles and a diaphragm prevents loss of hemolymph (reported for crickets by Graber as early as 1874).

According to Bordage (1905), only the jumping legs autotomize. However, in *Acheta*, autotomy occurs just more frequent in jumping legs than in the other legs (Brousse-Gaury 1958), and in *Scudderia texensis*, autotilly was found in the first two pairs of legs and autospasy in the jumping legs (Dixon 1989).

The regenerative power varies considerably within the orthopterans. In general, according to Lakes & Mücke (1989), the power of regeneration decreases within the Orthoptera, from crickets to tettigoniids to locusts. It is very low in *Ephippiger ephippiger* (Lakes & Mücke 1989). Interestingly, in newly hatched specimens, the amputation of the forelegs at the joint between femur and tibia leads to slow regeneration extending through all six instars. Eventually, a dimerous or trimerous regenerate is produced, composed of a tibial segment, a tarsal pad and the terminal claws. This regenerate is also reduced in size (just one quarter of normal length), sensory structures and neuronal pathways. In *Gryllus domesticus* (= *Achaeta domesticus*), the growth rate of the regenerating appendage is higher than in undisturbed development, and there is complete regeneration of the leg up to the tarsus, independent of the level of amputation. However, the level of amputation does affect the size of the first regenerate and the number of regenerated spines, which, in any case, is lower than in the undamaged legs (Rościszewska & Urvoy 1989b). The number of spines and the spatial distribution of external sensory organs require a higher number of molts to reach full condition. In *Teleogryllus commodus*, regeneration is complete in gross morphology, but does not re-establish the full pattern of sensory structures such as subgenual and tympanal organs, and campaniform sensilla (Biggin 1981). According to Bordage (1905), in gryllids and in tettigoniids, the regenerate lacks the tympanal organ, but Huber (1987) reported complete regeneration of the tympanal organ in *Gryllus bimaculatus*, if amputation occurs at or distal to the femur-tibia joint.

In several tettigoniids, acridids and gryllids studied by Bordage (1905), the jumping legs never regenerated except for the tarsus. However, in the first two pairs of legs, extensive regeneration occurred if the specimen was young enough and the cuts separated trochanter from femur. By contrast, sectioning between coxa and trochanter resulted in a regenerate that was a more or less rudimentary stump. The tarsus, which was often lost during exuviation, always regenerated, although its growth was very slow and the resulting appendage was not completely functional (in general the tarsomeres were slightly different from those of an undamaged appendage). In the tettigoniids *Phylloptera laurifolia* and *Conocephalus differens* (= *Ruspolia differens*), the regenerate was tetramerous as the undamaged limb, in acridids and gryllids, trimerous as the undamaged limb (Bordage 1905). According to Chopard (1938), the regenerating tarsus never exceeds the number of four articles, irrespective of the number characteristic of the species (up to five articles). The first regenerate may still have a lower number of tarsal articles, and further articles are added during additional molts.

Chopard (1938) reported that regeneration of the orthopteran antenna is easily accomplished if part of the flagellum is removed, while it seems more difficult when the cut affects the two basal articles, which often results in anomalies, including heteromorphosis. The regenerated antenna is usually smaller, and in gryllids, individual articles are longer and different in shape. In *Acheta*, the size of regenerate and number of regenerated articles increase with subsequent molts (Rościszewska & Urvoy 1989a).

In earwigs, there is evidence of regeneration of the antennae (Przibram 1909; Chopard 1938).

2.4.10 *Phasmida*

Studies on autotomy and regeneration were carried out in *Bacillus rossius* by Godelmann (1901), in *Monandroptera inuncans*, *Raphiderus scabrosus* and *Eurycantha horrida* by Bordage (1905) and in *Carausius morosus* by Schindler (1979).

There is a PBP at the virtual (ankylosed) joint between trochanter and femur. Unlike the condition in cockroaches and mantids, no muscle crosses the PBP or is even inserted in the trochanter and there is a diaphragm crossed by nerves and tracheae only (Godelmann 1901; Bordage 1905).

Autotomy is triggered by excitation of the sensitive nerve of the leg. As in decapods, this process also occurs in beheaded animals. Sometimes autotomy does not work perfectly and the leg remains partly attached (Bordage 1905). Autotomy is very common in all instars, especially in nymphs after the third molt, but it can also occur in adults, though it is much more difficult to trigger (Bordage 1905).

The regenerative power seems to be higher in young nymphs than in more aged ones and depends on the positional identity of the amputated limb: midlegs have the highest regenerative power, hindlegs the lowest (Godelmann 1901).

After the first post-operative molt, the leg regenerates with all final segments, even though it is usually reduced in size and ornamentation, and sometimes even ill-formed and not clearly segmented (Bordage 1905).

In *Bacillus rossius*, ablation of the last tarsomere or the last two tarsomeres does not trigger regeneration, while other cuts along the tarsus result in a regenerate of one or two segments. Amputations within the length of the first tarsomere cause the remaining tarsal stump to fall off, and the subsequent regeneration from the undamaged tibia produces a three- or four-segmented tarsus while the undamaged tarsus is five-segmented. Cuts at any level along the tarsus of the hindlegs are usually followed by appendotomy. Amputation in the distal part of the tibia leads either to appendotomy or to direct regeneration of a three- or four-segmented tarsus. Amputations along the femur always lead to autotomy. Regeneration from the PBP usually produces a four-segmented tarsus, but for amputations in the first instar a five-segmented tarsus was regenerated in seven cases out of 50. In one case, a nymph autotomized in the first instar regenerated at first a tarsus with four tarsomeres that changed into a five-segmented tarsus during the next molt (Godelmann 1901).

In *M. inuncans*, *R. scabrosus* and *Phyllium crurifolium*, the regenerated leg always has a four-segmented tarsus; cuts in the distal part of the tibia, or more distal to it, except for the last two tarsomeres, regenerate directly, while proximal cuts lead to autotomy. Interestingly, in *R. scabrosus*, amputation along the tibia produced a three-segmented tarsus that became four-segmented only after an additional molt (Bordage 1905).

If regeneration is induced from the distal part of the tibia, the femur also seems to contribute to the regenerative process since it becomes shorter and this shortening is not compensated for during subsequent molts (Godelmann 1901). Similarly, Bordage (1905) reported that coxa and trochanter both become shorter during regeneration following appendotomy. It is also noteworthy that regeneration from the PBP is faster than from any point distal to it (Bordage 1905).

The regenerative power of cerci and antennae is poor (Godelmann 1901). Urvoy (1970) carried out studies on regeneration of the antenna in *Sypyloidea sypylus*. Antennae were sectioned at various levels, using nymphs of various stages at different times during the intermolt. The regeneration potential decreased with age of the operated specimen, as indicated by smaller size and less numerous sensilla in the regenerate. Individual variation was observed, including differences between the two antennae of the same specimen. The nature of the regenerate depends on the level of sectioning. If it occurs proximal to the middle of the first antennal segment (the scape), there is no regenerate. If sectioning takes place between mid-scape and the articulation between second (pedicel) and third segment, the regenerate is a heteromorphic tarsus, more or less developed depending on the section level - the more distally the section is applied, the more developed is the appendage. In the heteromorphic appendage, the terminal part of the leg is always present, and the number of articles is fixed after the first molt. In some cases, there is a later increase in the size of the appendage and the number of sensilla. Finally, if the cut is distal to the proximal articulation of the third segment, the regenerate is an antenna in which the number of flagellar articles depends on age and level of the cut.

2.4.11 Hemiptera

A PBP has not been reported in this group.

Regeneration of lost appendages is poor and in several groups, especially in the homopterans, there is no regeneration at all. Experimental evidence, limited to a few heteropterans, showed that the regenerated leg is always reduced in size and often possesses an incomplete number of segments. No correlation with the nymphal stage at the time of operation has been noticed and, interestingly, the growth rate of the regenerating leg is not different from that of the opposite undamaged one (Lüscher 1948; Shaw & Bryant 1974).

There seems to be no regeneration following amputation proximal to the femur-tibia articulation. What remains of an amputated segment can be lost, e.g., following amputation at some level along the femur in the reduviid *Rhodnius prolixus*. In this species, the occasional appearance of a new, small terminal segment has been reported (Lüscher 1948). In the lygaeid *Oncopeltus fasciatus*, amputation in the middle of the femur results in a longer and deformed femur, sometimes followed by a new, small terminal segment (Shaw & Bryant 1974). Amputations at the femur-tibia articulation can result in a regenerative process or follow the same pattern as proximal cuts (Lüscher 1948; Shaw & Bryant 1974).

More distal amputations always start a regeneration process producing a leg with reduced segmentation. The final result depends on the level of the amputation and the number of molts ahead, since the regenerating leg will slowly increase the number of segments from instar to instar. In *R. prolixus* and *O. fasciatus*, where the number of tarsal segments should increase from two to three during the last molt, the final (adult) regenerate is sometimes a 'nymphal leg', i.e., one with two tarsomeres only. Regenerating legs have been documented following early amputation (before the first nymphal molt) in the middle of the tibia or more distal to it (Lüscher 1948; Shaw & Bryant 1974). In these cases, the number of tarsomeres increases from two to three during the last molt, as in an undamaged leg.

Cutting off the three distal segments of the antenna of *O. fasciatus* (the undamaged antenna has four segments) results in the regeneration of one segment, sometimes two. When regeneration is limited to one segment, the original bristle patterns of the two distal segments of the antenna is very often maintained. Sometimes, the regenerate is incompletely divided in two segments, and a complete division will not be attained during the remaining molts. In the regenerated antenna, the segments (including those left after amputation) usually become thicker and longer than expected (Shaw & Bryant 1974). The same was reported in the pentatomids *Raphigaster nebulosa* and *Euchistus variolarius* and several lygaeids (Wolsky 1957 and references therein). Removal of the two terminal segments of the antenna was followed by regeneration of only one segment (Wolsky 1957).

2.4.12 Endopterygota

There is no evidence that the larvae of holometabolans possess a PBP.

Evidence of regeneration is limited and the insect's response to the loss or breakage of appendages is far from uniform even within one order, as in Coleoptera. In this order, for example, regeneration of larval legs has been recorded in the tenebrionid *Tenebrio molitor*, the dynastid *Oryctes nasicornis*, the cerambycid *Rhagium indagator* (= *Hargium inquisitor*) (Megušar 1907), and to a limited extent also in the chrysomelid *Leptinotarsa decemlineata*, but in this case only following amputation in the first larval instar (Patay 1937; Poisson & Patay 1938). Lack of regeneration, however, has been reported for species of hydrophilids and dytiscids (Megušar 1907), and also in the chrysomelid *Timarcha* (Abeloos 1933;

Bourdon 1937).

The larval antennae of the moth *Lymantria dispar* regenerate their three articles passing through a stage of unsegmented bud (Kopeć 1913).

3 EVO-DEVO PERSPECTIVES ON ARTHROPOD APPENDAGE REGENERATION

3.1 *Mechanisms*

Before addressing specific questions about the diversity of regeneration processes in arthropod appendages and its possible evolutionary significance, it seems worthwhile to ask whether the comparative data summarized in the previous pages can help understanding regeneration and its mechanisms in more general terms.

The idea that regeneration is but a copy of 'normal' development has often been raised (e.g., Przibram 1909; Needham 1965). Taken literally, this idea is at the best grossly naive, nevertheless, it might be worth reconsidering, to some limited extent. We must first qualify this concept, however, since what we might call 'normal' development refers to two different processes, the equivalence of which is a matter for further speculation. In the vast majority of arthropods (a peculiar exception are polyembryonic wasps), normal development means embryogenesis, but in other metazoans, instead of embryogenesis - or in addition to it - it could mean blastogenesis. Indeed, several authors (e.g., Dehorne 1916; Berrill 1952; Herlant-Meewis 1953) highlighted asexual reproduction, when comparing regeneration with 'normal' development. This is obviously relevant for groups such as annelids, for which this comparison has recently been revived. Bely & Wray (2001) interpreted fission as a derived condition from pre-existing regeneration mechanisms recruited for a new role. For arthropods, that lack comparable reproductive mechanisms, we could adopt a speculation suggested by Sánchez Alvarado (2000). This author proposed that embryonic limb buds might be phylogenetically derived from the same, locally acting 'genetic organization' originally deployed in the regeneration of damaged body parts. This idea comes close to Minelli's (2000) notion of axis paramorphism, according to which the appendages are evolutionarily divergent copies of the main body axis.

When comparing regeneration to undisturbed developmental processes, one may be tempted to distinguish between adaptive regeneration following autotomy and regeneration as a general property of multicellular organisms. Such a clear-cut distinction, however, is unwarranted. Regeneration for certain traits has likely been shaped by natural selection. These include sophisticated mechanisms of autotomy with specialized autotomizer muscles and PBPs with protecting diaphragms and efficient re-arrangement of muscles, blood spaces and, when present, tracheae (e.g. crabs, cockroaches). In the case of the fiddler crabs, regeneration following autotomy has even become a developmental mechanism required to break up the initial body symmetry, which allows one of the male chelipeds to grow to its characteristic enormous size. In other cases, however, regeneration fails to produce a fully functional appendage, suggesting simple reactivation of growth and morphogenetic processes along unspecific pathways. For example, a dragonfly nymph may starve to death during the slow regeneration of the mask.

The fact that this local growth triggered by the loss of the appendage gives rise to a patterned regenerate rather than to a shapeless clump of cells has nourished the widespread belief in the existence of a directed program. However, we should keep in mind that the regenerate is not produced in a vacuum of gene expression, but in a spatially and temporally well-defined and information-rich setting. In this context, it may be significant to note that regenerating a lost small, apical part of an appendage may be much more difficult than regenerating a whole appendage from a proximal PBP, in terms of time (or molts) required to obtain the regenerate as well as the completeness of the latter. In some decapods at least, vastly different mechanisms seem to be at work in the two cases, possibly due either to different availability of metabolic supply, or to different positional information, or to both.

3.1.1 *Regeneration... of what?*

Defective regeneration shows how much morphogenesis depends on epigenetic interactions (*sensu* Schlichting & Pigliucci 1998) between nerves, muscles, tracheae and epidermis during undisturbed growth and differentiation of arthropod appendages. During regeneration, nerves, muscles and tracheae often seem to not keep pace with the rapidly growing and differentiating epidermis (but sometimes, as in the hemipteran antenna, it seems to be the other way around). This was remarked as early as 1903 by Child and Young in their study of regenerating legs in damselflies and is reflected in the frequent defects found in the arrangement of sensilla along regenerated appendages.

We should perhaps say that the regeneration of an arthropod appendage is not the product of a distinct 'modular' process. In terms of mechanisms, it would be more appropriate to distinguish between regeneration of epidermis and cuticle, regeneration of nerves, regeneration of tracheae, and regeneration of muscles. The extent to which these processes are actually synchronized, eventually coupled together and limiting each other, will obviously determine the anatomical and functional quality of the regenerate. However, the latter should not be regarded as the product of a 'local program' or even of the 'local re-deployment of a limb-producing program', because such program do probably not exist.

The common occurrence of a mismatch between the segmentation of the new appendage and the patterning of the sensilla along its proximo-distal axis also indicates independence of developmental events during the regeneration of the same appendage. A good example is the regenerated antenna of the isopod *Idotea,* which, following total ablation (a small part of the head also included), consists of epidermis and cuticle only, without any nerve or muscle (Bossuat 1958).

Quite often, the regenerate obtains the full number of segments, but these have an irregular set of sensilla (e.g., on the palps of the silverfish *Thermobia*; Larink 1983). In other cases, a more or less complete number of sensilla are developed on the regenerated appendage. However, when the regenerate is incompletely segmented, one 'double' segment can bear all the sensilla usually distributed over two segments. We discussed an example of such a condition in the section on the regenerating antenna of heteropterans. It is also similar to the regular condition in the antennule of adult males of some calanoid copepods (Boxshall & Huys 1998), and in the antennule of the isopod genus *Mancasellus* (= *Lirceus*) (Racovitza 1925).

3.1.2 *Local growth and competition*

The study of regeneration of arthropod appendages suggests that an organism is, to some extent, a mosaic of independent developmental domains (Paulian 1938). In other terms, growth and patterning of a regenerating appendage are largely autonomous from the remaining body. However, the regenerating appendage must compete for resources (Klingenberg & Nijhout 1998; Nijhout & Emlen 1998). The local rate of growth during regeneration can be exceptionally high, to the extent that the neighboring appendages are negatively affected by it (Hopkins 1985). Not surprisingly, therefore, production of a regenerate is sometimes accompanied by regression in size of segments of the appendage proximal to the level of the cut, as mentioned above for some species of stick insects. In decapod crustaceans, strong regression of ganglia and muscles proximal to the PBP is also common.

The convergence of the regenerate in size and complexity with the opposite appendage is extensive, and not easy to explain only in terms of competition for a shared pool of resources. In addition, this convergence is certainly not an absolute rule, as shown by the opposite (divergent) trend observed in heterochelous decapods such as *Uca*. In fiddler crabs, the loss of an appendage becomes a signal for a break of symmetry, starting a special allometric pathway of appendage growth.

3.1.3 *Regeneration, cuticle and mitosis*

Growth and differentiation are both involved in regenerating an appendage. Minimum requirements for growth are the detachment of the epithelium from the cuticle in a region proximal to the cut or the PBP, the recruitment of additional cells that will eventually result in a local blastema and the activation of mitosis therein. In mites other than ticks, lack of regeneration correlates positively with the lack of post-embryonic mitoses, and a similar explanation does perhaps apply to other small arthropods such as copepods, which apparently have very poor regeneration capability. Current evidence, however, does not cover a sufficient range of taxa to allow further speculations.

The detachment of the epidermis from the cuticle is probably required for morphogenesis no less than for cell proliferation. Similar to apolysis during the molting cycle, detaching the epidermal cells from the cuticle may release their mitotic potential, and, in addition, releases them from the morphostatic role of the cuticle (Minelli 2003b).

Generally speaking, regeneration is more effective in arthropods with higher and/or indeterminate number of molts than in those with a tight post-embryonic developmental schedule including a small, fixed number of molts: for example, regeneration is more conspicuous in isopods than in copepods, in cockroaches than in hemipterans. This point would require more systematic investigation.

3.1.4 *Appendotomy and regeneration*

The most efficient performances in regeneration are generally a follow-up to appendotomy, in particular autotomy. The two processes, however, are not necessarily interconnected. Most conspicuous is the lack of regeneration following autotomy of the jumping legs in the Orthoptera. This is surprising because the chances of survival (not to mention reproduction) of an autotomized specimen are likely to be dramatically small. The lack of regeneration of the appendotomized legs of some spiders and harvestmen is probably of lesser consequence for the fitness of the animal, if appendotomy is limited to one or a few legs.

3.2 Segmentation

3.2.1 Polarity of segment differentiation

In the new appendage, the tip is usually formed first. In scorpions, regeneration usually does not proceed further than the pretarsus. In this case, the tip of the appendage seems to generate a morphogenetic control over more proximal, original segments. For example, spine-like hairs typical of basitarsus and tarsus do appear on conserved pretarsal segments following amputation of the whole tarsal section, or more (Rosin 1964).

Unfortunately, evidence is poor as to the temporal order with which segmentation progresses in the regenerating appendage: literature data are often unreliable and sometimes conflicting, e.g., for decapod crustaceans. It seems safe to say, however, that in non-flagellar appendages segmentation proceeds neither in proximal to distal nor in distal to proximal sequence. The same was found by Norbeck & Denburg (1991) in the embryonic development in *Periplaneta*. For example, in the spider *Coelotes terrestris*, the regenerating equivalent to four distal-most leg articles (patella, tibia, metatarsus, tarsus) is at first represented by one segment, which later splits into two segments. Subsequently, the proximal segment is subdivided into patella and tibia, the distal segment into metatarsus and tarsus. This sequence agrees with the temporal sequence in undisturbed development (Vachon 1967).

3.2.2 Post-embryonic developmental schedules and regeneration: appendages with 'true segments' and 'flagella'

Post-embryonic development varies widely within arthropods, and a comparison between developmental schedules and regeneration processes suggests intriguing relationships.

Segmentation of the main body axis can follow two different modes. In epimorphic development, all body segments are already present at the end of embryonic development, whereas in anamorphic development juveniles hatch with an incomplete complement of segments. In the latter case, the final adult number of segments is reached later in ontogeny through a specific schedule of post-embryonic segment addition (Enghoff et al. 1993). Similarly, full segmentation of the developing appendages can be complete at their first appearance, or only later in ontogeny.

In the regeneration of appendages, the definitive number of segments in the regenerate (sometimes lower than the full number) is often complete within the first post-operative molt (e.g., in the legs of cockroaches and the spider *Dolomedes*). However, sometimes the number of segments increases according to an 'anamorphic' schedule (e.g., in the legs of zygopteran dragonflies). To some extent at least, the 'anamorphic' versus 'epimorphic' mode of regeneration matches the developmental mode, especially when comparing antennal development with antennal regeneration. The increase in the number of segments in the regenerating appendage recorded by Vachon (1956, 1967) for the spider *Coelotes* might seem an exception. However, although spiders are generally classified as epimorphic arthropods, their first free stage is embryo-like, and its appendages are still incompletely segmented. It was exactly this kind of 'larva' that Vachon investigated.

When different appendages of an arthropod or different parts of the same appendage can be contrasted as "truly segmental" versus "flagellar", their behavior during regeneration is generally also distinctly different. The number of flagellar units in the regenerate increases from molt to molt, whereas all 'true segments' are usually formed as soon as the regenerate emerges first. In appendages with flagellar organization, the number of flagellar units in the regenerate is variable around a modal value that is mostly lower than in the undamaged appendage, but sometimes higher. An example is the flagellar tibia and tarsus of the whip-like forelegs of the Amblypygi.

3.3 Serial homology

What we know about regeneration of appendages in arthropods is largely based on legs and, to a much lesser extent, on antennae. However, whenever studies have been carried out, evidence of regeneration has been found for most kinds of appendages. For example, some spiders regenerate legs, palps, chelicerae, the labium and even the spinnerets (Bonnet 1930; Mikulska et al. 1975) and some decapods regenerate eyes and their stalks, gills, copulatory organs and uropods (Vernet & Charmantier-Daures 1994).

Evidence for the mouthparts of arthropods other than chelicerates is very limited. It has been claimed (von Buddenbrock 1954) that those of decapod crustaceans do not regenerate at all. As for insects, evidence is limited to palps and the mask of dragonfly nymphs.

Experiments on both antennae and legs of the same species (and possibly by the same author) are rare. Available studies do not show major differences in regenerative behavior between legs and other appendages: i.e., they either show regeneration or no regeneration at all.

Mechanism of appendotomy and regenerative power, however, are not necessarily uniform even among similar appendages of the same animal, e.g. among the thoracic legs of an insect.

Differences in regeneration power do not correlate with the degree of specialization of the appendage: in crabs, chelipeds and walking legs may exhibit similar performances in regeneration, while the three pairs of legs of stick insects perform differently despite their broadly similar morphology. Mechanisms of appendotomy, however, are often different in appendages with different specialization, for example jumping versus walking legs in orthopterans or raptorial versus walking legs in mantises. However, while in orthopterans the walking legs do not autotomize, the jumping legs do. In mantises, on the other hand, the specialized raptorial legs do not autotomize, while the walking legs do. We wonder whether this difference reflects a plesiomorphic positional gradient of increasing autotomy from fore to mid to hindlegs rather than the divergent adaptive specializations of these appendages.

3.4 Phylogenetic patterns

Due to the limited and irregularly scattered taxonomic sampling of our current database on autotomy and regeneration of arthropod appendages, it is not yet possible to study these phenomena using standard phylogenetic methods. There is abundant evidence that regeneration performances are sometimes very different even among closely related arthropods. The best example of this variability are probably the beetles: larval legs, as mentioned before, do not seem to regenerate in Hydrophilidae and Dytiscidae, but regenerate in other families.

Nevertheless, the lack of PBP and thus of any kind of appendotomy in millipedes, xiphosurans, scorpions, and mites (with the possible exception of *Opilioacarus*, which represents a rather isolated clade within this group; cf. Vitzthum 1943; Brignoli 1967) is perhaps of phylogenetic significance. Some spiders also lack a PBP. In other spiders, the position of the PBP is variable (and there are even species that seem to possess more than one PBP). In centipedes the PBP is between the coxa and the trochanter. It seems significant that all the mentioned groups (a clade of chelicerates with myriapods, the Myriochelata of Pisani et al. 2004) lack real autotomy. Even the house centipede *Scutigera*, the only myriapod that loses its obviously fragile legs fairly easily in nature, seems to have autospasy or autotilly at most.

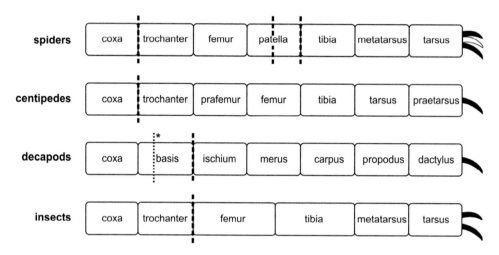

Figure 1. Position of the PBP along the leg in representatives of the major arthropod lineages. The appendages are drawn in very schematic and simplified way; in particular, possible secondary divisions of primary articles, e.g. in the tarsus, are ignored. The asterisk (*) marks the position of the PBP in the proximal part of the basis in isopods. Conditions in spiders are variable (see text).

The lack of PBP in the crustaceans other than malacostracans is probably also of phylogenetic significance. Only the remaining groups, malacostracans and hexapods, at least sometimes show real autotomy. The position of the PBP along the appendage is a further argument in favor of a Myriochelata/Pancrustacea hypothesis of higher relationships within the Arthropoda (Fig. 1), although caution is needed, owing to the incomplete data set available at the moment. Our comparison gives indirect support to Walossek & Müller's hypothesis (1997), that the most proximal articles of chelicerate (and trilobite) limbs are not homologous to the 'true' coxae of crustaceans. The PBP in insects seems to correlate well with the basis/ischium location in malacostracan crustaceans. Alternative locations of the PBP are probably a derived feature, especially when the PBP is not between two segments of the appendage, but within a segment, for example within the basis of the isopod leg. Another trait supporting a close relationship between malacostracans and hexapods is the presence of a PBP with a diaphragm that is not crossed by muscles. This condition has probably been lost at least twice in insect evolution. The Dictyoptera have retained a trochantero/femoral PBP, but apparently lost the diaphragm. In the Hemipteroidea and the (larval) Endopterygota, a true PBP has disappeared. On the other hand, a PBP with diaphragm, in a probably not-homologous position, has evolved in *Scutigera* within the Myriochelata. This is certainly not the only character in *Scutigera* that presents a convergence with insects.

ACKNOWLEDGEMENTS

We are grateful to many friends and colleagues for their advice on appendage regeneration in different arthropod groups, to Stefan Koenemann and two anonymous referees for their precious comments on a previous draft of this paper and to Ariel Chipman for a careful linguistic check of the final manuscript. We thank also Angela Mazzon, Monica Ortolan and Donata Pieri (Vallisneri Library, University of Padova) for their generous assistance with our extensive literature.

REFERENCES

Abeloos, M. 1933. Sur la régénération des pattes chez le coléoptère *Timarcha violaceo-nigra* De Geer. *C.R. hebdom. Séanc. Acad. Sci. Paris* 113: 17-19.

Adiyodi, R.G. 1972. Wound healing and regeneration in the crab *Paratelphusa hydrodromous* [sic]. *Int. Rev. Cyt.* 32: 257-285.

Agar, W.E. 1930. A statistical study of regeneration in two species of Crustacea. *J. Exp. Biol.* 7: 349-369.

Ameer Hamsa, K.M.S. 1982. Observationts [sic!] on moulting of crab *Portunus pelagicus* Linnaeus reared in the laboratory. *J. Mar. Biol. Ass. India* 24: 69-71.

Bain, B.A. 2003. Postembryonic development in the pycnogonid *Austropallene cornigera*. *Invertebr. Repr. Dev.* 43: 181-192.

Bateson, W. 1894. *Materials for the Study of Variation Treated with Especial Regard to Discontinuity in the Origin of Species*. London: Macmillan.

Belozerov, V.N. 2001. Regeneration of limbs and sensory organs in ixodid ticks (Acari, Ixodoidea, Ixodidae and Argasidae). *Russian J. Dev. Biol.* 32: 129-142.

Bely, A.E. & Wray, G.A. 2001. Evolution of regeneration and fission in annelids: insights from *engrailed-* and *orthodenticle-*class gene expression. *Development* 128: 2781-2791.

Berrill, N.J. 1952. Regeneration and budding in worms. *Biol. Rev.* 27: 401-438.

Biggin, R.J. 1981. Pattern re-establishment - transplantation and regeneration of the leg in the cricket *Teleogryllus commodus* (Walker). *J. Embryol. Exp. Morphol.* 61: 87-101.

Bliss, D.E. 1960. Autotomy and regeneration. In: Waterman, T.H. (ed.), *The Physiology of Crustacea, 1*: 561-589. New York: Academy Press.

Bonnet, P. 1930. La mue, l'autotomie et la régénération chez les araignées avec une étude des Dolomès d'Europe. *Bull. Soc. Hist. Nat. Toulouse* 59: 237-700.

Bordage, E. 1905. Recherches anatomiques et biologiques sur l'anatomie et régénération chez diverses arthropodes. *Bull. Sci. France Belgique* 39: 307-454.

Bossuat, M. 1958. Territoire de régénération des antennes de l'isopode *Idotea baltica* (Aud.). *C.R. hebdom. Séanc. Acad. Sci. Paris* 246: 2530-2532.

Bourdon, J. 1937. Sur la régénération des ébauches de quelques organes imaginaux chez le coléoptère *Timarcha goettingensis* L. *C.R. hebdom. Séanc. Acad. Sci. Paris* 124: 872-874.

Boxshall, G.A. & Huys, R. 1998 The ontogeny and phylogeny of copepod antennules. *Phil. Trans. R. Soc. London, Ser. B,* 353: 765-786.

Brignoli, M.P. 1967. Su *Opilioacarus italicus* (With). *Fragm. Entomol.* 5: 111-121.

Brousse-Gaury, P. 1958. Contribution a l'étude de l'autotomie chez *Acheta domestica* L. *Bull. Sci. France Belgique* 92: 55-85.

Bullière, D. & Bullière, F. 1985. Regeneration. In: Kerkut, G.A. & Gilbert, L.I. (eds.), *Comprehensive Insect Physiology, Biochemistry and Pharmacology,* 2: 371-424. Oxford: Pergamon Press.

Calman, W.T. 1909. *A Treatise on Zoölogy, Part 7, Appendiculata, 3rd Fascicle, Crustacea.* London: Adam & Charles Black.

Cameron, J.A. 1926. Regeneration in *Scutigera forceps. J. Exp. Zool.* 46: 169-179.

Charniaux-Cotton, H. 1957. Croissance, régénération et déterminisme endocrinien des caractères sexuels d'*Orchestia gammarella* (Pallas) crustacé amphipode. *Ann. Sci. Nat. Zool. Biol. Anim.* 19: 413-559.

Child, C.M. & Young, A.N. 1903. Regeneration of appendages in nymphs of the Agrionidae. *Arch. Entwicklungsmech. Organ.* 15: 543-602.

Chopard, L. 1938. *La Biologie des Orthoptères.* Paris: Paul Lechevalier.

Clare, A., Lumb, G., Clare, P.A. & Costolow, J.D. 1990. A morphological study of wound response and telson regeneration in postlarval *Limulus polyphemus* (L.). *Invertebr. Repr. Dev.* 17: 77-87.

Cloudsley-Thompson, J.L. 1961. A new sound-producing mechanism in centipedes. *Entomol. Monthly Mag.* 96: 110-113.

Coineau, Y. & Legendre, R. 1975. Sur un mode de régénération che les Arthropodes: la régénération des pattes marcheuses che les Opilioacarines (Acari: Notostigmata). *C.R. hebdom. Séanc. Acad. Sci. Paris, Sér. D, Sci. Nat.,* 280: 41-43.

Condé, B. 1955. Matériaux pour une monographie des Diploures Campodéidés. *Mém. Mus. Nat. Hist. Nat. Paris, Sér. A: Zool.,* 12: 1-201.

Darwin, C. 1854. *A monograph on the sub-class Cirripedia.* London: Ray Society.

Degrange, C. & Seassau, M. 1974. Sur la régénération du masque (= labium) des larves des odonates anisoptéres *Anax imperator* Leach et *Aeschna cyanea* (Müller). *C.R. hebdom. Séanc. Acad. Sci. Paris, Sér. D: Sci. Nat.,* 278: 281-284.

Dehorne, L. 1916. Les naïdimorphes et leur reproduction asexuée. *Arch. Zool. Exp. Gén.* 56: 25-157.

Desai, U.M. & Achuthankutty, C.T. 2000. Complete regeneration of ablated eyestalk in penaeid prawn, *Penaeus monodon. Curr. Sci.* 79: 1602-1603.

Dixey, L. 1938. Effects of repeated regeneration of the antenna in *Gammarus chevreuxi. Proc. Zool. Soc. London* 108: 289-296.

Dixon, K.A. 1989. Effect of leg type and sex on autotomy in the Texas bush katydid, *Scudderia texensis. Can. J. Zool.* 67: 1607-1609.

Dohrn, A. 1881. Die Pantopoden des Golfes von Neapel. *Fauna und Flora des Golfes von Neapel, Monographie 3.*

Edwards, J.S. 1972. Limb loss and regeneration in two crabs: the king crab *Paralithodes camtschatica* and the tanner crab *Chionoecetes bairdi. Acta Zool.* 53: 105-112.

Enghoff, H., Dohle, W. & Blower, J.G. 1993. Anamorphosis in millipedes (Diplopoda). The present state of knowledge and phylogenetic considerations. *Zool. J. Linn. Soc.* 109: 103-234.

Ernsting, G. & Fokkema, D.S. 1983. Antennal damage and regeneration in springtails (Collembola) in relation to predation. *Netherl. J. Zool.* 33 (4): 476-484.

Gaubert, P. 1892. Autotomie chez les Pycnogonides. *Bull. Soc. Zool. France* 17: 224-225.

Godelmann, R. 1901. Beiträge zur Kenntnis von *Bacillus Rossii* Fabr. Mit besonderer Berücksichtigung der bei ihm vorkommenden Autotomie und Regeneration einzelner Gliedmassen. *Arch. Entwicklungsmech. Organ.* 12: 265-301.

Govind, C.K. & Read, A.T. 1994. Regenerate limb bud sufficient for claw reversal in adult snapping shrimps. *Biol. Bull.* 186: 241-246.

Graber, V. 1874. Ueber eine Art fibrilloiden Bindegewebes der Insektenhaut und seine lokale Bedeutung als Tracheensuspensorium. *Arch. Mikrosk. Anat.* 10.

Hartnoll, R.G. 1988. Growth and molting. In: Burgreen, W.W. & McMahon, B.R. (eds.), *Biology of Land Crabs*: 186-210. Cambridge: Cambridge Univ. Press.

Herbst, C. 1891. Beiträge zur Kenntnis der Chilopoden. *Bibl. Zool.* 3: 1-42.

Herlant-Meewis, H. 1953. Contribution à l'étude de la régénération chez les oligochètes Aeolosomatidae. *Ann. Soc. R. Zool. Belg.* 84: 117-161.

Hopkins, P.M. 1985. Regeneration and relative growth in the fiddler crab. *Crustacean Issues* 3: 265-275.

Hopkins, P.M. 1993. Regeneration of the walking legs in the fiddler crab *Uca pugilator*. *Am. Zool.* 33: 348-356.

Huber, F. 1987. Plasticity in the auditory system of crickets: phonotaxis with one ear and neuronal reorganization within the auditory pathway. *J. Comp. Physiol.* 161: 583-604.

Igelmund, P. 1987. Morphology, sense organs, and regeneration of the forelegs (whips) of the whip spider *Heterophrynus elaphus* (Arachnida, Amblypygi). *J. Morphol.* 193: 75-89.

Juanes, F. & Smith, D.L. 1995. The ecological consequences of limb damage and loss in decapod crustaceans: a review and prospectus. *J. Exp. Mar. Biol. Ecol.* 193: 197-223.

Kaars, C., Greenblatt, S. & Fourtner, C.R. 1984. Patterned regeneration of internal femoral structures in the cockroach, *Periplaneta americana* L. *J. Exp. Biol.* 230: 141-144.

Kaestner, A. 1968. *Invertebrate Zoology, 2*. New York, London, Sidney: Interscience Publishers.

Klepal, W. & Barnes, H. 1974. Regeneration of the penis in *Balanus balanoides* (L.). *J. Exp. Mar. Biol. Ecol.* 16: 205-211.

Klingenberg, C.P. & Nijhout, H.F. 1998. Competition among growing organs and developmental control of morphological asymmetry. *Proc. Zool. Soc. London, Ser. B,* 265: 1135-1139.

Knoflach, B. & van Harten, A. 2000. Palpal loss, single palp copulation and obligatory mate consumption in *Tidarren cuneolatum* (Tullgren, 1910) (Araneae, Theridiidae). *J. Nat. Hist.* 34: 1639-1659.

Kocian, V. 1930. Un cas d'hétéromorphose chez *Argulus foliaceus* L. *Arch. Zool. Exp. Gén.* 70: 23-27.

Kopeć, S. 1913. Untersuchungen über die Regeneration von Larvalorganen und Imaginalscheiben bei Schmetterlingen. *Arch. Entwicklungsmech. Organ.* 37: 440-472.

Lakes, R. & Mücke, A. 1989. Regeneration of the foreleg tibia and tarsi of *Ephippiger ephippiger* (Orthoptera: Tettigoniidae). *J. Exp. Zool.* 250: 176-187.

Larink, O. 1983. Regeneration des Sensillenmusters der Labialpalpen bei Lepismatidae (Insecta: Zygentoma). *Verh. Deutsch. Zool. Ges.* 1983: 302.

Lawrence, R.F. 1953. *The Biology of the Cryptic Fauna of Forest, with Special Reference to the Indigenous Forest of South Africa*. Cape Town, Amsterdam: A.A. Balkema.

Lewis, J.G.E. 1968. Individual variations in a population of the centipede *Scolopendra amazonica* from Nigeria and its implications for methods of taxonomic discrimination in Scolopendridae. *J. Linn. Soc. London, Zool.,* 47: 315-326.

Lewis, J.G.E. 2000. Centipede antennal characters in taxonomy with particular reference to scolopendromorphs and antennal development in pleurostigmomorphs (Myriapoda, Chilopoda). In: Wytwer, J. & Golovatch, S. (eds.), *Progress in Studies on Myriapoda and Onychophora. Fragm. Faun. Warszawa* 43 (suppl.): 87-96.

Lüscher, M. 1948. The regeneration of legs in *Rhodnius prolixus* (Hemiptera). *J. Exp. Biol.* 25: 334-343.

Mariappan, P., Balasundaram, C. & Schmitz, B. 2000. Decapod crustacean chelipeds: an overview. *J. Bioscience* 25: 301-313.

McConaugha, J.R. 1991. Limb regeneration in juveniles of the mud crab *Rhithropanopeus harrisii* following removal of developing limb buds. *J. Exp. Zool.* 257: 64-69.

Megušar, F. 1907. Die Regeneration der Koleopteren. *Arch. Entwicklungsmech. Organ.* 26: 148-234.

Mellon, D. & Tewari, M.K. 2000. Heteromorphic antennules protect the olfactory midbrain from atrophy following chronic antennular ablation in freshwater crayfish. *J. Exp. Zool.* 286: 90-96.

Mikulska, I., Jacuński, L. & Weychert, K. 1975. The regeneration of appendages in *Tegenaria artica* C.L. Koch (Agelenidae, Araneae). *Zool. Poloniae* 25: 99-109.

Minelli, A. 2000. Limbs and tail as evolutionarily diverging duplicates of the main body axis. *Evol. Dev.* 2: 157-165.

Minelli, A. 2003a. The origin and evolution of appendages. *Int. J. Dev. Biol.* 47: 573-581.

Minelli, A. 2003b. *The Development of Animal Form*. Cambridge: Cambridge Univ. Press.

Minelli, A., Foddai, D., Pereira, L.A. & Lewis, J.G.E. 2000. The evolution of segmentation of centipede trunk and appendages. *J. Zool. Syst. Evol. Res.* 38: 103-117.

Morgan, T.H. 1920. Variations in the secondary sexual characters of the fiddler crab. *Am. Nat.* 54: 220-246.

Morgan, T.H. 1923. The development of asymmetry in the fiddler crab. *Am. Nat.* 57: 269-273.

Murakami, Y. 1958. The life history of *Bothropolys asperatus* Koch. *Zool. Mag. Tokyo* 67: 217-223 [in Japanese, with English summary].

Mykles, D.L. 2001. Interactions between limb regeneration and molting in decapod crustaceans. *Am. Zool.* 41: 399-406.

Myles, T.G. 1986. Evidence of parental and/or sibling manipulation in three species of termites in Hawaii (Isoptera). *Proc. Hawai. Entomol. Soc.* 27: 129-136.

Needham, A.E. 1947. Local factors and regeneration in Crustacea. *J. Exp. Biol.* 24: 220-226.

Needham, A.E. 1965. Regeneration in the Arthropoda and its endocrine control. In: Kiortsis, V. & Trampusch, H.A.L. (eds.), *Regeneration in Animals and Related Problems*: 283-323. Amsterdam: North-Holland Publ. Comp.

Newport, G. 1844. On the reproduction of lost part in Myriapoda and Insecta. *Phil. Trans. R. Soc. London* 1844: 283-294.

Nijhout, H.F. & Emlen, D.J. 1998. Competition among body parts in the development and evolution of insect morphology. *Proc. Nat. Acad. Sci. Un. St.* 95: 3685-3689.

Nilsson, C. 1986. The occurrence of lost and malformed legs in mayfly nymphs as a result of predator attacks. *Ann. Zool. Fenn.* 23: 57-60.

Niwa, K. & Kurata, H. 1964. Limb loss and regeneration in the adult king crab *Paralithodes camtschatica. Bull. Hokkaido Reg. Fish. Res. Lab.* 28: 51-55.

Norbeck, B.A. & Denburg, J.L. 1991. Pattern formation during insect leg segmentation: studies with a prepattern of a cell surface antigen. *Roux's Arch. Dev. Biol.* 199. 476-491.

Noulin, G. 1962. Contribution à l'étude d'un plan de rupture et des mécanismes d'amputation des appendices ches les Oniscoïdes (Crustacés Isopodes). *87 Congrès des Sociétés Savantes, Limoges*: 1159-1163.

Noulin, G. 1984. *Morphogenèse régénératrice chez les Crustacé Isopode* Porcellio dilatatus: *histologie et cytologie, interactions avec la mue, formations surnuméraires.* Thèse de Doctorat d'État, Université de Poitiers.

Nouvel-van Rysselberge, L. 1937. Contribution à l'étude de la mue, de la croissance et de la régénération chez les Crustacés Natantia. *Rec. Inst. Zool. Torley-Rousseau, Bruxelles,* 6: 5-161.

Patay, R. 1937. Sur la régénération des pattes chez le coléoptère chrysomélide *Leptinotarsa decemlineata* Say. *C.R. hebdom. Séanc. Acad. Sci. Paris, Sér. D, Sci. Nat.,* 126: 283-285.

Paulian, R. 1938. Contribution à l'étude quantitative de la régeneration chex les Arthropodes. *Proc. Zool. Soc. London, Ser. A,* 108: 297-383.

Penzlin, H. 1963. Über die Regeneration bei Schaben (Blattaria). I. Das Regenerationsvermögen und die Genese des Regenerats. *W. Roux's Arch. Entwicklungsmech. Organ.* 154: 434-465.

Pisani, D., Poling, L.L., Lyons-Weiler, M. & Hedges, S.B. 2004. The colonization of land by animals: molecular phylogeny and divergence times among arthropods. *BMC Biology* 2004, 2: 1. On-line available at: http://www.biomedcentral.com/1741-7007/2/1.

Poisson, R. & Patay, R. 1938. Sur quelques modalités de la régénération des pattes et des ailes chez la larve du doryphore: *Leptinotarsa decemlineata* Say. *C.R. hebdom. Séanc. Acad. Sci. Paris* 129: 126-128.

Przibram, H. 1909. *Regeneration.* Leipzig, Wien: Franz Deuticke.

Przibram, H. & Werber, E.J. 1907. Regenerationsversuche allgemeiner Bedeutung bei Borstenschwänzen (Lepismatidae). *Arch. Entwicklungsmech. Organ.* 27: 615-631.

Rabes, O. 1907. Regeneration der Schwanzfäden bei *Apus cancriformis. Zool. Anz.* 24: 753-755.

Racovitza, E. 1925. Notes sur les Isopodes. 13. Morphologie et phylogénie des antennes II, Le fouet. *Arch. Zool. Exp. Gén.* 63: 533-622.

Randall, J.B. 1981. Regeneration and autotomy exhibited by the black widow spider, *Latrodectus variolus* Walckenaer. *W. Roux's Arch. Dev. Biol.* 190: 230-232.

Read, A.T. & Govind, C.K. 1997a. Regeneration and sex-biased transformation of the sexually dimorphic pincer claw in adult snapping shrimps. *J. Exp. Zool.* 279: 356-366.

Read, A.T. & Govind, C.K. 1997b. Claw transformation and regeneration in adult snapping shrimp: test of the inhibition hypothesis for maintaining bilateral asymmetry. *Biol. Bull.* 193: 401-409.

Rockett, C.L. & Woodring, J.P. 1972. Comparative studies of acarine limb regeneration, apolysis, and ecdysis. *J. Insect Physiol.* 18: 2319-2336.

Rogers, D.C. 2001. Revision of the Nearctic *Lepidurus* (Notostraca). *J. Crust. Biol.* 21: 911-1006.

Rościszewska, M. & Urvoy, J. 1989a. Contribution à l'étude de la régénération d'appendices chez *Gryllus domesticus* L. (Orthoptera), I, Étude de la régénération de l'antenne. *Acta Biol. Cracov., Ser. Zool.,* 31: 125-135.

Rościszewska, M. & Urvoy, J. 1989b. Contribution à l'étude de la régénération d'appendices chez *Gryllus domesticus* L. (Orthoptera), II, Étude de la régénération des pattes. *Acta Biol. Cracov., Ser. Zool.,* 31: 136-143.

Rosin, R. 1964. On regeneration in scorpions. *Israel J. Zool.* 13: 177-183.

Roth, V.D. & Roth, M.R. 1984. A review of appendotomy in spiders and other arachnids. *Bull. Brit. Arachnol. Soc.* 6: 137-146.

Sánchez Alvarado, A. 2000. Regeneration in the metazoans: why does it happen? *BioEssays* 22: 578-590.

Schafer, R. 1973. Postembryonic development in the antenna of the cockroach, *Leucophaea maderae*: growth, regeneration and the development of the adult pattern of sense organs. *J. Exp. Zool.* 183: 353-364.

Scheffel, H. 1987. Häutungsphysiologie der Chilopoden: Ergebnisse von Untersuchungen an *Lithobius forficatus* (L.). *Zool. Jahrb., Abt. Allgem. Zool. Physiol. Tiere*, 91: 257-282.

Scheffel, H. 1989. Zur wechselseitigen Beeinflussung von Regeneration und Häutung bei Larven des Chilopoden *Lithobius forficatus* (L.). *Zool. Jahrb., Abt. Allgem. Zool. Physiol. Tiere*, 93: 436-505.

Schindler, G. 1979. Funktionsmorphologische Untersuchungen zur Autotomie der Stabheuschrecke *Carausius morosus* Br. (Insecta: Phasmida). *Zool. Anz.* 303: 316-326.

Schlichting, C.D. & Pigliucci, M. 1998. *Phenotypic evolution: a reaction-norm perspective*. Sunderland, Massachusetts: Sinauer.

Shaw, V.K. & Bryant, P.J. 1974. Regeneration of appendages in the large milkweed bag, *Oncopeltus fasciatus*. *J. Insect Physiol.* 20: 1849-1857.

Skinner, D.M. 1985. Molting and regeneration. In: Bliss, D.E. & Mantel, L.H. (eds.), *The Biology of Crustacea, 9*: 43-146. New York: Academic Press.

Spivak, E.D. & Politis, M.A. 1989. High incidence of limb autotomy in a crab population from a coastal lagoon in the province of Buenos Aires, Argentina. *Can. J. Zool.* 67: 1976-1985.

Sweetman, H.L. 1934. Regeneration of appendages and molting among the Thysanura. *Bull. Brooklyn Entomol. Soc.* 29: 158-161.

Tanaka, A., Akahane, H. & Ban, Y. 1992. The problem of the number of tarsomeres in the regenerated cockroach leg. *J. Exp. Zool.* 262: 61-70.

Toyota, K., Yamauchi, T. & Miyajima, T. 2003. A marking method of cutting uropods using malformed regeneration for kuruma prawn *Marsupenaeus japonicus*. *Fisheries Science*, 69, 161-169.

Truby, P.R. 1985. Separation of wound healing from regeneration in the cockroach leg. *J. Embryol. Exp. Morphol.* 85: 177-190.

Urvoy, J. 1963. Étude anatomo-functionelle de la patte et de l'antenne de la blatte *Blabera craniifer* B. *Ann. Sci. Nat. Zool. Biol. Anim.* 12: 287-413.

Urvoy, J. 1970. Étude des phénomènes de régénération après section d'antenne chez le phasme *Sypyloidea sypylus* W. *J. Embryol. Exp. Morphol.* 23: 719-728.

Urvoy, J. & Les Bris, R. 1968. Étude de la régénération des cerques, des palpes labiaux et des palpes maxillaires chez *Blabera craniifer* (Orth. Blattidae). *Ann. Soc. Entomol. France* 4: 371-383.

Vachon, M. 1956. Remarques sur la morphogenèse au cours de la régénération des pattes ches les araignées. *Proceedings of the XIV International Congress of Zoology, Copenhagen, 1953*.

Vachon, M. 1957. La régénération appendiculaire chez les scorpions (Arachnides). *C.R. hebdom. Séanc. Acad. Sci. Paris* 244: 2556-2559.

Vachon, M. 1967. Nouvelles remarques sur la régénération des pattes chez l'araignée: *Coelotes terrestris* Wid. (Agelenidae). *Bull. Soc. Zool. France* 92: 417-428.

Verhoeff, K.W. 1902-1925. Chilopoda. In: *H.G. Bronn's Klassen und Ordnungen des Tier-Reichs, 5 (2)*: 1-725. Leipzig: C.F. Winter.

Vernet, G. & Charmantier-Daures, M. 1994. Mue, autotomie et régénération. In: Grassé, P.P. (ed.), *Traité de Zoologie, Tome 7, Fascicule 1*: 153-194. Paris: Masson.

Vitzthum, G. 1943. Acarina. In: *H. G. Bronn's Klassen und Ordnungen des Tier-Reichs, Bd. 5, Abt. 4, Buch 5*: 1-1011. Leipzig: C.F. Winter.

Vollrath, F. 1990. Leg regeneration in web spider and its implications for orb spider exploration and web-building behaviour. *Bull. British Arachnol. Soc.* 8: 177-184.

Vollrath, F. 1995. Lyriform organs on regenerated spider legs. *Bull. British Arachnol. Soc.* 10: 115-118.
von Buddenbrock, W. 1954. Physiologie der Decapoden. Die Regeneration. In: *H.G. Bronn's Klassen und Ordnungen des Tier-Reichs, Bd. 5, Abt. 1, Buch 7, Lfg. 9*: 1231-1246. Leipzig: Akademische Verlagsgesellschaft.
Walossek, D. & Müller, K.J. 1997. Cambrian 'Orsten'-type arthropods and the phylogeny of Crustacea. In: Fortey, R.A. & Thomas, R.H. (eds.), *Arthropod Relationships*: 139-153. London: Chapman & Hall.
Wege, W. 1911. Morphologische und experimentelle Studien an *Asellus aquaticus*. *Zool. Jahrb., Abt. Allgem. Zool. Physiol. Tiere,* 30: 217-320.
Weis, J.S. 1982. Studies on limb regeneration in the anomuran *Pagurus longicarpus* and *Emerita talpoidea*. *J. Crust. Biol.* 2: 227-231.
Weise, R. 1991. Regeneration of antennae in the third larvae of *Lithobius forficatus*. *Zool. Anz.* 227: 343-355.
Weygoldt, P. 1984. L'autotomie chez les Amblypyges. *Rev. Arachnol.* 5: 321-327.
Wolsky, A. 1957. "Compensatory hyper-regeneration" in antennae of Hemiptera. *Nature* 180: 1144-1145.
Wood, F.D. 1926. Autotomy in Arachnida. *J. Morphol. Physiol.* 42: 143-195.
Wood, F.D. & Wood, H.E. 1932. Autotomy in decapod Crustacea. *J. Exp. Zool.* 62: 1-55.
Wygodzinsky, P. 1941. Beiträge zur Kenntnis der Dipluren und Thysanuren der Schweiz. *Denkschr. Schweiz. Naturf. Ges.* 74: 113-227.
Yamaguchi, T. 2001. Dimorphism of chelipeds in the fiddler crab, *Uca arcuata*. *Crustaceana* 74: 913-923.

V ARTHROPOD PHYLOGENETICS

What are Ostracoda? A cladistic analysis of the extant superfamilies of the subclasses Myodocopa and Podocopa (Crustacea: Ostracoda)

DAVID J. HORNE[1], ISA SCHÖN[2], ROBIN J. SMITH[3] & KOEN MARTENS[2,4]

1 Department of Geography, Queen Mary, University of London, London, U.K.
2 Freshwater Biology Section, Royal Belgian Institute of Natural Sciences, Brussels, Belgium
3 Department of Earth Sciences, Kanazawa University, Kakuma, Japan
4 University of Ghent, Department Biology, Ghent, Belgium

ABSTRACT

The higher phylogeny and classification of the Ostracoda have always been topics of contention and this has hampered inclusion of this group in general analyses of crustacean or arthropod phylogeny. Also, the inclusion of hard part characters in earlier attempts to investigate ostracod phylogeny has introduced a large degree of homoplasy in the resulting trees. Here, we present an analysis of the phylogeny of the extant ostracod subfamilies, using nearly exclusively morphological soft part characters. The homologies of these limb characters in a crustacean context are extensively discussed. The resulting maximum parsimony and distance analyses show a good resolution of the phylogeny of the myodocope subfamilies, well-supported by bootstrap values, but in neither of the two analyses are the podocope phylogenies supported by bootstrap statistics. We test the hypothesis that this discrepancy in resolution between the two subclasses is due to a difference in evolutionary tempo (punctuated in Podocopa, gradual in Myodocopa) by comparing mean distances, average branch lengths and by applying relative rate tests. To the best of our knowledge, this is the first time ever that Li's relative rate tests have been applied to a morphological data matrix. None of the tests yields significant differences in past evolutionary tempo within the two groups. However, there is a significant difference in mean relative distance between the lineages leading up to the two subclasses. Puncioidea, here considered to constitute Recent representatives of the otherwise Palaeozoic Palaeocopida, have significantly different evolutionary rates from other groups, but cluster within the Podocopida. The slowdown in evolution indicates that this lineage might have experienced, and might still be experiencing, period(s) of morphological stasis.

The present paper is dedicated to Frederick R. Schram, on the occasion of his retirement as professor in animal systematics and zoogeography at the University of Amsterdam. Fred Schram has conducted pioneering work on phylogeny of Crustacea, using both fossil and extant groups and we owe much of our present knowledge of this group to him, either through his own research or that of his students and collaborators, or due to the intriguing questions he has asked at various stages of his career and which provoked other scientists, including ourselves, to tackle a variety of problems. We wish him a happy retirement!

1 INTRODUCTION

In spite of their excellent fossil record, arguably the best of any arthropod group, the Ostracoda remain problematic with regard to both their own phylogeny and their position in crustacean phylogeny. The failure of ostracod specialists to agree on the higher classification of the group has been a source of difficulty and frustration for those seeking to consider the Ostracoda in a crustacean context (e.g., Schram & Hof 1998). Schram (1986) observed that this could be partly due to the difficulty of comparing fossil (without soft-parts) and living (with soft-parts) forms, and commented on the almost complete lack of rigid character analyses of living ostracod groups. Presenting a hybrid ostracod classification based on earlier works, he stated: "The higher classification of ostracodes may not stabilise until careful analysis of characters is attempted across all groups" (*op. cit.*, p. 416). Much more recently, Martin & Davis (2001) have expressed rather similar views about ostracod classification and the relationship of the group to the Maxillopoda. Here, we present the preliminary results of an analysis of characters across all extant ostracod superfamilies, as a first step towards resolving these issues.

The vast majority of ostracod fossils provide information only on carapace ('hard-part') morphology, which is at the same time rich in data and awash with homoeomorphies. Robust schemes compatible with other crustacean phylogenies must inevitably be based on 'soft-part' morphology, a proviso which immediately limits us to using extant ostracods plus those very rare and special instances of fossilisation of limbs. Unfortunately, such occurrences either seem to offer too little new information or too much. Some, such as the Early Cretaceous cypridoidean *Pattersoncypris* (Smith 2000) and the Silurian cylindroleberidoid *Colymbosathon* (Siveter et al. 2003) convey much about the surprising antiquity of extant lineages, but supply almost no new insights into the evolution of limb morphologies. Others, such as 'Orsten' phosphatocopines like *Hesslandona* and the Chengjiang bradoriid *Kunmingella*, show limb morphologies and arrangements so different from those of extant groups that they have had to be removed altogether from the Ostracoda (Hou et al. 1996; Walossek & Müller 1998). Really useful fossil ostracods with soft-part preservation (such as a Silurian beyrichioid or a Carboniferous carbonitoid) that would answer significant questions about the affinities of extinct groups within the Ostracoda *s. str.* remain undiscovered (although we have high hopes that this will not be the case for too much longer).

The discovery of what is possibly a living example of a major ostracod order that otherwise became extinct by the end of the Permian has had surprisingly little impact. To accept the puncioid *Manawa staceyi* Swanson, 1989 as an ostracod and an extant representative of the Palaeocopida requires us to expand our concepts of ostracod anatomy and ontogeny to allow for an extra pair of limbs and univalved early instars (Swanson 1989, 1991; Horne et al. 2002). The alternative is to remove palaeocopids, too, from the Ostracoda. Certain carapace features of the extinct Palaeozoic Leperditicopida suggest affinities with the Myodocopa, but in the absence of soft-part information it is impossible to confirm or deny such a relationship (Vannier et al. 2001). The fundamental question of ostracod monophyly is answered differently by morphological (e.g., Cohen et al. 1998) and molecular (e.g., Spears & Abele 1997) approaches. We are increasingly of the opinion that the two major groups, the Myodocopa and Podocopa, are not closely related and that the Ostracoda are consequently polyphyletic, but such a hypothesis has yet to be subjected to, let alone confirmed by, rigorous analyses.

A major impediment to such studies based on morphological features has been the lack of agreement on the homologies of ostracod limbs and their components (Cohen et al. 1998), something that Horne et al. (2002) and Horne (in press) have sought to rectify. Cohen & Morin (2003) observed that Ostracoda appear to lack a synapomorphy, but suggested a combination of five characters (in all instars) that unites all Ostracoda:

(1) Each pair of limbs uniquely different from the others.
(2) Whole body can be entirely enclosed by a bivalved carapace lacking growth lines (excepting univalved early instars in puncioids).
(3) A maximum of 9 pairs of limbs (including a pair of copulatory limbs; maximum of 10 if podocopan furca is accepted to be a pair of uropods, maximum 11 if male brush-shaped organ is accepted as a vestigial limb).
(4) The adult male limb just anterior to the furca is a copulatory limb/organ, usually paired, complex and conspicuously large.
(5) The body shows little or no segmentation, with a maximum of 10 trunk segments.

Leaving aside the contradiction that (4) is an adult character only, so not applicable to all instars, and noting that some authors at least consider 11 trunk segments plus a telson to be present in podocopans, but probably no more than 7 in myodocopans (Schulz 1976; Cohen et al. 1998; Tsukagoshi & Parker 2000; Horne et al. 2002), it must be pointed out that, even counting a pair of male copulatory limbs, no ostracod has more than 8 pairs of limbs plus a furca (discounting brush-shaped organs, possible limb vestiges present in males of some podocopan groups). The puncioid *Manawa* is the only ostracod to possess an eighth pair of limbs in both sexes, termed "uropods" by Swanson (1989), one of which in the male incorporates the single copulatory appendage. Complex copulatory appendages in other ostracod groups are almost certainly derived from limb pairs (probably more than one pair; see Martens 2003; Matzke-Karasz & Martens in press), but their homologies are far from clear, and Horne et al. (2002) pointed out that "eighth limbs" in different groups are not necessarily homologous. Ancestors of ostracods may have had a limb pair for each trunk segment, and different groups may have lost or retained different posterior limb pairs as a consequence of reduction of the trunk to fit inside a bivalved carapace. It appears that Ostracoda may be defined as *bivalved arthropods, which in the adult stage have up to 8 pairs of unambiguous limbs, plus copulatory limbs and a furca, all of which can be totally enclosed by a bivalved carapace, which lacks growth lines.*

Over the past few decades, several authors have presented evolutionary scenarios, almost invariably based on the fossil record, in which they picture the putative branching of various ostracod lineages superimposed on Phanerozoic time scales (Scott 1961; Swain 1976; Maddocks 1982; McKenzie et al. 1983; Whatley & Moguilevsky 1998; Horne et al. 2004). Such scenarios are useful in that they summarise information contained in the extensive fossil record of the Ostracoda. However, all of these evolutionary scenarios are highly subjective, as they were not the results of a rigorous phylogenetic analysis using a clearly defined criterion (parsimony, distance). They are also not repeatable, as no clearly defined character matrix is given, nor was a defined algorithm used. Finally, these analyses were mostly based on autapomorphic, not on synapomorphic features. Yamaguchi & Endo (2003) have presented an overview of the evolutionary hypotheses that were produced by these narrative analyses.

Vannier & Abe (1992) used multivariate and cluster analysis of 17 hard-part and 9 soft-part characters to assess the similarity between selected extant myodocopan superfamilies (Cypridinoidea, Polycopoidea (= Cladocopoidea in our classification), Thaumatocypridoidea and Halocypridoidea), and between these and three extinct myodocopan superfamilies using only 16 hard-part characters. They also presented a cladogram illustrating phylogenetic relationships between four extant and two extinct superfamilies, and a phyletic tree, shown against the Phanerozoic time-scale, based on a combination of their cladogram with stratigraphic ranges. It is not clear, however, how their cladogram was obtained, since no indication of method (e.g., parsimony), let alone bootstrap support, was given.

Park & Ricketts (2003) presented a primer to a cladistic analysis of the Ostracoda based on morphological features, but as will be argued below, their data matrix (and hence the subsequent analyses) is seriously flawed. Recently, a few papers (Oakley & Cunningham 2002; Yamaguchi & Endo 2003) have attempted to provide a molecular phylogeny of extant Ostracoda (using 18S and part of the 28S rDNA). The results are far from unequivocal, especially with regard to the arrangement of the podocope groups and of the Puncioidea.

Here, we present a cladistic analysis of 16 extant ostracod superfamilies (together forming the subclasses Podocopa and Myodocopa *sensu* Horne et al. 2002), using three out-groups and 27 morphological characters, all parsimony-informative.

2 MATERIAL AND METHODS

2.1 *Choice of taxa for analysis*

We have included in our analysis all of the extant superfamilies of the Ostracoda (Figs. 1+2), following the classification scheme of Horne et al. (2002), these being the major ostracod taxa for which information is available on 'soft-part' morphology. Extinct groups for which no information is available on limb morphology, such as the Leperditicopida, are excluded. The superfamily Puncioidea, represented by a single living species (*Manawa staceyi*), is included as the only extant representative of the Palaeocopida, the dominant ostracod group in the Early Palaeozoic. The extent to which it is truly representative of that group's morphology is far from certain, but until a Palaeozoic palaeocopid with fossilised limbs is discovered, it remains the best information we have.

Three out-group taxa were selected. The Cambrian Phosphatocopina were chosen because they were once considered to be ancestors of Ostracoda, and although now regarded as a sister group of the Eucrustacea (Maas et al. 2003), they are bivalved crustaceans *s. lat.* and as such a useful basis for comparison. Cephalocarida were chosen as an extant representative of the crown-group Crustacea, considered by some authors to possess plesiomorphic limb characters, and *Martinssonia*, a Cambrian representative of the stem-group Crustacea, was chosen for broadly similar reasons.

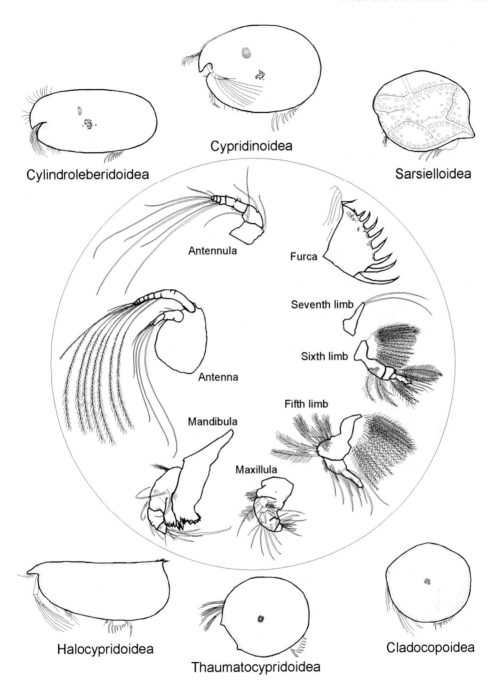

Figure 1. The superfamilies of the Myodocopa (Ostracoda); central inset shows example of limbs of Thaumatocypridoidea.

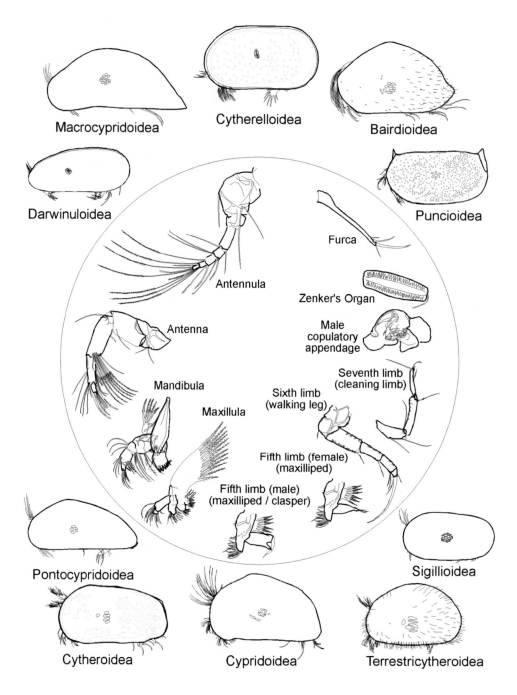

Figure 2. The superfamilies of the Podocopa (Ostracoda); central inset shows example of limbs of Cypridoidea.

2.2 Morphological characters and character states

The present analyses use morphological characters only (Figs. 1-3). The rationale for the choice of characters is straightforward: wherever we could be reasonably certain that a structures was homologous across the different groups, the morphology of that structure was used. This criterion lead to a surprisingly (and disturbingly) low number of characters that could be used. All but one of the characters used here are based on limb morphology; only one hard part (valve) character was used. This is so, because either structures are not shared between Podocopa and Myodocopa, or they show such evolutionary plasticity that they would have introduced a large degree of homoplasy into the analysis. Therefore, shape, dimorphism, external sculpturing and internal anatomy of the valves, including muscle scar patterns, were deemed not useful for an analysis at the present taxonomic level (see below for a discussion of an earlier attempt to use hard part characters for phylogenetic analyses of the Ostracoda). Furthermore, by using limb and other soft-part characters, we facilitate the inclusion of crustacean taxa without carapaces as out-groups (and pave the way for future inclusion of Ostracoda in analyses of the full range of crustacean groups).

Characters and character states are summarised in Appendix 1. The soft-part characters and character states used in the present analysis are discussed below, approximately in order from anterior to posterior. The data matrix drafted for the present analysis is given in Appendix 2.

Zenker's Organ (character 1). The Zenker's Organ, a muscular sperm pump external to the hemipenis of the male, has been considered to be a synapomorphy of the Cypridocopina (Cypridoidea, Macrocypridoidea, Pontocypridoidea) and the Sigillioidea (see, e.g., Horne 2003). The question of whether or not the ancient asexual Darwinuloidea ever had a Zenker's Organ is controversial; only a single rare male record of an extant darwinuloidean exists (see Martens 1998, for discussion) and we do not accept Maddocks' (1973) interpretation of the original illustrations as showing Zenker's Organs to be present. In the absence of evidence to the contrary it is most parsimonious to score this character as absent in darwinuloideans.

Antenna (characters 2, 3). We restrict the number of articulated endopodite podomeres to females because of sexual dimorphism in some taxa, with males showing subdivision of one of the antennal podomeres (e.g., cypridoidean Candonidae).

The antennal exopodite is one of the most useful characters for distinguishing the major groups (Fig. 3). Myodocopa have a multisegmented antennal exopodite, while in the Podocopa this ramus has a maximum of two podomeres (in Cytherelloidea). In Puncioidea, it is a single elongate podomere bearing terminal setae, in Cytheroidea a special seta for secreting a sticky thread from a spinneret gland, and in other podocopan superfamilies it is reduced to a small 'scale', bearing setae.

Mandibula (characters 4-7). A multisegmented mandibular exopodite may be regarded as the plesiomorphic condition (see, e.g., Walossek, 1999), as seen in Phosphatocopina and *Martinssonia*; in Cephalocarida the naupliar mandible is biramous with a multisegmented exopodite but that of the adult has lost the palp (basis, endopodite and exopodite).

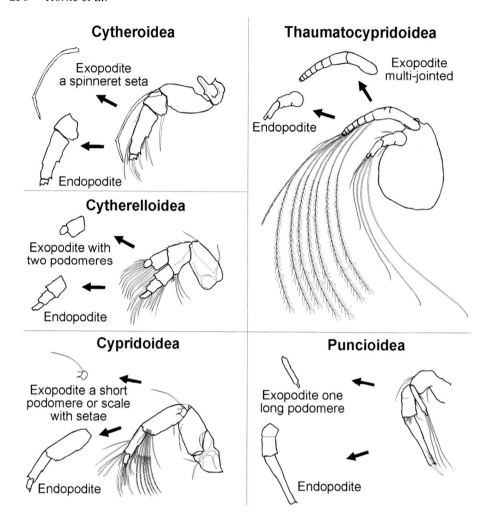

Figure 3. Examples of antennae of some ostracod superfamilies and their exopodite character states.

The mandibular palp filter screen comprises a comb or grille of closely-spaced, flattened, setuliferous setae. In the Darwinuloidea and Terrestricytheroidea there are typically eight setae, in the Cytherelloidea typically more than 30. No other taxa have this feature.

Phosphatocopina were once considered to have only a proximal movable endite on the basis, not a true coxa, in antenna and mandibula. Recent work has shown, however, that a coxa is present in both limbs in earlier ontogenetic stages, becoming fused with the basis in later stages (Maas et al. 2003).

A toothed endite on the mandibular basis is unusual in the Crustacea (Boxshall 2004), but it occurs in the ostracod superfamilies Halocypridoidea and Thaumatocypridoidea.

Maxillula (characters 8-11). Comparisons of morphology and musculature have led to the conclusion that no subdivision of the protopodite is recognisable in post-mandibular limbs in podocopan ostracods (Horne in press); subdivision into a basis, a coxa, and in some taxa a clear precoxa, is characteristic of the myodocopan maxillula, however (Boxshall 1997).

In considering the number of rays on the maxillular branchial plate, it is necessary to use overlapping ranges in order to allow for the variation seen within the various groups, but the ranges themselves have been chosen to reflect natural groupings in which most taxa are centred on 'typical' values (Smith et al. in press). Thus, while the number of branchial plate rays in Cytheroidea ranges from 10-29, most cytheroideans have, typically, 15-17 rays, while those with more than 20 belong to the families Entocytheridae and Bythocytheridae, which are arguably more ancient in origin. In character state 1, we group the Cytheroidea, Cypridoidea (16-27 rays), Pontocypridoidea (15-25 rays), Sigillioidea (17 rays), Macrocypridoidea (16-18 rays) and Terrestricytheroidea (20 rays). Character state 2 applies to Darwinuloidea (24-32 rays), Bairdioidea (25-34 rays) and Cytherelloidea (27-36 rays).

The leg-like endopodite of the puncioid maxillula, similar to those of the succeeding three pairs of thoracic limbs (Swanson, 1989), is considered to be a plesiomorphy; in all other podocopan ostracods the endopodite forms a palp lying alongside the three maxillular endites. The general morphology of the myodocopan maxillulae is typically quite different from that of podocopans, with the endopodite very short and compact although up to three podomeres may be discernable (Horne et al. 2003).

Whether ostracod branchial plates (which may occur on mandibula, maxillula, fifth limb and sixth limb) are exopodites or epipodites has been the subject of controversy in the literature (see, e.g., Cohen et al. 1998). Schram (1986: 401), apparently favouring phylogenetic hypothesis over morphological evidence, argued that they are exopodites because "to use the term epipod would imply an independent evolution of such a specialised exite in the ostracodes". Schram & Hoff (1998), in an analysis of the major crustacean groups utilising 90 characters, scored Ostracoda as having no epipodal development on the fifth limb. Myodocopans lack a maxillular branchial plate, but Boxshall (1997) has demonstrated that their fifth and sixth limb branchial plates are epipodites. Horne (in press) has concluded that the podocopan maxillular branchial plate is not an epipodite, as some authors have argued, but an exopodite arising from an undifferentiated protopodite. In adult myodocopans the maxillular exopodite is reduced to a single podomere, but in at least some Cladocopoidea there is a suggestion of a subdivision, seen more clearly in early instars and perhaps a vestige of the plesiomorphic multisegmented state (see Horne in press, for discussion).

Fifth limb (characters 12-15). The homology of the podocopan fifth limb is one of the most significant and contentious issues in ostracod phylogeny. For the purposes of the present study, we have assumed that for all of the fifth limb characters used, the fifth limb is homologous in all taxa analysed. We note, however, the view of some authors that, in podocopans at least, the true fifth limb (the posterior-most crustacean head limb, the maxilla) is missing and the so-called fifth limb is in fact the first thoracic limb (and therefore a homologue of the sixth limb in other Crustacea); see, e.g., Smith & Martens (2000), Horne et al. (2002) and Horne (in press). If that view were to be accepted, it would follow that none of the post-maxillular limbs could automatically be regarded as homologous. Here, we adopt the view that the fifth limb in all the analysed taxa is homologous, but consider that in Podocopa (as in Phosphatocopina and *Martinssonia*) it is the anterior-most trunk limb, not the hindmost head limb. In other words, unlike in other crown-group crustaceans, the maxillary segment in podocopans has not been encephalised (for further discussion see remarks on character 26). This interpretation allows us to consider fifth (and subsequent) limbs as homologous without contradicting the evidence that in podocopans they are trunk limbs; as such it is an operational compromise, but one which is not inconsistent with at least some phylogenetic scenarios (see Horne in press, for discussion).

A fifth limb endopodite is present in males only of the Cytherelloidea (this state being scored here) while in females the limb comprises only the protopodite and exopodite. Myodocopan fifth limbs are very complex and so extensively modified and shortened as to make the identification of rami and the determination of the number of podomeres extremely difficult (see, e.g., Kornicker, 2002), hence the use of the term "compacted".

The branchial plates on the fifth limbs of some podocopans (e.g., Bairdioidea) are exopodites, while those of myodocopans are coxal or precoxal epipodites (Boxshall, 1997; Horne in press) (see also discussion of Maxillula above); there is no evidence for the existence of epipodites in podocopan ostracods or in stem-group crustaceans such as *Martinssonia* and the Phosphatocopina, and the 'pseudepipodite' of cephalocarid post-maxillular limbs is here regarded as part of the exopodite, following Horne (in press).

The general morphological similarity of the leg-like fifth and sixth limbs of some podocopan (e.g., Cytheroidea) and myodocopan (e.g., Halocypridoidea) taxa has led some authors to regard them as (directly) homologous in form. Comparisons of morphology and musculature have shown, however, that the knee joints of the podocopan limbs are not in the same position as those of myodocopans (Horne in press).

Sixth limb (characters 16-19). The same argument applies to the sixth limb knee joint as to that of the fifth limb (see above).

It is not easy to decide whether the sixth limb exopodite of certain taxa should be regarded as a reduced branchial plate or simply a bunch of setae (in females of Sigillioidea, for example, there is a short stem or plate branching into three or four setae); accordingly several possible states, including the two-segmented cephalocarid exopodite, are placed together in character state 2.

As discussed above with regard to the maxillula and fifth limb, among the taxa analyzed only myodocopans have epipodites on the sixth limb.

As with the fifth limb, some myodocopan sixth limbs are so extensively modified and shortened as to make determination of the number of podomeres difficult or impossible, hence the use of the term "compacted". In the podocopan Cytherelloidea only the male has a sixth limb endopodite (the character state scored here) while in the female the limb is reduced to a small branchial plate, presumed to be the exopodite.

Seventh limb (character 20). The cleaning limb of Cypridoidea, Pontocypridoidea and Macrocypridoidea is generally regarded as a modified protopodite and endopodite, but that of the Myodocopida is a long, flexible and annulated appendage which defies homologisation.

Eighth limb (character 21). Of all the ostracod superfamilies, only the Puncioidea have an unambiguous eighth limb in both sexes (in the male one of them bears the single copulatory appendage). In others it is probable that the male (and sometimes female) copulatory appendages are modified limbs, but since it is unclear whether these represent the eighth limbs or a combination of limbs (Martens 2003; Matzke-Karasz & Martens in press), and because they may derive from different trunk segments and so not be homologous, we have declined to attempt any inclusion of these characters in our analysis.

Furca (characters 22, 23). It remains unclear whether the so-called furca is homologous in all of the groups analysed. Some have argued that the myodocopan furca is a modified telson, while that of the Podocopa represents a pair of uropods (Meisch 2000; Horne et al. 2002), a distinction which is encompassed by character 23. The position of the anus relative to the furca is a major distinguishing feature between Podocopa (behind) and Myodocopa (in front).

Lateral compound eyes (character 24). The possession of a naupliar (median) eye is a plesiomorphy of all Ostracoda, but lateral compound eyes may be an apomorphy of the Myodocopida (see Oakley & Cunningham 2002), notwithstanding arguments to the contrary, that compound eyes are plesiomorphic in Crustacea (see Fryer (1996) for discussion).

Carapace (character 25). Possession of a bivalved carapace, without growth lines, capable of completely enclosing the limbs, has previously been regarded as a possible synapomorphy of the Ostracoda *s. lat*. However, detailed comparisons show that some former 'ostracods', such as the Cambrian bivalved Phosphatocopina and Bradoriida, have quite different appendage morphologies, although some still argue for their retention in the Ostracoda (see, e.g., Hou et al. 1996; Siveter & Williams 1997; Walossek & Müller 1998; Shu et al. 1999; McKenzie et al. 1999).

Tagmosis - cephalic segments (character 26). As discussed above in connection with the fifth limb, we have adopted the view that podocopans have only four cephalic limb-bearing segments. We realise that it is important to distinguish between encephalisation of the fifth limb (i.e., its adaptation as a mouth-part, acting in concert with the mandibula and maxillula) and the encephalisation of the fifth body segment, which are two separate processes. In Cephalocarida, for example, the fifth body segment is encephalised, but the fifth limb retains the morphology of the trunk limbs. The converse seems possible for podocopids, which may (in some superfamilies) have an encephalised fifth limb while the segment to which it is attached remains part of the trunk; for example in the Cytheroidea the fifth limb has the morphology of succeeding trunk limbs, but in the Cypridoidea it is a cephalic feeding limb. Our argument is not based on the morphology of the fifth limb, however. Podocopans retain only traces of body segmentation, but in our view the limited evidence offered by the endoskeletal structure (see, e.g., Schulz 1976 for podocopids and platycopids, and Swanson 1989 for puncioids) and the location of the maxillary glands (Kesling 1951) offers little support for the notion of an encephalised fifth segment, but can be interpreted as supporting the idea of the fifth segment as a trunk segment. A detailed exploration of these arguments is in preparation; for now, as stated above, our decision to regard the fifth segment as non-cephalic is an operational one.

Bellonci organ (character 27). This anterior sensory organ is a synapomorphy of the myodocopan superfamilies.

2.3 Analyses

The above characters and characters states (see Appendix 1) were scored for all investigated taxa and compiled in a data matrix of 19 OTU (16 in-group, 3 out-group) and 27 characters (all parsimony informative, see Appendix 2). Analyses were carried out with the computer programme PAUP (version 4.0b10, Swofford 1998). Character types were all set at 'unordered', all character weights were set at '1'. Inapplicables were scored as '?'. Maximum parsimony trees were calculated via stepwise input with furthest addition sequence using the 'branch and bound' algorithm. Both strict and 50% majority rule consensus trees were computed when multiple, equally parsimonious trees were found. Bootstrap values were calculated for 10,000 replicates with the same algorithm. Distance trees were calculated with Neighbor Joining using mean (relative) character difference; bootstraps were calculated for 10,000 replicates with the same algorithm.

In order to test if different lineages have experienced different evolutionary rates, mean relative distances (p) were calculated for different comparisons of branch lengths within the various groups. Student t-tests were one-tailed and assuming equal variances. Relative rate tests were carried out following Li (1997). The difference (d) between the rates K_{AC} and K_{BC} is calculated assuming independent evolution of A and B from C. $d = K_{AC} - K_{BC}$ and under equal rates, d should be 0 or close to it. Only the longest and shortest branches of the two subclasses (Podocopa and Myodocopa) were compared. Finally, mean average branch lengths were compared between Podocopa and Myodocopa; statistical significance was tested with the F-test in Microsoft Excel.

3 RESULTS

A 'branch and bound' analysis using parsimony as criterion calculated 21 equally parsimonious trees. A 50% majority rule consensus tree, with bootstrap values over the branches leading to the pertinent node, is given in Figure 4. A strict consensus of the same 21 equally parsimonious trees did not retain any of the internal branching in Podocopa, except for the basal split of the Puncioidea. A neighbour joining tree, with bootstrap values over the branches leading to the pertinent node, is given in Figure 5. Both trees have very similar topologies:

(1) Both subclasses (Podocopa and Myodocopa) form well-supported clusters of superfamilies: bootstrap values 70 and 95 respectively in the parsimony tree, 82 and 96 respectively in the distance tree.
(2) In both trees, the out-groups cluster well out of the in-groups. The phosphatocopines and *Martinssonia* - cluster appears in both trees and even has acceptable bootstrap support (75 and 81 respectively in parsimony and distance trees). Cephalocarida are always the most basal of the out-groups.
(3) Within the subclass Podocopa, the distinction between the Palaeocopida (Puncioidea) and the rest of the group is supported in both analyses, with bootstrap values of 75 (parsimony) and 84 (distance) respectively. There is no bootstrap support for the resolution of the remainder of the podocopan groups, the superfamilies in the Platycopida (Cytherelloidea) and the Podocopida (the remaining 8 superfamilies) in either analyses,

except for the Macrocypridoidea/Pontocypridoidea cluster in the MP tree (57%) and (marginally) for the cluster Cytherelloidea/Darwinuloidea/Terrestricytheroidea in the distance tree (51%). Note that in both trees the Cytherelloidea (the only extant lineage of the Platycopida) cluster *within* the Podocopida. Several attempts to manipulate the dataset by changing either character types or character weights could not improve the phylogenetic resolution.

(4) The 6 superfamilies of the Myodocopa, on the other hand, are better resolved into the two orders (Halocyprida and Myodocopida), with bootstrap values of 66 and 92 respectively for the parsimony analysis and 88 and 97 respectively for the distance analysis.

(5) The positions of the three superfamilies within each of the myodocopan orders are resolved in both trees.

The difference in resolution in the two subclasses could have been due to a different evolutionary history of the two groups, for example when Myodocopa would have had a more gradual cladogenesis, while the Podocopa could have had an early burst of cladogenesis followed by a long period of stasis. To test this hypothesis, mean relative distances, relative rate tests and branch length comparisons were carried out.

(1) Mean relative distances (MRD) between Podocopa and Myodocopa were not significantly different (p = 0.0.4837, Table 1). Significances for MRD comparisons between Myodocopa and the out-groups and Podocopa and the out-groups, respectively, were highly significant (p = 0.0008).

(2) Table 2 gives the relative rate tests between longest and shortest branches (taxa A and B) within the two subclasses, using either of the three out-groups as reference taxon (C). Values in bold are larger than twice the Standard Error (SE) and can be considered significant at the 5% level (Li 1997), although they provide no real statistical tests. Of the 7 pairwise taxon (A, B) comparisons, only 3 provide 'significantly' different relative rates, all three cases include Puncioidea, while using the Phosphatocopina as reference taxon; none of the other two out-groups provide significant comparisons.

(3) Branch length comparisons (Table 3) used branch length descriptions of the distance tree as provided by the program PAUP (Swofford 1998). We computed mean and standard deviation for branch lengths in Podocopa and Myodocopa from the node where both subclasses are branching off (indicated by arrow in Fig. 5); from this node to the last internal nodes and for all the tip-branches. The general branch length comparison was performed with and without the Puncioidea. The results (Table 3) show that none of the general comparison of the branch lengths is significantly different, indicating that there is no evidence for different evolutionary rates in either group.

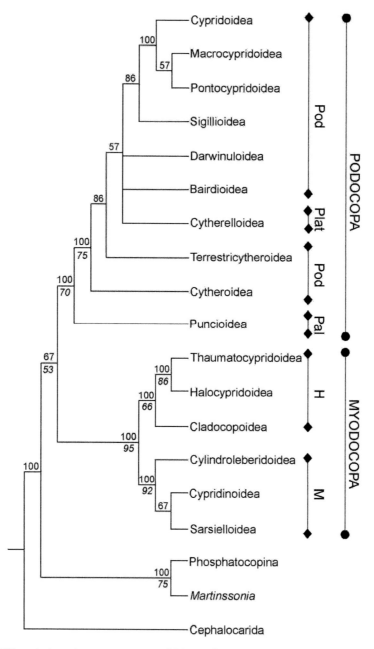

Figure 4. 50% majority rule consensus tree of 21 equally parsimonious trees using stepwise input with furthest addition sequence and the 'branch and bound' algorithm (PAUP) with morphological dataset in appendix 2 (19 taxa, 27 characters, all parsimony informative). Bootstrap values (for 10,000 replicas) higher than 50% are shown below the branches leading up to the pertinent nodes in italic. Figures above the branches are % of the 21 equally parsimonious trees showing these nodes. Tree length = 77 steps. CI = 0.6494, RI = 0.7840. Abbreviations: Pod = Podocopida; Plat = Platycopida; Pal = Palaeocopida; H = Halocyprida; M = Myodocopida.

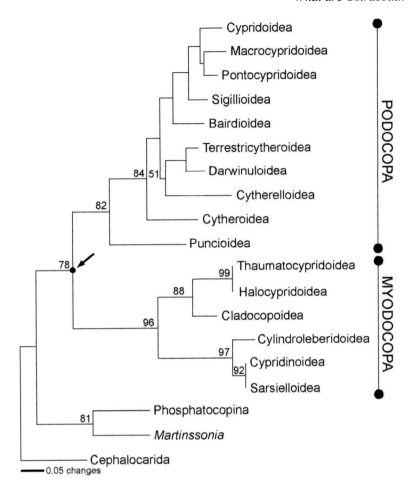

Figure 5. Distance tree, calculated with Neighbor Joining algorithm using mean character difference (PAUP) with morphological dataset in appendix 2 (19 taxa, 27 characters). Bootstrap values (for 10,000 replicas) higher than 50% are shown above the branches leading up to the pertinent nodes. Arrow indicates node from where branch lengths are calculated (see Table 3).

Table 1. Mean relative distances (p) calculated for different comparisons. Student t-tests were 1-tailed and assuming equal variances.

Comparison	Mean	SD	Student t-test
Within Podocopa	0.2324	0.0982	$p = 0.4837$ NS
Within Myodocopa	0.2338	0.1515	
Podocopa versus out-groups	0.5782	0.0583	$p = 0.0008$***
Myodocopa versus out-groups	0.6257	0.0364	

Table 2. Relative rate tests according to Li (1997). The difference (d) between the rates K_{AC} & K_{BC} is calculated, assuming independent evolution of A and B from C. $d = K_{AC} - K_{BC}$. Under equal rates, d should be 0 or close to it. Only the longest and shortest branches of the two superfamilies were compared. Values of 'd' in bold are larger than twice the SE and can be considered significant at the 5% level (Li 1997) although they provide no real, statistical test.

Taxon A	Taxon B	comparison	d
Puncioidea	Cypridinoidea	Cephalocarida	0.077
		Phosphatocopina	**0.200**
		Martinssonia	0
	Cytherelloidea Cylindroleberidoidea	Cephalocarida	0
		Phosphatocopina	0.040
		Martinssonia	0.039
Puncioidea	Cylindroleberidoidea	Cephalocarida	0.077
		Phosphatocopina	**0.240**
		Martinssonia	0.039
Cypridinoidea	Cytherelloidea	Cephalocarida	0
		Phosphatocopina	0
		Martinssonia	0
Puncioidea	Cytherelloidea	Cephalocarida	0.077
		Phosphatocopina	**0.200**
		Martinssonia	0.000
Puncioidea	Cytheroidea	Cephalocarida	0.077
		Phosphatocopina	0.120
		Martinssonia	0
Cypridinoidea	Cylindroleberidoidea	Cephalocarida	0
		Phosphatocopina	0.040
		Martinssonia	0.040

Table 3. Comparison of branch lengths. Mean plus minus standard deviation is given. P values are calculated by F-tests; only general branch length comparisons with Puncioidea included is significant at the 5% level. Column **A**: Branch lengths from last shared node (28) to tips (including Puncioidea). Column **B**: Branch lengths from last shared node (28) to tips (no Puncioidea). Column **C**: Branch lengths from node 28 to last internal node. Column **D**: Branch lengths from last internal nodes to tips.

	A	B	C	D
Podocopa	0.2903 ± 0.0247	0.2969 ± 0.0138	0.2190 ± 0.0594	0.0712 ± 0.0427
Myodocopa	0.3405 ± 0.0279	0.3405 ± 0.0279	0.3249 ± 0.0040	0.0157 ± 0.0243
F-test	$F = 0.705$ NS	$F = 0.076$ NS	$F = 0.410$ NS	$F = 0.227$ NS

4 DISCUSSION

4.1 *Monophyly of Ostracoda?*

Several studies have discussed the position of the different ostracod lineages within the Crustacea and within the Arthropoda, but this is beyond the scope of the present paper. Our present interest in those approaches is limited to the question whether or not the ostracod lineages (mostly limited to podocopes and myodocopes) come out as sister clades, or whether the Ostracoda are a polyphyletic conglomerate of unrelated clades. The results are conflicting. In various analyses of their molecular data, Spears & Abele (1998), for example, obtain different results according to the type of analysis they performed, although in most cases myodocopans and podocopans do not cluster together.

In the present study, we have analysed podocopan and myodocopan phylogeny under the assumption that the Ostracoda (as defined by Horne et al. 2002) are indeed monophyletic, as this was the best way to compare evolutionary tempos in both groups. The three out-groups used here always appeared in the trees outside of the podocopan-myodocopan cluster (the Ostracoda), also when they are not constrained to be out-groups, but this of course offers no clear-cut proof that Ostracoda are monophyletic. The use of other out-groups might have yielded different results.

4.2 *Previous analyses of the phylogeny of the Ostracoda*

Park & Ricketts (2003) carried out a cladistic analysis of nine major higher taxa (orders, suborders), using just seven hard-part characters and obtaining a poorly resolved consensus (majority rule) of four most parsimonious trees. This poor resolution (at which they expressed surprise) is probably a consequence, not only of the low number of characters used (which they admitted), but also of ambiguities and lack of precision in the definition of character states. For example, their character 3 (Park & Ricketts, *op. cit.*, Appendix 1) "Dimorphic structures" has three states: "none" (0), "posterior swelling" (1) and "relative convexity" (2); the last, especially, is difficult to understand, let alone apply to major ostracod taxa. Character 6, "Muscle scars" includes two states that overlap: "unknown" (0) and "circular spots or unknown" (2). The fundamental problem, one that we also encountered when we embarked on the present study, is the extreme difficulty of determining and applying suitable hard-part character states for ostracod carapaces. Park & Ricketts (2000) also carried out a more extensive analysis of 12 ostracod taxa (at order or suborder level) using a mixture of hard- and soft-part characters (14 of each). Once again, the validity of their results is called into question by ambiguities and errors in their determination of character states. For example, their character 18 (male reproductive organ) combines two pairs of character states that should be kept separate: paired or single copulatory limbs, and presence or absence of a Zenker's Organ. According to their scheme, Cypridocopina are scored as '2' because they have a Zenker's Organ, but they could just as well be scored as '0' because they have paired copulatory limbs (they also scored Cytherocopina as '2', which is incorrect since no cytherocopines (= Cytheroidea and Terrestricytheroidea) have Zenker's Organs). For their character 27, the presence or absence of compound eyes, they have incorrectly scored as '1' (= present) for Platycopina, Cytherocopina, Bairdiocopina,

Cypridocopina, Darwinulocopina and Sigilliocopina. The limited value of many of the hard part characters is highlighted by the multiple scores necessarily assigned to some taxa (for example, for carapace/valve shape, four character states are defined; only four out of the twelve taxa are assigned a single state, four are assigned two states, and the remaining four, three states). Examples of adductor muscle scar patterns are illustrated (*op. cit.*, fig. 3) and named for the purposes of character state definition; unfortunately the scores assigned in their appendix 3 do not match the illustrations: their fig. 3d, described as a "pawprint" pattern, is correctly ascribed to the Kirkbyocopina (Puncioidea), but in appendix 3, the "pawprint" score (3) goes to the Cypridocopina, while Kirkbyocopina is incorrectly scored as '0' (= no muscle scars). The typical cypridoidean adductor muscle scar pattern is indeed commonly described in the literature as a "pawprint", but it is quite distinct from that of the puncioids and certainly no basis for considering a close affinity between the two groups.

In view of the above, as well as rather numerous other errors and confusions, and considering the absence of any bootstrap (or other) support for their trees, we consider the phylogenetic analysis of Park & Ricketts (2003) to be seriously flawed. Comparisons of the present results with these unsupported trees are therefore not useful.

Recently, molecular phylogenetic analyses of (part of the) Ostracoda, based on 18S (and partial 28S) ribosomal DNA sequences, have been published by Oakley & Cunningham (2002) and Yamaguchi & Endo (2003) The results of these two studies are not fully congruent with each other, nor with our present results. However, because of certain unsatisfactory aspects of the molecular analyses (for example, only a single genomic region was used) and since we are not yet fully satisfied with our morphological analysis (see below), we feel that it would be premature to start making comparisons between them and our results.

4.3 Topologies of the present phylogenetic trees

With some notable exceptions, the topologies of the two trees based on morphological data and presented above are largely congruent. The most notable aspect of these trees is the difference in supported resolution between the two ostracod subclasses, Podocopa and Myodocopa. The former group is not at all resolved in the parsimony strict consensus tree, only partly resolved in the 50% majority consensus tree, while the branching pattern in the distance tree fails to generate bootstrap support. The analyses do not even show a distinction between the Platycopida and the Podocopida, only the Palaeocopida are clearly branched off the remainder of the Podocopa. The myodocopans consist of two well-supported clades, which are moreover congruent with the extant orders, while in three of the four cases (two orders in two analyses) the resolution is supported down to the superfamily level.

There are various ways in which this difference in resolution within the two main ostracod clades can be interpreted. (1) The data matrix has insufficient characters to resolve the podocopan phylogeny. (2) The lineages leading up to the clades and within the clades could have had different modes of (past) evolution. For example, Myodocopa could have had a more gradual cladogenesis, while the Podocopa could have had an early burst of cladogenesis followed by a long period of stasis.

We tested the latter hypothesis with relative rate tests, comparisons of mean relative distances and comparisons of mean average branch lengths in the two groups, all computed from the distance tree. Because of the said failure to obtain statistical support in the podocope part of the distance tree, we will refrain from discussing the putative meaning of the branching pattern of these superfamilies. We have, however, used this tree to compute mean relative distances and relative rate tests because distances of the OTU's and relative rates are only partly dependent on topologies, even if extant branches are used to compute these figures.

4.4 A different mode of evolution for Podocopa and Myodocopa?

There are no significant differences between mean relative differences within Podocopa and within Myodocopa, but the branches leading up to the two clades are significantly different (Table 1). The latter result is difficult to interpret, especially in light of the unresolved phylogeny of the clades within the Podocopa. The are several possible causes for the difference in evolutionary rate between these ancestral lineages. For example, it could be that the Myodocopa constitute the only descendants of their ancestral lineage, while it could be that the Podocopa are not. This could be so because more podocopan than myodocopan lineages have gone extinct (for which the fossil record does not really offer much support) or because we have not considered other extant crustacean groups as sister clades to Podocopa, i.e. descending from the same ancestral lineage. The latter assumption could be tested by including, for example, Branchiura or certain Cirripedia.

Subsequent evolution within the two clades was then investigated by pairwise comparisons of the relative evolutionary rates (Table 2) and by comparisons of the branch lengths (Table 3). To the best of our knowledge, this is the first time ever that relative rate tests have been used on morphological data.

Only pairwise comparisons with the Puncioidea yield some results that can, according to Li (1997) be considered significantly different. Puncioids would thus have had a different evolutionary rate than some of the other superfamilies, namely Cypridinoidea, Cylindroleberidoidea and Cytherelloidea. Moreover, these pairwise comparisons are only significant when compared to the Phosphatocopina, not to any of the other two out-groups. Why this would be so remains unclear. The fact that punciids are the only clade with deviating evolutionary rate is also shown by branch length comparison. Mean all-round branch length comparisons between Myodocopa and Podocopa are only significantly different (and than only marginally so) if Puncioidea are included.

A final test of different evolutionary tempos in the history of the Myodocopa and the Podocopa compared the earlier (internal) to the tip branches within the two groups. None of these comparisons yielded significant differences, although it must be pointed out that the small number of internal branches in the Myodocopa (only 2 versus 7 in the Podocopa) and the large variation in length of tip branches in the Myodocopa (four branches with zero-length) will have skewed these calculations. Based on the three sets of tests, there seem to be indications that different evolutionary rates have been experienced in the lineages leading up to the two ostracod subclasses, but the hypothesis that the clades within these groups have evolved at different evolutionary tempo must at present be rejected.

5 OUTLOOK

(1) Since the hypothesis that Podocopa cannot be resolved because they experienced early and rapid cladogenesis must be rejected on the basis of several tests presented above, a new character set must be explored to attempt resolution of the podocopan groups. Most likely, such a character set could use hard part morphology to some extent.

(2) New molecular phylogenies, using representatives of all ostracod superfamilies must be drafted using a variety of molecular markers. It appeared from the earlier molecular phylogenies referred to above that 18S might not be the best marker to use for analyses at this level. This is not surprising: if 18S can be used successfully at the level of the crustacean classes (and above), than chances are small that the same marker will be useful for lower taxonomic levels, such as superfamilies.

(3) The present analysis nevertheless shows that a number of soft part characters can be used to resolve part of the ostracod phylogeny; it is hoped that they will be useful for incorporating the two ostracod subclasses in general morphological analyses of Crustacea and Arthropoda. In such analyses, Myodocopa and Podocopa must be used as separate groups in order to finally test ostracod mono- or polyphyly.

ACKNOWLEDGEMENTS

Renate Matzke-Karasz (Munich, Germany), Geoff Boxshall (London, UK) and Ronald Jenner (Cambridge, UK) offered valuable advice. The editors of this book are gratefully acknowledged for their patience. Isa Schön acknowledges OSTC grant OSTC (MO-36-005). Robin Smith thanks the Royal Society, Koen Martens and Dave Horne acknowledge the Belgian FWO for contract G.0118.03 which allowed several visits of the DJH to the lab of KM. All authors acknowledge EU Marie Curie RTN grant MRTN-CT-2004-512492 (SEXASEX) for facilitating mobility and transfer of knowledge.

APPENDIX 1

Characters and character states (in square brackets) used in the present phylogenetic analysis of the Ostracoda.

1. Zenker's Organ: absent [0], present [1].
2. Maximum number of antennal endopodite podomeres: 2 [0], 3 [1], 4 [2].
3. Antennal exopodite: with > 2 podomeres [0], with 1-2 podomeres [1], with a scale bearing setae [2], a spinneret seta [3].
4. Mandibular exopodite: multisegmented [0], unsegmented (seta or branchial plate) or absent [1].
5. Mandibular palp filter screen of > 7 setae: absent [0], present [1].
6. Mandibular coxa: a 'normal' podomere with an endite or secondarily fused with basis [0], a laterally produced podomere with a gnathobase [1].
7. Mandibular basis: with toothed endite [0], without toothed endite [1].
8. Maxillular protopodite: not subdivided or with proximal endite only [0], subdivided into coxa and Basis [1], subdivided into precoxa, coxa and basis [2].
9. Number of rays on maxillular branchial plate: inapplicable (no unsegmented branchial plate) [?], 15-27 [0], 24-36 [1].
10. Maxillular endopodite: leg-like [0], a palp [1], compacted [2].
11. Maxillular exopodite: multisegmented [0], an unsegmented branchial plate [1], single-segmented [2], reduced / absent [3].
12. Fifth limb endopodite: leg-like with 4 or more podomeres [0], leg-like with 3 podomeres [1], a palp / leg with 1-2 podomeres [2], compacted [3].
13. Fifth limb exopodite: multisegmented [0], a branchial plate [1], single-segmented [2], reduced / absent [3].
14. Fifth limb knee joint: absent [0], present at protopod-endopod junction [1], present at coxa-basis junction [2].
15. Epipodite on 5th limb: absent [0], present [1].
16. Sixth limb knee joint: inapplicable (limb absent) [?], limb without knee joint [0], knee joint present at protopod-endopod junction [1], knee joint present at coxa-basis junction [2].
17. Sixth limb exopodite: inapplicable (limb absent) [?], multisegmented [0], a branchial plate / 1-2 segments / setae / absent [1].
18. Epipodite on 6th limb: inapplicable (limb absent) [?], absent from limb [0], present [1].
19. Sixth limb endopodite: inapplicable (limb absent) [?], leg-like with 4 or more podomeres [0], leg-like with 3 podomeres [1], leg-like with 1 or 2 podomeres [2], compacted [3].
20. Seventh limb: leg-like with multisegmented exopodite [0], leg-like, with exopodite reduced to setae or absent [1], an articulated cleaning limb [2], a vermiform cleaning limb [3], reduced / absent [4].
21. Eighth limb in female: present [0], absent [1].
22. Furca: well-developed, two rami + claws / setae [0], reduced to setae / absent [1].
23. Furcal position terminal, in line with anus [0], posterior to anus [1], anterior to anus [2].
24. Lateral compound eyes: absent [0], present [1].
25. Carapace: a dorsal shield [0], bivalved, totally enclosing limbs [1].
26. Tagmosis - cephalic segments: 4 cephalic segments with limb pairs [0], 5 cephalic segments with limb pairs [1].
27. Bellonci organ: absent [0], present [1].

APPENDIX 2

Matrix of 27 character states for Ostracoda superfamilies and out-groups: Cephalocarida, Phosphatocopina and *Martinssonia*.

Taxa	Characters
Cypridoidea	1 1 2 1 0 1 1 0 0 1 1 2 1 1 0 1 1 0 0 2 1 0 2 0 1 0 0
Macrocypridoidea	1 2 2 1 0 1 1 0 0 1 1 0 2 1 0 1 1 0 0 2 1 0 2 0 1 0 0
Pontocypridoidea	1 1 2 1 0 1 1 0 0 1 1 1 2 1 0 1 1 0 0 2 1 0 2 0 1 0 0
Cytheroidea	0 2 3 1 0 1 1 0 0 1 1 1 2 1 0 1 1 0 1 1 1 1 2 0 1 0 0
Terrestricytheroidea	0 1 2 1 1 1 1 0 0 1 1 1 2 1 0 1 1 0 0 1 1 1 2 0 1 0 0
Darwinuloidea	0 1 2 1 1 1 1 0 1 1 1 1 1 1 0 1 1 0 0 1 1 1 2 0 1 0 0
Sigillioidea	1 2 2 1 0 1 1 0 0 1 1 1 1 1 0 1 1 0 0 1 1 0 2 0 1 0 0
Bairdioidea	0 2 2 1 0 1 1 0 1 1 1 0 1 1 0 1 1 0 0 1 1 0 2 0 1 0 0
Cytherelloidea	0 1 1 1 1 1 1 0 1 1 1 2 1 1 0 1 1 0 2 4 1 0 2 0 1 0 0
Puncioidea	0 1 1 1 0 1 1 0 ? 0 3 1 2 1 0 0 1 1 1 1 0 0 2 0 1 0 0
Thaumatocypridoidea	0 1 0 1 0 1 0 1 ? 2 2 1 2 2 1 2 1 1 1 4 1 0 1 0 1 1 1
Cylindroleberidoidea	0 1 0 1 1 0 1 1 ? 2 2 3 2 0 1 0 1 0 3 3 1 0 1 1 1 1 1
Cypridinoidea	0 1 0 1 0 0 1 2 ? 2 2 3 2 0 1 0 1 0 3 3 1 0 1 1 1 1 1
Halocypridoidea	0 1 0 1 0 1 0 1 ? 2 2 1 2 2 1 2 1 1 1 4 1 0 1 0 1 1 1
Cladocopoidea	0 1 0 1 0 1 1 2 ? ? 2 2 2 2 2 1 ? ? ? ? 4 1 0 1 0 1 1 1
Sarsielloidea	0 1 0 1 0 0 1 2 ? 2 2 3 2 0 1 0 1 0 3 3 1 0 1 1 1 1 1
Cephalocarida	0 0 0 1 0 1 1 1 ? 0 1 0 0 0 0 0 1 0 0 0 0 0 0 0 0 1 0
Phosphatocopina	0 0 0 0 0 0 0 0 ? 0 0 1 0 0 0 0 0 0 1 0 0 0 ? 0 1 0 0
Martinssonia	0 2 0 0 0 0 0 0 ? 0 0 0 0 0 0 0 0 0 0 4 1 0 1 0 0 0 0

REFERENCES

Boxshall, G. 1997. Comparative limb morphology in major crustacean groups: the coxa-basis joint in postmandibular limbs. In: Fortey, R.A. & Thomas, R.H. (eds.), *Arthropod Relationships*. Syst. Assoc. Spec. Vol., Ser. 55: 155-167.

Boxshall, G. 2004. The evolution of arthropod limbs. *Biol. Rev.* 79: 253-300.

Cohen, A.C., Martin, J.W. & Kornicker, L.S. 1998. Homology of Holocene ostracode biramous appendages with those of other crustaceans: the protopod, epipod, exopod and endopod. *Lethaia* 31: 251-265.

Cohen, A.C. & Morin, J.G. 2003. Sexual morphology, reproduction and the evolution of bioluminescence in Ostracoda. In: Park, L.E. & Smith, A.J. (eds.), *Bridging the Gap: Trends in the Ostracode Biological and Geological Sciences*. Paleontol. Soc., Pap. 9: 37-70.

Fryer, G. 1996. Reflections on arthropod evolution. *Biol. J. Linn. Soc.* 58: 1-55.

Horne, D.J. 2003. Key events in the ecological radiation of the Ostracoda. In: Park, L.E. & Smith, A.J. (eds.), *Bridging the Gap: Trends in the Ostracode Biological and Geological Sciences*. Paleontol. Soc., Pap. 9: 181-201.

Horne, D.J. In press. Homology and homoeomorphy in ostracod limbs. *Hydrobiologia*.

Horne, D.J., Cohen, A. & Martens, K. 2002. Taxonomy, morphology and biology of Quaternary and living Ostracoda. In: Holmes, J.A. & Chivas, A.R. (eds.), *The Ostracoda: Applications in Quaternary Research*. Amer. Geophys. Monogr. 131: 5-36.

Horne, D.J., Smith, R.J., Whittaker, J.E. & Murray, J.W. 2004. The first British record and a new species of the superfamily Terrestricytheroidea (Crustacea, Ostracoda): morphology, ontogeny, lifestyle and phylogeny. *Zool. J. Linn. Soc.* 142: 253-288

Hou, Xianguag, Siveter, D.J., Williams, M., Walossek, D. & Bergstrom, J. 1996. Appendages of the arthropod *Kunmingella* from the early Cambrian of China: its bearing on the systematic position of the Bradoriida and the fossil record of the Ostracoda. *Phil. Trans. R. Soc. London, Ser. B*, 351: 1131-1145.

Kesling, R.V. 1951. The morphology of ostracod molt stages. *Illin. Biol. Monogr.* 21 (1/3): 1-324.

Kornicker, L.S. 2002. Comparative morphology of the fifth limb (second maxilla) of myodocopid Ostracoda. *J. Crust. Biol.* 22: 797-818.

Li, W-H. 1997. *Molecular Evolution*: 487 pp. Sunderland: Sinauer.

Maas, A., Waloszek, D. & Müller, K.J. 2003. Morphology, ontogeny and phylogeny of the Phosphatocopina (Crustacea) from the Upper Cambrian "Orsten" of Sweden. *Fossils & Strata* 49: 1-238.

Maddocks, R.F. 1973. Zenker's Organ and a new species of *Saipanetta* (Ostracoda). *Micropaleontology* 19: 193-208.

Maddocks, R.F. 1982. Part 4: Ostracoda. In: Abele, L.G. (ed.), *The Biology of the Crustacea; Volume 1, Systematics, the Fossil Record and Biogeography*: 221-239. New York; London: Academic Press.

Martens, K. 1992 On *Namibcypris costata* n.gen. n.sp. (Crustacea, Ostracoda, Candoninae) from a spring in northern Namibia, with the description of a new tribe and a discussion on the classification of the Podocopina. *Stygologia* 7: 27-42.

Martens, K. 1998. General morphology of non-marine ostracods. In: Martens, K. (ed.), *Sex and Parthenogenesis: Evolutionary Ecology of Reproductive Modes in Non-Marine Ostracods*: 57-75. Leiden: Backhuys Publishers.

Martens, K. 2003. On a remarkable South African giant ostracod (Crustacea, Ostracoda, Cyprididae) from temporary pools, with additional appendages. *Hydrobiologia* 500: 115-130.

Martin, J.W. & Davis, G.E. 2001. An Updated Classification of the Recent Crustacea. *Nat. Hist. Mus. L.A. County, Sci. Ser.* 39.

Matzke-Karasz, R. & Martens, K. In press. The female reproductive organ in podocopid ostracods is homologous to five appendages: histological evidence from *Liocypris grandis* (Crustacea, Ostracoda). *Hydrobiologia*.

McKenzie, K.G., Müller, K.J. & Gramm, M.N. 1983. Phylogeny of Ostracoda. In: Schram, F.R. (ed.), *Crustacean Phylogeny*: 29-46. Rotterdam: A.A. Balkema.

McKenzie, K.G., Angel, M.V., Becker, G., Hinz-Schallreuter, I., Kontrovitz, M., Parker, A.R., Schallreuter, R.E.L. & Swanson, K.M. 1999. Ostracods. In: Savazzi, E. (ed.), *Functional Morphology of the Invertebrate Skeleton*: 459-507. John Wiley & Sons, Ltd.

Meisch, C. 2000. Freshwater Ostracoda of Western and Central Europe. In: Schwoerbel, J. & Zwick, P. (eds.), *Süßwasserfauna von Mitteleuropa 8/3*: 1-522. Heidelberg: Spektrum Akad. Verlag.

Oakley, T.H. & Cunningham, C.W. 2002. Molecular phylogenetic evidence for the independent evolutionary origin of an arthropod compound eye. *Proc. Nat. Acad. Sci. USA* 99 (3): 1426-1430.

Park, L.E. & Ricketts, R.D. 2003. Evolutionary history of the Ostracoda and the origin of nonmarine faunas. In: Park, L.E. & Smith, A.J. (eds.), *Bridging the Gap: Trends in the Ostracode Biological and Geological Sciences*. Paleontol. Soc., Pap. 9: 11-35.

Schram, F.R. 1986. *Crustacea*. Oxford, New York: Oxford Univ. Press.
Schram, F.R. & Hof, C.H.J. 1998. Fossils and the interrelationships of major crustacean groups. In: Edgecombe, G.D. (ed.), *Arthropod Fossils and Phylogeny*: 233-302. New York: Columbia Univ. Press.
Schulz, K. 1976. *Das Chitinskelett der Podocopida (Ostracoda, Crustacea) und die Frage der Metamerie dieser Gruppe*. Doctoral thesis, Fachber. Biologie, Univ. Hamburg: 167 pp.
Scott, H.W. 1961. Classification of Ostracoda. In: Moore, R.C. (ed.), *Treatise on Invertebrate Paleontology, Part Q, Arthropoda 3, Crustacea, Ostracoda*: Q74-Q92. Boulder; Lawrence: Geol. Soc. Amer. and Univ. Kansas Press.
Shu, D., Vannier, J, Luo, H., Chen, L., Zhang, X. & Hu, S. 1999. Anatomy and lifestyle of *Kunmingella* (Arthropoda, Bradoriida) from the Chenjiang fossil Lagerstätte (lower Cambrian, southwest China). *Lethaia* 32: 279-298.
Siveter, D.J., Sutton, M.D., Briggs, D.E.G. & Siveter, D.J. 2003. An ostracode crustacean with soft parts from the Lower Silurian. *Science* 302: 1749-1751.
Siveter, D.J. & Williams, M. 1997. Cambrian bradoriid and phosphatocopid arthropods of North America. *Spec. Pap. Palaeontol.* 57: 1-69.
Smith, R.J. 2000. Morphology and ontogeny of Cretaceous ostracods with preserved appendages from Brazil. *Palaeontology* 43: 63-98.
Smith, R.J., Kamiya, T., Horne, D.J. & Tsukagoshi, A. In press. Evaluation of a new character for the phylogenetic analysis of Ostracoda (Crustacea): the podocopan maxillular branchial plate. *Zool. Anz.*
Smith, R.J. & Martens, K. 2000. The ontogeny of the cyprid ostracod *Eucypris virens* (Jurine, 1820) (Crustacea, Ostracoda). *Hydrobiologia*, 419: 31-63.
Spears, T. & L.G. Abele 1998. Crustacean phylogeny inferred from 18S rDNA. In: Fortey; R.A. & Thomas, R.H. (eds.), *Arthropod Relationships*. Syst. Assoc. Spec. Vol., Ser. 55: 169-188.
Swain, F.M. 1976. Evolutionary development of cypridopsid Ostracoda. *Abh. Verh. Naturwiss. Ver. Hamburg, N.F. (Suppl.)*, 18/19: 103-118.
Swanson, K.M. 1989. *Manawa staceyi* n. sp. (Punciidae, Ostracoda), soft anatomy and ontogeny. *Cour. Forsch.-Inst. Senckenb.* 113: 235-249.
Swanson, K.M. 1991. Distribution, affinities and origin of the Punciidae (Crustacea: Ostracoda). *Mem. Queensland Mus.* 31: 77-92.
Swofford, D.L. 1998. *PAUP. Phylogenetic analysis using parsimony (and other methods), Ver. 4.0*. Sunderland: Sinauer Assoc.
Tsukagoshi, A. & Parker, A.R. 2000. Trunk segmentation of some podocopine lineages in Ostracoda. *Hydrobiologia* 419: 15-30.
Vannier, J. & Abe, K. 1992. Recent and early Palaeozoic myodocope ostracods: functional morphology, phylogeny, distribution and life-styles. *Palaeontology* 35: 485-517.
Vannier, J., Wang, S.Q. & Coen, M. 2001. Leperditicopid arthropods (Ordovician-Late Devonian): functional morphology and ecological range. *J. Paleontol.* 75: 75-95.
Walossek, D. 1999. On the Cambrian diversity of Crustacea. In: Schram, F.R. & von Vaupel Klein, J.C. (eds.), *Crustaceans and the Biodiversity Crisis; Proc. 4th Int. Crust. Congr., Amsterdam, Netherlands, 1998, Vol. 1*: 3-27. Leiden: Brill.
Walossek, D. & Müller, K.J. 1998. Early arthropod phylogeny in the light of the Cambrian "Orsten" fossils. In: Edgecombe, G.D. (ed.), *Arthropod Fossils and Phylogeny*: 185-231. New York: Columbia Univ. Press.

Whatley, R.C. & Moguilevsky, A. 1998. The origins and early evolution of the Limnocytheridae (Crustacea, Ostracoda). In: Crasquin-Soleau, S., Braccini, E. & Lethiers, F. (eds.), *What about Ostracoda! Act. 3e Congr. Europ. Ostracodologists, Paris-Bierville, France, 1996*. Bull. Centre Rech. Elf Explor. Prod., Mem. 20: 271-285.

Yamaguchi, S. & Endo, K. 2003. Molecular phylogeny of Ostracoda (Crustacea) inferred from 18S ribosomal DNA sequences: implications for its origin and diversification. *Mar. Biol.* 143: 23-38.

Relationships within the Pancrustacea: Examining the influence of additional Malacostracan 18S and 28S rDNA

COURTNEY C. BABBITT[1] & NIPAM H. PATEL[2]

[1] *Committee on Evolutionary Biology, University of Chicago, Chicago, Illinois, U.S.A.*

[2] *Departments of Integrative Biology and Molecular Cell Biology, Howard Hughes Medical Institute, University of California, Berkeley, California, U.S.A.*

ABSTRACT

Recent molecular and morphological phylogenetic analyses have lent support to the Pancrustacea hypothesis, which argues that the Crustacea is either the sister group to the Hexapoda, or is paraphyletic in relation to the Hexapoda. Developmental evidence has been used to argue for a sister relationship between the Malacostraca and the Hexapoda, while the molecular evidence has been equivocal. These data provide four scenarios that require testing with increased taxon sampling: 1) hexapods and crustaceans are sister groups; 2) crustaceans are paraphyletic with hexapods most closely related to branchiopod crustaceans; 3) crustaceans are paraphyletic with hexapods most closely related to malacostracan crustaceans; or 4) hexapods are more closely related to myriapods, and that the Pancrustacea hypothesis is incorrect. The focus of this chapter is on the monophyly of the Crustacea, relationships within the Malacostraca, and the relationship of the Crustacea and Hexapoda. To this end, we examine the developmental and molecular evidence for the above hypotheses by testing them with new crustacean, specifically malacostracan, 18S and 28S sequence data. Our data suggest that branchiopod crustaceans may be the sister group to the Hexapoda, but they are not resolved on the placement of the Myriapoda. Within the Malacostraca, the Eucarida and Peracarida are found not to be monophyletic groups due to changes in the affinities of mysids, krill, and caridean shrimp.

1 PREVIOUS STUDIES OF THE ARTHROPODA

The complexity of arthropod relationships is mirrored by the diverse literature attempting to decipher it; most of the possible permutations in relationships between the major groups have been proposed at least once. The extraordinary diversity and long evolutionary history of all of these groups has led to a continuing controversy over arthropod relationships in studies based on morphological, molecular, developmental, and genomic evidence. Ancient and possibly explosive radiations are thought to be involved in the early history of the arthropods, with subsequent periods of adaptation and convergence in new environments (Schram 1986; Briggs & Fortey 1989; Briggs et al. 1992; Averof & Akam 1995). This complicated history may be the reason for the contradictory phylogenetic signals within the Arthropoda.

1.1 Arthropod monophyly and groups within the Arthropoda

There is a long history of considering the Arthropoda to be monophyletic (Snodgrass 1935; Field et al. 1988; Turbeville et al. 1991; Adoutte & Philippe 1993; Wheeler et al. 1993; Boore et al. 1995; Boore et al. 1998; Friedrich & Tautz 1995; Giribet et al. 1996; Giribet et al. 2001; Wheeler 1997; Giribet & Ribera 2000; Hwang et al. 2001; Mallatt et al. 2004). However, some authors have argued in the past that the Arthropoda is either diphyletic or polyphyletic (Manton 1977; Anderson 1973). This hypothesis was based on a few characters that were thought to have critical importance, characters such as cleavage patterns that were described using a framework of 'Articulata' fate maps and spiralian cleavage, and argued similarities in adult limb and mandible functional morphology. However, using systematic techniques, the majority of studies using molecular and/or morphological data have reaffirmed the arthropod monophyly sensu Snodgrass (Field et al. 1988; Wheeler et al. 1993; Giribet et al. 1996; Giribet et al. 2001; Wheeler 1997; Giribet & Ribera 2000; Hwang et al. 2001; Mallatt et al. 2004). Historical reviews of the different hypotheses concerning the relationship between the five major arthropod groups (the Cheliceriformes, Myriapoda, Trilobitomorpha, Hexapoda and Crustacea) have been reviewed extensively in previous volumes of arthropod relationships (Fortey & Thomas 1997; Deuve 2001).

Morphological studies (Wheeler et al. 1993; Wheeler 1997) have supported an Atelocerata, combining the myriapods and hexapods based on a number of synapomorphies, such as the possession of a single pair of antennae, the presence of ectodermal Malpighian tubules, and the presence of tracheae (Snodgrass 1935; Wheeler et al. 1993; Wheeler 1997). However, it has also been argued that these morphological and functional similarities between myriapods and hexapods are convergent adaptations to life on land (Averof & Akam 1995).

Molecular evidence, more recently, has provided new light on this problem. Combined analyses with morphology using either 18S and 28S (Wheeler 1997) or histone H3 and U2 (Edgecombe et al. 2000) support the traditional Mandibulata and Atelocerata, but the latter only weakly. The Pancrustacea hypothesis of Hexapoda + Crustacea was first described in Zrzavý & Štys (1997), and has been supported by molecular analyses (Giribet et al. 2001; Hwang et al. 2001; Mallatt et al. 2004).

First, we will review different sub-groups of crustaceans, and then examine the recent evidence placing the Hexapoda within a polyphyletic Crustacea (Fig. 1).

1.2 Groups within the Crustacea

As is the case with the Arthropoda as a whole, the enormous morphological diversity and long evolutionary history of the Crustacea make the phylogenetic affinities within the Crustacea very controversial. Multiple lines of evidence support this idea of complex relationships and convergences towards similar body plans (Cunningham et al. 1992; Morrison et al. 2001). Morphologically, crustacean monophyly is based on a unique series of head appendages with two pairs of antennae, a mandibular segment, and two pairs of maxillae (Schram 1986), though it is important to note that in the Cephalocarida and in some fossil crustaceans outside of the crown-group the presence and structure of the maxillae is variable (Schram & Hof 1998; Walossek & Müller 1998). Naupliar stages of development are another general characteristic of the Crustacea, in which only the first three segments are formed before hatching, and it has been argued that this is the primitive state for crustacean development (Scholtz 2000). Other characters that suggest crustacean monophyly, such as the tendency to fuse segments to form a carapace, or repeated trends of tagmatization, have been argued to be more generalized trends in different crustacean groups, than they are static characters (Schram 1986). However, these hypotheses of primary character homology may be difficult to implement and/or interpret in studies with only extant taxa. Schram (1986) performed the first systematic treatment of the Crustacea based on morphological characters which presented remipedes as the basal group, and the Malacostraca (without phyllocarids) as a sister group to the Maxillopoda + Phyllopoda (with the Phyllopoda including phyllocarids, cephalocarids, and branchiopods). A more recent study that included fossil representatives (Wills 1997), defines six major crustacean groups: the Malacostraca, Phyllocarida, Maxillopoda, Cephalocarida, Branchiopoda, and Remipedia. This is the general classification discussed below, with the exception that the Phyllocarida is considered here to be part of the Malacostraca. Molecular studies of the Crustacea with broad taxon sampling have been limited to date. Spears & Abele (1997) produced one of the first studies, based on 18S rDNA. This study found that the Crustacea, Eumalacostraca (with the inclusion of the Phyllocarida), and the Branchiopoda were well-supported monophyletic groups, and that there was weak support for a clade containing representatives from the Maxillopoda, Cephalocarida, and Remipedia.

The Branchiopoda is considered to be a monophyletic group with two main branches: the Anostraca and the bivalved branchiopods, currently classified as the Phyllopoda (Spears & Abele 2000; Martin & Davis 2001; Braband et al. 2002), and this classification is strongly supported by molecular data (Spears & Abele 2000; Braband et al. 2002). The Anostraca appears to have diverged early from the main branchiopod line, and retains many primitive features, such as biramous, foliaceous thoracic limbs used for swimming and feeding, but are unique within the Branchiopoda because they lack a carapace (Spears & Abele 2000).

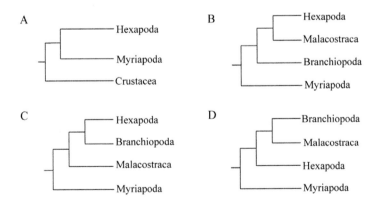

Figure 1. Competing hypotheses about relationships within the Mandibulata. (A) A monophyletic Atelocerata. (B) Hexapoda + Malacostraca grouping to make the Crustacea paraphyletic. (C) Hexapoda + Branchiopoda grouping to make the Crustacea paraphyletic. (D) monophyletic Pancrustacea and Crustacea.

Although they contain the bounty of the morphological, ecological, and developmental diversity of crustaceans, the Malacostraca is also considered to be a stable monophyletic group (Bowman & Abele 1982; Spears & Abele 1997; Watling 1999; Shultz & Regier 2000; Richter & Scholtz 2001) with the first fossil record from this group is a phyllocarid species (Briggs 1978) from the Cambrian. The Malacostraca has five major divisions: Eucarida, Syncarida, Peracarida, Hoplocarida, and the Phyllocarida (Calman 1904, 1909). This last group, the Phyllocarida, have been placed in their own class (Schram 1986; Martin et al. 1996), though recent morphological (Dahl 1987; Richter & Scholtz 2001) and molecular analyses (Spears & Abele 1997) argue for placement of the Phyllocarida as a sister group to the rest of the Malacostraca. The monophyly of the Hoplocarida, Peracarida and Syncarida was supported in a large morphological analysis by Richter & Scholtz (2001). However, there is still controversy about the monophyly of the Eucarida: specifically, the placement of the Euphausiacea (krill) either with the Peracarida (Richter & Scholtz 2001) or with specific peracarids, such as the mysids (Jarman et al. 2000).

Relationships between the remaining crustacean groups are more problematic. Molecular attempts to resolve these relationships have argued that the Maxillopoda is an artificial assemblage, and that the similarities in spermatozoan and adult morphology are either symplesiomorphic characters or convergences (Spears & Abele 1997). The two remaining crustacean classes are the Remipedia and the Cephalocarida. The Remipedia are a recently discovered group (Yager 1981), of which still little is known. Morphological studies have placed them as the basal group of crustaceans (Schram 1986; Wills 1997; Schram and Hof 1997) based mainly on their homonomous segmentation and lack of tagmosis. However, a recent phylogenetic analysis employing characters from the remipede brain argues for a sister group relationship between the Malacostraca and the Remipedia. This problem is not easily solvable as the sequence data for remipedes, as well as for cephalocarids is problematic in that both branch lengths are long (Spears & Abele 1997), which may mislead different types of phylogenetic analyses (Felsenstein 1985; Siddall 1998).

1.3 *Evidence of affinities between crustacean groups and insects*

Recently, multiple lines of evidence have suggested that either malacostracan or branchiopod crustaceans may be more closely aligned with the hexapods than they are to each other (Boore et al. 1995; Boore et al. 1998; Friedrich & Tautz 1995; Giribet et al. 1996; Giribet et al. 2001; Wheeler 1997; Giribet & Ribera 2000; Hwang et al. 2001; Mallatt et al. 2004). Morphologically, this is difficult to determine due to the diversity and specialization of some crustacean forms, as well as to the modifications made in the insect lineage for life on land (Schram & Jenner 2001). However, both neurological and developmental data have been used to look for synapomorphies between these groups.

Comparative data from studies on neurogenesis and axonogenesis point to affinities between malacostracans and hexapods in the structure of the nervous system and the placement and organization of the neurons, though there are differences in the axonal pathways of specific cells in the CNS (Thomas et al. 1984; Whitington et al. 1991; Whitington et al. 1993; Whitington 1996; Duman-Scheel & Patel 1999). There is also a marked similarity between malacostracan and hexapod compound eyes, particularly in ommatidial structure, which has not been found in either chelicerates or myriapods (Averof & Akam

1995; Osorio et al. 1995; Dohle 2001).

Characters during development can also contribute to these analyses, not necessarily as providers of support to homology arguments about adult morphology, but as characters themselves. One example is the precise and stereotypic formation of CNS precursor cells from neuroblasts (NBs) in malacostracan crustaceans and in insects. These NB cells are large stem cells, which divide unequally to form a stereotyped pattern of neurons in each segment (this process was described in Dohle & Scholtz 1988; Whitington 1996). After this, in both malacostracans and hexapods, the NBs then divide asymmetrically perpendicular to the ventral surface to produce ganglion mother cells (GMCs), and then the GMCs go on to produce neurons. NB-like cells have not been found to date in myriapods, and the processes of neurogenesis and axon formation in the centipede *Ethmostigmus rubripes* are different than the patterns seen in either the Malacostraca or the Hexapoda (Whitington et al. 1991; Whitington et al. 1993). In contrast to this, almost all branchiopod neural stem cells produce neurons through a general inward proliferation of cells (Weygoldt 1960; Benesch 1969). However, the expression patterns of some of the genes known to influence later neuronal fates are shared between the Malacostraca, Branchiopoda, and Hexapoda (Duman-Scheel & Patel 1999). Even-skipped (Eve) is an excellent neural marker for this, as the protein is expressed only in a small sub-set of neurons. As measured by location and expression levels, homologues of all Eve-expressing neurons have been found in members of the Branchiopoda, Malacostraca, and Hexapoda (Duman-Scheel & Patel 1999). Taken all together, this developmental data suggests a relationship between the Malacostraca and the Hexapoda, and supports the Pancrustacea to the exclusion of the myriapods.

Examining other types of developmental data, three levels of developmental genetic expression have been examined in some depth: mouthpart formation, segmentation, and the establishment of segment identity. The absence of *Distal-less* (*Dll*) expression has been examined as a marker for homologous development of the gnathobasic mandible. However, the transformations and loss of this character do not allow for any suggestions of relationships within the Mandibulata, although they do support a monophyletic Hexapoda (Scholtz et al. 1998). In the same vein, earlier phylogenetic arguments focusing on similarities in segmental (*engrailed*) and *Hox* (*Ultrabithorax*, etc.) gene expression between malacostracans and insects are now being filled in with more extensive taxon sampling. It now appears that, at least at the levels of segment polarity genes and segment identity genes as assayed by engrailed and Ultrabithorax protein expression, that similar developmental genes are being employed throughout the arthropods, and that their expression patterns vary within the major clades (Averof & Patel 1997; Davis & Patel 1999). Past authors had also predicted (Averof & Akam 1995) that *Hox* cluster structure would provide some insight, but with more extensive sequencing it appears that all major groups of arthropods contain one (with the exception of spiders) copy of each of the ten *Hox* genes (Cook et al. 2001; Hughes & Kaufman 2002). Therefore, *Hox* cluster structure does not seem to be informative in parceling apart the major arthropod groups. Yet, *Hox* clusters may be phylogenetically useful at lower levels: the Araneae possess two Ultrabithorax genes which differentiates them from the rest of the Chelicerata that only possess one (Damen et al. 1998), and there have been characteristic losses of *Hox* genes in that define clades within the nematodes (Aboobaker & Blaxter 2003). Neither Hox complex structure, nor protein expression of genes in the segmentation hierarchy, have been useful in differentiating between the possible relationships within the Mandibulata.

A separate line of evidence that there is a more complex relationship between crustaceans and hexapods than previously thought comes from mitochondrial gene order studies. Mitochondrial data argues for support for a clade of crustaceans and insects to the exclusion of myriapods (Boore et al. 1995; Wilson et al. 2000; Hwang et al. 2001; Nardi et al. 2003). However, these studies seem to be very dependent on the taxon sampling and the more specific relationships are variable between studies. The whole or partial mitochondrial studies with the largest number of crustacean representatives show support for the decapods as a sister clade to insects, with the branchiopods as sister group to that clade (Garcia-Machado et al. 1999; Wilson et al. 2000; Hwang et al. 2001). Yet, when sequences that could be causing uncertainty are removed, the Crustacea regains monophyly and is the sister group to insects, to the exclusion of Collembola (Nardi et al. 2003).

The molecular sequence data has been ambiguous about the relationships between the different crustacean groups and hexapods. In early combined analyses (Wheeler et al. 1993; Wheeler 1997) the molecular data weakly argue for a sister group relationship between crustaceans and hexapods, but this support collapsed in the combined (morphological + molecular) analyses, which supported a monophyletic Atelocerata. Recent ribosomal DNA analyses have continued to support the Pancrustacea (Friedrich & Tautz 1995; Giribet et al. 1996; Giribet et al. 2001; Giribet & Ribera 2000; Hwang et al. 2001; Mallatt et al. 2004). Hwang et al. (2001) also pointed out that the ribosomal genes gave higher support to the Pancrustacea than the mitochondrial data. Nuclear non-ribosomal genes, such as EF-1α and RNA Polymerase II have also been used for arthropod phylogeny reconstruction (Regier & Shultz 1997; Shultz & Regier 2000; Regier & Shultz 2001). Sections of RNA Pol II provides support for a crustacean + hexapod clade; yet, EF-1α does not support this grouping (Shultz & Regier 2000; Regier & Shultz 2001). Within all of these studies there are again varying levels of support for scenarios B-D shown in Figure 1.

One limitation of these previous studies has been the limited number of taxa represented from the Crustacea, especially the Malacostraca. With a large variety of data pointing to a paraphyletic Crustacea with respect to the Hexapoda, the inclusion of these taxa is crucial to teasing apart these ambiguous relationships. In this chapter, we attempt to fill in some of the missing data as we test the relationship of the Malacostraca, Hexapoda, and Branchiopoda, and the relationships within the Malacostraca by the addition of malacostracan 18S and 28S rDNA sequences.

2 MATERIALS AND METHODS

2.1 *Specimens and Sequencing*

Primer pairs used to amplify 18S (~1800 bp) from malacostracan species were: (1F, 5R), (3F, bi), and (5F, 9R). Primer sequences 5' to 3' are:
1F: ACCTGGTTGATCCTGCCAGTAG; 5R: CTTGGCAAATGCTTTCGC;
3F: GTTCGATTCCGGAGAGGGA; bi: GAGTCTCGTTCGTTATCGGA;
5F: GCGAAAGCATTTGCCAAGAA; and 9R: GATCCTTCCGCAGGTTCACCTAC;
(nomenclature and sequences as in Giribet et al. 1996; Giribet & Ribera 2000). The D3 region of 28S (~350 bp) was amplified with the primers described in Whiting et al. (1997). Tissue was extracted from either fresh tissue or from specimens preserved in 95% ethanol

and stored at –20° C. Genomic DNA was extracted using a Nucleospin tissue kit, and then phenol: chloroform extracted and ethanol precipitated. PCR was done on a Stratagene robocycler gradient 96, and the PCR program consisted of a 5 min denaturing step at 95° C, then 30 cycles of 30 sec at 95° C, 45 sec at 54° C, and 1 minute extension at 72° C, followed by 5 minutes at 72° C. The PCR product was cleaned up of any single stranded products using 2 U of Shrimp Alkaline Phosphatase and 20 U of Exonuclease I per each 25 µl PCR reaction. This mix was incubated at 37° C for one hour, and then the reaction was terminated by incubation at 80° C for 20 minutes. Sequencing reactions were done on both strands. Reactions used BigDye 3.0 (Applied Biosystems) and sequencing was done at the HHMI sequencing facility at the University of Chicago on an ABI 377. Taxa were chosen for sequencing that part of groups underrepresented in previous molecular studies, such as the Peracarida and Caridea. Sequences were deposited in Genbank with accession numbers AY743938-AY743958 (18S) and AY739181-AY739205 (28S). Accession numbers for all of the sequences used in the analyses are listed in Table 1 (Appendix 1), and new sequences are in bold.

2.2 Phylogenetic Analysis

Sequence alignments were done using ClustalX (Thompson et al. 1997), using the following parameters: equal character weighting, gap opening penalty = 10 and gap extension penalty = 4, delayed addition of sequences ≥ 40% divergence. Ambiguous alignment indels larger than five base pairs were removed in MacClade 4.03 (Maddison & Maddison 2001) and the 18S and 28S sequences were concatenated. All analyses used the combined 18S and 28S sequence data. Maximum Parsimony (MP) analysis was done in PAUP* ver. 4.0b.10 (Swofford 2003) using 100 random stepwise addition sequence replicates and TBR branch swapping. All MP searches were done three separate times to check for shorter trees. All characters were treated as equally weighted and unordered. Support for the trees was evaluated with 50% jackknife support (Felsenstein 1985) in 100 repetitions, as well as with Bremer support (Bremer 1988). Markov Chain Monte Carlo (MCMC) analysis of Maximum Likelihood was done using the program MrBayes ver. 3.0b4 (Huelsenbeck & Ronquist 2001). The commands for the Bayesian analysis were: lset nst = 6, rates = gamma, mcmc ngen = 5000000, burnin = 100000, temp = 0.5, printfreq = 1000, samplefreq = 100, nchains = 4, savebrlens = yes. Empirical nucleotide frequencies and the default priors were used. Stationarity of the Markov chain was established by plotting the sampled ln L scores against generation time. Analyses were run twice to ensure that different starting points did not bias the resulting tree topology. The percentage representation, or posterior probability, of each branch was also calculated. In all programs, the designated outgroups were the three representatives from the Chelicerata. To illustrate the variation in branch lengths, they are represented in Figures 2-5.

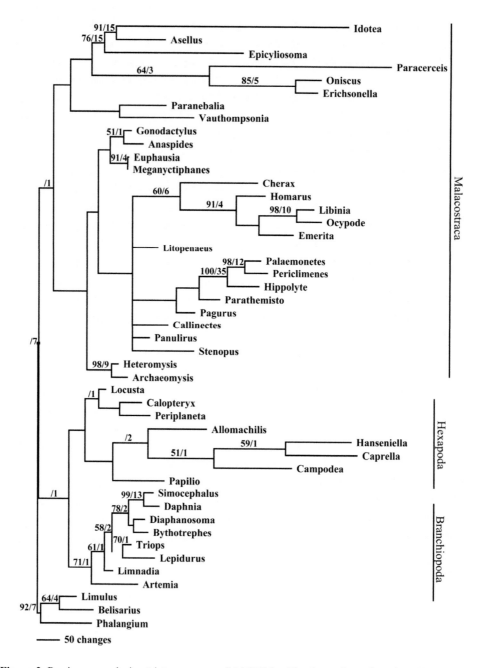

Figure 2. Parsimony analysis, strict consensus of 4 MPT for 47 arthropod taxa based on 18S and 28S RNA (tree length = 6508 steps). Values above the nodes are 50% jackknife values and Bremer decay indices (TreeRot 2; Sorenson 1999), respectively. The separate bar at the bottom represents the branch length for 50 changes.

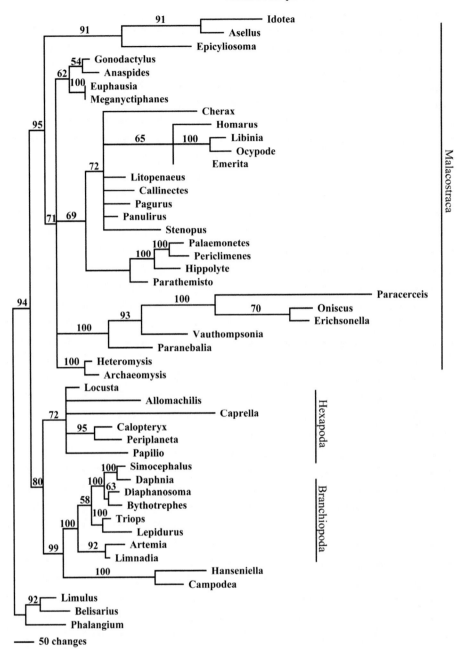

Figure 3. 50% Majority Rule Bayesian tree of 47 arthropod taxa based on GTR + Γ (-log likelihood = 29777.104 averaged over last 1000 generations). Analyses were run for 5,000,000 generations, sampled every 100, and the first 1000 trees were discarded as burn-in. Values above the branches indicate posterior probabilities. The separate bar at the bottom represents the branch length for 50 changes.

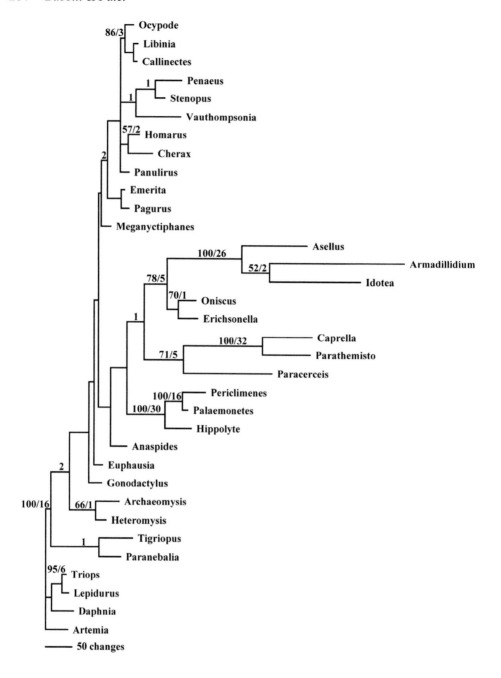

Figure 4. Parsimony analysis, strict consensus of 2 MPT for 34 malacostracan and outgroup taxa based on 18S and 28S RNA (tree length = 2667 steps). Values above the nodes are 50% jackknife and Bremer support values (TreeRot 2; Sorenson 1999), respectively. The separate bar at the bottom represents the branch length for 50 changes.

Relationships within the Pancrustacea 285

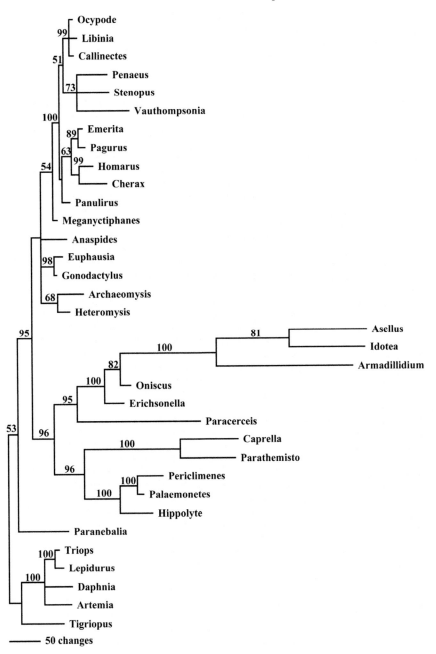

Figure 5. 50% Majority Rule Bayesian tree of 34 malacostracan representatives using a GTR + Γ model (-log likelihood = 13877.87 averaged over last 1000 generations). Analyses were run for 5,000,000 generations, sampled every 100, and the first 1000 trees were discarded as burn-in. Values above the branches indicate posterior probabilities. The separate bar at the bottom represents the branch length for 50 changes.

3 RESULTS

3.1 *Maximum Parsimony*

We first analyzed the data using MP as the optimality criterion. The strict consensus of 4 most parsimonious trees (MPTs), based on 47 taxa, is shown in Figure 2. The general trend in the parsimony analysis, even under different alignment regimes, was a relationship of the Branchiopoda and the Hexapoda to the exclusion of the Malacostraca and the rest of the Crustacea. However, the affinity of the myriapods cannot be ascertained. All trees had significant support for the monophyly of the Branchiopoda, which consistently was the strongest in the tree. The placement of the myriapod taxa and the malacostracan amphipod *Caprella* was problematic, as these taxa always associated with the longest branches of the tree, a problem that had been seen with one of these taxa (the symphylan myriapod *Hanseniella*) in previous studies (Mallatt et al. 2004).

The deeper branches of the tree are very short, suggesting that in the future sequencing other loci is necessary to decipher these relationships. An analysis with more myriapod and non-malacostracan crustaceans was performed (data not shown); however, the result were extremely difficult to interpret, indicating that these loci alone are not able to distinguish relationships for that many terminal taxa with such varied rates of change.

The malacostracan sequences were aligned under the same regime as described for the larger data set and were analyzed using parsimony. Four branchiopods and one additional copepod representative were used as outgroup taxa. These analyses contained 34 taxa, and consistently converged on two MPTs, with lengths of 2667 steps (the strict consensus is shown in Fig. 4). Within the Malacostraca, the phyllocarid representative (*Paranebalia*) groups with the copepod and is outside the Eumalacostraca. The Peracarida and Eucarida become polyphyletic as the representatives from the Mysida and the krill *Euphausia* fall outside as much more basal clades. In addition, in the analyses focusing on the Malacostraca, the caridean shrimp (*Periclimenes, Palaemonetes*, and *Hippolyte*) group with the Amphipoda + Isopoda, although they are with the Eucarida in the larger analyses. All of these findings were consistent through different alignment regimes, and the non-monophyly of the Peracarida is congruent with previous studies placing mysids outside of the Peracarida and nearer the Euphausiacea and Anaspidacea (Watling 1999; Jarman et al. 2000).

3.2 *Bayesian Analyses*

In order to explore the data set, we also used explicit models of sequence evolution in a Bayesian framework. Bayesian methods are known to behave similarly to Maximum Likelihood methods (ML), and ML algorithms are generally more robust to taxa that have long branches. Modeltest ver. 3.5 (Posada & Crandall 1998) was run on the data set and based on Akaike information criterion (AIC) criteria the best fit was a general-time-reversible (GTR+Γ) model, in which the number of substitution sites was set at six and the substitution rate was according to a gamma distribution with a shape parameter = 0.5. MrBayes was then run using four Markov chains and samples taken every 100 generations for 5,000,000 generations. The first 1,000 trees were eliminated as 'burn-in' and a 50% Majority Rule consensus tree was created in PAUP* ver. 4.0b10 (Swofford 2003).

The Bayesian consensus tree (Fig. 3) supports, as measured by Bayesian posterior probabilities, an association of the Branchiopoda and Hexapoda, but the relationship of the myriapod representative remains unclear. The monophyly of the Branchiopoda is strongly supported in the Bayesian analysis and the Malacostraca (with the exclusion of *Caprella*) receives strong support.

Within the Malacostraca (Fig. 5), the Bayesian analyses places the Phyllocarida as the sister group to the remainder of the Malacostraca, though not with high support. The Eumalacostraca is highly supported. The Amphipoda, Isopoda, and Caridea are the sister group to the rest of the Eumalacostraca. The relationship between amphipods and isopods has been contentious, but a close relationship has been supported by the morphological study of Richter and Scholtz (2001). The cumacean *Vauthompsonia* also consistently groups within the Eucarida in both the parsimony and Bayesian analyses.

4 DISCUSSION

Long branches have affected analyses in previous studies of deeper arthropod relationships (Spears and Abele 1997; Giribet et al. 2001), and so we chose to address this problem by adding taxa (Graybeal 1998) from a group under-represented in previous analyses (the Malacostraca); as well as to investigate the data using Bayesian methods that might be more resilient to these problems. In both the parsimony and Bayesian studies (Figs. 2, 3) a sister-group relationship between the branchiopod crustaceans and the hexapods was consistently recovered, but with weak support, and the position of the Myriapoda remains unclear. A monophyletic Crustacea was never recovered in any of the analyses. Within the Crustacea, monophyly of the Branchiopoda seems clear-cut and robust.

The position of the Myriapoda is uncertain in these analyses, so an Atelocerata hypothesis cannot be ruled out. However, most recent molecular work has strongly supported a Myriapoda + Chelicerata clade (Hwang et al. 2001; Mallatt et al. 2004). A sister group relationship between the Hexapoda and Malacostraca was never recovered. Focusing within the Malacostraca, the results for the both the parsimony and Bayesian analyses show support for the Leptostraca as outside of, or as a sister group, to the rest of the Eumalacostraca, supporting previous findings based on both DNA sequences (Spears & Abele 1997) and morphology (Richter & Scholtz 2001). Within the Eumalacostraca, the mysids consistently fall outside of the Peracarida to a basal position in the Eumalacostraca. A previous study showed support for the Mysida having a sister group relationship with the Euphausiacea (Jarman et al. 2000). In addition, at least one, if not both of the taxa in the Euphausiacea fall outside of the Eucarida in all of the analyses. This change in relationships rejects the monophyly of both the Eucarida and Peracarida. The position of the Hoplocarida is variable, but this group is variably allied with the Euphausiacea and Syncarida, as seen in Figures 2, 3, and 5. There is also support for a close relationship between the Amphipoda and Isopoda, which are sister groups in the parsimony analysis (Fig. 4). This is a node that had been controversial in previous morphological studies (Schram & Hof 1997; Richter & Scholtz 2001). The placement of the cumacean *Vauthompsonia* within the Eucarida has not been proposed before. This sequence has been isolated multiple times from different samples and we are confident that this is not a case of contamination. Therefore, its unusual placement may be due to either effects from multiple substitutions or the presence of pseudogenes, which have been found in other crustacean taxa (Spears & Abele 1997). Overall, the monophyly of either the Peracarida or the Eucarida is not supported and these analyses suggest that the affinities between the Syncarida, Hoplocarida, Euphausiacea, and Mysida need to be investigated further.

Though the developmental evidence may suggest a close relationship between the Hexapoda and the Malacostraca, sequence data here suggest otherwise. More thorough taxon sampling of developmental evidence may reveal more commonalities between crustacean groups and the Hexapoda, developmental synapomorphies of the Branchiopoda + Hexapoda, or confirm the convergence of developmental mechanisms in the Hexapoda and Malacostraca. This last hypothesis also leads to truly interesting questions about homoplasy and the evolution of development between the Malacostraca and the Hexapoda.

5 CONCLUSIONS

The outcome of this study is that the monophyly of the Crustacea, Peracarida, and Eucarida is not supported, and that many of the relationships within the Malacostraca need to be investigated thoroughly with more molecular data. The monophyly of the Branchiopoda is strongly supported. In addition, the association of the Hoplocarida, Euphausiacea, and Syncarida warrants further investigation. Overall, inclusion of more Malacostracan taxa does little to confidently resolve the monophyly and position of the Hexapoda within the Pancrustacea. One would hope that in the future the addition of other sequences, such as complete 28S rDNA, which has been shown to be robustly informative at higher taxonomic levels (Mallatt et al. 2004), would work to resolve the relationships between the major groups of the Crustacea and the Hexapoda. However, it is possible that the taxa at the base of long branches, which would be most informative to resolve this trichotomy, (Greybeal 1998) are extinct. If this is the case, then fossil evidence would be the best avenue to try to resolve these deep arthropod relationships. The differences between the morphological and molecular topologies of the Malacostraca hint at the complexity of malacostracan relationships and possible morphological convergences that have yet to be untangled.

ACKNOWLEDGEMENTS

We would like to thank Todd Haney, Kenneth Macdonald, and Dan Cadien for assistance with identification. The *Gonodactylus viridus* tissue was a gift from Roy Caldwell. We would also like to thank two anonymous reviewers for detailed comments.

APPENDIX 1

Table 1. Species used in the analyses in this chapter. Accession numbers in bold represent new data and were sequenced by the authors: (AY743938-AY743958 for 18S; AY739181-AY739205 for 28S sequences).

	Order	Family	Species	18S	28S
Crustacea					
Malacostraca	Isopoda	Armadillidae	*Armadilidlium vulgarae*	AJ287061	**AY739196**
		Asellidae	*Asellus aquaticus*	AJ287055	**AY739195**
		Sphaeromatidae	*Paracerceis glynni*	**AY743958**	**AY739201**
		Idoteidae	*Idotea baltica*	AF279603	**AY739187**
		Idoteidae	*Erichsonella attenuata*	**AY743948**	**AY739191**
		Oniscidae	*Oniscus asellus*	AF255699	
	Amphipoda	Caprellidae	*Caprella equilibria*	**AY743950**	**AY739193**
		Hyperiidae	*Parathemisto gaudichaudi*	**AY743940**	**AY739182**
	Cumacea	Bodotriidae	*Vaunthompsonia minor*	**AY743938**	**AY739181**
	Mysidacea	Mysidae	*Heteromysis* sp.	**AY743946**	**AY739189**
		Mysidae	*Archaeomysis kokuboi*	AJ566085	
	Decapoda	Penaeidae	*Litopenaeus vannamei*	AF340220	
		Palaemonidae	*Palaemonetes vulgaris*	**AY743941**	**AY739184**
		Palaemonidae	*Periclimenes pedersoni*	**AY743954**	**AY739198**
		Hippolytidae	*Hippolyte pleurocanthus*	**AY743956**	
		Stenopoidae	*Stenopus hispidus*	**AY743957**	**AY739199**
		Parastacidae	*Cherax quadricarinatus*	AF235966	AY211962
		Nephropidae	*Homarus americanus*	**AY743945**	**AY739205**
		Palinuridae	*Panilurus argus*	**AY743955**	**AY739199**
		Paguridae	*Pagurus longicornus*	AF436018	**AY739185**
		Hippidae	*Emerita talpoida*	**AY743949**	**AY739192**
		Majidae	*Libinia emarginata*	**AY743953**	**AY739186**
		Portunidae	*Callinectes sapidus*	**AY743951**	**AY739194**
		Ocypodidae	*Ocypode quadrata*	**AY743942**	**AY739202**
	Euphausiacea	Euphausiidae	*Meganyctiphanes norvegica*	AF296707	AF296700
		Euphausiidae	*Euphausia pacifica*	AY141010	AF169700
	Stomatopoda	Gonodactylidae	*Gonodactylus viridis*	**AY743947**	**AY739190**
	Anaspidacea	Anaspididae	*Anaspides tasmaniae*	AF169703	AF169703
		Neballiidae	*Paranebalia belizensis*	**AY743952**	**AY739183**
Copepoda	Podoplea	Harpactidae	*Tigriopus californicus*	AF363324	AF363324
Branchiopoda	Anostraca	Artemiidae	*Artemia franciscana*	AJ238061	
		Artemiidae	*Artemia* sp.		AY210805
	Notostraca	Triopsidae	*Lepidurus packardi*	L34048	AF209047
		Triopsidae	*Triops longicaudatus*	AF144219	
		Triopsidae	*Triops* sp.		AY210844
	Diplostraca	Daphniidae	*Daphnia pulex*	AF014011	
		Daphniidae	*Simocephalus serrulatus*	AF144216	AF346519
		Cercopagididae	*Bythotrephes cederstroemi*	AF144207	

Table 1 continued.

	Order	Family	Species	18S	28S
		Sididae	*Diaphanosoma* sp.	AF144210	
	Conchostraca	Limnadiidae	*Limnadia lenticularis*	L81934	
Hexapoda	Archarognatha	Meinertellidae	*Allomachilis froggarti*	AF370788	AF370806
		Campodeidae	*Campodea tillyardi*	AF173234	
	Odonata	Calopterygidae	*Calopteryx aequabilis*	AY338716	AY338673
	Orthoptera	Acrididae	*Locusta migratoria*	AF370793	AF370809
	Blattaria	Blattidae	*Periplaneta americana*	AF370792	
	Lepidoptera	Papilionoidae	*Papilio troilus*	AF286299	U65199
Myriapoda		Scutigerellidae	*Hanseniella* sp.	AF173237	AF173268
	Sphaerotheriida	Sphaerotheriidae	*Epicyliosoma* sp.	AF370785	
Chelicerata	Scorpiones	Brotheinae	*Belisarius xambeui*	AF005442	AF124954
	Opiliones	Phalangiidae	*Phalangium opilio*	AF124937	AF124965
	Xiphosura	Limulidae	*Limulus polyphemus*	AF062973	U91492

REFERENCES

Aboobaker, A.A. & Blaxter, M.L. 2003. Hox gene loss during dynamic evolution of the nematode cluster. *Curr. Biol.* 13: 37-40.

Adoutte, A. & Philippe, H. 1993. The major lines of metazoan evolution: summary of traditional evidence and lessons from ribosomal RNA sequence analysis. In: Pichon, G. (ed.), *Comparative Molecular Neurobiology*: 1-30. Basel: Birkhauser Verlag.

Anderson, D.T. 1973. *Embryology and Phylogeny in Annelids and Arthropods*. New York: Pergamon Press.

Averof, M. & Akam, M. 1995. Insect-crustacean relationships: insights from comparative developmental and molecular studies. *Phil. Trans. R. Soc. London, B,* 347: 293-303.

Averof, M. & Patel, N.H. 1997. Crustacean appendage evolution associated with changes in *Hox* gene expression. *Nature* 388: 607-608.

Benesch, R. 1969. Zur Ontogenie und Morphologie von *Artemia salina*. L. *Zool. Jahrb., Abt. Anat. Ontog. Tiere,* 86: 307-458.

Boore, J.L., Collins, T.M., Stanton, D., Daehler, L.L. & Brown, W.M. 1995. Deducing the pattern of arthropod phylogeny from mitochondrial DNA rearrangements. *Nature* 376: 163-165.

Boore, J.L., Lavrow, D.V. & Brown, W.M. 1998. Gene translocation links insects and crustaceans. *Nature* 392: 667-668.

Bowman, T.E. & Abele, L.G. 1982. Classification of the Recent Crustacea. In: Abele, L.G. (ed.), *Systematics, the Fossil Record, and Biogeography, Vol. 1*: 1-27. New York: Academic Press.

Braband, A., Richter, S., Hiesel, R. & Scholtz, G. 2002. Phylogenetic relationships within the Phyllopoda (Crustacea, Branchiopoda) based on mitochondrial and nuclear markers. *Mol. Phyl. Evol.* 25: 229-244.

Bremer, K. 1988. The limits of amino acid sequence data in angiosperm phylogenetic reconstruction. *Evolution* 42: 795-803.

Briggs, D.E.G. 1978. The morphology, mode of life, and affinities of *Canadaspis perfecta* (Crustacea: Phyllocarida), Middle Cambrian, Burgess Shale, British Columbia. *Phil. Trans. R. Soc. London, B,* 281: 439-487.

Briggs, D.E.G. & Fortey, R.A. 1989. The early radiation and relationships of the major arthropod groups. *Science* 246: 241-243.

Briggs, D.E.G., Fortey, R.A. & Wills, M.A. 1992. Morphological disparity in the Cambrian. *Science* 256: 1670-1673.

Calman, W.T. 1904. On the classification of the Crustacea Malacostraca. *Ann. Mag. Nat. Hist.* 7 (13): 144-158.

Calman, W.T. 1909. *A Treatise on Zoology, 7: Appendiculata, Crustacea.* London: Adam & Charles Black.

Cook, C.E., Smith, M.L., Telford, M.J., Bastianello, A. & Akam, M. 2001. Hox genes and the phylogeny of the arthropods. *Curr. Biol.* 11: 759-763.

Cunningham, C.W., Blackstone, N.W. & Buss, L.W. 1992. Evolution of king crabs from hermit crab ancestors. *Nature* 355: 539-542.

Dahl, E. 1987. Malacostraca maltreated - the case of the Phyllocarida. *J. Crust. Biol.* 7: 721-726.

Damen, W.G.M., Hausdorf, M., Seyfarth, E.-A. & Tautz, D. 1998. A conserved mode of head segmentation on arthropods revealed by the expression patterns of *Hox* genes in a spider. *Proc. Nat. Acad. Sci. USA* 95: 10665-10670.

Davis, G. & Patel, N.H. 1999. The origin and evolution of segmentation. *Trends Genet.* 24 (12): M68-M72.

Deuve, T. (ed.) 2001. *The Origin of the Hexapoda. Ann. Soc. Entomol. France, N.S.,* 37 (1-2).

Dohle, W. 2001. Are insects terrestrial crustaceans? A discussion of some new facts and arguments and the proposal of the proper name 'Tetraconata' for the monophyletic unit Crustacea + Hexapoda. In: Deuve, T. (ed.), *The Origin of the Hexapoda. Ann. Soc. Entomol. France, N.S.,* 37 (1-2): 85-103.

Dohle, W. & Scholtz, G. 1988. Clonal analysis of the crustacean segment: the discordance between genealogical and segmental borders. *Development Suppl.* 104: 147-160.

Duman-Scheel, M. & Patel, N.H. 1999. Analysis of molecular marker expression reveals neuronal homology in distantly related arthropods. *Development* 126: 2327-2334.

Edgecombe, G.D., Wilson, G.D.F., Colgan, D.J., Gray, M.R. & Cassis, G. 2000. Arthropod cladistics: combined analysis of histone H3 and U2 snRNA sequences and morphology. *Cladistics* 16: 155-203.

Felsenstein, J. 1985. Confidence limits on phylogenies: An approach using the bootstrap. *Evolution* 39: 783-791.

Field, K.G., Olsen, G.J., Lane, D.J., Giovanni, S.J., Ghiselin, M.T., Raff, E.C., Pace, N.R. & Raff, R.A. 1988. Molecular analysis of the animal kingdom. *Science* 239: 748-753.

Fortey, R.A. & Thomas, R.H. (eds.) 1997. *Arthropod Relationships.* London: Chapman & Hall.

Friedrich, M. & Tautz, D. 1995. Ribosomal DNA phylogeny of the major extant arthropod classes and the evolution of myriapods. *Nature* 376: 165-167.

Garcia-Machado, E., Pempera, M., Dennebouy, N., Oliva-Suarez, M., Mounolou, J-C. & Monnerot, M. 1999. Mitochondrial genes collectively suggest the paraphyly of Crustacea with respect to Insecta. *J. Mol. Evol.* 49: 142-149.

Giribet, G., Carranza, S., Baguna, J., Riutort, M. & Ribera, C. 1996. First molecular evidence for the existence of a Tartigrada + Arthropoda clade. *Mol. Bio. Evol.* 13: 76-84.

Giribet, G., Edgecombe, G.D., & Wheeler, W.C. 2001. Arthropod phylogeny based on eight molecular loci and morphology. *Nature* 413: 157-161.

Giribet, G. & Ribera, C. 2000. A review of arthropod phylogeny: new data based on ribosomal DNA sequences and direct character optimization. *Cladistics* 16: 204-231.

Graybeal, A. 1998. Is it better to add taxa or characters to a difficult phylogenetic problem? *Syst. Biol.* 47: 9-17.

Huelsenbeck, J.P. & Ronquist. F. 2001. MRBAYES: Bayesian inference of phylogeny. *Bioinformatics* 17: 754-755.

Hughes, C.L. & Kaufman, T.C. 2002. Hox genes and the evolution of the arthropod body plan. *Evol. Devel.* 4: 459-499.

Hwang, U.W., Friedrich, M., Tautz, D., Park, C.J. & Kim, W. 2001. Mitochondrial protein phylogeny joins myriapods with chelicerates. *Nature* 413: 154-157.

Jarman, S.N., Nicol, S., Elliot, N.G. & McMinn, A. 2000. 28S rDNA evolution in the Eumalocostraca and the phylogenetic position of krill. *Mol. Phyl. Evol.* 17: 26-36.

Maddison, D.R. & Maddison, W.P. 2001. MacClade version 4.03 PPC. Sunderland, Massachusetts: Sinauer Assoc.

Mallatt, J.M., Garey, J.R. & Shultz, J.W. 2004. Ecdysozoan phylogeny and Bayesian inference: first use of nearly complete 28S and 18S rDNA gene sequences to classify the arthropods and their kin. *Mol. Phyl. Evol.* 31: 179-191.

Manton, S.M. 1977. *The Arthropoda*. Oxford: Clarendon Press.

Martin, J.W. & Davis, G.E. 2001. *An Updated Classification of the Recent Crustacea*. Los Angeles: Nat. Hist. Mus. Los Angeles County.

Martin, J.W., Vetter, E.W. & Cash-Clark, C.E. 1996. Description, external morphology, and natural history observations of *Nebalia hessleri*, new species (Phyllocarida: Leptostraca), from southern California, with a key to the extant families and genera of the Leptostraca. *J. Crust. Biol.* 16: 347-372.

Morrison, C.L., Harvey, A.W., Lavery, S., Tieu, K., Huang, Y. & Cunningham, C.W. 2001. Mitochondrial gene rearrangements confirm the parallel evolution of the crab-like form. *Proc. R. Soc. London, B,* 269: 345-350.

Nardi, F., Spinsanti, G., Boore, J.L., Carapelli, A. Dalli, R. & Frati, F. 2003. Hexapod origins: monophyletic or paraphyletic? *Science* 299: 1887-1889.

Osorio, D., Averof, M. & Bacon, J.P. 1995. Arthropod evolution: great brains, beautiful bodies. *Trends Ecol. Evol.* 10: 449-454.

Posada, D. & Crandall, K.A. 1998. Modeltest: testing the model of DNA substitution. *Bioinformatics* 14: 817-818.

Regier, J.C. & Shultz, J.W. 1997. Molecular phylogeny of the major arthropod groups indicates polyphyly of crustaceans and a new hypothesis for the origin of hexapods. *Mol. Biol. Evol.* 14: 902-913.

Regier, J.C. & Shultz, J.W. 2001. Elongation factor-2: a useful gene for arthropod phylogenetics. *Mol. Phyl. Evol.* 20: 136-148.

Richter, S. & Scholtz, G. 2001. Phylogenetic analysis of the Malacostraca (Crustacea). *J. Zool. Syst. Evol. Res.* 39: 113-136.

Scholtz, G. 2000. Evolution of the nauplius stage in malacostracan crustaceans. *J. Zool. Syst. Evol.Res.* 38: 175-187.

Scholtz, G., Mittmann, B. & Gerberding, M. 1998. The pattern of *Distal-less* expression in the mouthparts of crustaceans, myriapods and insects: new evidence for a gnathobasic mandible and the common origin of the Mandibulata. *Int. J. Dev. Biol.* 42: 801-810.

Schram, F.R. 1986. *Crustacea*. New York: Oxford Univ. Press.

Schram, F.R. & Hof, C.H.J. 1998. Fossils and the interrelationships of major crustacean groups. In: Edgecombe, G.D. (ed.), *Arthropod Fossils and Phylogeny*: 233-302. New York: Columbia Univ. Press.

Schram, F.R. & Jenner, R.A. 2001. The origin of Hexapoda: a crustacean perspective. In: Deuve, T. (ed.), *Origin of the Hexapoda*. Ann. Soc. Entomol. France, N.S., 37: 243-264.

Shultz, J.W. & Regier, J.C. 2000. Phylogenetic analysis of arthropods using two nuclear protein encoding genes supports a crustacean + hexapod clade. *Proc. R. Soc. London, B,*. 267: 1011-1019.

Siddall, M.E. 1998. Success of parsimony in the four-taxon case: long-branch repulsion by likelihood in the Farris zone. *Cladistics* 14: 209-220.

Snodgrass, R.E. 1935. *Principles of Insect Morphology*. New York: McGraw-Hill Book Company.

Sorenson, M.D. 1999. TreeRot, version 2. Program online available at: http://people.bu.edu/msoren/TreeRot.html

Spears, T. & Abele, L.G. 1997. Crustacean phylogeny inferred from 18S rRNA. In: Fortey, R.A. & Thomas, R.H. (eds.), *Arthropod Relationships*: 169-187. London: Chapman & Hall.

Spears, T. & Abele, L.G. 2000. Branchiopod monophyly an interordinal phylogeny inferred from 18S ribosomal DNA. *J. Crust. Biol.* 20: 1-24.

Swofford, D.L. 2003. *PAUP*, Phylogenetic Analysis Using Parsimony (* and Other Methods), Version 4.0b10*. Sunderland, Massachusetts: Sinauer Assoc.

Thomas, J.B., Bastiani, M.J., Bate, M. & Goodman, C.S. 1984. From grasshopper to *Drosophila*: a common plan for neuronal development. *Nature* 310: 203-207.

Thompson, J.D., Gibson, T.J., Plewniak, F., Jeanmougin, F. & Higgins, D.G. 1997. The ClustalX windows interface: flexible strategies for multiple sequence alignment aided by quality analysis tools. *Nucl. Acids Res.* 24: 4876-4882.

Turbeville, J.M., Pfeiffer, D.M., Field, K.G. & Raff, R.A. 1991. The phylogenetic status of arthropods, as inferred from 18S rDNA sequences. *Mol. Biol. Evol.* 8: 669-686.

Walossek, D. and Müller, K.J. 1998. Cambrian "Orsten" fossils. In: Edgecombe, G.D. (ed.), *Arthropod Fossils and Phylogeny*: 185-231. New York: Columbia Univ. Press.

Watling, L. 1999. Toward understanding the relationships of the peracaridean orders: the necessity of determining exact homologies. *Proc. 4th Int. Crust. Congr., Amsterdam 1998*: 73-89.

Weygoldt, P. 1960. Embryologische Untersuchungen an Ostrakoden: die Entwicklung von *Cyprideis litoralis*. *Zool. Jahrb., Abt. Anat. Ontog. Tiere* 78: 369-426.

Wheeler, W.C. 1997. Sampling, groundplans, total evidence and the systematics of arthropods. In: Fortey, R.A. & Thomas, R.H. (eds.), *Arthropod Relationships*: 87-96. London: Chapman & Hall.

Wheeler, W.C., Cartwright, P., & Hayashi, C.Y. 1993. Arthropod phylogeny: a combined approach. *Cladistics* 9: 1-39.

Whiting, M.F., Carpenter, J.C., Wheeler, Q.D., & Wheeler, W.C. 1997. The Strepsiptera problem: phylogeny of the holometabolous insect orders inferred from 18S and 28S ribosomal DNA sequences and morphology. *Syst. Biol.* 46: 1-68.

Whitington, P.M. 1996. Conservation versus change in early axonogenesis in arthropod embryos: a comparison between myriapods, crustaceans and insects. In: Breidbach, O. & Kutch, W. (eds.), *The Nervous System of Invertebrates: An Evolutionary and Comparative Approach, Vol. 72*: 181-219. Basel: Experientia Supplementum.

Whitington, P.M., Leach, D. & Sandeman, R. 1993. Evolutionary change in neural development within the Arthropoda: axonogenesis in the embryos of two crustaceans. *Development* 118: 449-461.

Whitington, P.M., Meier, T. & King, P. 1991. Segmentations, neurogenesis and formation of early axonal pathways in the centipede *Ethmostigmus rubripes* (Brandt). *Roux's Arch. Dev. Biol.* 199: 349-363.

Wills, M.A. 1997. A phylogeny of recent and fossil Crustacea derived from morphological characters. In: Fortey, R.A. & Thomas, R.H. (eds.), *Arthropod Relationships*: 189-209. London: Chapman & Hall.

Wilson, K., Cahill, V., Ballment, E. & Benzie, J. 2000. The complete sequence of the mitochondrial genome of the crustacean *Penaeus monodon*: are malacostracan crustaceans more closely related to insects than to branchiopods? *Mol. Bio. Evol.* 17: 863-874.

Yager, J. 1981. Remipedia, a new class of crustaceans from a marine cave in the Bahamas. *J. Crust. Biol.* 1: 328-333.

Zrzavý, J. & Štys, P. 1997. The basic body plan of arthropods: insights from evolutionary Morphology and developmental biology. *J. Evol. Biol.* 10: 353-367.

Relationships between hexapods and crustaceans based on four mitochondrial genes

ANTONIO CARAPELLI[1], FRANCESCO NARDI[1], ROMANO DALLAI[1], JEFFREY L. BOORE[2], PIETRO LIÒ[3] & FRANCESCO FRATI[1]

[1] *Department of Evolutionary Biology, University of Siena, Siena, Italy*
[2] *U.S. Department of Energy Joint Genome Institute and Lawrence Berkeley National Laboratory, and University of California, Berkeley, U.S.A.*
[3] *Computer Laboratory, University of Cambridge, Cambridge, U.K.*

ABSTRACT

The ever-increasing use of molecular data in phylogenetic studies have revolutionized our view of the phylogenetic relationships among the major lineages of arthropods. In this context, an important contribution is offered by mitochondrial genes, and the now widely available sequences of entire mitochondrial genomes. One of the most debated issues in arthropod phylogeny is the relationship between crustaceans and hexapods, and particularly, whether the traditional taxa Crustacea and Hexapoda are mono- or paraphyletic. A key role is played by basal hexapodan taxa, the entognathan apterygotans (Protura, Collembola, Diplura), whose phylogenetic position as the sister taxa of the Insecta *s. str.* is not totally convincing. The phylogenetic analysis based on mitochondrial protein-coding genes suggests that there are crustacean taxa which are more closely related to the Insecta *s. str.* than Collembola and Diplura, therefore suggesting non-monophyly of the taxon Hexapoda as traditionally defined. Hence, Collembola and Diplura might have differentiated from different pancrustacean ancestor(s) than those from which the remaining hexapods (Insecta) arose. These results also imply a new scenario for the evolution of several morphological and physiological features of hexapods, including terrestrialization.

1 INTRODUCTION

In the last ten years, mitochondrial genomics (the analysis of sequence and structural features of the mitochondrial genome) has had a considerable impact on the reconstruction of higher-level phylogeny among Arthropods. For example, Boore et al. (1995, 1998) showed that the translocation of a tRNA gene links crustaceans and hexapods (= Pancrustacea; Zrzavý & Štys 1997). These studies favored the Pancrustacea concept (= Tetraconata; Dohle 2001), and the exclusion of myriapods (and others) contributed to the widely accepted dismissal of the Atelocerata (hexapods + myriapods) concept (Telford & Thomas 1995). Other studies using molecular (Friedrich & Tautz 1995; Regier & Shultz 1997) and developmental (Averof & Akam 1995; Panganiban et al. 1995) data have supported similar relationships. Subsequently, a considerable bulk of data have been produced in support of the Pancrustacea hypothesis, owing to renewed interest and the collection of new phylogenetic evidence from different perspectives: developmental genetics (Cook et al. 2001; Deutsch 2001), neurobiology (Duman-Scheel & Patel 1999; Dohle 2001; Simpson 2001), skeletal structures (Deuve 2001), the sequences of nuclear genes (Giribet & Ribeira 1998; Shultz & Regier 2000; Regier & Shultz 2001), and the analysis of combined molecular and morphological data sets (Giribet et al. 2001). Mitochondrial genes have continued to contribute extensively to phylogenetic studies given the signal that could be recovered from the nucleotide and the putative amino acid sequences of mitochondrial protein-coding genes (Garcia-Machado et al. 1999; Wilson et al. 2000; Nardi et al. 2001; Lavrov et al. 2004; Negrisolo et al. 2004). The considerable amount of molecular and developmental data supporting the Pancrustacea (Richter 2002) is in contrast with the widely accepted evidence supporting the Atelocerata coming from morphological data sets (Koch 2001; Kraus 2001), as well as from combined (morphology + molecules) analyses (Wheeler et al. 1993; Wheeler, 1998; Edgecombe et al. 2000). However, it now seems that the Pancrustacea concept is favored over that of the Atelocerata by most molecular systematists.

The question of the monophyly of Crustacea has received renewed attention, and evidence has been collected to suggest their paraphyly, either in the context of the Atelocerata (Moura & Christoffersen 1996), or in the context of the Pancrustacea (Regier & Shultz 1997; Garcia-Machado et al. 1999). Regarding this latter hypothesis, many efforts now attempt to identify which crustacean lineage should be considered the sister taxon of the Hexapoda (Schram & Jenner 2001). In this respect, a crucial role is played by the most basal lineages of six-legged arthropods, which comprise five major taxa of quite neglected, soil-dwelling animals, collectively known as apterygotans: Protura, Collembola, Diplura, Microcoryphia and Zygentoma. According to the classical view (Kristensen 1981), the first three of these taxa, which share entognathan mouthparts, are included in the taxon Entognatha, a lineage which would have branched off earlier along the hexapod lineage, before their closest relatives acquired ectognathan mouthparts. Due to the many peculiar features they possess, entognathan taxa are not usually granted the status of insects, and the taxon Insecta is formally limited to the ectognathan orders (Kristensen 1981). While the monophyly of Hexapoda has hardly been questioned at all (Bitsch & Bitsch 2000; Wheeler et al. 2001), considerable debate has grown over the phylogenetic relationships of the entognathan groups (Carapelli et al. 2000), either challenging the monophyly of the taxon Entognatha (Kukalová-Peck 1987; Kristensen 1997), or the monophyly of some of its taxa, such as the Diplura (Štys & Bilinski 1990). The monophyly of Hexapoda has been broadly accepted, but a closer look to the pertinent literature shows that the number of shared features is small (Bitsch & Bitsch 1998; Klass & Kristensen 2001), and the support in favor of their monophyly arguably weak (Friedrich & Tautz 2001). Therefore, this question merits further testing.

One possible reason which would make it difficult to reconstruct the splitting events at the origin of the hexapod lineage is the fact that they might have occurred in a short period of time, leaving a very long period to each lineage to differentiate its own autapomorphic features, and masking useful synapomorphic characters, both morphological and molecular. A sudden radiation in a short period of time is indeed the way that basal hexapod evolution is often represented (e.g., Engel & Grimaldi 2004). Again, molecular data have been important in stimulating the discussion on this subject. Several data sets have been produced including apterygotan taxa in the analysis: some of them support a monophyletic Hexapoda with strong or moderate support (Friedrich & Tautz 1995; Shultz & Regier 2000; Regier & Shultz 2001; Kjer 2004), others reject hexapod monophyly (Giribet & Ribeira 1998; Nardi et al. 2001, 2003a; Negrisolo et al 2004). To add uncertainties to the whole picture, different methods of analysis of the same data set produce different reconstructions (Delsuc et al. 2003; Nardi et al. 2003b). The aim of this work is to extend the previous data set (Nardi et al. 2003a) by including a representative of the apterygotan order Diplura, *Japyx solifugus*.

2 THE DATA

The complete sequence of the mitochondrial genome is available for quite a number of arthropod taxa, so we have concentrated our sequencing efforts in the apterygotan hexapods that have been generally neglected thus far. Currently, two sequences are available from Collembola: the onychiurid *Tetrodontophora bielanensis* (Nardi et al. 2001) and the hypogastrurid *Gomphiocephalus hodgsoni* (Nardi et al. 2003a). Also available and of special interest is the sequence of the zygentoman *Tricholepidion gertschi* (Nardi et al. 2003a), considered to be one of the most basal taxa of the Dicondylia (Pterygota + Zygentoma).

Adding to this, we recently determined the complete mtDNA sequence of a dipluran (the japygid *Japyx solifugus* - GenBank acc. nr. AY771989). The sequence of the complete genome was obtained with a combination of primer walking and shotgun sequencing approaches. First, we amplified with PCR two long fragments encompassing the regions between the *cytochrome oxidase I* (*cox1*) and *cytochrome oxidase III* (*cox3*) genes (with the universal primers C1-J-1751 and C3-N-5460: Simon et al. 1994), and the region between the *cox3* and *NADH dehydrogenase 4L* (*nad4L*) genes (with the primers 5'-CTCCCATAG-GCATTTCACCATTCAA-3' and 5'-GCTTTCGGGGGTGTGTGTGGTTATTT-3'). These two fragments were completely sequenced via primer walking. We also amplified and sequenced a small fragment encompassing the large (*rrnL*) and small (*rrnS*) ribosomal RNA subunits using the universal primers LR-J-13417 and SR-N-14588 (Simon et al. 1994). Then, we designed primers specific for *J. solifugus* using known sequences as follows:

cox1 = 5'-AAAGCCCAGTGCTCACAGAATGGACG-3',
nad4 = 5'-GACCAATAACCATTCTACGACTACCAACACG-3',
trnV = 5'-GAATTGCACAGATCCTACTCAGTGTA-3',
rrnS = 5'-GGTGTGTACATATCGCCCGTCACTCTC-3'.

We used these primers to amplify the remaining part of the genome in two long fragments (*trnV-cox1* and *nad4-rrnS*). This was achieved with a long-PCR approach producing two fragments of about 5.7 Kb and 4.2 Kb, respectively. The long-PCR products were then purified with Microcon PCR (Amicon-Millipore), and sheared into ~1.5 Kb fragments by running them through a Hydroshear (GeneMachines). The fragments were subsequently pooled together and cloned. Over 350 clones were sequenced from this library, and the sequences were automatically assembled using the software Sequencher. Due to the considerable number of clones sequenced, we obtained a minimum 5-fold coverage on each position of the mtDNA sequence.

The mitochondrial genome of *Japyx solifugus*, 15,785 bp long, shows the same gene content typical of most metazoans (Boore 1999), and the same gene order as *Gomphiocephalus hodgsoni*, *Tricholepidion gertschi* and *Drosophila yakuba* (Clary & Wolstenholme 1985). This gene order, shared also with *Daphnia pulex* (Crease 1999) and differing by only the position of one tRNA gene from the mtDNA of *Limulus polyphemus* (Lavrov et al. 2000), is believed to be the ancestral arrangement of the Pancrustacea (Crease 1999).

Interestingly, it differs from the gene order of one of the two collembolans, *Tetrodontophora bielanensis*, for two tRNA translocations (Nardi et al. 2001), which, therefore, are likely to be autapomorphic features of an internal lineage of Collembola. Hence, gene arrangement does not seem to provide useful information to test whether hexapods are monophyletic, nor to reconstruct between-order relationships, but it could still be useful for within-order phylogeny.

3 THE ANALYSIS

Since gene order is not informative as far as within-hexapod relationships, we turned our attention to comparisons of gene sequences. Changes in the nucleotide sequences are long saturated at this level of divergence, therefore, we focused on the amino acid sequences of protein-coding genes. To perform a preliminary survey of the rates of variability and the levels of confidence in the alignment (i.e., the establishment of homology across positions), we aligned the inferred amino acid sequences of the 13 mitochondrially-encoded proteins from all the species with complete mtDNA sequences available at the time of this analysis. The genes encoding for ribosomal subunits were not considered in this analysis due to difficulties in the alignment.

The alignment was performed using ClustalX (Thompson et al. 1997; default settings: gap opening = 10; gap extension = 0.20; Gonnet series matrix), followed by visual inspection. It soon became apparent that some parts of the genome, and even some entire genes, were very difficult to align, given the high levels of variability (both in primary sequence and length) at this taxonomic level. Many aligned genes had less than 15% of invariable sites, with gaps introduced at over 25% of positions (Nardi et al. 2003a). In order to minimize the phylogenetic noise generated by possible alignment errors, we took the conservative step of excluding from the analysis the most variable genes, and included only the four that are most conserved: *cytochrome oxidase I, II and III (cox1, cox2 and cox3,* respectively), and *cytochrome b (cob)*. The concatenated alignment of these amino acid sequences totalled 1413 positions.

To minimize systematic errors due to unequal base composition and uneven rates of evolution across sequences (Swofford et al. 1996), and to reduce the data set to manageable size, we excluded taxa when they failed tests for biased base composition and rates of evolution using the approach described in Nardi et al. (2003a). We also excluded congeneric taxa, as they would not add any significant information. A total of 17 taxa (Table 1) were subjected to phylogenetic analysis.

We performed a Bayesian analysis using MrBayes (Huelsenbeck & Ronquist 2001; Huelsenbeck et al. 2001), with the GTR model of evolution (Tavare 1986) (selected as the most general time-reversible model of evolution), assuming that a proportion of sites is invariant, and the remaining sites evolve according to a Γ-distribution (parameters estimated from the data). The analysis was run for 500,000 generations, from which the first 150,000 were removed as a burn-in due to the fact that likelihood values reached a plateau after that point. The tree is rooted by specifying the mollusk *Katharina tunicata* (Boore & Brown 1994) as an out-group. Other out-groups were found more difficult to use because of alignment problems, such as the nematodes, where rates of evolution are dramatically accelerated (Hwang et al. 2001). This tree (Fig. 1) supports, with high posterior probabilities, the taxon Pancrustacea, and the monophyly of Insecta, but rejects the monophyly of Hexapoda. Both the collembolans and the dipluran *Japyx solifugus*, in fact, are placed at the base of a clade joining all Insecta with the three crustaceans. The same analysis (MrBayes, GTR + I + Γ, with 500,000 generations, burn-in = 150,000) was also run on each of the four genes separately. While the *cob* gene gave the same results as the concatenated data set, with highly supported Pancrustacea and Insecta, but no support for Hexapoda, neither of the other three genes recovered these groups with high confidence. The better performance of the analysis of the concatenated data set over the analysis of single genes may be due to the fact that it averages the discrepancies across genes with different selective pressures and functional constraints, and it combines congruent phylogenetic signals (Cao et al. 1994).

Table 1. List of taxa used in the analysis, with GenBank accession numbers for the sequence of the mitochondrial genome.

Acc. no.	Taxon	Taxonomical assignment
NC002735	*Tetrodontophora bielanensis*	HEXAPODA, Collembola
NC005438	*Gomphiocephalus hodgsoni*	HEXAPODA, Collembola
AY771989	*Japyx solifugus*	HEXAPODA, Diplura, Japygidae
NC005437	*Tricholepidion gertschi*	INSECTA, Zygentoma
NC001712	*Locusta migratoria*	INSECTA, Orthoptera
NC002609	*Triatoma dimidiata*	INSECTA, hemipteroid, Heteroptera
NC003081	*Tribolium castaneum*	INSECTA, Endopterygota, Coleoptera, Tenebrionidae
NC003372	*Crioceris duodecimpunctata*	INSECTA, Endopterygota, Coleoptera, Crysomelidae
NC002084	*Anopheles gambiae*	INSECTA, Endopterygota, Diptera, Culicidae
NC003368	*Ostrinia furnacalis*	INSECTA, Endopterygota, Lepidoptera, Pyralidae
NC001620	*Artemia franciscana*	CRUSTACEA, Branchiopoda, Anostraca
NC003058	*Pagurus longicarpus*	CRUSTACEA, Malacostraca, Decapoda, Anomura
NC004251	*Panulirus japonicus*	CRUSTACEA, Malacostraca, Decapoda, Palinura
NC003343	*Narceus annularus*	MYRIAPODA, Diplopoda
NC003344	*Thyropygus* sp.	MYRIAPODA, Diplopoda
NC003057	*Limulus polyphemus*	CHELICERATA, Merostomata, Xiphosura
NC001636	*Katharina tunicata*	MOLLUSCA, Polyplacophora

We evaluated statistically the relative likelihood of seven possible relationships for the most crucial taxa using PAML (Yang 1997) with the mitochondrially-based mt-REV24 + Γ model of evolution (Fig. 2). Although this model was developed for mammalian mitochondrial genes, and was found to lack wide generality (Liò & Goldman 2002), it is still one of the best available models to study phylogenetic relationships using amino acid sequences of mitochondrially-encoded proteins. For this analysis, we assumed monophyly for each of Collembola, Diplura, Insecta, and Crustacea, based on the analysis in Fig. 1, and considered the relationships among these groups. Five of the seven topologies imply non-monophyly of Hexapoda, including the one that has the best score (nr. 3 in Fig. 2). This topology shows a clade uniting Collembola with Diplura and a clade uniting Insecta with Crustacea. The topologies where Hexapoda are monophyletic (nr. 1 and nr. 2 in Fig. 2) have significantly

300 Carapelli et al.

lower scores for both the Kishino-Hasegawa (KH: Kishino & Hasegawa 1989) and the Shimodaira-Hasegawa (SH: Shimodaira 2002) tests, except for topology nr. 2 for the SH test. We also used PAML (and the mtREV24 + Γ model) to compare the seven different topologies based on each of the four genes individually. Figure 3 shows the best topologies selected in each analysis, and their significance; as in the analysis shown before with GTR + I + Γ model and MrBayes, *cox1* gives the same results as the concatenated data set (hexapod non-monophyly, with both Collembola and Diplura outside the Hexapoda, significantly better supported than hexapod monophyly), while the remaining three genes did not provide significant resolution to distinguish between the two hypotheses.

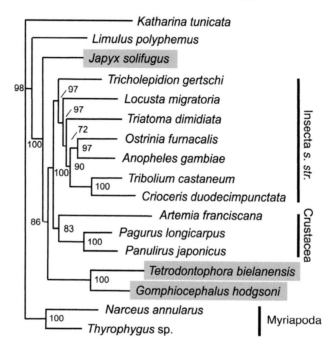

Figure 1. Phylogenetic tree obtained on the concatenated amino acid sequences of the four genes using a Bayesian Inference approach (MrBayes, 500,000 generations, burn-in = 150,000) with a GTR + I + Γ model of evolution. Numbers at the nodes indicate posterior probabilities. Taxa highlighted in gray indicate collembolan and dipluran species; out-group = *Katharina tunicata*.

4 EVOLUTIONARY IMPLICATIONS OF HEXAPOD NON-MONOPHYLY

The data shown here and the analysis of mitochondrial genomes collectively support non-monophyly of Hexapoda as traditionally defined, with both Collembola and Diplura placed outside the Hexapoda, and with Crustacea (or, rather, some crustacean groups) as the sister taxon of Insecta *s. str.* Therefore, collembolans and diplurans might have branched off very early from a pancrustacean ancestor, either before all present crustaceans differentiated from ectognathan insects, or as a derived group of one or more lineages of crustaceans. The best tree obtained in our analysis also does not cluster the collembolans with the dipluran, therefore suggesting non-monophyly of Entognatha. However, the support for the node dividing Diplura from Collembola is very low, and the result in partial contrast with the topological test performed with PAML, leaving the question of entognathan monophyly still unanswered. Our data, and particularly the lack of a denser sampling of crustacean groups, do not allow us to resolve whether crustaceans are mono- or paraphyletic. While the tree in Fig. 1 clusters the three crustacean taxa in a single clade (as in Negrisolo et al. 2004), the analysis of the larger data set, and also other molecular studies based on mitochondrial genes (Hwang et al. 2001; Wilson et al. 2001; Nardi et al. 2001, 2003a), suggest that some crustacean lineages (namely the Malacostraca) might be more closely related to insects than other crustacean groups. The hypothesis of paraphyly of Crustacea with respect to insects is also suggested by other molecular studies (Regier & Shultz 1997, 2001; Garcia-Machado et al. 1999; Shultz & Regier 2000), and corroborated by other lines of evidence (Schram & Jenner 2001; Fanenbruck et al. 2004).

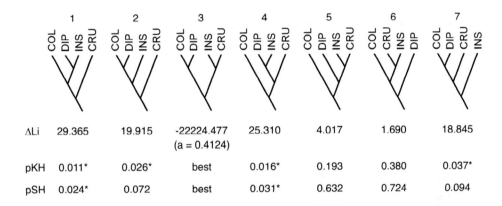

Figure 2. Test of significance of the difference in likelihood scores (mtREV24 + Γ model) among seven selected topologies depicting different potential relationships of the crucial taxa. pKH = Kishino-Hasegawa test (Kishino & Hasegawa 1989); pSH = Shimodaira-Hasegawa test (Shimodaira 2002); ΔLi = difference in likelihood from the best topology; asterisks indicate cases where the difference in likelihood is statistically significant; α = shape parameter of the gamma distribution of among-site rate variation; COL = Collembola (*G. hogdsoni* + *T. bielanensis*); DIP = Diplura (*J. solifugus*); INS = Insecta; CRU = Crustacea (*A. franciscana* + *P. longicarpus* + *P. japonicus*).

The hypothesis of hexapod non-monophyly implies that the features that were typically claimed to support the taxon Hexapoda, such as, for instance, characters resulting from terrestrialization, the type of body tagmosis (with head, thorax and abdomen), and a habitus with three pairs of legs, might have independently evolved in different lineages of pancrustacean arthropods. It is widely accepted that terrestrialization has occurred independently several times in different lineages of arthropods, and it should come as no surprise if one (or more) additional independent terrestrialization event is discovered among apterygotans (Negrisolo et al. 2004). Furthermore, the traditional association between terrestrialization and the tagmosis pattern (and the three pairs of legs) was challenged by the recent discovery of a presumed marine hexapod (Haas et al. 2003), with six legs and a tagmosis pattern intermediate between those of insects and some long-bodied crustaceans. Entognathan apterygotes also show a variety of peculiar characters with respect to 'true insects', for example: entognathous mouthparts, different numbers of abdominal segments (12 in proturans, six in collembolans), the absence of antennae in proturans, and the absence of accessory microtubules in the sperm of proturans and collembolans (Dallai & Afzelius 1999). However, some of these characters might well be autapomorphic or symplesiomorphic with respect to Hexapoda (Kristensen 1981), and the difficulties in interpreting such morphological characters makes the position of basal hexapods still doubtful (Klass & Kristensen 2001). In fact, the monophyly of Hexapoda is not well supported by either molecular (Friedrich & Tautz 2001) and morphological (Bitsch & Bitsch 1998) data.

In the scenario of non-monophyletic hexapods, alternative hypotheses may be considered for the phylogenetic relationships of Collembola and Diplura (leaving aside Protura, for which no molecular data are yet available). The monophyly of Collembola seems apparent and supported by all molecular analyses. Monophyly of Diplura has been questioned on the basis of the structure of ovarioles (Štys & Bilinski 1990) and other morphological characters. Evidence has also been collected for the placement of Diplura as the sister taxon of Ectognatha (= Insecta) (Bitsch & Bitsch 1998; Dallai & Afzelius 1999), therefore rejecting the monophyly of Entognatha (Protura + Collembola + Diplura). However, our phylogenetic analysis places Diplura, as well as Collembola, outside the Hexapoda.

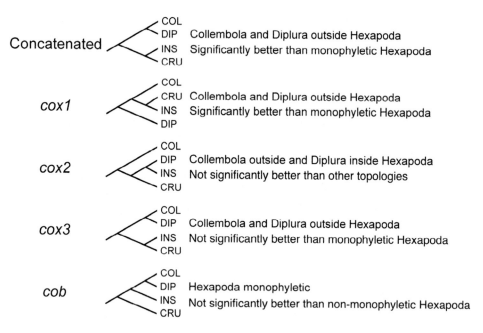

Figure 3. Level of significance of different tree topologies selected with the mtREV24 + Γ model for each single gene, and in the concatenated data set.

5 HOW IS IT LIKELY THAT NEW DATA WOULD SOLVE THE CURRENT-DAY PHYLOGENETIC ISSUES?

The scientist to whom this special issue is dedicated concluded his discussion concerning the Atelocerata-Pancrustacea controversy by pointing out that the two taxa were *"alternative hypotheses to be explored in the light of more information yet to be gathered"* (Schram & Jenner 2001). The same words could certainly be used for describing the controversy between monophyly *vs.* paraphyly of Hexapoda. Although the evidence available from mitochondrial genomics is not conclusive, the data presented here suggest that the hypothesis of the non-monophyly of hexapods deserves consideration and further scrutiny.

In order to complete the overall picture, new taxa need to be included in these analyses (i.e., additional mitochondrial genomes need to be sequenced), especially more crustaceans, and, among hexapods, proturans, campodeid diplurans, microcoryphians and more zygentomans. At the same time, methods of phylogenetic analysis are improving, following the exponential amount of molecular data gathered, and, concerning mitochondrial proteins, considerable improvements have been made by using models of evolution which take structural information into account (Liò & Goldman 2002). Finally, Bayesian inference more efficiently allows the combination of molecular and morphological data, and the incorporation into the phylogenetic analysis of prior information based on morphological characters. All these steps, more taxa, better methods, and more efficient combination of information will hopefully help in reconstructing this fundamental step in the evolution of arthropods.

REFERENCES

Averof, M. & Akam, M. 1995. *Hox* genes and the diversification of insect and crustacean body plans. *Nature* 376: 420-423.

Bitsch, C. & Bitsch, J. 1998. Internal anatomy and phylogenetic relationships among apterygote insect clades (Hexapoda). *Ann. Soc. Entomol. Fr.* 34: 339-363.

Bitsch, C. & Bitsch, J. 2000. The phylogenetic interrelationships of the higher taxa of apterygote hexapods. *Zool. Scripta* 29: 131-156.

Boore, J.L. 1999. Animal mitochondrial genomes. *Nucl. Acid Res.* 27: 1767-1780.

Boore, J.L. & Brown, W.M. 1994. Complete DNA sequence of the mitochondrial genome of the black chiton, *Katharina tunicata*. *Genetics* 138: 423-443.

Boore, J.L., Collins, T.M., Stanton, D., Daheler, L.L. & Brown, W.M. 1995. Deducing the pattern of arthropod phylogeny from mitochondrial DNA rearrangements. *Nature* 376: 163-165.

Boore, J.L., Lavrov, D.V. & Brown, W.M. 1998. Gene translocation links insects and crustaceans. *Nature* 392: 667-668.

Cao, Y., Adachi, J., Janke, A., Pääbo, S. & Hasegawa, M. 1994. Phylogenetic relationships among eutherian orders estimated from inferred sequences of mitochondrial proteins: instability of a tree based on a single gene. *J. Mol. Evol.* 39: 519-527.

Carapelli, A., Frati, F., Nardi, F., Dallai R. & Simon, C. 2000. Molecular phylogeny of the apterygotan insects based on nuclear and mitochondrial genes. *Pedobiologia* 44: 361-373.

Clary, D.O. & Wolstenholme, D.R. 1985. The mitochondrial DNA molecule of *Drosophila yakuba*: Nucleotide sequence, gene organization, and genetic code. *J. Mol. Evol.* 22: 252-271.

Cook, C.E., Louise Smith, M., Telford, M.J., Bastianello, A. & Akam, M. 2001. *Hox* genes and the phylogeny of the arthropods. *Curr. Biol.* 11: 759-763.

Crease, T.J. 1999. The complete sequence of the mitochondrial genome of *Daphnia pulex* (Cladocera: Crustacea). *Gene* 233: 89-99.

Dallai, R. & Afzelius, B.A. 1999. Accessory microtubules in insect spermatozoa: structure, function and phylogenetic significance. In: Gagnon, C. (ed.), *The Male Gamete: From Basic Science to Clinical Applications*: pp. 333-350. Cache River Press.

Delsuc, F., Phillips, M.J. & Penny, D. 2003. Comment on "Hexapod origins: monophyletic or paraphyletic?". *Science* 301: 1482.

Deutsch, J.S. 2001. Are Hexapoda members of the Crustacea? Evidence from comparative developmental genetics. *Ann. Soc. Entomol. Fr.* 37: 41-49.

Deuve, T. 2001. The epipleural field in hexapods. *Ann. Soc. Entomol. Fr.* 37: 195-231.

Dohle, W. 2001. Are the insects terrestrial crustaceans? A discussion of some new facts and arguments and the proposal of the proper name "Tetraconata" for the monophyletic unit Crustacea + Hexapoda. *Ann. Soc. Entomol. Fr.* 37: 85-103.

Duman-Scheel, M. & Patel, N.H. 1999. Analysis of molecular marker expression reveals neuronal homology in distantly related arthropods. *Development* 126: 2327-2334.

Edgecombe, G.D., Wilson, G.D.F., Colgan, D.J., Gray, M.R. & Cassis, G. 2000. Arthropod cladistics: combined analysis of Histone H3 and U2 snRNA sequences and morphology. *Cladistics* 16: 155-203.

Engel, M.S. & Grimaldi, D.A. 2004. New light shed on the oldest insect. *Nature* 427: 627-630.

Fanenbruck, M., Harzsch, S. & Wägele, J.W. 2004. The brain of the Remipedia (Crustacea) and an alternative hypothesis on their phylogenetic relationships. *Proc. Nat. Acad. Sci. USA* 101: 3868-3873.

Friedrich, M. & Tautz, D. 1995. Ribosomal DNA phylogeny of the major extant arthropod classes and the evolution of myriapods. *Nature* 376: 165-167.

Friedrich, M. & Tautz, D. 2001. Arthropod rDNA phylogeny revisited: a consistency analysis using Monte Carlo simulation. *Ann. Soc. Entomol. Fr.* 37: 21-40.

Garcia-Machado, E., Pempera, M., Dennebouy, N., Oliva-Suarez, M., Mounolou, J.C. & Monnerot, M. 1999. Mitochondrial genes collectively suggest the paraphyly of Crustacea with respect to Insecta. *J. Mol. Evol.* 49: 142-149.

Giribet, G., Edgecombe, G.D. & Wheeler, W. 2001. Arthropod phylogeny based on eight molecular loci and morphology. *Nature* 413: 157-161.

Giribet, G. & Ribeira, C. 1998. The position of arthropods in the animal kingdom: a search for a reliable outgroup for internal arthropod phylogeny. *Mol. Evol. Phylog.* 9: 481-488.

Haas, F., Waloszek, D. & Hartenberger, R. 2003. *Devonohexapodus bocksbergensis*, a new marine hexapod from the lower Devonian Hunsrück Slates, and the origin of Atelocerata and Hexapoda. *Org. Divers. Evol.* 3: 39-54.

Huelsenbeck, J.P. & Ronquist, F. 2001. MrBayes: Bayesian inference of phylogenetic trees. *Bioinformatics* 17: 754-755.

Huelsenbeck, J.P., Ronquist, F.R., Nielsen, R. & Bollback, J.P. 2001. Bayesian inference of phylogeny and its impact on evolutionary biology. *Science* 294: 2310-2314.

Hwang, U.W., Friedrich, M., Tautz, D., Park, C.J. & Kim, W. 2001. Mitochondrial protein phylogeny joins myriapods with chelicerates. *Nature* 413: 154-157.

Kishino, H. & Hasegawa, M. 1989. Evaluation of the maximum likelihood estimate of the evolutionary tree topologies from DNA sequence data, and the branching order in Hominoidea. *J. Mol. Evol.* 29: 170-179.

Klass, K.-D. & Kristensen, N.P. 2001. The ground plan and affinities of hexapods: recent progress and open problems. *Ann. Soc. Entomol. Fr.* 37: 265-298.

Koch, M. 2001. Mandibular mechanisms and the evolution of hexapods. *Ann. Soc. Entomol. Fr.* 37: 129-174.

Kraus, O. 2001. "Myriapoda" and the ancestry of Hexapoda. *Ann. Soc. Entomol. Fr.* 37: 105-127.

Kristensen, N.P. 1981. Phylogeny of insect orders. *Ann. Rev. Entomol.* 26: 135-157.

Kristensen, N.P. 1997. The groundplan and basal diversification of the hexapods. In: Fortey, R.A. & Thomas, R.H. (eds.), *Arthropod Relationships*: pp. 281-293. London: Chapman & Hall.

Kukalová-Peck, J. 1987. New Carboniferous Diplura, Monura and Thysanura, the hexapod groundplan, and the role of thoracic side lobes in the origin of wings (Insecta). *Can. J. Zool.* 65: 2327-2345.

Kjer, K.M. 2004. Aligned 18S and insect phylogeny. *Syst. Biol.* 53: 506-514.

Lavrov, D., Boore, J.L. & Brown, W.M. 2000. The complete mitochondrial DNA sequence of the horseshoe crab *Limulus polyphemus*. *Mol. Biol. Evol.* 17: 813-824.

Lavrov, D., Brown, W.M. & Boore, J.L. 2004. Phylogenetic position of the Pentastomida and (pan)-crustacean relationships. *Proc. R. Soc. London, B,* 271: 537-544.

Liò, P. & Goldman, N. 2002. Modeling mitochondrial protein evolution using structural information. *J. Mol. Evol.* 54: 519-529.

Moura, G & Christoffersen, M.L. 1996. The system of the mandibulate arthropods: Tracheata and Remipedia as sister groups, "Crustacea" non-monophyletic. *J. Comp. Biol.* 1: 95-113.

Nardi, F. Carapelli, A., Fanciulli, P.P., Dallai, R. & Frati, F. 2001. The complete mitochondrial DNA sequence of the basal hexapod *Tetrodontophora bielanensis*: evidence for heteroplasmy and tRNA translocations. *Mol. Biol. Evol.* 18: 1293-1304.

Nardi, F., Spinsanti, G., Boore, J.L., Carapelli, A., Dallai, R. & Frati, F. 2003a. Hexapod origins monophyletic or polyphyletic? *Science* 299: 1887-1889.

Nardi, F., Spinsanti, G., Boore, J.L., Carapelli, A., Dallai, R. & Frati, F. 2003b. Response to comment on "Hexapod origins: monophyletic or paraphyletic?". *Science* 301: 1842.

Negrisolo, E., Minelli, A. & Valle, G. 2004. The mitochondrial genome of the house centipede *Scutigera* and the monophyly versus paraphyly of myriapods. *Mol. Biol. Evol.* 21: 770-780.

Panganiban, G., Sebring, A., Nagy, L. & Carroll, S. 1995. The development of crustacean limbs and the evolution of arthropods. *Science* 270: 1363-1366.

Regier, J.C. & Shultz, J.W. 1997. Molecular phylogeny of the major arthropod groups indicates polyphyly of crustaceans and a new hypothesis for the origin of hexapods. *Mol. Biol. Evol.* 14: 902-913.

Regier, J.C. & Shultz, J.W. 2001. Elongation Factor-2: a useful gene for arthropod phylogenetics. *Mol. Phylog. Evol.* 20: 136-148.

Richter, S. 2002. The Tetraconata concept: hexapod-crustacean relationships and the phylogeny of crustaceans. *Org. Divers. Evol.* 2: 217-237.

Schram, F.R. & Jenner, R.A. 2001. The origin of Hexapoda: a crustacean perspective. *Ann. Soc. Entomol. Fr.* 37: 243-264.

Shimodaira, H. 2002. An approximately unbiased test of phylogenetic tree selection. *Syst. Biol.* 51: 492-508.

Shultz, J.W. & Regier, J.C. 2000. Phylogenetic analysis of arthropods using two nuclear protein-encoding genes supports a crustacean + hexapod clade. *Proc. R. Soc. London, B*, 267: 1011-1019.

Simon, C., Frati, F., Beckenbach, A., Crespi, B., Liu, H. & Flook, P. 1994. Evolution, weighting, and phylogenetic utility of mitochondrial gene sequences and a compilation of conserved Polymerase Chain Reaction primers. *Ann. Ent. Soc. Amer.* 87: 651-701.

Simpson, P. 2001. A review of early development of the nervous system in some arthropods: comparison between insects, crustaceans and myriapods. *Ann. Soc. Entomol. Fr.* 37: 71-84.

Štys, P. & Bilinski, S. 1990. Ovariole types and the phylogeny of hexapods. *Biol. Rev.* 65: 401-429.

Swofford, D.L., Olsen, G.J., Waddell, P.J. & Hillis, D.M. 1996. Phylogenetic inference. In: Hillis, D.M., Moritz, C. & Mable, B.K. (eds.), *Molecular Systematics*: pp. 407-514. Sunderland: Sinauer Assoc.

Tavare, S. 1986. Some probabilistic and statistical problems on the analysis of DNA sequences. *Lect. Math. Life Sci.* 17: 368-376.

Telford, M.J. & Thomas, R.H. 1995. Demise of the Atelocerata? *Nature* 376: 123-124.

Thompson, J.D., Gibson, T.J., Plewniak, F., Jeanmougin, F. & Higgins, D.G. 1997. The ClustalX windows interface: flexible strategies for multiple sequence alignment aided by quality analysis tools. *Nucl. Acid Res.* 24: 4876-4882.

Wheeler, W. 1998. Sampling, groundplans, total evidence and the systematics of arthropod. In: Fortey, R.A. & Thomas, R.H. (eds.), *Arthropod Relationships*: pp. 87-96. London: Chapman & Hall.

Wheeler, W., Cartwright, P. & Hayashi, C.Y. 1993. Arthropod phylogeny: a combined approach. *Cladistics* 9: 1-39.

Wheeler, W.C., Whiting, M., Wheeler, Q.D. & Carpenter, J.M. 2001. The phylogeny of the extant hexapod orders. *Cladistics* 17: 113-169.

Wilson, K., Cahill, V., Ballment, E. & Benzie, J. 2000. The complete sequence of the mitochondrial genome of the crustacean *Penaeus monodon*: are malacostracan crustaceans more closely related to insects than branchiopods. *Mol. Biol. Evol.* 17: 863-874.

Yang, Z. 1997. PAML: a program package for phylogenetic analysis by maximum likelihood. *CABIOS* 13: 555-556.

Zrzavý, J. & Štys, P. 1997. The basic body plan of arthropods: insights from evolutionary morphology and developmental biology. *J. Evol. Biol.* 10: 353-367.

The position of crustaceans within Arthropoda - Evidence from nine molecular loci and morphology

GONZALO GIRIBET[1], STEFAN RICHTER[2], GREGORY D. EDGECOMBE[3] & WARD C. WHEELER[4]

[1] *Department of Organismic and Evolutionary Biology, Museum of Comparative Zoology, Harvard University, Cambridge, Massachusetts, U.S.A.*

[2] *Friedrich-Schiller-Universität Jena, Institut für Spezielle Zoologie und Evolutionsbiologie, Jena, Germany*

[3] *Australian Museum, Sydney, NSW, Australia*

[4] *Division of Invertebrate Zoology, American Museum of Natural History, New York, U.S.A.*

ABSTRACT

The monophyly of Crustacea, relationships of crustaceans to other arthropods, and internal phylogeny of Crustacea are appraised via parsimony analysis in a total evidence framework. Data include sequences from three nuclear ribosomal genes, four nuclear coding genes, and two mitochondrial genes, together with 352 characters from external morphology, internal anatomy, development, and mitochondrial gene order. Subjecting the combined data set to 20 different parameter sets for variable gap and transversion costs, crustaceans group with hexapods in Tetraconata across nearly all explored parameter space, and are members of a monophyletic Mandibulata across much of the parameter space. Crustacea is non-monophyletic at low indel costs, but monophyly is favored at higher indel costs, at which morphology exerts a greater influence. The most stable higher-level crustacean groupings are Malacostraca, Branchiopoda, Branchiura + Pentastomida, and an ostracod-cirripede group. For combined data, the Thoracopoda and Maxillopoda concepts are unsupported, and Entomostraca is only retrieved under parameter sets of low congruence. Most of the current disagreement over deep divisions in Arthropoda (e.g., Mandibulata versus Paradoxopoda or Cormogonida versus Chelicerata) can be viewed as uncertainty regarding the position of the root in the arthropod cladogram rather than as fundamental topological disagreement as supported in earlier studies (e.g., Schizoramia versus Mandibulata or Atelocerata versus Tetraconata).

1 INTRODUCTION

Crustaceans show a remarkable amount of morphological disparity with respect to tagmosis, limb morphology, and internal anatomy (Schram 1986). Improving our knowledge of the phylogenetic relationships of crustaceans is, therefore, always a challenge. For more than a century, phylogenetic scenarios have shown as much diversity as the taxon itself.

308 Giribet et al.

The position of Crustacea within the arthropods is a matter of incessant debate. Paleontologists often favored a closer relationship between Crustacea and Chelicerata (Cisne 1974; Bergström 1979), whereas most neontologists (with the exception of the Manton school) argued for a monophyletic Mandibulata with Crustacea as a sister group to Tracheata or Atelocerata (myriapods + hexapods). During the last decade, however, increasing evidence supports a closer relationship between crustaceans and hexapods, leaving the myriapods outside (Fig. 1). Early molecular analyses (e.g., Turbeville et al. 1991) favored a crustacean-hexapod clade, named Pancrustacea (Zrzavý & Štys 1997) or Tetraconata (Dohle 2001), which has subsequently been substantiated in almost all molecular analyses of arthropods (e.g., Wheeler et al. 1993: fig. 8; Friedrich & Tautz 1995; Giribet et al. 1996, 2001; Hwang et al. 2001; Regier & Shultz 2001; Kusche et al. 2003; Mallatt et al. 2004). In addition, mitochondrial genome rearrangements (Boore et al. 1998) and certain morphological characters were proposed as support for a crustacean-hexapod clade (Averof & Akam 1995; Whitington 1995; Dohle 2001). However, the traditional concept of a monophyletic Atelocerata still has advocates (Klass & Kristensen 2001; Kraus 2001).

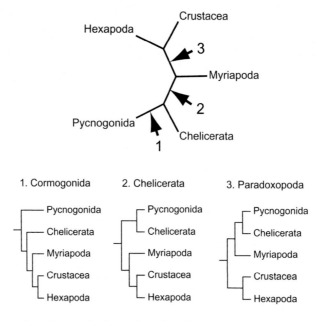

Figure 1. Alternative hypotheses of arthropod relationships and putative positions of the root. Rooting point 1 supports monophyly of Cormogonida; rooting point 2 implies a single origin of chelicerae or chelifores; rooting point 3 shows monophyly of Paradoxopoda as the sister group to Tetraconata.

Other aspects of arthropod relationships have also been debated intensively. Myriapods have been suggested to be paraphyletic, with Progoneata as a sister group to the Hexapoda, and Chilopoda as a sister group to Progoneata + Hexapoda (Kraus 1998, 2001). Others have considered myriapods as being the monophyletic sister taxon of Tetraconata (Edgecombe & Giribet 2002) (Fig. 1). However, some molecular studies favor a clade including Myriapoda and Chelicerata (e.g., Friedrich & Tautz 1995; Hwang et al. 2001; Mallatt et al. 2004), recently named Paradoxopoda (Mallat et al. 2004) or Myriochelata (Pisani et al. 2004). Some correspondences in the development of the nervous system might even support this relationship from a morphological point of view (Dove & Stollewerk 2003; Kadner & Stollewerk 2004), whereas recent studies on compound eye structure (Müller et al. 2003) and mandible morphology (Edgecombe et al. 2003) have added support for the Mandibulata hypothesis, contradicting Paradoxopoda (see a review in Richter & Wirkner 2004).

Independent of those questions, crustacean monophyly itself remains unresolved (see Schram & Koenemann 2004a, 2004b). Crustacean monophyly was suggested by Lauterbach (1983) after a careful analysis of the morphological evidence available at that time, and has been defended by Waloszek (Walossek 1999; Waloszek 2003) based on extant morphology and evidence from fossil taxa. Cladistic morphological analyses also support a monophyletic Crustacea (Wheeler et al. 1993; Wheeler 1998; Schram & Hof 1998; Wills 1998; Zrzavý et al. 1998; Edgecombe et al. 2000; Giribet et al. 2001; Edgecombe 2004; Wheeler et al. 2004). On the contrary, with few exceptions (e.g., Negrisolo et al. 2004), analyses based solely on molecular data and including at least two major crustacean taxa suggest non-monophyly (Friedrich & Tautz 1995 2001; Giribet & Ribera 2000; Hwang et al. 2001). Paraphyly with respect to hexapods is a recurring pattern in these analyses. Additionally, in light of the Tetraconata concept, an increasing number of morphological characters support crustacean paraphyly. In particular, Malacostraca is a good candidate as a potential sister group of Hexapoda because of the shared presence of certain neuroanatomical features such as specific pioneer neurons (Whitington 1996) and neuroblasts (Harzsch 2001). However, these characters might also be present in branchiopods (Duman-Scheel & Patel 1999) and they are not well studied in the maxillopodan taxa (see Richter 2002, for a review of these characters). Recently, a close relationship between hexapods, malacostracans and remipedes has been suggested based on brain characters (Fanenbruck et al. 2004). A hexapod-malacostracan clade is also supported by the presence of two optic chiasmata between the optic ganglia (Harzsch 2002; Sinakevitch et al. 2003). Some of the mentioned molecular analyses indeed favor a closer relationship between malacostracans and hexapods (Wilson et al. 2000; Hwang et al. 2001), sometimes including remipedes (Regier & Shultz 2001: fig. 1A), but others support a closer relationship of branchiopods to hexapods (Friedrich & Tautz 2001).

Concerning the higher crustacean groups, there seems to be no doubt that the two smaller taxa, Cephalocarida and Remipedia, with their unique morphology, are each monophyletic. There is also increasing evidence from both morphology and molecular data that Branchiopoda and Malacostraca are each monophyletic (Walossek 1999; Giribet & Ribera 2000; Spears & Abele 2000; Giribet et al. 2001; Richter & Scholtz 2001). Concerning Maxillopoda, molecular analyses failed to show its monophyly (Giribet et al. 2001; Mallatt et al. 2004) and morphological support for the group seems to be weak (but see Walossek 1999 for a different opinion).

The relationships among these main clades, Remipedia, Cephalocarida, Branchiopoda, Malacostraca, and the putative Maxillopoda have been discussed for a long time, although mostly in the context of crustacean monophyly. In particular, Thoracopoda - including Cephalocarida, Branchiopoda and Malacostraca (see Hessler 1992; Ax 1999; Edgecombe et al. 2000; Richter 2002) and Entomostraca - including Cephalocarida, Branchiopoda and Maxillopoda - as sister taxon to Malacostraca (Walossek 1999; Waloszek 2003) represent reasonable alternatives (Fig. 2).

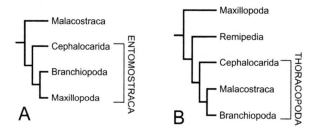

Figure 2. Two alternative hypotheses showing suggested relationships of Crustacea. (A) According to Walossek (1999), Entomostraca is monophyletic. (B) Relationships based on Edgecombe et al. (2000) supporting the Thoracopoda concept of Hessler (1992).

One obvious issue that has become evident is that crustacean phylogeny can be only sensibly discussed in the framework of global arthropod relationships (see also Schram & Jenner 2001; Richter 2002). However, only a few phylogenetic analyses that deal with arthropods using a broad sample of crustacean taxa are available based on molecular data (Giribet & Ribera 2000; Hwang et al. 2001; Mallatt et al. 2004), or using a combined approach of molecules and morphology (Edgecombe et al. 2000; Giribet et al. 2001; Wheeler et al. 2004). Here we present a re-analysis of arthropod relationships based upon the most comprehensive data set yet assembled to study such problem, the 52-taxon, 303-morphological character, eight-gene data set of Giribet et al. (2001). To that analysis we have added 15 taxa, mostly crustaceans (24 crustaceans are represented in the new analyses versus the previous 13), refined the terminal taxa by decreasing the number of composite taxa, refined morphological codings and added another 50 morphological characters, and added ca. 25% new sequences for the eight previous loci. Moreover, we have added a new gene (elongation factor-2), and increased the previous sampling of the D3 region of 28S rRNA to the complete gene sequence. We also doubled the length of the COI gene fragment used in our previous analyses. With the new data, we attempt to shed new light on the long-standing controversies in arthropod phylogenetics, especially for those aspects that concern the position of crustaceans, their monophyly, and the relationships among its constituent lineages.

2 METHODOLOGY

We analyze data for 67 taxa (Appendix 1, Tables 1 and 2), 352 morphological characters (Appendices 2 and 3), and 9 molecular loci for over 10 Kb per complete taxon. The data are analyzed as in several of our previous studies by using the direct optimization method (Wheeler 1996) and in the computer software POY version 3.0 (Wheeler et al. 2002).

2.1 *Taxa*

The analyses here presented include data from 67 terminal taxa, 2 Onychophora, 2 Tardigrada, and 63 Arthropoda (3 Pycnogonida, 10 Chelicerata, 13 Myriapoda, 13 Hexapoda, 1

Remipedia, 1 Cephalocarida, 8 Malacostraca, 6 Maxillopoda [including 1 Pentastomida], and 8 Branchiopoda).

We agree with Prendini (2001) that one should use species as terminals for phylogenetic studies. In that respect, we have attempted to select our terminals based on well-studied species both at morpho-anatomical and molecular levels, and a large proportion of our taxa are based on an exemplar approach. Whenever this was not possible, and always with the aims of representing the maximum amount of available data per terminal, we combined representatives of several species within the same genus; sometimes this combination includes unidentified species within the genus. In a few cases, suprageneric terminals were created by pooling data from different species only if the evidence for the monophyly of such taxa is undisputed. That way, we represent the following suprageneric terminals in our analyses: Peripatidae, Buthidae, Scorpionidae, Scolopendridae, Geophilidae, Pauropodinae, Polyxenidae, Sphaerotheriidae, Acerentomidae, Campodeidae, Japygidae, Meinertellidae, Machilidae, Lepismatidae, Remipedia, Oniscidea, Calanoida, Stomatopoda, and Pentastomida. The exact species pooled to constitute these terminals are available in Appendix 1 (Table 1).

2.2 Morphological characters

The data we present, 352 characters, are based on the data set presented by Giribet et al. (2001), later more explicitly described and updated (Edgecombe 2004). Characters have been refined since then, especially for crustaceans, and new data available since the publication of the previous articles were incorporated into the new matrix (Appendix 2). The matrix in NDE (Page 2000) format is available upon request from the authors.

All morphological characters were considered unordered, with the exception of characters 3, 28, 92, 153, 173, 185 and 232. Character 300, referring to the male gonopods, includes 13 character states to reflect its broad variability across arthropods. This character required independent coding because it needed special analysis (most programs cannot handle more than 10 character states). As such, character 300 was coded as all '?' in the file used for analyses (file **artmor1.ss**), and a second file containing a single Sankoff character with the 13 states was generated (file **artmor2.ss**). This improves upon our previous analyses (Giribet et al. 2001; Edgecombe 2004), in which the gonopod character presented 11 states, but the 11th state was coded as '?' for Remipedia. Treatment of more than 10 character states in the program POY required novel implementation by one of us (WCW).

2.3 Molecular sampling

The data presented here include information on nine molecular loci. Compared to our previous analysis (Giribet et al. 2001), we were able to add DNA sequence information for several extra taxa, but we also made an effort to fill in many sequence data that did not amplify in our previous analysis. For 28S rRNA, we are now using the complete gene for a large proportion of the samples, although some remain partial (but longer than 2 Kb), and a few more remain at ca. 350 bp for the D3 expansion fragment. We also added the novel data on elongation factor-2 published by Regier & Shultz (2001), which were not available when we published our previous studies. Therefore, we use molecular data for the following markers (approximate number of bp used indicated in parentheses): 18S rRNA (ca. 1,800 bp), 28S rRNA (ca. 3,600 bp), U2 sn rRNA (130 to 133 bp), histone H3 (327 bp), elongation factor-1α (1,092 bp), elongation factor-2 (2,178 to 2,184 bp), RNA polymerase II (1,161 bp), cytochrome *c* oxidase subunit I (1,230 bp), and 16S rRNA (437 to 540 bp). This amounts to over 12 Kb of genetic information per complete taxon and includes genes evolving in very different manners, including nuclear coding and non-coding markers, and mitochondrial coding and non-coding. No comparable molecular data set has ever been compiled to study the evolution of any group of invertebrate animals.

2.4 *Analytical methods*

All data were analyzed simultaneously under direct optimization using the program POY version 3.0. Our choice for analyzing data in such a fashion is straightforward; we do not think that DNA-based homology can be assigned independently of topology.

Homology is the relationship between features that is derived from their shared, unique origin on a cladogram. Two features are homologous if their origin can be traced back to a specific transformation on a specific branch of a specific cladogram. The same features may or may not be homologous on alternate cladograms. Therefore, homology is entirely cladogram dependent and the relative optimality of alternate cladograms determines whether or not features have this relationship. The dynamic homology framework (Wheeler 2001) extends through optimization of transformations to the correspondences among features (often referred to as 'putative' or primary homology) themselves. The joint scenario of correspondence and transformation is chosen such that the overall cladogram cost is minimal. In this framework, there is no distinction between 'putative' or 'primary' and 'secondary' homology - all variation is optimized de novo for each cladogram (contra Simmons 2004).

In our study morphological data are analyzed using a static homology framework, and several fragments of DNA sequences of protein-coding genes were analyzed this way. Not all coding fragments could be analyzed that way, and they pose identical (or even harder) challenges to homology recognition as in non-coding regions. Therefore, 'alignment-issues' should not be a reason for preferring coding versus non-coding genes, despite the claim of some authors.

All data were analyzed using a combination of branch swapping and refining techniques that included 20 random addition replicates followed by tree fusing (Goloboff 1999), ratcheting (Nixon 1999), TBR branch swapping, and another round of tree fusing. The data were run under 20 parameter sets, and the resulting trees for each parameter set were pooled into a file and submitted to tree fusing once more, for each of the parameter sets analyzed. Variables studied were different indel costs and transversion/transition ratios (Wheeler 1995). The relative weight of the morphological matrix was increased as the molecular costs increased, so its signal would not be obliterated by the much larger costs assigned to indel events or transversions (i.e., morphological transformation weight = indel weight). Nodal support was evaluated by jackknifing (Farris et al. 1996), as implemented in POY.

Splitting of large DNA sequences into smaller fragments speeds up analyses (Giribet 2001), and it helps to incorporate information on secondary structure or for refining original homology statements (Giribet & Wheeler 2001). In that respect, we have partitioned several of the markers employed in our analyses as follows:

- 18S rRNA was divided into 29 sequence fragments, four of which (fragments 4, 9, 21 and 28) were deactivated due to extreme length variation (even within species) that would cause random matching of bases at no cost.
- 28S rRNA was divided into 31 sequence fragments from which 10 were deactivated (fragments 1, 7, 9, 10, 16, 18, 22, 25, 30, and 31) due to extreme length variation or to the presence of missing data for most taxa.
- 16S rRNA was divided into 9 sequence fragments from which fragment 1 was deactivated.
- Cytochrome *c* oxidase subunit I (COI hereafter) was divided into 12 sequence fragments using the translation to amino acids, and fragment 9 was deactivated due to the lack of sequence data for most taxa (this fragment connects our previously used fragment between primers LCO and HCO and a new COI fragment used in this analysis). The COI fragments showed length variation and therefore were analyzed under a dynamic homology framework.
- Elongation factor-2 was divided into six sequence fragments, analyzed under dynamic homology due to presence of sequence length variation.
- Elongation factor-1α was analyzed in five prealigned sequence fragments.
- RNA polymerase II was divided into six sequence fragments, from which fragments 2 and 5 were deactivated (due to lack of sequence data for most taxa) and was analyzed under static homology.
- The other two loci were analyzed as single fragments, U2 sn rRNA under dynamic homology, and histone H3 under static homology.

3 RESULTS

3.1 *Morphological analysis*

The analysis of the morphological data set resulted in three trees at 740 steps. The strict consensus of these trees, with jackknife proportions over 50%, is presented in Figure 3.

This tree places pentastomids outside of Arthropoda, and shows monophyly of the main arthropod lineages; Pycnogonida, Chelicerata[1], Myriapoda, Hexapoda, and Crustacea, as well as the higher groups Pycnogonida + Chelicerata, Mandibulata and Tetraconata. While the major arthropod taxa receive jackknife support values above 50%, none of the relationships among these taxa receives jackknife support above 50%. The monophyly of Arthropoda also has low jackknife support, and only when pentastomids are considered arthropods does one get jackknife support above 50% (Fig. 3).

---

[1] For many authors, the taxon Chelicerata is formed by Pycnogonida and by a clade formed of Xiphosura + Arachnida, which is often referred to as Euchelicerata. Due to the possible paraphyly of Chelicerata *s. lat.*, we prefer to adopt the terms Pycnogonida and Chelicerata, the latter referring to what other authors term Euchelicerata. We follow this nomenclature throughout this chapter.

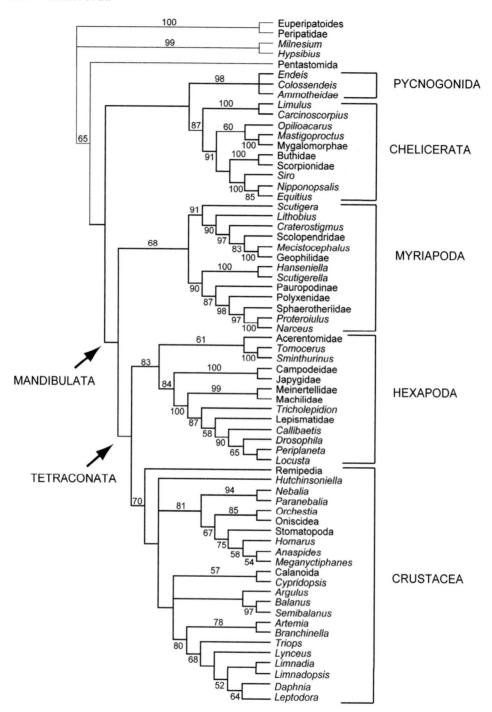

Figure 3. Strict consensus of 3 trees at 740 steps for the parsimony analysis of the morphological data set from Appendix 3 (see Appendix 2 for character descriptions). Numbers on branches indicate jackknife support values above 50%.

Relationships within Crustacea show Remipedia as the sister group to the remaining taxa, which form a polytomy of three clades, Cephalocarida, Malacostraca, and a clade of Maxillopoda + Branchiopoda, although maxillopodans do not form a resolved clade in the strict consensus tree (they are monophyletic in one of the three optimal trees). The branching pattern among these clades conflicts, and the two possible relationships have been summarized in Figure 4. Basically, the morphological data are unable to generate a stable hypothesis of the main crustacean groups, other than suggesting a sister group relationship of remipedes to other crustaceans, and the Maxillopoda + Branchiopoda clade, although jackknife support is below the 50% threshold in both cases.

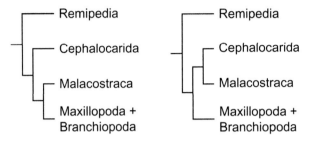

Figure 4. Alternative minimal length resolutions of the crustacean relationships summarized in Figure 3. See Appendix 1 (Table 1) for detailed overview of taxa.

3.2 Combined analysis

The combined analysis of all nine genes plus morphology (see Fig. 5 for the results under the optimal parameter set) is best summarized in Figures 6 and 7, which represent the sensitivity plots ('Navajo rugs') for several major arthropod clades. When pentastomids are considered as arthropods (as favored by most molecular systematists), arthropod monophyly is found under most parameter sets, but not under equal weights because tardigrades appear within the progoneate myriapods. Tardigrades also nest within this group in one other case (indel cost = 2; transversions/transitions = 1).

Within arthropods, Chelicerata + Pycnogonida is supported by most parameter sets, wheras the alternative Cormogonida occurs only under two (see complementary Navajo rugs in Fig. 6). A monophyletic Mandibulata is found under most analyses, and the alternative Paradoxopoda hypothesis appears only under two parameter sets. Myriapoda, Tetraconata, and Hexapoda appear to be monophyletic under most analytical conditions, whereas Crustacea is found under 13 parameter sets, especially for those with higher indel costs (and higher weight for the morphological partition). In almost all cases in which Crustacea is monophyletic, pentastomids are included within Crustacea except for two of the analyses (shaded squares in the Crustacea plot in Fig. 6). In three of the analyses without support for Crustacea, a clade comprising Cephalocarida + Ostracoda + Cirripedia is sister to Hexapoda (+ Remipedia in the analysis which minimizes the ILD [Incongruence Length Difference]; appendix 1, Table 3). There is no support for a Malacostraca-Hexapoda clade. In three of the analyses in which Crustacea is not monophyletic, hexapods are also not monophyletic.

316 *Giribet et al.*

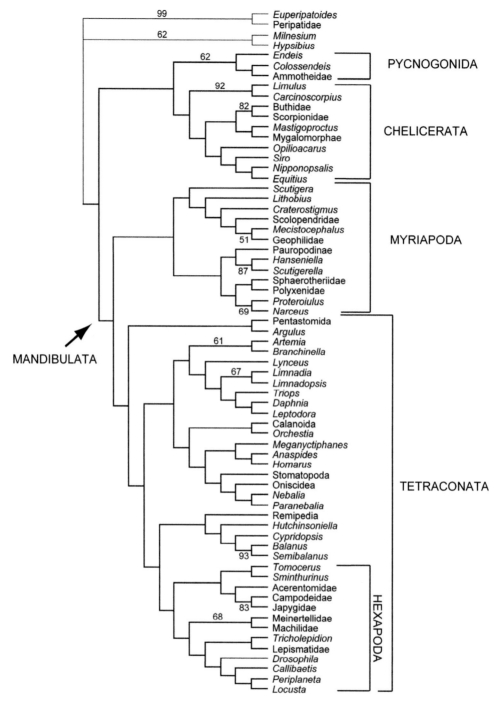

Figure 5. Most parsimonious cladogram at cost 109,957 for the optimal parameter set (indel/transversion cost of 1 and transversion/transition cost of 4) for nine genes and morphology. Numbers on branches represent jackknife support values above 50%.

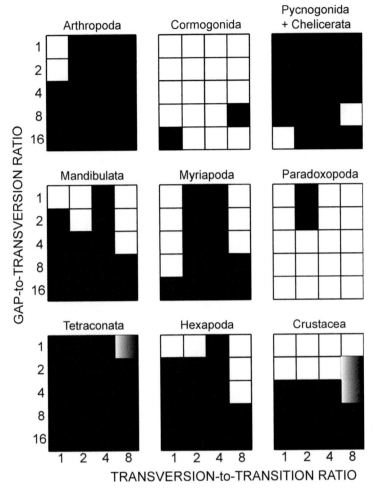

Figure 6. Navajo rugs (sensitivity plots) for higher arthropod relationships based on combined analysis of nine genes and morphology. Black squares indicate monophyly; white squares indicate non-monophyly. For Tetraconata and Crustacea shaded square indicates that Pentastomida is not included.

Internal relationships of the crustacean clades are summarized in Figure 7. Within Crustacea (or Tetraconata if Crustacea is not supported), Malacostraca and Branchiopoda are monophyletic under most parameter sets. In the single case in which Branchiopoda is not supported, a malacostracan species is included within Branchiopoda. In the analyses where Malacostraca is not monophyletic, four show the copepod terminal included in the otherwise monophyletic clade. The internal relationships within Branchiopoda and Malacostraca vary throughout the parameter space of the combined analysis.

Maxillopoda does not find any support in our analyses. In the morphological analyses maxillopodans are unresolved, and only one of the three optimal trees shows maxillopodan monophyly. However, one clade within Maxillopoda, an ostracod-cirripede clade is supported under almost all parameter sets in the combined analyses (Fig. 7). As well, a pentastomid-branchiuran clade is supported in 17 of 20 analyses.

318 *Giribet et al.*

None of the major hypotheses of internal crustacean relationship is well supported by our data. The Thoracopoda concept is not supported by a single parameter set, and even the core group Malacostraca + Branchiopoda finds only support in one analysis. The alternative Entomostraca concept is supported in seven analyses (but also including Pentastomida), and one additional analysis excluding the pentastomid-branchiuran clade. Concerning the position of Remipedia, three alternatives are represented in 16 of the 20 analyses: Remipedia + Cephalocarida, Remipedia + Calanoida, and Remipedia + Malacostraca.

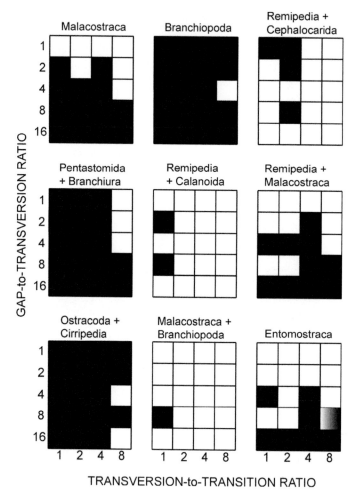

Figure 7. Navajo rugs (sensitivity plots) for crustacean relationships based on combined analysis of nine genes and morphology. Black squares indicate monophyly; white squares indicate non-monophyly. For Entomostraca black squares include Pentastomida; shaded square exludes Pentastomida and Branchiura.

For the parameter set that minimizes incongruence between the nine genes and morphology (indel/transversion = 1, transversion/transition = 4), the groupings Pycnogonida + Chelicerata, Mandibulata and Tetraconata are supported by the data, but with jackknife support values below 50% (Fig. 5). The fundamental divisions within Chelicerata (= Xiphosura and Arachnida) and Myriapoda (= Chilopoda and Progoneata) are congruent with the morphological cladogram (Fig. 3). Crustaceans are resolved as paraphyletic with respect to a monophyletic Hexapoda, the latter dividing into Entognatha and Ectognatha. A branchiuran-pentastomid group resolves basal to other Tetraconata. The ostracod-cirripede clade is represented in this tree forming a clade with cephalocarids and remipedes. Malacostraca and Branchiopoda are resolved as sister groups but malacostracans include copepods. Jackknife support values for this tree are in general below the 50% threshold, and only a few clades receive meaningful values, such as Onychophora (99%), Tardigrada (62%), Pycnogonida (62%), Xiphosura (92%), Scorpiones (82%), Geophilomorpha (51%), Symphyla (87%), Juliformia (69%), Anostraca (61%), Spinicaudata (67%), Cirripedia (93%), Diplura (83%), and Archaeognatha (68%). This low proportion of nodes with high jackknife values contrasts with the large number of clades supported under most parameter sets (see Giribet 2003).

4 DISCUSSION

Our analysis represents the most comprehensive account in terms of taxa studied and amount of data (morphological and molecular) used to understand arthropod relationships to date. Nevertheless, many clades are still not clearly resolved and disagreements with several hypotheses based on morphological data alone are obvious; some of the resulting clades in our analyses even might appear as 'nonsense clades'. Nevertheless, we are convinced that a combined approach using as much data as possible is a more sensible way to deal with the information explored from nature than ignoring some particular data because of the seemingly nonsensical results they might create. To distinguish between 'nonsense clades' and evolutionary patterns that we just do not understand might be hazardous. However, what is most important is to test the stability of certain assumptions. Particular clades that show stability throughout an entire parameter space have a much higher degree of corroboration that those favored by only a few parameter sets (or even by a single evolutionary model) (Giribet 2003). Therefore, we refer to the 'Navajo rugs' as a summary of our results. Based on those, we can accept clades with a high degree of confidence.

4.1 *Arthropod relationships*

Arthropod relationships in general, and the position of crustaceans in particular, have remained contentious for several decades. While current analyses including molecular data tend to support Tetraconata - including crustaceans and hexapods - other issues remain controversial. Although fossils may have a fundamental role in elucidating arthropod relationships (e.g., see Edgecombe 1998) as shown in the few combined analyses of morphology and molecules including several fossil arthropod taxa (Giribet et al. 2002; Wheeler et al. 2004), no fossil species has been included in this study. Therefore, we caution the reader to interpret our results and conclusions in the absence of fossils. Disregarding the possibility of monophyly of Atelocerata for the sake of simplicity, most molecular and combined analyses tend to view arthropod relationships as a conflicting set of hypotheses as summarized in Figure 1. Different authors have argued more or less strongly for the hypotheses based on their data, such as Mandibulata versus Paradoxopoda; both supported by the same data in at least one case (Giribet et al. 1996). Another current debate shows conflict in the monophyly of arthropods with the first appendage in the form of chelicerae or chelifores (pycnogonids, xiphosurans and arachnids) versus the Cormogonida concept (Zrzavý et al. 1998), which considers pycnogonids as the sister group to all other arthropods, thus rendering chelicerae as the plesiomorphic condition for the first pair of arthropod appendages. In reality all these hypotheses can be reduced easily to a rooting issue, as illustrated in Figure 1. Rooting problems are common in phylogenetic analyses when distant outgroups are selected, massive extinction may have occurred in the stem lineage of a taxon, or molecular and morphological change is fast (e.g., Wheeler 1990; Giribet et al. 2000). In the case of arthropods, which were subjected to massive extinction since they diverged from a common ancestor shared with Cambrian lobopodians (e.g., Budd 2002), rooting with onychophorans and tardigrades has been problematic in some cases due to the autapomorphic degree of molecular data in the two outgroups. It is in this respect that we prefer to view arthropod relationships as one possibly well-solved topology (Fig. 1) with alternative rooting places.

As for our data, rooting position 2 (arthropods with chelicerae as the sister group to Mandibulata) is supported in most analyses. However, a few parameter sets also support the Cormogonida hypothesis (Zrzavý et al. 1998; Giribet et al. 2001) or the Paradoxopoda hypothesis (Turbeville et al. 1991; Friedrich & Tautz 1995; Giribet et al. 1996; Hwang et al. 2001; Mallatt et al. 2004). Besides the rooting issue at the base of the arthropod tree, our data support myriapod monophyly in 13 of 20 parameter sets, with a division into Chilopoda and Progoneata (Edgecombe 2004). Noteworthy as well is the monophyly of Hexapoda across most of the explored parameter space (15 out of 20 parameter sets). Hexapod monophyly has moderate morphological support (Fig. 3) but has not been retrieved in most previous molecular analyses (see a review in Giribet et al. 2004).

4.2 Crustacean relationships

Crustacean relationships will remain controversial. Nevertheless, a few landmarks become manifest. This concerns in particular the monophyly of both Malacostraca and Branchiopoda. Our analysis supports the monophyly of Malacostraca, which is found under most parameter sets. This is also found in previous molecular (Spears & Abele 1998; Giribet & Ribera 2000; Shultz & Regier 2000) and combined (Giribet et al. 2001) analyses, although now with a more extensive taxon sampling (8 terminals). Malacostraca is also well supported by our morphological data as shown in previous morphological analyses (Richter & Scholtz 2001). Previous assumptions of a closer relationship of Leptostraca and Branchiopoda (e.g., Schram 1986; Schram & Hof 1998) cannot be substantiated. The relationships within Malacostraca are not stable and are outside the focus of the present study; though at least our morphological analysis and some of the combined analyses support the sister group relationship between Leptostraca and Eumalacostraca.

Branchiopod monophyly is still controversially discussed by morphologists. Whereas monophyly is advocated, e.g., by Walossek (1993), Schram & Hof (1998), and Richter (2004) using different characters, paraphyly has been suggested by Ax (1999). In our present analysis, the monophyly of Branchiopoda is found under almost all parameter sets for our extended taxon sampling (8 terminals), which is also in agreement with previous molecular (Spears & Abele 1998, 1999, 2000; Giribet & Ribera 2000; Shultz & Regier 2000) and combined (Giribet et al. 2001) analyses. Within Branchiopoda the two cladocerans form a clade (18 of 20 analyses) which is in agreement with previous morphological and molecular analyses (Olesen et al. 1996; Spears & Abele 2000; Braband et al. 2002). A close relationship between Spinicaudata and Cladocera is found in some of the analyses (8 of 20), a relationship which has been previously suggested by Braband et al. (2002). However, *Cyclestheria hislopi* (Cyclestherida) as potential closest cladoceran relative was not considered in our analysis.

Martin & Davis (2001: 20) stated that "the Maxillopoda continues to be a terribly controversial assemblage". Maxillopodan monophyly has been suggested by Boxshall & Huys (1989) and by Walossek (1993, 1999) and Ax (1999). It is also found in one of our most parsimonious trees based on morphology alone. However, as in previous molecular analyses (see for example Spears & Abele 1998; Giribet & Ribera 2000), there is no support for Maxillopoda in our combined analyses.

What appears to be very interesting is the strong support for an ostracod-cirripede clade, found for the first time for molecular or combined analyses. Within Maxillopoda, Boxshall & Huys (1989) suggested a separation between a thecostracan-branchiuran-ostracod clade and a mystacocarid-copepod clade. Within the first clade they proposed a sister group relationship between Thecostraca + Branchiura with Ostracoda as the sister group to both taxa, but this particular relationship has not been found in any of our analyses. However, a sister group relationship of ostracods and thecostracans has been proposed by Ax (1999) under the name Thecostracomorpha, although the single apomorphy mentioned by him, a bivalved carapace with adductor muscle, is certainly problematic because it also occurs in Malacostraca and Branchiopoda.

Remipede relationships also remain mysterious. Previously suggested close relationships to copepods (Ito 1989), to cephalocarids (Spears & Abele 1998; Giribet et al. 2001) and to malacostracans (Fanenbruck et al. 2004) have been also found in some of our analyses. However, none of these hypotheses can be considered as much more favored than any other although the remiped-malacostracan clade occurs especially in those analyses with higher indel costs (and higher weight for the morphological partition).

4.3 *Pentastomid relationships*

The unusual morphology of pentastomids (tongue worms), a group composed entirely of parasitic species, has long intrigued zoologists. Two main hypotheses are currently favored. A crustacean affinity with a possible relationship to maxillopodan branchiurans has been suggested based on sperm ultrastructure (Wingstrand 1972; Riley et al. 1978; Jamieson & Storch 1992), 18S rRNA sequence data (Abele et al. 1989; Zrzavý 2001), and amino acid analysis of 12 mitochondrial genes (Lavrov et al. 2004) (but see a relationship to nematodes suggested by the same data in their figures 1a and 1b). A second hypothesis suggests that pentastomids branched off the arthropod stem lineage between tardigrades and the (eu)arthropod stem group (Walossek & Müller 1994), rather than being especially closely related to crustaceans. Interestingly, our data also support both of these hypotheses. The morphological data set suggests a sister group relationship between pentastomids and the other arthropods when rooted with onychophorans and tardigrades (Fig. 3), although further testing should include more distantly related outgroups. In fact, Jenner (2004) suggested that the extreme morphological modification of parasitic groups such as pentastomids may mislead morphological cladistic analyses. Our molecular and combined analyses, which include novel 18S and 28S rRNA data for one pentastomid species, mostly support a relationship between pentastomids and branchiurans as in previous molecular analyses (Abele et al. 1989; Zrzavý 2001) (Figs. 5, 7). A three parameter sets do not find a sister group relationship between Pentastomida and *Argulus* (Fig. 7).

5 CONCLUDING REMARKS

Arthropod relationships and especially the relationships within the clade Tetraconata remain contentious in many respects, but application of the most strict taxonomic sampling and inclusion of evidence from external morphology, internal anatomy, mitochondrial gene order data and DNA sequence data under a wide range of parameter sets allows us to scientifically evaluate previous hypotheses formulated in more restricted studies. Our studies differ from those of other students of arthropod relationships in that we are not interested in presenting a single hypothesis derived from a more or less inclusive set of data. On the contrary, we are interested in exploring the variation of phylogenetic hypotheses in a sensitivity analysis framework (variation to parameter choice) because this is the only way that we will be able to trace where and how transitions between topological representations occur, regardless of data set or data partition. We cannot justify either eliminating information by applying arbitrary decisions to exclude sequence data (e.g., Nardi et al. 2003) or taxa (e.g., Regier and Schultz 2001; Nardi et al. 2003). We will continue to attempt to overcome problems of taxon and data deficiencies in expanded versions of the data sets presented in this study.

ACKNOWLEDGEMENTS

We are indebted to Stefan Koenemann for inviting this contribution and for his careful editorial work, and to Buz Wilson and Ron Jenner for their reviews of an early version of this chapter. Wolfgang Boeckeler kindly supplied tissue samples of the pentastomid species used in this study, Mikael Agolin advised on pentastomid morphology, and Jon Mallatt kindly provided the 18S and 28S rRNA sequence data for the pentastomid as well as important criticism on a previous version of this chapter. Kelly DeMeo assisted in generating novel sequence data presented in this article. This material is based upon work supported by the NASA Fundamental Biology Program (Award NCC2-1169) to WCW and GG, and through a Lerner Gray Research Fellowship from the AMNH to SR.

APPENDIX 1

Table 1. Taxonomy of the non-crustacean terminals employed in the analyses. When single species are used per line, that species is coded; if more than one species is listed per genus, a generic groundplan is implied. When more that one genus is listed, the higher taxon name underlined is used in the analyses.

Phylum			Family	Species
Onychophora			Peripatopsidae	*Euperipatoides leuckartii*, *E. rowelli*
			Peripatidae	*Peripatus* sp., *Oroperipatus corradei*
Tardigrada	Eutardigrada		Milnesiidae	*Milnesium tardigradum*
			Hypsibiidae	*Hypsibius dujardini*
Phylum?	Pentastomida		Raillietiellidae	*Raillietiella* sp.
Arthropoda	Pycnogonida		Endeidae	*Endeis laevis*, *E. spinosa*, *E. flaccida*
			Colossendeidae	*Colossendeis* sp.
			Ammotheidae	*Achelia echinata*, *Ammothella* sp., *Tanystylum* sp.
	Chelicerata	Xiphosura	Limulidae	*Limulus polyphemus*
			Limulidae	*Carcinoscorpius rotundicauda*
		Scorpiones	Buthidae	*Androctonus australis*, *Lychas marmoreus*, *L. mucronatus*, *Centruroides exilicauda*
			Scorpionidae	*Heterometrus spinifer*, *Pandinus imperator*
		Uropygi	Thelyphonidae	*Mastigoproctus giganteus*
		Araneae	Mygalomorphae	*Atrax* sp., *Aphonopelma* sp.
		Opiliones	Sironidae	*Siro acaroides*, *S. valleorum*
			Nipponopsalididae	*Nipponopsalis abei*
			Triaenonychidae	*Equitius doriae*
		Acari	Opilioacaridae	*Opilioacarus texanus*
	Myriapoda	Chilopoda	Scutigeridae	*Scutigera coleoptrata*
			Lithobiidae	*Lithobius obscurus*, *L. variegatus*, *L. forficatus*
			Craterostigmidae	*Craterostigmus tasmanianus*
			Scolopendridae	*Cormocephalus monteithi*, *Scolopendra polymorpha*, *Ethmostigmus rubripes*
			Mecistocephalidae	*Mecistocephalus* sp., *M. tahitiensis*
			Geophilidae	*Ribautia* sp., *Pachymerium ferrugineum*

Table 1 continued.

Phylum		Family	Species
	Symphyla	Scutigerellidae	*Hanseniella* sp.
		Scutigerellidae	*Scutigerella* sp.
	Pauropoda	Pauropodidae	pauropodine spec.
	Diplopoda	Polyxenidae	*Polyxenus fasciculatus*, *Unixenus* sp.
		Sphaerotheriidae	*Epicyliosoma* sp., sphaerotheriid spec.
		Blaniulidae	*Proteroiulus fuscus*
		Spirobolidae	*Narceus americanus*
Hexapoda	Protura	Acerentomidae	*Acerentulus traegardhi*, *Nipponentomon* sp., *Acerentomon* sp.
	Diplura	Campodeidae	*Campodea tillyardi*, *Eumesocampa frigilis*
		Japygidae	*Heterojapyx* sp., *Metajapyx* sp.
	Collembola	Tomoceridae	*Tomocerus* sp., *T. minor*
		Katiannidae	*Sminthurinus bimaculatus*
	Archaeognatha	Meinertellidae	*Allomachilis froggatti*, *Machiloides* sp.
		Machilidae	*Dilta littoralis*, *Pedetontus saltator*, petrobiine spec.
	Zygentoma	Lepidotrichidae	*Tricholepidion gertschi*
		Lepismatidae	*Thermobia domestica*, *Ctenolepisma lineata*
	Ephemeroptera	Baetidae	*Callibaetis ferrugineus ferrugineus*
	Blattodea	Blattidae	*Periplaneta americana*
	Orthoptera	Acrididae	*Locusta migratoria*
	Diptera	Drosophilidae	*Drosophila melanogaster*
Crustacea			
Branchiopoda	Anostraca	Artemiidae	*Artemia salina*, *A. franciscana*
		Thamnocephalidae	*Branchinella* sp., *B. occidentalis*
	Notostraca	Triopsidae	*Triops cancriformis*, *T. longicaudus*, *T. australiensis*
	Laevicaudata	Lynceidae	*Lynceus gracilicornis*, *L. tatei*, *L. brachyurus*
	Spinicaudata	Limnadiidae	*Limnadia lenticularis*
		Limnadiidae	*Limnadopsis birchii*
	Cladocera	Leptodoridae	*Leptodora kindtii*
		Daphniidae	*Daphnia* sp., *D. pulex*, *D. galeata*

Table 1 continued.

Phylum			Family	Species
	Malacostraca	Leptostraca	Paranebaliidae	*Paranebalia longipes*
			Nebaliidae	*Nebalia* sp., *N. longicornis*, *N. hessleri*
		Stomatopoda		*Kempina mikado*, *Squilla empusa*, *Gonodactylus smithii*
		Anaspidacea	Anaspididae	*Anaspides tasmaniae*
		Reptantia	Nephropidae	*Homarus americanus*
		Euphausiacea	Euphausiidae	*Meganyctiphanes norvegica*
		Amphipoda	Talitridae	*Orchestia cavimana*
		Isopoda, Oniscidea		*Armadillidium vulgare*, *Porcellio scaber*
	Maxillopoda	Copepoda, Calanoida		*Calanus finmarchicus*, *C. pacificus*, *C. marshallae*, *Eurytemora affinis*
		Branchiura	Argulidae	*Argulus* sp., *A. nobilis*
		Cirripedia	Balanidae	*Balanus balanus*
			Balanidae	*Semibalanus balanoides*
		Ostracoda	Cypridopsidae	*Cypridopsis* sp., *C. vidua*
	Cephalocarida		Hutchinsoniellidae	*Hutchinsoniella macracantha*
	Remipedia		Speleonectidae	*Speleonectes gironensis*, *S. tulumensis*, *Lasionectes exleyi*

Table 2. GenBank accession numbers for the non-crustacean terminals utilized in the analyses. Loci abbreviations are: 18S (18S rRNA), 28S (28S rRNA), H3 (histone H3), U2 (sn U2 rRNA), EF1 (elongation factor-1alpha), POL (RNA polymerase II), EF2 (elongation factor 2), COI (cytochrome c oxidase subunit I), 16S (16S rRNA).

	18S	28S	H3	U2	EF1	POL	EF2	COI	16S
Euperipatoides	U49910	-	AF110849	AF110880	AF137394	AF139015	-	U62426	-
Peripatidae	AY210837	AY210836	AF110848	AF110879	AF137395	AF139017	AF240835	U62429	-
Milnesium	U49909	AY210826	-	-	AF063419	AH010426	AF240833	-	-
Hypsibius	Z93337	-	CD449232	-	CK325900	CK325867	-	CK326084	-
Raillietiella	AY744887	AY744894	-	-	-	-	-	-	-
Endeis	AF005441	AF005462	-	-	AF063409	AH010424	AF240819	AF259656	AF448567
Colossendeis	AF005440	AF005461	-	-	AF063406	AF138974	-	AF259659	-
Ammotheidae	AF005438	AF005459	AF110874	AF110903	AF063417	AF139013-4	AF240831	AF259657	AF370854
Limulus	U91490	U91492	AF370813	AF110902	U90051	U90037	AF240821	AF370827	U09397
Carcinoscorpius	U91491	U91493	AF370814	AF370820	AF063407	AF138975	-	AF370828	U09396
Buthidae	X77908	AF124955	AF110876	AF110905	AF240840	AF240988	-	AF370829	AF370855
Scorpionidae	AY210831	AY156529	-	-	AF240847	AH010441	-	AY156574	AY156559
Mastigoproctus	AF005446	AF062989	-	-	U90052	U90038	AF240823	AF370830	AF370856
Mygalomorphae	AF370784	X90464	AF110877	AF110906	U90045	U90035	-	AF370831	AF370857
Siro	AY639490	AY639490	AY639458	-	AF240855	AH010449	-	AY639678	AY639551
Nipponopsalis	AF124948	AF124975	-	-	AF137391	AF138993-5	AF240824	-	-
Equitius	U37003	U91503	AF110875	AF110904	AF240867	AH010461	-	AY744908	-
Opilioacarus	AF124935	AF124963	-	-	AF240849	AH010443	-	-	-
Scutigera	AF173238	AF173269	AY744902	-	AH009886	AF240951	-	AF370834	AF370859

Table 2 continued.

	18S	28S	H3	U2	EF1	POL	EF2	COI	16S
Lithobius	AF334271	AF334292	AF110853	-	AF240799	AH010406	-	AF334311	AF334333
Craterostigmus	AF000774	AF000781	AF110850	AF110882	AF240793	AH010400	-	AF370835	AF370860
Scolopendridae	AF173249	AF173280	AF110855	AF110885	AF137393	AF139006	AF240828	AF370836	AF370861
Mecistocephalus	AF173254	AF173285	AF110852	-	-	-	-	AF370837	AF370862
Geophilidae	AF173263	AF370803	AF110851	AF110883	AF240807	AF240949	-	AF370838	AF370863
Hanseniella	AY210823	AF173268	AF110856	AF110886	U90049	AF138982	-	AF370839	AF370864
Scutigerella	AF007106	AF005464	-	-	AF137392	AF139003	AF240827	-	-
Pauropodinae	AF005451	AF005466	AF110857	AF110887	-	-	-	-	-
Polyxenidae	AF173235	AF173267	AF110859	AF110889	U90055	AF139001-2	AF240826	AF370840	-
Sphaerotheriidae	AF370785	-	AF110858	AF110888	AF24808	AH010416	-	AF370841	AF370865
Proteroiulus	AF173236	AF370804	-	-	AF063415	AF139000-240942	AF370842	AF370866	-
Narceus	AF370786-7	AF370805	-	-	U90053	U90039-AF240927	AF370843	AF370867	-
Acerentomidae	AF173233	AF005469	AF110861	-	AH009876	-	-	-	-
Campodeidae	AF173234	AF005471	AF110860	-	AF137388	AF138978	AF240818	AF370844	AF370868
Japygidae	AY555524	AY555537	AY555567	-	AF137389	AF138987-8	-	AF370845	AF370869
Tomocerus	AY555516	AY555530	AY555562	-	U90059	AF139011-2	AF240830	-	-
Sminthurinus	AY555522	AY555536	AY555566	-	-	-	-	AY555545	AY555555
Meinertellidae	AF370788	AF370806	AF110864	AF110893	AF137390	AF138990-2	AF240822	AF370846	AF370870
Machilidae	AF005457	AF005473	AF110865	-	U90056	U90041	-	AF370847	AF370871
Tricholepidion	AF370789	AF370807	AF110863	AF110892	-	-	-	AY191994	AY191994
Lepismatidae	AF370790	AF370808	AY555568	-	AF063405	AF138973	-	AF370848	AF370872
Callibaetis	AF370791	-	AF370815	AF370821	-	-	-	AF370849	AF370873
Periplaneta	AF370792	AF321248	AF370816	AF370822	U90054	U90040	-	AF370850	AF262620

Table 2 continued.

	18S	28S	H3	U2	EF1	POL	EF2	COI	16S
Locusta	AF370793	AF370809	AF370817	AF370823	AY077627	-	-	X80245	X80245
Drosophila	M21017	M21017	X81207	X04256	X06869	M27431	X15805	M57910	X53506
Artemia	X01723	X01723	-	-	X03349	U10331	AF240815	X69067	X69067
Branchinella	AY744888	AY744895	AF110871	AF110900	-	-	-	AF308964	AF527558
Triops	AF144219	AY744896	AF110870	AF110898	U90058	U90043	-	-	AF200964
Lynceus	AF144215	AY137136	-	-	AF526294	-	-	-	-
Limnadia	L81934	AY137135	-	-	AF063412	AF138989	-	-	-
Limnadopsis	AY744889	AY744897	AY744903	-	AF526290	-	-	-	-
Leptodora	AF144214	AY137130	-	-	AF526278	-	-	-	-
Daphnia	AF014011	AF532888	-	AF370824	-	-	-	AF117817	AF064189
Paranebalia	AY744891	AY744899	AY744905	-	-	-	-	-	AY744909
Nebalia	L81945	AF169699	AF110869	AF110897	AF063413	AF138996	-	-	-
Stomatopoda	AF370802	AY210842	AF110873	AF110901	-	-	-	AF205233	AF107615
Anaspides	L81948	AF169703	-	-	-	-	-	AF048821	AF133679
Homarus	AF235971	AF370812	AF370819	AF370826	-	-	-	AF370853	AF370876
Meganyctiphanes	AY744892	AY744900	AY744906	-	-	-	-	AF177191	AY744910
Orchestia	AY744893	-	AY744907	-	-	-	-	-	AY744911
Oniscidea	AJ267293	AY744901	-	-	U90046	AF138970	AF240816	AF255779	AJ419996
Calanoida	L81939	AF385466	-	-	AF063408	AF138977	-	AF259651	AF295335
Argulus	M27187	AY210804	-	-	-	-	-	-	-
Balanus	AY520628	AY520594	AY520696	-	-	-	-	-	AY520730
Semibalanus	AY520626	AY520694	AY520592	-	AF063404	AF138971-2	AF240817	AF242660	AY520728

Table 2 continued.

	18S	28S	H3	U2	EF1	POL	EF2	COI	16S
Cypridopsis	AY457057	AY455765	-	-	AF063414	AF138997-9	AF240825	-	-
Hutchinsoniella	AF370801	AF370811	AF110867	-	AF063411	AF138984	AF240820	AF370852	AF370875
Remipedia	AF370794	AF370810	AF110868	AF110895	AF063416	AF139008	AF240829	AF370851	AF370874

Table 3. Tree lengths and ILD values at 20 different parameter set combinations ranging from a gap to transversion ratio (g:v) of 1 to 16 and transversion to transition ratio (v:s) of 1 to 8. The parameter set that minimizes ILD is shown in bold font type. Abbreviations for the different partitions are as follows: 18S (18S rRNA), 28S (28S rRNA), H3 (histone H3), EF1 (elongation factor-1alpha), EF2 (elongation factor 2), POL (RNA polymerase II), U2 (sn U2 rRNA), COI (cytochrome *c* oxidase subunit I), 16S (16S rRNA), MOR (morphology), MOL (molecular: all loci analyzed simultaneously), TOT (total evidence: molecular + morphology).

g:v	v:s	18S	28S	H3	EF1	EF2	POL	U2	COI	16S	MOR	MOL	TOT	ILD
1	1	4111	4520	1484	7034	6495	6840	255	6627	4399	740	43338	44309	0.04071
1	2	5987	6649	2188	10288	9770	10283	373	10170	7063	1480	65008	66849	0.03886
1	4	**9530**	**10794**	**3502**	**16572**	**16060**	**16880**	**590**	**16963**	**12017**	**2960**	**106445**	**109957**	**0.03719**
1	8	16208	18373	6057	28988	28453	29892	1014	30220	14303	5920	184924	192594	0.06836
2	1	4468	5036	1484	7034	6588	6840	272	6707	5019	1480	45114	46898	0.04201
2	2	6625	7672	2188	10288	9873	10283	397	10300	8017	2960	67977	71521	0.04080
2	4	10798	12762	3502	16572	16267	16880	628	17172	13625	5920	112493	119146	0.04213
2	8	18339	21032	6057	28988	29012	29892	1084	30843	13541	11840	192503	207570	0.08162
4	1	4889	5637	1484	7034	6593	6840	289	6712	5432	2960	46681	50144	0.04535
4	2	7426	8819	2188	10288	9908	10283	432	10273	8810	5920	71035	77600	0.04192
4	4	12256	14920	3502	16572	16313	16880	691	17157	14240	11840	117545	130502	0.04698
4	8	20798	22791	6057	28988	29089	29892	1199	30791	13860	23680	198721	228266	0.09253
8	1	5376	6317	1484	7034	6593	6840	317	6716	5576	5920	48184	54832	0.04849
8	2	8200	9943	2188	10288	9904	10283	474	10281	9018	11840	73537	86276	0.04471
8	4	13647	16765	3502	16572	16306	16880	771	17173	14256	23680	121461	147018	0.05078
8	8	22746	24221	6057	28988	29050	29892	1359	30817	13770	47360	202788	257594	0.09058
16	1	5743	6770	1484	7034	6593	6840	352	6720	5625	11840	49253	61873	0.04642
16	2	8656	10670	2188	10288	9904	10283	533	10284	9080	23680	74926	99636	0.04085
16	4	14247	17952	3502	16572	16306	16880	872	17164	14160	47360	123756	173014	0.04623
16	8	23911	24932	6057	28988	29050	29892	1540	30817	13707	94720	204894	311186	0.08860

APPENDIX 2

Description of characters used in morphological analysis. See Edgecombe (2004) for key literature citations. New characters or those with revised character state descriptions are indicated by an asterisk after the character number.

1. Non-migratory gastrulation: (0) absent; (1) present.
2. Early cleavage: (0) total cleavage with radially oriented position of cleavage products; (1) intralecithal cleavage.
3. Blastokinesis: (0) absent; (1) open amnionic cavity; (2) closed amnionic cavity, amnioserosal fold fuses beneath the embryo.
4. Blastodermal cuticle: (0) absent; (1) present.
5. Dorsal closure of embryo: (0) definitive dorsal closure; (1) provisional dorsal closure.
6. Ectoteloblasts: (0) absent; (1) present at anterior border of blastopore.
7. Caudal papilla: (0) absent; (1) present.
8. Origin of fat body: (0) vitellophagal; (1) mesodermal.
9. Midgut developed within the yolk: (0) midgut cells enclose the yolk; (1) lumen of embryonic midgut lacking yolk globules.
10. Fate map ordering of embryonic tissues: (0) presumptive mesoderm posterior to presumptive midgut; (1) presumptive mesoderm anterior to midgut; (2) mesoderm midventral, cells sink and proliferate, midgut internalizes during cleavage; (3) mesoderm diffuse through ectoderm; (4) midgut develops from anterior / posterior rudiments at each end of midventral mesoderm band.
11. Embryological development: (0) with a growth zone giving rise to both the prosoma and opisthosoma; (1) with a growth zone giving rise to the opisthosoma.
12. *engrailed* expressed in mesoderm patterning: (0) present; (1) absent.
13. Epimorphic development: (0) absent; (1) present.
14. Nauplius larva (orthonauplius) or egg nauplius: (0) absent; (1) present.
15. Pupoid stage: (0) absent; (1) present.
16. Imaginal molt: (0) present; (1) absent.
17.* Cyclic parthenogenesis as part of the life cycle: (0) absent; (1) present.
18. Sclerotization of cuticle into hard, articulated exoskeleton: (0) absent; (1) present.
19. Cuticle calcification: (0) absent; (1) present.
20. Cilia: (0) present in several organ systems, including photoreceptor, nephridia, and genital tracts; (1) present in (at most) sperm.
21. Tendon cells with tonofilaments penetrating epidermis: (0) absent; (1) present.
22. Dorsal longitudinal ecdysial suture with forking on head: (0) absent; (1) present.
23. Transverse and antennocellar sutures on head shield: (0) absent; (1) present.
24. Resilin protein: (0) absent; (1) present.
25. Molting gland: (0) absent; (1) present.
26. Bismuth staining of Golgi complex beads: (0) not staining; (1) staining.
27. Metanephridia with sacculus with podocytes: (0) absent; (1) present.
28. Distribution of segmental glands: (0) on many segments; (1) on at most last four cephalic segments and first two post-cephalic segments; (2) on second antennal and maxillary segments only.
29. Maxillary nephridia: (0) absent in postembryonic stadia; (1) paired; (2) fused nephridia of both maxillary segments.

30. Coxal gland orifice, leg I: (0) absent; (1) present.
31. Tömösváry organ ('temporal organs' at side of head behind insertion of first antenna): (0) absent; (1) present.
32. Salivary gland reservoir: (0) absent; (1) present.
33. Malpighian tubules formed as endodermal extensions of the midgut: (0) absent; (1) present.
34. Malpighian tubules formed as ectodermal extensions of the hindgut: (0) absent; (1) single pair; (2) multiple pairs.
35. Form of ectodermal Malpighian tubules: (0) elongate; (1) papillate.
36. Neck organ: (0) absent; (1) present.
37. Hemoglobin: (0) absent; (1) present.
38. Subcutaneous hemal channels in body wall: (0) absent; (1) present.
39. Dorsal heart with segmental ostia and pericardial sinus: (0) absent; (1) present.
40. Internal valves formed by lips of ostiae: (0) absent; (1) present.
41. Circumesophageal circulatory loop with trumpet-shaped opening: (0) absent; (1) present.
42.* Aorta descendens connects heart and ventral vessel: (0) absent; (1) present.
43. Slit sensilla: (0) absent; (1) present.
44. Ganglion formation: (0) ganglia formed by invagination of the ventral organ; (1) neuroblasts.
45. Early differentiating neurons aCC, pCC, RP2, U-CQ, EL and AUN: (0) absent (1) present.
46. EC neurons: (0) Eca and Ecp only; (1) Eca, Ecp and EcI.
47.* Anterior pair of serotonergic neurons with neurites that cross to contralateral side: (0) absent; (1) present.
48.* Posterior pair of serotonergic neurons with neurites that cross to contralateral side: (0) absent; (1) present.
49.* Serotonergic somata clustered: (0) unclustered; (1) clusters of about 10 cells.
50.* Serotonergic cell group 'b' of Harzsch (2004): (0) absent; (1) present.
51.* Single median serotonergic neurons 'c' and 'd' of Harzsch (2004): (0) absent; (1) present.
52. Globuli cells: (0) confined mainly to brain, in massive clusters; (1) making up majority of neuropil and ventral layer of ventral nerve cord.
53. Corpora allata: (0) absent; (1) present.
54. Intrinsic secretory cells in protocerebral neurohemal organ: (0) absent; (1) present.
55. Enlarged epipharyngeal ganglia: (0) absent; (1) present.
56. Innervation of mouth area by anterior stomogastric nervous system: (0) absent; (1) present.
57. Ganglia of pre-esophageal brain: (0) protocerebrum; (1) protocerebrum and deutocerebrum; (2) proto-, deuto- and tritocerebra.
58. Ganglia of post-oral appendages fused into single nerve mass: (0) absent; (1) present.
59. Fan-shaped body in brain with neurons extending laterally into protocerebral lobes: (0) absent; (1) present.
60. Midline neuropil (ml1): (0) absent; (1) present.
61. Midline neuropil 2: (0) absent; (1) present.
62. Arcuate body in brain: (0) absent; (1) present.
63. Ellipsoid body in brain: (0) absent; (1) present.
64. Noduli in brain: (0) absent; (1) present.
65. Protocerebral bridge: (0) absent; (1) present.
66. Mushroom body calyces: (0) absent; (1) present.
67.* Deutocerebral olfactory lobe with glomeruli: (0) absent; (1) present.
68.* Deutocerebral olfactory-globular tract: (0) absent; (1) uncrossed; (2) with chiasma.

69.* Deutocerebrum with bipartite antennular neuropils: (0) absent; (1) present.
70. Cephalon composed of one pair of pre-oral appendages and three or more pairs of post-oral appendages: (0) absent (1) present.
71.* Cephalic tagma with four post-pedipalpal locomotory limbs: (0) fewer locomotory limbs; (1) present.
72. Prosomal shield: (0) absent; (1) present.
73. Transverse furrows on prosomal carapace corresponding to margins of segmental tergites: (0) absent; (1) present.
74. Cephalic kinesis: (0) absent; (1) present.
75.* Carapace adductor: (0) absent; (1) present.
76. Flattened head capsule, with head bent posterior to the clypeus, accommodating antennae at anterior margin of head: (0) absent; (1) present.
77. Clypeofrontal suture: (0) absent; (1) present.
78. Lateral eyes: (0) absent; (1) simple lens with cup-shaped retina; (2) stemmata; (3) compound (mostly facetted); (4) onychophoran eye.
79. Compound eyes medial margins: (0) separate; (1) medially contiguous; (2) fused.
80. Compound eye stalked, basally articulated: (0) absent (eye sessile); (1) present.
81. Compound eyes internalized early in ontogeny, shifted dorsally into a cuticular pocket: (0) absent; (1) present.
82. Ophthalmic ridges: (0) absent; (1) present.
83.* Number of corneagenous cells: (0) many; (1) two.
84.* Corneagenous cells containing pigment grains: (0) corneagenous cells lacking pigment grains; (1) corneagenous cells are primary pigment cells.
85.* Bipartite distal pigment cells with an inner pigment free portion, and an outer pigment bearing portion separated by an extracellular space: (0) distal pigment cells not bipartite; (1) distal pigment cell bipartite.
86.* Interommatidial pigment cells attached to cornea and basement membrane: (0) absent; (1) present.
87. Ommatidium with crystalline cone: (0) cone absent; (1) cone present.
88. Crystalline cone cells: (0) tetrapartite crystalline cone, lacking accessory cells; (1) cone bipartite, with two accessory cells; (2) pentapartite cone; (3) tripartite cone.
89. Reduction of processes of crystalline cone-producing cells: (0) all cells have processes that pass through clear zone and rhabdom; (1) only accessory cells have processes.
90. Distally displaced nuclei of accessory crystalline cone cells: (0) absent; (1) present.
91. Clear zone between dioptric apparatus and retina: (0) absent (apposition eye); (1) present (superposition eye).
92.* Optic chiasma between lamina and medulla: (0) absent (uncrossed axons); (1) present.
93.* Medulla divided into two layers by Cuccati bundle: (0) undivided; (1) divided.
94.* Lobula or protolobula receiving crossed axons from medulla: (0) absent; (1) present.
95.* Third optic neuropil (lobula) separated from protocerebrum: (0) protolobula contiguous with protocerebrum or absent; (1) lobula separated from protocerebrum.
96.* Fourth optic neuropil (lobula plate), receiving uncrossed axons from medulla: (0) absent; (1) present.
97. Lateral eye rhabdoms with quadratic network: (0) absent; (1) present.
98. Number of median eyes: (0) none; (1) four; (2) three; (3) two; (4) one, embryonic.
99. Inverted median eye: (0) absent; (1) present.

100. Median eyes fused to naupliar eyes: (0) absent; (1) present.
101. Type of naupliar eye: (0) inverse; (1) everse.
102. Tapetal cells in cups of naupliar eye: (0) absent; (1) present.
103. Dorsal frontal organ (malacostracan type): (0) absent; (1) present.
104. Ocular tubercle: (0) absent; (1) present.
105. Trichobothria innervated by several sensory cells: (0) absent; (1) present.
106. Basal bulb in trichobothria: (0) absent; (1) present.
107. Head/mouth orientation: (0) prognathous, mouth directed anteroventrally; (1) hypognathous, mouth directed ventrally; (2) mouth directed posteriorly.
108. Labrum: (0) absent; (1) present.
109. Fleshy labrum: (0) absent; (1) present.
110. Entognathy (overgrowth of mandibles and maxillae by cranial folds): (0) absent; (1) present.
111. Admentum differentiated lateroventrally on each side of head capsule: (0) absent; (1) present.
112. Sclerotic sternum formed by antennal to maxillulary sternites: (0) absent; (1) present.
113. Tritosternum: (0) absent; (1) present.
114. Hypopharynx: (0) absent or only median lingua; (1) complete hypopharynx of lingua and paired superlinguae.
115. Fulturae: (0) absent or limited to a hypopharyngeal suspensor; (1) present, in a groove between arthrodial membrane of maxilla and labium.
116. Posterior process of tentorium fused anteriorly with hypopharyngeal bar and transverse bar: (0) absent; (1) present.
117. Triradiate pharyngeal lumen: (0) absent; (1) present.
118.*Flexible buccal tube and stylet apparatus: (0) absent; (1) present.
119. Three-branched epistomal skeleton supporting pharyngeal dilator muscles: (0) absent; (1) present.
120. Stomothecae: (0) absent; (1) present.
121. Post-cephalic filter feeding apparatus with sterniitic food groove: (0) absent; (1) present.
122. Appendage of second (deutocerebral) head segment: (0) locomotory leg 1; (1) antenna (antennula in crustaceans); (2) chelicera / chelifore; (3) jaw.
123.*Antennal rami: (0) uniramous; (1) polyramous; (2) one ramus + scale. Remipedia scored after Boxshall (2004).
124. Antennal apical cone sensilla: (0) absent; (1) present.
125. Two lateral areas bearing club-like sensilla on terminal antennal article: (0) absent; (1) present.
126. Intrinsic muscles of antennae: (0) present; (1) absent. Remipedia scored after Boxshall (2004).
127. Scape and pedicel differentiated, with Johnson's organ: (0) absent; (1) present.
128. Antennal circulatory vessels: (0) joined; (1) separate; (2) absent.
129. Ampullo-ampullary dilator and ampullo-aortic dilator: (0) absent; (1) present.
130. Statocyst in basal segment of first antenna: (0) absent; (1) present.
131. Cheliceral segmentation: (0) three segments, the last two forming a chela; (1) two segments, subchelate.
132. Plagula ventralis: (0) absent; (1) present.
133. Cheliceral tergo-deutomerite muscle: (0) absent; (1) present.
134. Appendage on third (tritocerebral) head segment: (0) unspecialized locomotory limb; (1) second antenna; (2) absent (intercalary segment); (3) pedipalp; (4) oral papilla with slime glands and adhesive glands.
135. Antennal scale: (0) absent; (1) present.

136. Antennal naupliar protopod: (0) short; (1) long.
137.* Scorpionid-type chelate pedipalps: (0) absent; (1) present.
138. *Distal-less* expressed in mandible (or positionally equivalent limb): (0) present; (1) absent.
139. Mandible: (0) absent; (1) present.
140. Mandibular base plate: (0) absent; (1) present.
141. Telognathic mandible with musculated gnathal lobe, flexor arising dorsally on cranium: (0) absent; (1) present.
142. Pectinate lamellae: (0) absent; (1) present.
143.* Mandibular gnathal edge: (0) consisting of molar and incisor process; (1) only ellipsoid molar present; (2) number of teeth arranged in a row; (3) shovel with terminal teeth; (4) group of paired teeth and hair pad.
144. Mandibular cranial articulation: (0) absent; (1) present.
145. Ball-and-socket mandibular articulation: (0) absent; (1) present, formed between clypeal condyle and mandibular ridges.
146. Mandibular scutes: (0) absent; (1) present.
147.* Mandibular palp: (0) present; (1) absent in adults
148. 'Movable appendage' between pars incisivus and pars molaris of mandible: (0) absent; (1) present in the adults.
149. Posterior tentorial apodeme: (0) absent; (1) present as metatentorium.
150. Pre- and metatentorium fused: (0) absent; (1) present.
151. Anterior tentorial arms: (0) absent; (1) cuticular tentorium developed as ectodermal invaginations; (2) cuticular fulcro-tentorium.
152. Posterior suspension of anterior apodemes to cranial wall: (0) absent; (1) present.
153. Anterior tentorium: (0) separate, rod-like anterior tentorial apodemes; (1) anterior part of tentorial apodeme forms arched, hollow plates that approach each other mesially but remain separate; (2) anterior tentorium an unpaired roof.
154. Swinging tentorium: (0) absent; (1) present.
155. Mandibular articulation with tentorium: (0) gnathal lobe articulates with epipharyngeal bar; (1) mandible articulates with hypopharyngeal bar.
156. Suspensory bar: (0) absent; (1) present.
157. Intergnathal connective lamina: (0) present; (1) absent.
158. Mandibulo-hypopharyngeal muscle: (0) absent; (1) present.
159. Cephalic post-occipital ridge: (0) absent; (1) present.
160. Ovigers: (0) absent; (1) present.
161.* Enlarged apodemes on coxae of legs 1 and 2: (0) absent; (1) present.
162.* Walking leg II longer than adjacent legs and modified as a feeler: (0) absent; (1) present.
163. Salivary gland position: (0) ectodermal, on second maxilla; (1) mesodermal segmental organs on first maxilla.
164. Maxillary salivary gland opening: (0) pair of openings at base of second maxilla; (1) median opening in midventral groove on labium; (2) median opening on salivarium, between labium and hypopharynx.
165. First maxilla: (0) absent; (1) present.
166. First maxillary precoxa: (0) absent; (1) present.
167. Number of medially directed lobate endites on first maxilla: (0) two endites; (1) one endite.
168. First maxillary palps: (0) present (including telopodite of positionally equivalent limb in chelicerates); (1) absent.

169. First maxillary palp hypertrophied: (0) absent; (1) present.
170.*First maxillary palp an elongate cleaning organ: (0) absent; (1) present.
171. First maxilla divided into cardo, stipes, lacinia, and galea, with similar musculation and function: (0) absent; (1) present.
172. Interlocking of galea and superligua: (0) absent; (1) present.
173. First maxilla coalesced with sternal intermaxillary plate: (0) absent; (1) present, with unfused stipital and intermaxillary components; (2) mental elements of gnathochilarium consolidated.
174. Second maxillae on fifth metamere: (0) appendage developed as trunk limb; (1) well developed maxilla differentiated as mouthpart; (2) vestigial appendage; (3) appendage lacking; (4) well developed, not a mouth part.
175. Egg tooth on second maxilla: (0) absent (no embryonic egg tooth on cuticle of fifth limb-bearing metamere); (1) present.
176. Maxillary plate: (0) absent; (1) present.
177. Coxae of second maxillae medially fused: (0) absent (coxae of fifth metamere not fused); (1) present.
178. Symphylan-type labium: (0) absent; (1) present.
179. Linea ventralis: (0) absent; (1) present.
180. Divided glossae and paraglossae: (0) undivided pair of glossae and paraglossae; (1) glossae and paraglossae bilobed.
181. Rotation of labial *Anlagen*: (0) absent; (1) present.
182. Widened apical segment of labial palp: (0) absent; (1) present.
183. Collum covering posterior part of head capsule and part of segment II: (0) absent; (1) present.
184. Direct articulation between first and fourth articles of telopodite of maxilliped: (0) absent (first and fourth articles of telopodite of sixth metamere lack a common hinge); (1) present.
185. Coxosternite of maxilliped sclerotized in midline: (0) coxae separated medially, with sternite present in adult; (1) coxosternal plates meeting medially, with flexible hinge; (2) coxosternal plates meeting medially, hinge sclerotized and non-functional.
186. Coxosternite of maxilliped deeply embedded into cuticle above second trunk segment: (0) not embedded; (1) embedded.
187. Maxilliped segment with pleurite forming a girdle around coxosternite: (0) small lateral pleurite; (1) large girdling pleurite.
188. Sternal muscles truncated in maxilliped segment, not extending into head: (0) sternal muscles extended into head; (1) sternal muscles truncated.
189. Maxilliped tooth plate (anteriorly-projecting, serrate coxal endite): (0) absent; (1) present.
190. Maxilliped poison gland: (0) absent; (1) present.
191. Maxilliped distal segments fused as a tarsungulum: (0) separate tarsus and pretarsus; (1) tarsus and pretarsus fused as tarsungulum.
192. Oblique muscle layer in body wall: (0) absent; (1) present.
193. Longitudinal muscles: (0) united sternal and lateral longitudinal muscles; (1) separate sternal and lateral longitudinal muscles, with separate segmental tendons.
194. Superficial pleural muscles: (0) absent; (1) present.
195. Crossed, oblique dorsoventral muscles: (0) absent; (1) present.
196. Deep dorsoventral muscles in trunk: (0) absent; (1) present.
197. Circular body muscle: (0) present; (1) suppressed.
198. Discrete segmental cross-striated muscles attached to cuticular apodemes: (0) absent; (1) present.

199. Trunk muscles: (0) straight; (1) twisted.
200. Proventriculus: (0) absent; (1) present.
201. Lateralia and inferolateralia anteriores in the cardiac chamber: (0) absent; (1) present.
202. Unpaired superomedianum: (0) absent; (1) present.
203. Inferomedianum anterius (midventral cardiac ridge): (0) absent; (1) present.
204. Inferomedianum posterius (midventral pyloric ridge): (0) absent; (1) present.
205. Atrium between inferomediana connecting cardiac primary filter grooves with pyloric filter grooves: (0) absent; (1) present.
206. Gut caecae: (0) absent; (1) present along midgut; (2) restricted to the anterior part of midgut.
207. Proctodeal dilation: (0) posterior section of hindgut simple, lacking a dilation; (1) proctodeum having a rectal ampulla with differentiated papillae.
208.* Pyloric region with ring of flattened cells with thick intima: (0) absent; (1) present.
209. Peritrophic membrane: (0) absent; (1) present.
210. Tubular diverticula: (0) absent; (1) present.
211. Fusion of all (opisthosomal) tergites behind opercular tergite into a thoracetron: (0) absent; (1) present.
212. Opisthosoma greatly reduced, forming a slender tube emerging from between posterior-most legs, with a terminal anus: (0) absent; (1) present.
213. Lamellate respiratory organs derived from posterior wall of opisthosomal limb buds: (0) absent; (1) present.
214. Position of lamellate respiratory organs: (0) on opisthosomal segments 3-7; (1) on opisthosomal segments 4-7; (2) on opisthosomal segments 2-3.
215. Type of lamellate respiratory organs: (0) book gills; (1) book lungs.
216. Appendage on first opisthosomal segment: (0) appendage present on eighth limb-bearing metamere in post-embryonic stages; (1) appendage absent on eighth metamere.
217. Limb VII as chilaria: (0) absent; (1) present.
218.* Pectines: (0) absent; (1) present.
219. First opisthosomal segment: (0) broad; (1) narrow.
220.* Claspers as modified anterior thoracopods (applicable for taxa with phyllopodous limbs only): (0) absent; (1) one pair of claspers (at least movable finger); (2) two pairs of claspers.
221. Abdomen (limb-free somites between the terminal segment and limb-bearing trunk segments; if known posterior to expression domain of *Ubx*, *abdA* and *abdB*): (0) absent; (1) present.
222. Limb bearing trunk: (0) not divided; (1) divided into thorax and pleon (with different locomotory functions)
223. Thorax with three limb-bearing segments: (0) absent; (1) present.
224. Meso- and metathorax in mature stages bearing wings: (0) absent; (1) present.
225. Wing flexion: (0) absent; (1) present.
226. Segmentation of pleon: (0) seven segments (1) six segments.
227. Diplosegments: (0) absent; (1) present.
228. Endosternum (ventral tendons fused into prosomal endosternum): (0) absent; (1) present.
229. Dorsal endosternal suspensor of fourth post-oral segment with anterolateral carapacal insertion: (0) absent; (1) present.
230. Tergal scutes extend laterally into paratergal folds: (0) absent; (1) present.
231. Paramedian sutures: (0) absent; (1) present.
232. Intercalary sclerites: (0) absent; (1) developed as small rings; (2) developed as pre-tergite and pre-sternite.

233. Trunk heterotergy: (0) absent; (1) present.
234. Trunk sternites: (0) large sternum; (1) sternal area divided into two hemisternites by linea ventralis; (2) sternum mostly membranous, with pair of small sternites; (3) sternal plate bears Y-shaped ridge/apodeme; (4) sternites extended rearwards to form substernal laminae; (5) thoracic sternal areas reduced and partly invaginated along median line; (6) sternal plate absent.
235. Endoskeleton of trunk in each segment: (0) pair of lateral connective plates; (1) pair of sternocoxal rods (ventral apodemes); (2) complex connective endosternite; (3) mainly cuticular, composed of two intrasegmental furcal arms and intersegmental spinal process.
236. Pleural part of trunk segments: (0) pleurites absent; (1) supracoxal arches on each segment; (2) pleural part of thoracic segments II and III consisting of a single sclerite with a large pleural process; (3) pleuron in each thoracic segment composed of a single sclerite divided into anterior and posterior parts by pleural suture.
237. Procoxal and metacoxal pleurites surround coxa: (0) pleurites absent or incompletely surrounding coxa; (1) procoxa and metacoxa surround coxa.
238. Elongate coxopleurites on anal legs: (0) absent; (1) present.
239. Pleuron filled with small pleurites: (0) absent; (1) present.
240. Complete body rings: (0) absent (sternites and/or pleurites free); (1) present (sternites, pleurites and tergites fused).
241. Longitudinal muscles attach to intersegmental tendons: (0) absent; (1) present.
242. Lobopods with pads and claws: (0) absent; (1) present.
243. Limbs (mostly articulated) with intrinsic muscles: (0) absent; (1) present.
244.*Telescopic legs: (0) absent; (1) present.
245. Biramy: (0) absent; (1) present.
246. Paddle-like epipods: (0) absent; (1) present.
247. Trunk limbs with lobate endites formed by folds in limb bud: (0) absent; (1) present.
248. Coxal swing: (0) coxa mobile, promotor-remotor swing between coxa and body; (1) coxa with limited mobility or immobile, promotor-remotor swing between coxa and trochanter.
249. Coxopodite articulation: (0) arthrodial membrane; (1) pleural condyle; (2) sternal condyle; (3) sternal and pleural condyles; (4) internal plate.
250. Coxal vesicles: (0) absent; (1) present at limb base on numerous trunk segments; (2) on distal part of first abdominal segment (modified as *Ventraltubus*).
251. Styli: (0) absent; (1) present.
252.*Furcula: (0) absent; (1) present.
253. Musculi lateralis: (0) absent; (1) present.
254. Coxotrochanteral joint: (0) simple; (1) complex.
255. Trochanteronotal muscle: (0) absent; (1) present.
256. Trochanter distal joint: (0) mobile; (1) short, ring-like trochanter lacking mobility at joint with prefemur.
257. Trochanterofemoral joint of walking legs: (0) transverse bicondylar; (1) vertical bicondylar.
258. Unique trochanteral femur-twisting muscle: (0) absent; (1) present.
259. Unique femur-tibia pivot joint: (0) absent; (1) present.
260. Patella/tibia joint: (0) free; (1) fused.
261. Patellotibial joint of walking legs: (0) dorsal monocondylar; (1) simple bicondylar; (2) vertical bicondylar; (3) dorsal hinge.
262. Femoropatellar joint: (0) transverse dorsal hinge; (1) bicondylar articulation.

263. Origin of posterior transpatellar muscle: (0) arises on distodorsal surface of femur, traverses femoropatellar joint ventral to axis of rotation, receives fibers from wall of patella; (1) arises on distal process of femur, traverses femoropatellar joint dorsal to axis of rotation, does not receive fibers from patella
264. Tibiotarsus: (0) separate tibia and tarsus; (1) unjointed tibiotarsus.
265. Elastic arthrodial sclerites spanning the tibia-tarsus joints: (0) absent; (1) present.
266. Tarsus segmentation: (0) not subsegmented; (1) subsegmented.
267. Tarsal organ: (0) absent; (1) present.
268. Pretarsal depressor muscle origin: (0) on tarsus; (1) on tibia or patella.
269. Pretarsal levator muscle: (0) present; (1) absent (depressor is sole pretarsal muscle).
270. Pretarsal claws: (0) paired; (1) unpaired.
271. Pretarsal claw articulation: (0) on pretarsal base; (1) on distal tarsomere.
272. Plantulae: (0) absent; (1) present.
273. Tracheae/spiracles: (0) absent; (1) pleural spiracles; (2) spiracles at bases of walking legs, opening into tracheal pouches; (3) single pair of spiracles on head; (4) dorsal spiracle opening to tracheal lungs; (5) open-ended tracheae with spiracle on second opisthosomal segment; (6) many spiracles scattered on body; (7) pair of spiracles in the collar region; (8) four pairs of opisthosomal stigmata with irregular unprotected opening.
274. Longitudinal and transverse connections between segmental tracheal branches: (0) tracheae not connected; (1) tracheae connected.
275. Pericardial tracheal system with chiasmata: (0) dendritic tracheae; (1) long, regular pipe-like tracheae with specialized molting rings.
276. Abdominal spiracles: (0) present (pleural spiracles on posterior part of trunk); (1) absent on first abdominal segment; (2) absent on all abdominal segments.
277. Abdominal segmentation (in hexapods): (0) six segments; (1) ten segments; (2) eleven segments; (3) twelve segments.
278. Annulated caudal filament: (0) absent; (1) present.
279. Abdominal segment XI modified as cerci: (0) absent; (1) present.
280. Articulate furcal rami: (0) absent; (1) present.
281. Uropods: (0) absent; (1) present.
282. Tail fan escape reaction: (0) absent; (1) present.
283. Telson shape: (0) round; (1) laterally depressed; (2) dorsoventrally depressed.
284. One pair of dorsal telsonal setae: (0) absent; (1) present.
285. Styliform post-anal telson: (0) absent; (1) present.
286. Paired terminal spinnerets: (0) absent; (1) present.
287. Anal segment with pair of large sense calicles, each with a long sensory seta: (0) absent; (1) present.
288. Egg cluster guarded until hatching, female coiling around egg cluster: (0) absent; (1) female coils ventrally around cluster; (2) female coils dorsally around egg cluster.
289. Peripatoid and foetoid stages protected by mother: (0) absent; (1) present.
290. Female gonopod used to manipulate single eggs: (0) absent; (1) present.
291. Female abdomen with ovipositor formed by gonapophyses of segments VIII and IX: (0) absent; (1) present.
292. Gonangulum sclerite fully developed as ovipositor base, articulating with tergum IX and attached to 1st valvula/valvifer: (0) not developed; (1) fully developed.
293. Ovipositor: (0) absent; (1) present.

294. Legs of seventh trunk segment transformed into gonopods: (0) absent; (1) present.
295. Penes: (0) absent; (1) present.
296. Penis (spermatopositor): (0) absent; (1) present.
297. Penis form: (0) short, membranous, undivided; (1) long, chitinous, divided into shaft and glans.
298. Male parameres: (0) undifferentiated; (1) pair of lateral plates on segment XI; (2) pair of parameres on segment IX; (3) incorporated into phallic apparatus as sclerites.
299. Penis on abdominal segment IX: (0) absent; (1) present.
300. Male gonopore location: (0) posterior end (opisthogoneate); (1) somite 11 (sixth trunk segment); (2) somite 12 (seventh trunk segment); (3) somite 8 (first opisthosomal segment); (4) behind legs of somite 8 (second pair of trunk legs); (5) somite 13 (eighth trunk segment); (6) somite 17 (twelfth trunk segment); (7) somite 16; (8) on multiple leg bases; (9) between segments VIII and IX, more or less hidden by hind border of sternum VIII; (10) somite 19; (11) somite 9 (fourth trunk segment); (12) dorsally.
301. Female gonopore position: (0) on same somite as male; (1) two segments anterior to male; (2) six segments anterior to male; (3) seven segments anterior to male.
302.*Female gonopore parity: (0) paired; (1) median, unpaired.
303. Genital operculum divided, incorporated into pedicel: (0) absent; (1) present.
304. Genital operculum overlapping third opisthosomal sternite: (0) absent; (1) present.
305. Postgenital appendages: (0) opercular and/or lamellar; (1) poorly sclerotized or eversible; (2) absent.
306. Embryonic gonoduct origin: (0) gonoduct arising as a mesodermal coelomoduct; (1) gonoduct arising as a secondary ectodermal ingrowth; (2) gonoduct arising in association with splanchnic mesoderm.
307. Lateral testicular vesicles linked by a central, posteriorly-extended deferens duct: (0) absent; (1) present.
308. Testicular follicles with pectinate arrangement: (0) absent (elongated testicular sac or sacs); (1) several pectinate follicles present.
309. Spermatophore web produced by 'Spingriffel' structure: (0) absent; (1) present.
310. 'By-passing' foreplay, spermatophore transfer on web, 'waiting' ritual by female: (0) absent; (1) present.
311. Sperm dimorphism: (0) absent; (1) present (microsperm and macrosperm).
312. Acrosomal complex in sperm: (0) bilayered (filamentous actin perforatorium present); (1) monolayered (perforatorium absent); (2) acrosome absent.
313.*Pseudoacrosome with dorsal ribbon, granulosome, apical membrane and pseudoacrosomal granular material: (0) absent; (1) present.
314. Perforatorium bypasses nucleus: (0) absent (perforatorium penetrates nucleus); (1) present.
315. Periacrosomal material: (0) absent; (1) present.
316. Striated core in subacrosomal space: (0) absent; (1) present.
317. Centrioles in sperm: (0) proximal and distal centrioles present, not coaxial; (1) coaxial centrioles; (2) single centriole; (3) centrioles absent; (4) doublet centrioles with radial 'foot'.
318. Centriole adjunct: (0) absent; (1) present.
319. Sperm 'accessory bodies' developed from centriole: (0) absent; (1) present.
320. Cristate, non-crystalline mitochondrial derivatives in sperm: (0) absent; (1) present.
321. Three filamentous mitochondria symmetrically disposed between nucleus and axoneme: (0) absent; (1) present.
322. Connecting bands between axoneme and mitochondria: (0) absent; (1) present.

323. Axoneme parallels entire length of nucleus: (0) absent; (1) present.
324. Supernumerary axonemal tubules (peripheral singlets): (0) absent; (1) present, formed from the manchette; (2) present, formed from axonemal doublets.
325. Number of protofilaments in wall of accessory tubules: (0) 13; (1) 16.
326.*Mediodorsal peripheral doublet (doublet 1) connected to dorsal ribbon by an obliquely oriented membrane: (0) absent; (1) present.
327. Axonemal endpiece 'plume': (0) endpiece not extended; (1) endpiece extended, plume-like.
328. Sperm flagellum: (0) present; (1) absent.
329. Nucleus of sperm forms spiral ridge: (0) absent; (1) present.
330. Sperm nucleus with manchette of microtubules: (0) absent; (1) present.
331. Coiling of spermatozoa flagellum: (0) absent (filiform); (1) present.
332. Medial microtubules in spermatozoan axoneme: (0) 9 + 2; (1) 9 + 3; (2) 9 + 0; (3) 12 + 0.
333. Sperm conjugation: (0) absent; (1) present.
334. Female spermathecae formed by paired lateral pockets in mouth cavity: (0) absent; (1) present.
335. Ovary shape: (0) sac- or tube-shaped, entire; (1) divided into ovarioles; (2) ovarian network.
336.*Location of ovary germarium: (0) germarium forms elongate zone in the ventral or lateral ovarian wall; (1) germarium in the terminal part of each egg tube; (2) single, median mound-shaped germarium on the ovarian floor; (3) paired germ zones on ovarian wall; (4) median germ zone on ovarian roof.
337. Site for oocyte growth: (0) in ovarian lumen; (1) on outer surface of ovary, in hemocoel, connected by egg stalk.
338.*Ventral marsupium formed by oostegites: (0) absent; (1) present.
339. Coxal organs on last pair of legs: (0) absent; (1) present.
340. Crural glands: (0) absent; (1) present.
341. Pair of repugnatorial glands in the carapace: (0) absent; (1) present.
342. Pleural defense glands with benzoquinones: (0) absent; (1) present.
343.*Aculeus with sting / opisthosomal venom glands: (0) absent; (1) present.
344. *Labial* expression domain: (0) expressed over multiple segments; (1) expression confined to second antennal/intercalary segment.
345. *proboscipedia* expression domain: (0) collinear with *labial* and *Deformed* domains; (1) anterior boundary of main expression domain of *proboscipedia* behind anterior boundary of *Deformed*.
346. *Deformed* expression domain: (0) expressed over three or more segments; (1) expression confined to mandibular and first maxillary segments.
347. *Antennapedia* expression domain: (0) strong throughout trunk; (1) restricted from the posterior of embryo.
348. Relative position of COI and COII: (0) COI/COII; (1) COI/L2/COII.
349.*Relative position of tRNA-L: (0) lsu rRNA/L1/L2/NADH 1; (1) lsu rRNA/L1/NADH 1; (2) NADH 1/H'/lsu rRNA/L1; (3) lsu rRNA/NADH 1; (4) lsu rRNA/L1/L2/Cytb.
350.*Relative position of tRNA-R and tRNA-N: (0) R/N; (1) R/K/N; (2) N/E/R; (3) N/A/S1/R; (4) R/S1.
351.*Relative position of tRNA-C and tRNA-Y: (0) Y/C; (1) Y/Q/C; (2) C/Q; (3) Q/Y/C; (4) Q/Y/F; (5) Q/I/C.
352.*Relative position of tRNA-P and tRNA-T: (0) T/-P; (1) -P/T; (2) T between W and Cytb; (3) P between NADH 4L and NADH 1; (4) T between S2 and NADH 1.

APPENDIX 3

Morphological data matrix for the terminals studied. Special characters used: A = character state 10; B = character state 11; C = character state 12; D = [0,1]

	1	11	21	31	41	51	61
Peripatidae	??????-???	-?100?0000	00-00010-?	0100-0011?	?-0???????	?10-0?00??	?????????0
Milnesium	0?00?0-???	-?100?000?	00-0010--?	0000-0?00-	?-0???????	??0-0100??	?????????0
Hypsibius	0?00?0-???	-?100?000?	00-0010--?	0000-0?00-	?-0???????	??0-0100??	?????????0
Endeis	000???-???	-?000?0101	100?010--?	0000-0001?	?-00??????	??0?0?10??	?????????1
Colossendeis	??????-???	-??00?0101	100?010--?	0000-0001?	?-00??????	??0?0?10??	?????????1
Ammotheidae	0?0???-???	-?000?0101	100?010--?	0000-0001?	?-00??????	?00?0?100?	?10000???1
Limulus	0101?0-???	0?10000101	100?0?11-0	0000-0001?	??00??0010	00000?1101	0100000001
Carcinoscorpius	0101?0-??2	0?10000101	100?0?11-0	0000-0?01?	??00??????	??000?11??	?????????1
Buthidae	0100?0-???	0?10010101	10010?11-0	0010-0001?	??10??????	?001011101	?10000???1
Scorpionidae	0100?0-???	0?10010101	10010?11-0	0010-0001?	??10??0010	0001011101	?10000???1
Mastigoproctus	??????-???	1?10010101	100?0?11-1	0010-0001?	??1???????	?001011101	010000???1
Mygalomorphae	0100?0-??2	1?100D0101	100?0?11-1	0010-0001?	??10??????	?201011101	?????????1
Siro	??????-???	??100?0101	100?0?11-0	0000-0?01?	??1???????	??000?11??	?????????1
Nipponopsalis	??????-???	??100?0101	100?0?11-0	0000-0?01?	??1???????	??000?11??	?????????1
Equitius	??????-???	??100?0101	100?0?11-0	0000-0?01?	??1???????	??000?11??	?????????1
Opilioacarus	??????-???	??100?0101	100?0?11-0	0010-0?01?	??1???????	?????11??	?????????1
Scutigera	??01?0-???	-?000?0101	100?1?1120	1001000010	0000??????	??010120??	?????????1
Lithobius	??????-???	-?000?0101	101?1?1120	1001000010	100???0101	1001012001	?00001???1
Craterostigmus	??????-???	-?000?0101	101??21100	1001000011	000???????	??0?0?20??	?????????1
Scolopendridae	010000-100	-010000101	1001??1100	0001000011	000???????	?201012001	100001???1
Mecistocephalus	??????-???	-?100?0101	100???1100	000100?01?	?00???????	??0102?0??	?????????1
Geophilidae	??????-???	-?100?0101	100?0?1100	0001000011	000???????	??010?20??	?????????1
Hanseniella	000100-013	-?000?0101	100?0?1110	100100?01?	?000??????	??000?20??	?????????1
Scutigerella	??????-???	-?000?0101	100?0?1110	100100?01?	?00???????	??0?0?20??	?????????1
Pauropodinae	000100-013	-?001?0101	100?0?1110	100100?00-	?-00??????	??0?0?20??	?????????1
Polyxenidae	??????-??3	-?00100101	100?1?1110	100100?01?	?-0???????	??000?20??	?????????1
Sphaerotheriida	0001?0-013	-?000?0111	100???1110	100100?01?	?-00??????	??000?20??	?????????1
Proteroiulus	??????-???	-?000?0111	100???1110	0001000001?	?-0??0001	100?0?2000	000000???1
Narceus	??01??-???	-?000?0111	100???1110	000100001?	?-0???????	?00?0120??	?????????1
Acerentomidae	??????-???	-?00?0101	100???1100	100210?01?	0-0???????	??101?20??	?????????1
Tomocerus	000110-103	-?10000101	100???1110	1000-0001?	1-0?11????	?0101?1011	100000???1
Sminthurinus	00011?-103	-?10000101	100??????0	1000-0001?	?-0???????	?0101?10??	?????????1
Campodeidae	010110-103	-?10000101	110???1110	000210?01?	1-0???????	??000?10??	?????????1
Japygidae	??????-???	-?10?0101	110???1110	0000-0?01?	1-0???????	??100?10??	?????????1
Meinertellidae	??????-???	-?10000101	110???1110	000200?01?	?-0???????	??1?0?20??	?????????1
Machilidae	001110-104	-?10000101	110?1?11?0	000200001?	1-0???????	??1?0?20??	?????????1
Tricholepidion	??????-???	-??00?0101	110???1110	000200?01?	?-0???????	??????????	?????????1
Lepismatidae	012110-104	-110000101	110?1?1110	000200001?	1-01111100	00110?2011	1000001111
Callibaetis	012???-1??	-?10000101	110?1?1100	000200001?	0-0???????	??1?0?20??	?????????1
Periplaneta	?22???-1??	-?10000101	11011?1100	000200?01?	0-0???1100	0011012011	1011111111
Locusta	012?10-104	-?10010101	110111?1100	000200001?	0-01111100	0011012011	1011111111
Drosophila	0120?0-10?	-110010101	11011?1100	000200001?	0-0111????	?011012011	1011111111
Remipedia	??????????	-?????010?	100???1210	0000-0?01?	?20???????	??0?012012?	?000?01211
Hutchinsoniella	??????????	-?000?0101	100???1210	0000-0?01?	0-0???1100	000?0110??	?0000????1
Artemia	0000?00??1	-1010?0101	10-?0?1210	0000-1101?	0-0??21100	0?00010??	?????0001
Branchinella	??????????	-?010?0101	10-???1210	0000-1101?	0-0???????	??0?0?10??	?????????1
Triops	0?????0???	-1010?0101	100?0?1210	0000-1101?	0-011?1100	0?0?0110??	?????????1
Lynceus	??????0???	-?01010101	100?0?1210	0000-1?01?	0-0???????	??0?0?0??	?????????1
Limnadia	??????0???	-?01010101	100?0?1210	0000-1101?	0-0???????	??0?0?10??	?????????1
Limnadopsis	??????0???	-?01010101	100???1210	0000-??01?	0-0???????	??0?0?0??	?????????1
Daphnia	0000?00???	-?000?1101	100?0?1210	0000-1101?	0-0???????	??0?0?10??	?????????1
Leptodora	0100?00???	-?000?1101	100?0?1210	0000-1?01?	0-01??????	??0?0?0??	?????????1
Calanoida	0000?0??1	-?010?0101	1001?1220	0000-1101?	?-0???????	??0?0120??	?????0001
Balanus	0000?00??1	-?01000111	10-???1210	0000-0100-	--0??1100	-0?0120??	?????????1
Semibalanus	0000?00??1	-?01000111	10-???1210	0000-0100-	--0??1100	0?0?0120??	?????????1
Cypridopsis	?0???????	-?01010101	100???1210	0000-??00-	0-0???????	??0?0?0??	?????????1
Argulus	??????????	-?000?0101	100???1210	0000-??01?	0-0???????	??0?0?20??	?????????1
Pentastomida	?001??????	-?000?0001	10-???0--0	0000-0?00-	--0???????	??????????	?????????0
Nebalia	0100?11??1	-?0?0?0101	100?1?1210	0000-1?01?	0-0??1000	0?0?0?20??	????0???1
Paranebalia	??????????	-?0?0?0101	100???1210	0000-??01?	??0???????	????0?????	?????????1

The position of crustaceans within the Arthropoda

	1	11	21	31	41	51	61
Stomatopoda	0100?11??1	-?01000101	100?1?1210	0000-1001?	000???????	??000120??	?????????1
Anaspides	0000?11?01	-?010?0101	100?1?1210	0000-1?01?	010???1000	000?0?201?	?0000-???1
Orchestia	0000?01?0?	-100000101	100?1?1200	0000-1?01?	0-01??????	??0?0?20??	?????????1
Oniscidea	0100?10?0?	-100000111	100?111210	0000-0001?	0-01101000	0000012011	10001-1211
Meganyctiphanes	0000?11?0?	-?010?0101	100?1?1200	0000-1?01?	010???????	??0?0?20??	?????????1
Homarus	0100?11?01	-101000111	100?111200	0000-1001?	0101101000	000?012011	10000-1211

	71	81	91	101	111	121	131
Peripatidae	------?4--	00--?-----	-------0--	---00-?0-0	--??-?10--	03--?-----	---4----0-
Milnesium	------?0--	-0--?-----	-------0--	---00-?0-0	--??-?11--	00--?-----	---0----0-
Hypsibius	------?0--	-0--?-----	-------0--	---00-?0-0	--??-?11--	00--?-----	---0----0-
Endeis	101---?0--	-0--?-----	-------100	---10-?0-0	-???-010--	02--------	000---0-0-
Colossendeis	101---?0--	-0--?-----	-------100	---10-?0-0	-???-010--	0---------	-003--0-0-
Ammotheidae	10?---?0--	-0--?-----	-------100	---10-?0-0	-???-010--	02--------	0003--0-0-
Limulus	110---?300	010??00---	?100-?0300	--?00-2100	-00?-00000	02--------	000---000-
Carcinoscorpius	110---?300	01?????---	?100-?0300	--?00-2100	-00?-00000	02--------	00?0--0?0-
Buthidae	111---?1--	000??-0--	?-----0310	--?0100100	-00?-00011	02--------	0013--1?0-
Scorpionidae	111---?1--	000??-0--	?-----0310	--?0100100	-00?-00011	02--------	0013--1?0-
Mastigoproctus	110---?1--	000??-?--	?-----1310	--?0100100	-01?-00000	02--------	1103--0?0-
Mygalomorphae	110---?1--	000??-0--	?-----1310	--?0100100	-01?-00000	02--------	1103--000-
Siro	11?---?0--	-0--?-----	-------0--	--?00-0100	-00?-00011	02--------	0013--0?0-
Nipponopsalis	111---?0--	-0--?-----	-------310	--?10-0100	-00?-000?1	02--------	0013--0?0-
Equitius	110---?0--	-0--?-----	-------310	--?10-0100	-00?-00011	02--------	0013--0?0-
Opilioacarus	11?---?1--	00??-?---	?-----10--	--?00-0100	-00?-000?0	02--------	00?3--0?0-
Scutigera	0--0-0?300	0001?110??	?000-000--	--000-0100	00?00100--	01000100-0	---2---?10
Lithobius	0--0-102--	000??-0--	?-----00--	--000-0100	00?00100--	01000000-0	---2---?10
Craterostigmus	0--0-102--	00?-?-?---	???????0--	--000-0100	00?00100--	0100000?0	---2---?10
Scolopendridae	0--0-102--	000??-0--	?-----00--	--000-0100	00?00100--	01000000-0	---2---?10
Mecistocephalus	0--0-100--	-0--?-----	-------0--	--000-0100	00?00000--	01001000-0	---2---?10
Geophilidae	0--0-100--	-0--?-----	-------0--	--000-0100	00?00000--	01001000-0	---2---?10
Hanseniella	0--0-000--	-0--?-----	-------0--	--00110100	00?10100--	01000000-0	---2---?11
Scutigerella	0--0-000--	-0--?-----	-------0--	--00110100	00?10100--	01000000-0	---2---?11
Pauropodinae	0--0-0?0--	-0--?-----	-------0--	--00110100	00?10100--	01000002-0	---2---?10
Polyxenidae	0--0-002--	000??10??	?-----00--	--00110100	00?0100--	0101000?0	---2---?11
Sphaerotheriida	0--0-002--	000??-0-??	?-----?0--	--000-1100	00?0100--	01010000-0	---2---011
Proteroiulus	0--0-002--	000??-0-??	?-----?0--	--000-0100	00?0100--	0101000?0	---2---?11
Narceus	0--0-002--	000??-0-??	?-----?0--	--000-0100	00?0100--	0101000?0	---2---?11
Acerentomidae	0--0-000--	-0--?-----	-------0--	--000-0101	00?11000--	0---------	---2---?10
Tomocerus	0--0-00300	0011??10??	?0????0100	--?00-?101	00?11000--	01000002-0	---2---110
Sminthurinus	0--0-00300	0011??10??	??????0100	--?00-1101	00?11000--	01000002-0	---2---?10
Campodeidae	0--0-000--	-0--?-----	-------?0--	--000-0101	10?11000--	01000000-0	---2---?10
Japygidae	0--0-000--	-0--?-----	-------?0--	--000-0101	10?11000--	01000000-0	---2---?10
Meinertellidae	0--0-00310	0011??1???	?111100?200	--?00-1100	00?11000--	0100011100	---2---?10
Machilidae	0--0-00310	0011??1000	0111000200	--000-1100	00?11000--	0100011100	---2---?10
Tricholepidion	0--0-01300	0011??10??	?1????0200	--000-1100	00?00000--	0100011??0	---2---?10
Lepismatidae	0--0-01300	0011??10??	011??0200	--000-1100	00?00000--	0100011100	---2---110
Callibaetis	0--0-01300	0011??10??	?1????0200	--000-1100	00?10000--	01000111-0	---2---?10
Periplaneta	0--0-00300	0011??10??	?1111?0300	--000-1100	00?00000--	0100011110	---2---?10
Locusta	0--0-01300	0011??10??	0111??0200	--000-1100	00?00000--	0100011110	---2---110
Drosophila	0--0-01300	0011??10??	0111110200	--000-1100	00?00000--	0100011110	---2---10-
Remipedia	0--0-0?0--	-0--------	-------0--	--?00-2110	01??-000--	0100000??0	---10?-?10
Hutchinsoniella	0--0-0?0--	-0--------	-------0--	--000-2110	01??-000--	0100000?-0	---100-?10
Artemia	0--0-0?301	00100?1000	00000?0201	01000-2110	01??-000--	1100000?-0	---101-010
Branchinella	0--0-0?301	00???????	0?????0201	01000-2110	01??-000--	1100000?-0	---101-?10
Triops	0--000?300	10100?1000	0000000101	01000-2110	01??-000--	1100000?-0	---101-?10
Lynceus	0--010?300	10??0?1000	0000???101	01000-2110	????-00?--	1100000?-0	---101-?10
Limnadia	0--010?320	10100?1200	00000??101	01000-2110	01??-000--	1100000?-0	---101-?10
Limnadopsis	0--010?320	10??0?1200	00000??101	01000-2110	????-00?--	1100000?-0	---101-?10
Daphnia	0--010?320	10100?1200	00000?0101	0?000-2110	01??-000--	1100000?-0	---10--?10
Leptodora	0--000?320	10??0?1200	00000?0201	??000-2110	????-00?--	0100000?-0	---101-?10
Calanoida	0--0-0?0--	-0--------	-------201	01000-2110	01??-000--	0100000?-0	---10?-?10
Balanus	0--0?0?300	00??101300	0?-----201	01000-?110	????-000--	01--------	-----0-?10
Semibalanus	0--0?0?300	00??????1??	??-----201	01000-?110	????-000--	01--------	-----0-?10
Cypridopsis	0--010?0--	-0--------	-------201	0?000-?110	????-000--	0100000?-0	---10?-?10
Argulus	0--0-0?300	00??101000	00??0?0201	0?000-?110	????-000--	0100000?-0	---10?-?10

	71	81	91	101	111	121	131
Pentastomida	0--0-0?0--	----?-----	-------0--	--000-?0-0	????-000--	0?--------	---?----0-
Nebalia	0--110?301	00100?1000	01010100--	--000-2110	?1??-000--	01200000-0	---10--?10
Paranebalia	0--110?301	00????????	????????0--	--?00-2110	????-000--	0120000?-0	---10--?10
Stomatopoda	0--100?301	0010011000	01011?0201	10100-2110	?1??-000--	0110000?-1	---11--?10
Anaspides	0--0-0?301	0010011111	11011?0401	10100-2110	01??-000--	01100000-1	---11--?10
Orchestia	0--0-0?300	0010001111	01??1?00--	--000-2110	0???-000--	0100000?-0	---10--010
Oniscidea	0--0-0?300	00100111-1	01011100--	--000-2110	01??-000--	01000000-0	---10--010
Meganyctiphanes	0--000?301	0010011111	11??1?0201	10000-2110	????-000--	0110000?-0	---11?-?10
Homarus	0--000?301	0010011000	1101110201	10100-2110	?1??-000--	01100000-1	---11--010

	141	151	161	171	181	191	201
Peripatidae	--------0-	0-----?--0	-0?-0-----	----------	----------	-1-?--000-	-----00?10
Milnesium	--------0-	0-----?--0	-0?-0-----	----------	----------	-0-?0?110-	-----00?00
Hypsibius	--------0-	0-----?--0	-0?-0-----	----------	----------	-0-?0?110-	-----00?00
Endeis	--------0-	0-----?--1	00?-0-?000	--00000---	---0---?00	-0??-?1100	-----10?01
Colossendeis	--------0-	0-----?--1	00?-0-?000	--00000---	---0---?00	-0??-?1100	-----10?01
Ammotheidae	--------0-	0-----?--1	00?-0-?000	--00000---	---0---?00	-0??-?1100	-----10?01
Limulus	--------0-	0-----?--0	00?-0-?000	--00000---	---0---?00	-0??1?1100	-----10?01
Carcinoscorpius	--------0-	0-----?--0	00?-0-?000	--00000---	---0---?00	-0??1?1100	-----10?01
Buthidae	--------0-	0-----?--0	10?-0-?000	--00000---	---0---?00	-0??0?1100	-----10?01
Scorpionidae	--------0-	0-----?--0	10?-0-?000	--00000---	---0---?00	-0??0?1100	-----10?01
Mastigoproctus	--------0-	0-----?--0	00?-0-?000	--00000---	---0---?00	-0??0?1100	-----10?01
Mygalomorphae	--------0-	0-----?--0	00?-0-?000	--00000---	---0---?00	-0??0?1100	-----10?01
Siro	--------0-	0-----?--0	00?-0-?000	--00000---	---0---?00	-0??0?1100	-----10?01
Nipponopsalis	--------0-	0-----?--0	01?-0-?000	--00?00---	---0---?00	-0??0?1100	-----10?01
Equitius	--------0-	0-----?--0	01?-0-?000	--00?00---	---0---?00	-0??0?1100	-----10?01
Opilioacarus	--------0-	0-----?--0	00?-0-?000	--00000---	---0---?00	-0??0?1100	-----1???1
Scutigera	110--1100-	10000100-0	00001-?000	0-011000--	-?-0000001	0000101100	-----00?10
Lithobius	114--1100-	10010100-0	00001-?000	0-011010--	-?-0100001	1000101100	-----00010
Craterostigmus	114--1100-	10010100-0	00001-?000	0-01?010--	-?-0211111	1011111100	-----00??0
Scolopendridae	114--1100-	10010100-0	00001-?000	0-011010--	-?-1211111	1011111100	-----00010
Mecistocephalus	01?--0100-	10000000-0	00001-?000	0-01?010--	-?-1211101	1011111100	-----00?10
Geophilidae	01?--0100-	10000000-0	00001-?000	0-011010--	-?-1211101	1011111100	-----00?10
Hanseniella	100--0110-	1001?000-0	001-1-?1--	0-0101010-	---00-0000	00??1?1100	-----00?10
Scutigerella	100--0110-	10010000-0	001-1-?1--	0-0101010-	---00-0000	0000101100	-----00?10
Pauropodinae	00?--0100-	100110?0-0	001-1-?1--	0-13-0----	--00--0000	00001?1100	-----00??0
Polyxenidae	110--0100-	10011000-0	001-1-?1--	0-13-0----	--00--0?00	00001?1100	-----00?10
Sphaerotheriida	110--0100-	1001?000-0	001-1-?1--	0-23-0----	--00--0?00	000???1100	-----00?10
Proteroiulus	110--0100-	10011000-0	001-1-?1--	0-23-0----	--10--0?00	000???1100	-----00??0
Narceus	110--0100-	10011000-0	001-1-?1--	0-23-0----	--10--0?00	000???1100	-----00??0
Acerentomidae	00?0-0100-	2000?10000	00011-?000	1001?10010	?0-0--0000	00????1100	-----00000
Tomocerus	0000-0100-	1000?10000	00011-?000	100101001?	0--0--0000	001?1?1100	-----00110
Sminthurinus	00?0-0100-	1000?10000	00011-?000	100101001?	?--0--0000	001?1?1100	-----0?1?0
Campodeidae	0030-0100-	0---?00000	00011-?000	110101000?	10-0--0?00	00-?1?1100	-----01010
Japygidae	0030-0100-	0---?00000	00011-?000	1101?10000	10-0--0?00	000??1?1100	-----0?010
Meinertellidae	0000-01010	1110?00000	00021-?010	1001?10001	?0-0--0?00	000?1?1110	-----21?10
Machilidae	0000-01010	1110?00000	00021-?010	1001010001	00-0--0?00	000?1?1110	-----21010
Tricholepidion	0001001010	1120?00010	00021-?000	1001?10000	?1-0--0?00	000?1?1100	-----???10
Lepismatidae	0001001010	1120?01110	00021-?000	1001010000	01-0--0?00	000?1?1101	????21010
Callibaetis	00?1001011	1120?01110	00021-?000	1001010000	?0-0--0?00	000?1?1101	????21010
Periplaneta	00?1101011	1120?01110	00021-?000	1001010000	?0-0--0?00	000?1?1101	????21010
Locusta	00?1101011	1120?01110	00021-?000	1001010000	00-0--0?00	000?1?1101	????21010
Drosophila	--------11	1120?-1-10	00021-?000	1001010000	00-0--0?00	000?1?1100	-----01010
Remipedia	000--0100-	0----0??-0	00?-110000	0-01?000--	-?-0---?00	-0??1?1100	-----10??0
Hutchinsoniella	000--0100-	0----0??-0	00?-100000	0-000000--	-?-0---?00	-0??1?1100	-----20??0
Artemia	001--0100-	0----00?-0	00?-10-1--	0-020000--	-?-0---?00	-0??1?1100	-----20?10
Branchinella	001--0100-	0----00?-0	00?-10-1--	0-020000--	-?-0---?00	-0??1?1100	-----20?10
Triops	002--0100-	0----00?-0	00?-10-1--	0-020000--	-?-0---?00	-0??1?1100	-----20?10
Lynceus	002--0100-	0----0??-0	00?-??-1--	0-020000--	-?-0---?00	-0??1?1100	-----20??0
Limnadia	001--0100-	0----00?-0	00?-1?-1--	0-020000--	-?-0---?00	-0??1?1100	-----20?10
Limnadopsis	001--0100-	0----00?-0	00?-??-1--	0-020000--	-?-0---?00	-0??1?1100	-----20??0
Daphnia	001--0100-	0----00?-0	00?-10-1--	0-020000--	-?-0---?00	-0??1?1100	-----20?10
Leptodora	00?--0100-	0----00?-0	00?-10-1--	0-020000--	-?-0---?00	-0??1?1100	-----20??0
Calanoida	00?--0000-	0----0??-0	00?-110000	0-010000--	-?-0---?00	-0??1?1100	-----00??0
Balanus	00?--0100-	0----0??-0	00?-1?-1--	0-010000--	-?-0---?00	-0????1100	-----20?10

The position of crustaceans within the Arthropoda   345

	141	151	161	171	181	191	201
Semibalanus	00?--0100-	0----0??-0	00?-1?-1--	0-010000--	-?-0---?00	-0????1100	-----20?10
Cypridopsis	00?--00?0-	0----0??-0	00?-1??000	0-010000--	-?-0---?00	-0????1100	-----20??0
Argulus	00?--0100-	0----0??-0	00?-1--000	0-?40000--	-?-0---?00	-0????1100	-----20??0
Pentastomida	--------0-	0----??-0	0-?-0-----	----------	-------?--	-0?????0?00	-----00?00
Nebalia	000--0000-	0----00?-0	00?-101001	0-010000--	-?-0---?00	-0??1?1101	0100000??0
Paranebalia	000--0000-	0----0??-0	00?-101001	0-010000--	-?-0---?00	-0??1?1101	??????0??0
Stomatopoda	000--0000-	0----00?-0	00?-101000	0-010000--	-?-0---?00	-0??1?1101	1011120?10
Anaspides	000--0000-	0----00?-0	00?-101000	0-010000--	-?-0---?00	-0??1?1101	1110020??0
Orchestia	000--0110-	0----0??-0	00?-1011--	0-010000--	-?-0---?00	-0??1?1101	1011120??0
Oniscidea	000--0110-	0----0??-0	00?-1011--	0-020000--	-?-0---?00	-0??1?1101	1011120?00
Meganyctiphanes	000--0000-	0----0??-0	00?-101000	0-010000--	-?-0---?00	-0??1?1101	11???20??0
Homarus	000--0000-	0----0??-0	00?-101000	0-010000--	-?-0---?00	-0??1?1101	1111120?10

	211	221	231	241	251	261	271
Peripatidae	--0----0--	0-00--00--	---???----	0100000--0	00???-???-	?--??-?--?	--6----000
Milnesium	--0----0--	0-00--00-0	---???----	0011000--0	00???-???-	?--??-?--?	--0----000
Hypsibius	--0----0--	0-00--00-0	---???----	0011000--0	00???-???-	?--??-?--?	--0----000
Endeis	-10--0000-	0-00--00-0	000???----	0010000000	00?0?00???	0???01?00?	?-0----000
Colossendeis	-10--0000-	0-00--00-0	000???----	0010000000	00?0?00???	0???01?00?	?-0----000
Ammotheidae	-10--0000-	0-00--00-0	000???---0	0010000000	00?0?00???	0???01?00?	?-0----000
Limulus	101000100-	0-00--0101	000???---0	1010100000	0000?00???	000?00?001	?-0----000
Carcinoscorpius	101000100-	0-00--0101	000???---0	1010100000	0000?00???	000?00?001	?-0----000
Buthidae	001111010-	0-00--0101	000???---0	10100001-0	0000?00???	111?110100	?-0----000
Scorpionidae	001111010-	0-00--0101	000???---0	10100001-0	0000?00???	111?110100	?-0----000
Mastigoproctus	001211001-	0-00--0110	000???---0	10100001-0	0011?00???	000?011100	?-0----000
Mygalomorphae	001211001-	0-00--0110	000???---0	10100001-0	0011?00???	000?011100	?-0----000
Siro	000--1000-	0-00--0100	000???---0	10100001-0	0000?01???	211?110101	?-5----000
Nipponopsalis	000--1000-	0-00--0100	000???---0	10100001-0	0000?01???	211?110101	?-5----000
Equitius	000--1000-	0-00--0100	000???---0	10100001-0	0000?01???	211?110101	?-5----000
Opilioacarus	000--1000-	0-00--0100	000???---0	10100001-0	0000?00???	300?0?0100	?-80---000
Scutigera	0-0---00--	0-00--00-0	0010?10000	1010000020	00??01??0	??-0?1??11	0-400--000
Lithobius	0-0---00--	0-00--00-0	0010010000	1010000020	00??01??0	??-0?1??11	0-1000-000
Craterostigmus	0-0---00--	0-00--00-0	0110?10100	1010000020	00??01??0	??-0?0??11	0-1000-000
Scolopendridae	0-0---00--	0-00--00-0	1110011100	1010000020	00??01??0	??-0?1??11	0-1100-000
Mecistocephalus	0-0---00--	0-00--00-0	1200011110	1010000020	00???1??0	??-0?0??11	0-1110-000
Geophilidae	0-0---00--	0-00--00-0	1200011110	1010000020	00???1??0	??-0?1??11	0-1110-000
Hanseniella	0-0---00--	0-00--00-0	0002100000	1010000021	10??00??0	??-0?0??10	0-3----000
Scutigerella	0-0---00--	0-00--00-0	000210-0-0	1010000021	10??00??0	??-0?0??10	0-3----000
Pauropodinae	0-0---00--	0-00--00-0	000610-0-0	1010000000	00??00??0	??-0?1??11	0-0----000
Polyxenidae	0-0---00--	0-00--10-0	000?1?0000	1010000100	00??00??0	??-0?1??11	0-2----000
Sphaerotheriida	0-0---00--	0-00--10-0	00001?0000	1010000020	00??0?0	??-0?0??11	0-2----000
Proteroiulus	0-0---00--	0-00--10-0	00001?0001	1010000020	00??0?0	??-0?0??11	0-2----000
Narceus	0-0---00--	0-00--10-0	00001?0001	1010000020	00??0?0	??-0?0??11	0-2----000
Acerentomidae	0-0---00--	0-10--00-0	0003210000	1010000032	00??00?001	??-0?0??11	0-0---3000
Tomocerus	0-0---00--	0-10--00-0	0001210000	1010000042	01??00?001	??-1?0??11	0-0---0000
Sminthurinus	0-0---00--	0-10--00-0	0001210000	1010000042	01??00?001	??-1?0??11	0-70020000
Campodeidae	0-0---00--	0-10--00-0	0003210000	1010000021	10??00?111	??-0?0??10	0010021010
Japygidae	0-0---00--	0-10--00-0	0003210000	1010000021	10??00?111	??-0?0??10	0010001010
Meinertellidae	0-0---00--	0-10--00-1	0000220000	1010000011	10??10?001	??-0?1??10	1010012110
Machilidae	0-0---00--	0-10--00-1	0000220000	1010000011	10??10?001	??-0?1??10	1010012110
Tricholepidion	0-0---00--	0-10--00-1	0000210000	1010000011	10??10?001	??-0?1??10	1011002110
Lepismatidae	0-0---00--	0-10--00-1	0004210000	1010000010	10??10?001	??-0?1??10	1011002110
Callibaetis	0-0---00--	0-110--00-1	0000330000	1010000010	10??10?001	??-0?1??10	1011002110
Periplaneta	0-0---00--	0-111-00-1	0000330000	1010000010	10??10?001	??-0?1??10	1111002010
Locusta	0-0---00--	0-111-00-1	0000330000	1010000010	00??10?001	??-0?1??10	1111002010
Drosophila	0-0---00--	0-111-00-1	0005330000	1010000010	00??10?001	??-0?1??10	1011002000
Remipedia	0-0---00--	0-000--00-1	000??---0	1010100000	00???????	??-???????	?-0----001
Hutchinsoniella	0-0---00-0	1000--00-1	000???---0	1010110000	00???????	??-???????	?-0----001
Artemia	0-0---00-0	1000--00-1	000???---0	1010111000	00???????	??-???????	?-0----001
Branchinella	0-0---00-0	1000--00-1	000???---0	1010111000	00???????	??-???????	?-0----001
Triops	0-0---00-0	1000--00-1	000???---0	1010111000	00???????	??-???????	?-0----000
Lynceus	0-0---00-1	0000--00-?	000???---0	?010111000	00???????	??-???????	?-0----000
Limnadia	0-0---00-2	0000--00-?	000???---0	1010111000	00???????	??-???????	?-0----001
Limnadopsis	0-0---00-2	0000--00-?	000???---0	?010111000	00???????	??-???????	?-0----001
Daphnia	0-0---00-1	?000--00-?	000???---0	1010111000	00???????	??-???????	?-0----001

	211	221	231	241	251	261	271
Leptodora	0-0---00--	?000--00-?	000???---0	?010001000	00????????	??-???????	?-0----000
Calanoida	0-0---00--	1000--00-1	000???---0	1010100000	00????????	??-???????	?-0----001
Balanus	0-0---00--	1000--00--	000???---0	1010100000	00????????	??-???????	?-0----000
Semibalanus	0-0---00--	1000--00--	000???---0	1010100000	00????????	??-???????	?-0----000
Cypridopsis	0-0---00--	1000--00-?	000???---0	1010100000	00????????	??-???????	?-0----001
Argulus	0-0---00--	1000--00-?	000???---0	?010100000	00????????	??-???????	?-0----000
Pentastomida	0-0---00--	?000--00--	000???---0	?010000--0	00????????	??-???????	?-0----000
Nebalia	0-0---00-0	1100-000-1	000???---0	1010110000	00????????	??-???????	?-0----001
Paranebalia	0-0---00-0	1100-000-1	000???---0	?010110000	00????????	??-???????	?-0----001
Stomatopoda	0-0---00--	0100-100-1	0000??---0	1010110000	00????????	??-??????1	?-0----000
Anaspides	0-0---00--	0100-100-1	0000??---0	1010110000	00????????	??-?????01	?-0----000
Orchestia	0-0---00--	0100-100-1	0000??---0	?010110000	00????????	??-??????1	?-0----000
Oniscidea	0-0---00--	0100-100-1	0000??---0	1010110000	00????????	??-??????1	?-0----000
Meganyctiphanes	0-0---00--	0100-100-?	0000??---0	?010110000	00????????	??-??????1	?-0----000
Homarus	0-0---00--	0100-100-?	0000??---0	1010110000	00????????	??-?????01	?-0----000

	281	291	301	311	321	331	341
Peripatidae	--??00000-	0-0000-0-0	00---0000?	0?0???2000	0001?00000	0?00001001	000???????? ??
Milnesium	--??00000-	0-0000-0-0	01---?000?	000?00????	0000-0001?	000001?000	000??????? ??
Hypsibius	--??00000-	0-0000-0-0	01---?000?	000?00????	0000-0001?	000001?000	000???00? ??
Endeis	0-??00000-	0-0000-008	00---0000?	0?????????	????????0?	0?00001000	000??????? ??
Colossendeis	0-??00000-	0-0000-008	00---0000?	??????????	??????????	???00??000	000??????? ??
Ammotheidae	0-??00000-	0-0000-008	00---0000?	??????????	??????????	0?000??000	000??????? ??
Limulus	0-??10000-	0-0000-003	00000?000?	0000001000	0000-00000	0000201000	000???000 00
Carcinoscorpius	0-??10000-	0-0000-003	00000?000?	0000001000	0000-00000	0200201000	000??????? ??
Buthidae	0-??10000-	0-0000-003	0100000000	000?001000	0000-00000	0200201000	001??????? ??
Scorpionidae	0-??10000-	0-0000-003	0100000000	000?001000	0000-00000	0200201000	001?????00? ??
Mastigoproctus	0-??00000-	0-0000-003	011120000?	000?001000	0000-0-011	1100001000	000??????? ??
Mygalomorphae	0-??00000-	0-0000-003	0111100000	000?001000	0000-0-011	1100001000	0000000013 04
Siro	0-??00000-	0-10010003	01002?000?	000000?000	0000-00000	02000??000	100??????? ??
Nipponopsalis	0-??00000-	0-10011003	01002?000?	??????????	??????????	???00??000	100??????? ??
Equitius	0-??00000-	0-10011003	01002?000?	??????????	??????????	???00??000	100??????? ??
Opilioacarus	0-??00000-	0-1000-003	01001??0?	000000??0	-----?-100	--00??000	000??????? ??
Scutigera	0-??000001	0-0000-000	01---00000	1000002100	0000-01010	00000??000	000???044 00
Lithobius	0-??000001	0-0000-000	01---?0010	110-002100	0000-01010	0000000010	0000001000 50
Craterostigmus	0-??00010-	0-0000-000	01---?101?	?10-00?100	0?????1010	0??00??010	000??????? ??
Scolopendridae	0-??00011-	0-0000-000	01---01010	110-002100	0000-01010	0000000010	000??????? ??
Mecistocephalus	0-??000110	0-0000-000	01---?101?	??????????	??????????	???00??010	000??????? ??
Geophilidae	0-??000210	0-0000-000	01---?1010	010-002100	0000-01010	00000??010	000??????? ??
Hanseniella	0-??01100-	0-0000-004	01---1000?	??????????	??????????	???1020000	000??????? ??
Scutigerella	0-??01100-	0-0000-004	01---?0000	110-012000	0000-00000	0001020000	000??????? ??
Pauropodinae	0-??00000-	0-0010-004	01---1000?	020---2000	0000-01000	0000020000	000??????? ??
Polyxenidae	0-??00000-	0-0010-004	00---10000	010-003--0	-------100	--00020000	000??????? ??
Sphaerotheriida	0-??00000-	0-0000-004	00---10000	010-00??0	-------100	--00030000	000??????? ??
Proteroiulus	0-??00000-	0-0110-004	00---?000?	010-013--0	-------100	--0003?000	010??????? ??
Narceus	0-??00000-	0-0110-004	00---?0000	010-01??0	-------100	--1003?000	010???000 43
Acerentomidae	0-??00000-	0-0000-100	01---?000?	010-00?100	0000-00000	0300010000	000??????? ??
Tomocerus	0-??00000-	0-0000-000	01---20000	0000002001	-000-0-100	-000000000	000??????? ??
Sminthurinus	0-??00000-	0-0000-000	01---20000	0000002001	-000-0-100	-000000000	000??????? ??
Campodeidae	0-??00000-	0-0000-009	01---?0000	000?00?0?1	1002000000	0?00010000	000??????? ??
Japygidae	0-??00000-	0-0000-009	01---?000?	000?00?0?1	1002000000	0000110000	000??????? ??
Meinertellidae	0-??00000-	100000-010	11---?0100	??????????	??????????	???0110000	000??????? ??
Machilidae	0-??00000-	100000-210	11---?0101	000?102111	0002100000	0000110000	000??????? ??
Tricholepidion	0-??00000-	1?0000-210	11---?0101	000??02111	0002100000	0010110000	000???110 00
Lepismatidae	0-??00000-	110000-210	11---20101	000?102111	0002100000	0010110000	0001111??? ??
Callibaetis	0-??00000-	0-0000-?10	11---?010-	0?0?002001	0??2000000	0200110000	000??????? ??
Periplaneta	0-??00000-	110000-310	11---?010-	0?0?10?101	0102100000	0000110000	000??????? ??
Locusta	0-??00000-	110000-310	11---2010-	000010?101	0102100000	0000110000	000???100 00
Drosophila	0-??00000-	0-0000-310	11---2010-	010-0?0101	0002?00000	0000110000	0001111110 00
Remipedia	0-?0000??-	0-0000-00A	30---0000?	0000003--0	0000-0000?	00000??000	000???122 20
Hutchinsoniella	0-?000000-	0-0000-001	00---0000?	0000003--0	-------10?	--00010000	000???011 31
Artemia	0-0000000-	0-0000-006	01---0000?	020---2000	-------100	--00??000	000??1110 00
Branchinella	0-0000000-	0-0000-006	01---0000?	020---2000	-------100	--00??000	000??????? ??
Triops	0-0100000-	0-0000-007	00---0000?	020---2000	-------100	--00010000	000???110 00
Lynceus	0-1100000-	0-0000-007	00---0000?	020---2000	-------100	--00??000	000??????? ??
Limnadia	0-1100000-	0-0000-007	00---0000?	??????????	??????????	???0??000	000??????? ??

	281	291	301	311	321	331	341
Limnadopsis	0-1100000-	0-0000-007	00---0000?	??????????	??????????	???00??000	000??????? ??
Daphnia	0-1100000-	0-0000-00C	00---0000?	020---2000	-------100	--000?0000	000????110 00
Leptodora	0-0100000-	0-0000-00C	0?---0000?	020---2000	-------100	--000?0000	000??????? ??
Calanoida	0-0000000-	0-0000-002	00---0000?	0?0???3--0	-------100	--00010000	000??????? ??
Balanus	0-?000000-	0-0000-002	20---0000?	0?0?00?000	0010-00000	00000?0000	000????110 01
Semibalanus	0-?000000-	0-0000-002	20---0000?	0?0?00?000	0010-00000	00000?0000	000????110 01
Cypridopsis	0-?000000-	0-0000-00?	?0---0000?	0?0??????	-------110	--000?0000	000??????? ??
Argulus	0-?000000-	0-0000-00B	00---0000?	021---?000	1000-10100	-0000?1000	000????131 12
Pentastomida	0-?000000-	0-0000-00?	?1---?000?	021---2000	1000-10100	-000041000	000????131 10
Nebalia	000000000-	0-0000-005	10---0000?	020---2000	-------100	--000?0000	000??????? ??
Paranebalia	000000000-	0-0000-005	10---0000?	??????????	??????????	???00??000	000??????? ??
Stomatopoda	102000000-	0-0000-005	10---0000?	0000004?00	-------100	--000?0000	000??????? ??
Anaspides	112000000-	0-0000-005	10---0000?	000100??-0	-------100	--000?0000	000??????? ??
Orchestia	102000000-	0-0000-005	10---0000?	0?????????	??????????	???00??100	000??????? ??
Oniscidea	102000000-	0-0000-005	10---0000?	0001004000	-------100	--00010100	0001011??? ??
Meganyctiphanes	112000000-	0-0000-005	10---0000?	0?????????	??????????	???00??000	000??????? ??
Homarus	112000000-	0-0000-005	10---0000?	0000004000	-------100	--000?0000	000???1110 00

REFERENCES

Abele, L.G., Kim, W. & Felgenhauer, B.E. 1989. Molecular evidence for inclusion of the phylum Pentastomida in the Crustacea. *Mol. Biol. Evol.* 6: 685-691.

Averof, M. & Akam, M. 1995. Insect-crustacean relationships: Insights from comparative developmental and molecular studies. *Phil. Trans. R. Soc. London, B,* 347: 293-303.

Ax, P. 1999. *Das System der Metazoa II. Ein Lehrbuch der phylogenetischen Systematik.* Stuttgart: Gustav Fischer Verlag.

Bergström, J. 1979. Morphology of fossil arthropods as a guide to phylogenetic relationships. In: Gupta, A.P. (ed.), *Arthropod Phylogeny*: pp. 3-56. New York: Van Nostrand Reinhold.

Boore, J.L., Lavrov, D.V. & Brown, W.M. 1998. Gene translocation links insects and crustaceans. *Nature* 392: 667-668.

Boxshall, G.A. 2004. The evolution of arthropod limbs. *Biol. Rev.* 79: 253-300.

Boxshall, G.A. & Huys, R. 1989. New tantulocarid, *Stygiotantulus stocki*, parasitic on harpacticoid copepods, with an analysis of the phylogenetic relationships within the Maxillopoda. *J. Crustacean Biol.* 9: 126-140.

Braband, A., Richter, S., Hiesel, R. & Scholtz, G. 2002. Phylogenetic relationships within the Phyllopoda (Crustacea, Branchiopoda) based on mitochondrial and nuclear markers. *Mol. Phylogenet. Evol.* 25: 229-244.

Budd, G.E. 2002. A palaeontological solution to the arthropod head problem. *Nature* 417: 271-275.

Cisne, J.L. 1974. Trilobites and the origin of arthropods. *Science* 186: 13-18.

Dohle, W. 2001. Are the insects terrestrial crustaceans? A discussion of some new facts and arguments and the proposal of the proper name Tetraconata for the monophyletic unit Crustacea + Hexapoda. *Ann. Soc. Entomol. Fr., N.S.,* 37: 85-103.

Dove, H. & Stollewerk, A. 2003. Comparative analysis of neurogenesis in the myriapod *Glomeris marginata* (Diplopoda) suggests more similarities to chelicerates than to insects. *Development* 130: 2161-2171.

Duman-Scheel, M. & Patel, N.H. 1999. Analysis of molecular marker expression reveals neuronal homology in distantly related arthropods. *Development* 126: 2327-2334.

Edgecombe, G.D. 1998. *Arthropod Fossils and Phylogeny*. New York: Columbia Univ. Press.

Edgecombe, G.D. 2004. Morphological data, extant Myriapoda, and the myriapod stem-group. *Contrib. Zool.* 73: 207-252.

Edgecombe, G.D. & Giribet, G. 2002. Myriapod phylogeny and the relationships of Chilopoda. In: Llorente Bousquets, J.E. & Morrone, J.J. (eds.), *Biodiversidad, taxonomía y biogeografía de artrópodos de México; Hacia una síntesis de su conocimiento*: pp. 143-168. Mexico D.F.: Prensas de Ciencias, Univ. Nac. Autónoma México.

Edgecombe, G.D., Richter, S. & Wilson, G.D.F. 2003. The mandibular gnathal edges: Homologous structures throughout Mandibulata? *African Invertebrates* 44: 115-135.

Edgecombe, G.D., Wilson, G.D.F., Colgan, D.J., Gray, M.R. & Cassis, G. 2000. Arthropod cladistics: Combined analysis of Histone H3 and U2 snRNA sequences and morphology. *Cladistics* 16: 155-203.

Fanenbruck, M., Harzsch, S. & Wägele, J.W. 2004. The brain of the Remipedia (Crustacea) and an alternative hypothesis on their phylogenetic relationships. *Proc. Nat. Acad. Sci. USA* 101: 3868-3873.

Farris, J.S., Albert, V.A., Källersjö, M., Lipscomb, D. & Kluge, A.G. 1996. Parsimony jackknifing outperforms neighbor-joining. *Cladistics* 12: 99-124.

Friedrich, M. & Tautz, D. 1995. Ribosomal DNA phylogeny of the major extant arthropod classes and the evolution of myriapods. *Nature* 376: 165-167.

Friedrich, M. & Tautz, D. 2001. Arthropod rDNA phylogeny revisited: a consistency analysis using Monte Carlo simulation. *Ann. Soc. Entomol. Fr., N.S.*, 37: 21-40.

Giribet, G. 2001. Exploring the behavior of POY, a program for direct optimization of molecular data. *Cladistics* 17: S60-S70.

Giribet, G. 2003. Stability in phylogenetic formulations and its relationship to nodal support. *Syst. Biol.* 52: 554-564.

Giribet, G., Carranza, S., Baguñà, J., Riutort, M. & Ribera, C. 1996. First molecular evidence for the existence of a Tardigrada + Arthropoda clade. *Mol. Biol. Evol.* 13: 76-84.

Giribet, G., Distel, D.L., Polz, M., Sterrer, W. & Wheeler, W.C. 2000. Triploblastic relationships with emphasis on the acoelomates and the position of Gnathostomulida, Cycliophora, Plathelminthes, and Chaetognatha: A combined approach of 18S rDNA sequences and morphology. *Syst. Biol.* 49: 539-562.

Giribet, G., Edgecombe, G.D., Carpenter, J.M., D'Haese, C.A. & Wheeler, W.C. 2004. Is Ellipura monophyletic? A combined analysis of basal hexapod relationships with emphasis on the origin of insects. *Org. Divers. Evol.* 4: 319-340.

Giribet, G., Edgecombe, G.D. & Wheeler, W.C. 2001. Arthropod phylogeny based on eight molecular loci and morphology. *Nature* 413: 157-161.

Giribet, G., Edgecombe, G.D., Wheeler, W.C. & Babbitt, C. 2002. Phylogeny and systematic position of Opiliones: a combined analysis of chelicerate relationships using morphological and molecular data. *Cladistics* 18: 5-70.

Giribet, G. & Ribera, C. 2000. A review of arthropod phylogeny: new data based on ribosomal DNA sequences and direct character optimization. *Cladistics* 16: 204-231.

Giribet, G. & Wheeler, W.C. 2001. Some unusual small-subunit ribosomal RNA sequences of metazoans. *Am. Mus. Novitates* 3337: 1-14.

Goloboff, P.A. 1999. Analyzing large data sets in reasonable times: solutions for composite optima. *Cladistics* 15: 415-428.

Harzsch, S. 2001. Neurogenesis in the crustacean ventral nerve cord: homology of neuronal stem cells in Malacostraca and Branchiopoda? *Evol. Dev.* 3: 154-169.

Harzsch, S. 2002. The phylogenetic significance of crustacean optic neuropils and chiasmata: a re-examination. *J. Comp. Neurol.* 453: 10-21.

Hessler, R.R. 1992. Reflections on the phylogenetic position of the Cephalocarida. *Acta Zool.* 73: 315-316.

Hwang, U.W., Friedrich, M., Tautz, D., Park, C.J. & Kim, W. 2001. Mitochondrial protein phylogeny joins myriapods with chelicerates. *Nature* 413: 154-157.

Ito, T. 1989. Origin of the basis in copepod limbs, with reference to remipedian and cephalocarid limbs. *J. Crust. Biol.* 9: 85-103.

Jamieson, B.G.M. & Storch, V. 1992. Further spermatological evidence for including the Pentastomida in the Crustacea. *Int. J. Parasitol.* 22: 95-108.

Jenner, R.A. 2004. Quo Vadis? *The Systematist* 23: 12-16.

Kadner, D. & Stollewerk, A. 2004. Neurogenesis in the chilopod *Lithobius forficatus* suggests more similarities to chelicerates than to insects. *Dev. Genes Evol.* 214: 367-379.

Klass, K.D. & Kristensen, N.P. 2001. The ground plan and affinities of hexapods: Recent progress and open problems. *Ann. Soc. Entomol. Fr., N.S.,* 37: 265-298.

Kraus, O. 1998. Phylogenetic relationships between higher taxa of tracheate arthropods. In: Fortey, R.A. & Thomas, R.H. (eds.), *Arthropod Relationships*: pp. 295-303. London: Chapman & Hall.

Kraus, O. 2001. "Myriapoda" and the ancestry of the Hexapoda. *Ann. Soc. Entomol. Fr., N.S.,* 37: 105-127.

Kusche, K., Hembach, A., Hagner-Holler, S., Gebauer, W. & Burmester, T. 2003. Complete subunit sequences, structure and evolution of the 6 x 6-mer hemocyanin from the common house centipede, *Scutigera coleoptrata*. *Europ. J. Biochem.* 270: 2860-2868.

Lauterbach, K.-E. 1983. Zum Problem der Monophylie der Crustacea. *Verh. Naturwiss. Ver. Hamburg* 26: 293-320.

Lavrov, D.V., Brown, W.M. & Boore, J.L. 2004. Phylogenetic position of the Pentastomida and (pan)crustacean relationships. *Proc. R. Soc. London, B,* 271: 1471-2954.

Mallatt, J.M., Garey, J.R. & Shultz, J.W. 2004. Ecdysozoan phylogeny and Bayesian inference: first use of nearly complete 28S and 18S rRNA gene sequences to classify the arthropods and their kin. *Mol. Phylogenet. Evol.* 31: 178-191.

Martin, J.W. & Davis, G.E. 2001. An updated classification of the recent Crustacea. *Nat. Hist. Mus. Los Angeles Co., Contrib. Sci.* 39: 1-124.

Müller, C.H.G., Rosenberg, J., Richter, S. & Meyer-Rochow, V.B. 2003. The compound eye of *Scutigera coleoptrata* (Linnaeus, 1758) (Chilopoda: Notostigmophora): an ultrastructural reinvestigation that adds support to the Mandibulata concept. *Zoomorphology* 122: 191-209.

Nardi, F., Spinsanti, G., Boore, J.L., Carapelli, A., Dallai, R. & Frati, F. 2003. Hexapod origins: Monophyletic or paraphyletic? *Science* 299: 1887-1889.

Negrisolo, E., Minelli, A. & Valle, G. 2004. The mitochondrial genome of the house centipede *Scutigera* and the monophyly versus paraphyly of myriapods. *Mol. Biol. Evol.* 21: 770-780.

Nixon, K.C. 1999. The Parsimony Ratchet, a new method for rapid parsimony analysis. *Cladistics* 15: 407-414.

Olesen, J., Martin, J.W. & Roessler, E.W. 1996. External morphology of the male of *Cyclestheria hislopi* (Baird, 1859) (Crustacea, Branchiopoda, Spinicaudata), with a comparison of male claspers among the Conchostraca and Cladocera and its bearing on phylogeny of the "bivalved" Branchiopoda. *Zool. Scripta* 25: 291-316.

Page, R.D.M. 2000. NDE, Version 0.4.8. Program online available at: http://taxonomy.zoology.gla.ac.uk/rod/NDE/nde.html

Pisani, D., Poling, L.L., Lyons-Weiler, M. & Hedges, S.B. 2004. The colonization of land by animals: molecular phylogeny and divergence times among arthropods. *BMC Biology* 2: 1-10.

Prendini, L. 2001. Species or supraspecific taxa as terminals in cladistic analysis? Groundplans versus exemplars revisited. *Syst. Biol.* 50: 290-300.

Regier, J.C. & Shultz, J.W. 2001. Elongation factor-2: a useful gene for arthropod phylogenetics. *Mol. Phylogenet. Evol.* 20: 136-148.

Richter, S. 2002. The Tetraconata concept: hexapod-crustacean relationships and the phylogeny of Crustacea. *Org. Divers. Evol.* 2: 217-237.

Richter, S. 2004. A comparison of the mandibular gnathal edges in branchiopod crustaceans: implications for the phylogenetic position of the Laevicaudata. *Zoomorphol.* 123: 31-44.

Richter, S. & Scholtz, G. 2001. Phylogenetic analysis of the Malacostraca (Crustacea). *J. Zool. Syst. Evol. Res.* 39: 113-136.

Richter, S. & Wirkner, C. 2004. Kontroversen in der phylogenetischen Systematik der Euarthropoda. In: Richter, S. Sudhaus, W. (eds.), *Kontroversen in der Phylogenetischen Systematik der Metazoa*: pp. 73-102. Sber. Ges. Naturf. Freunde Berlin 43.

Riley, J., Banaja, A.A. & James, J.L. 1978. The phylogenetic relationships of the Pentastomida: The case for their inclusion within the Crustacea. *Intern. J. Parasitol.* 8: 245-254.

Schram, F.R. 1986. *Crustacea*. New York: Oxford Univ. Press.

Schram, F.R. & Hof, C.H.J. 1998. Fossils and the interrelationships of major crustacean groups. In: Edgecombe, G.D. (ed.), *Arthropod Fossils and Phylogeny*: pp. 233-302. New York: Columbia Univ. Press.

Schram, F.R. & Jenner, R.A. 2001. The origin of Hexapoda: a crustacean perspective. *Ann. Soc. Entomol. Fr., N.S.,* 37: 243-264.

Schram, F.R. & Koenemann, S. 2004a. Are the crustaceans monophyletic? In: Cracraft, J. & Donoghue, M.J. (eds.), *Assembling the Tree of Life*: pp. 319-329. New York: Oxford Univ. Press.

Schram, F.R. & Koenemann, S. 2004b. Developmental genetics and arthropod evolution: On body regions of Crustacea. In: Scholtz, G. (ed.), *Evolutionary Developmental Biology of Crustacea*: pp. 75-92. Lisse: Balkema.

Shultz, J.W. & Regier, J.C. 2000. Phylogenetic analysis of arthropods using two nuclear protein-encoding genes supports a crustacean + hexapod clade. *Proc. R. Soc. London, B,* 267: 1011-1019.

Simmons, M.P. 2004. Independence of alignment and tree search. *Mol. Phylogenet. Evol.* 31: 874-879.

Sinakevitch, I., Douglass, J.K., Scholtz, G., Loesel, R. & Strausfeld, N.J. 2003. Conserved and convergent organization in the optic lobes of insects and isopods, with reference to other crustacean taxa. *J. Comp. Neurobiol.* 467: 150-172.

Spears, T. & Abele, L.G. 1998. Crustacean phylogeny inferred from 18S rDNA. In: Fortey, R.A. & Thomas, R.H. (eds.), *Arthropod Relationships*: pp. 169-187. London: Chapman & Hall.

Spears, T. & Abele, L.G. 1999. Phylogenetic relationships of crustaceans with foliaceous limbs: an 18S rDNA study of Branchiopoda, Cephalocarida, and Phyllocarida. *J. Crust. Biol.* 19: 825-843.

Spears, T. & Abele, L.G. 2000. Branchiopod monophyly and interordinal phylogeny inferred from 18S ribosomal DNA. *J. Crust. Biol.* 20: 1-24.

Turbeville, J.M., Pfeifer, D.M., Field, K.G. & Raff, R.A. 1991. The phylogenetic status of arthropods, as inferred from 18S rRNA sequences. *Mol. Biol. Evol.* 8: 669-686.

Walossek, D. 1993. The Upper Cambrian *Rehbachiella* and the phylogeny of Branchiopoda and Crustacea. *Fossils & Strata* 32: 1-202.

Walossek, D. 1999. On the Cambrian diversity of Crustacea. In: Schram, F.R. & Vaupel-Klein, J.C. (eds.), *Crustaceans and the Biodiversity Crisis. Proc. 4th Int. Crust. Congr., Amsterdam 1998, Vol. I*: pp. 3-27. Leiden: Brill.

Waloszek, D. 2003. Cambrian 'Orsten'-type preserved arthropods and the phylogeny of Crustacea. In: Legakis, A., Sfenthourakis, S., Polymeni, R. & Thessalou-Legaki, M. (eds.), *The New Panorama of Animal Evolution*: pp. 69-87. Sofia: Pensoft.

Walossek, D. & Müller, K.J. 1994. Pentastomid parasites from the Lower Palaeozoic of Sweden. *Trans. R. Soc. Edinburgh, Earth Sci.*, 85: 1-37.

Wheeler, W.C. 1990. When is an outgroup not an outgroup and how to root DNA sequence based topologies without an outgroup. *Cladistics* 6: 363-367.

Wheeler, W.C. 1995. Sequence alignment, parameter sensitivity, and the phylogenetic analysis of molecular data. *Syst. Biol.* 44: 321-331.

Wheeler, W.C. 1996. Optimization alignment: the end of multiple sequence alignment in phylogenetics? *Cladistics* 12: 1-9.

Wheeler, W.C. 1998. Sampling, groundplans, total evidence and the systematics of arthropods. In: Fortey, R.A. & Thomas, R.H. (eds.), *Arthropod Relationships*: pp. 87-96. London: Chapman & Hall.

Wheeler, W.C. 2001. Homology and the optimization of DNA sequence data. *Cladistics* 17: S3-S11.

Wheeler, W.C., Cartwright, P. & Hayashi, C.Y. 1993. Arthropod phylogeny: a combined approach. *Cladistics* 9: 1-39.

Wheeler, W.C., Gladstein, D. & DeLaet, J. 2002. POY version 3.0. Program and documentation online available at American Museum of Natural History: ftp.amnh.org/pub/molecular

Wheeler, W.C., Giribet, G. & Edgecombe, G.D. 2004. Arthropod systematics. The comparative study of genomic, anatomical, and paleontological information. In: Cracraft, J. & Donoghue, M.J. (eds.), *Assembling the Tree of Life*: pp. 281-295. New York: Oxford Univ. Press.

Whitington, P.M. 1995. Conservation *versus* change in early axonogenesis in arthropod embryos: A comparison between myriapods, crustaceans and insects. In: Breidbach, O. & Kutsch, W., (eds.), *The Nervous System of Invertebrates; An Evolutionary and Comparative Approach*: pp. 181-219. Basel: Birkhäuser.

Whitington, P.M. 1996. Evolution of neural development in the arthropods. *Seminars Cell Dev. Biol.* 7: 605-614.

Wills, M.A. 1998. A phylogeny of recent and fossil Crustacea derived from morphological characters. In: Fortey, R.A. & Thomas, R.H. (eds.), *Arthropod Relationships*: pp. 189-209. London: Chapman & Hall.

Wilson, K., Cahill, V., Ballment, E. & Benzie, J. 2000. The complete sequence of the mitochondrial genome of the crustacean *Penaeus monodon*: are malacostracan crustaceans more closely related to insects than to branchiopods? *Mol. Biol. Evol.* 17: 863-874.

Wingstrand, K.G. 1972. Comparative spermatology of a pentastomid, *Raillietiella hemidactyli*, and a branchiuran crustacean, *Argulus foliaceus*, with a discussion of pentastomid relationships. *K. Dan. Vidensk. Selsk. Biol. Skr.* 19: 1-72.

Zrzavý, J. 2001. The interrelationships of metazoan parasites: a review of phylum-and higher-level hypotheses from recent morphological and molecular phylogenetic analyses. *Folia Parasitologica* 48: 81-103.

Zrzavý, J. & Štys, P. 1997. The basic body plan of arthropods: insights from evolutionary morphology and developmental biology. *J. Evol. Biol.* 10: 353-367.

Zrzavý, J., Hypša, V. & Vlaskova, M. 1998. Arthropod phylogeny: taxonomic congruence, total evidence and conditional combination approaches to morphological and molecular data sets. In: Fortey, R.A. & Thomas, R.H. (eds.), *Arthropod Relationships*: pp. 97-107. London: Chapman & Hall.

VI METAZOAN PHYLOGENETICS

Playing another round of metazoan phylogenetics: Historical epistemology, sensitivity analysis, and the position of Arthropoda within the Metazoa on the basis of morphology

RONALD A. JENNER[1] & GERHARD SCHOLTZ[2]

[1] *Section of Evolution and Ecology, University of California, Davis, California, U.S.A.*
[2] *Institut für Biologie/Vergleichende Zoologie, Humboldt-Universität zu Berlin, Berlin, Germany*

ABSTRACT

Morphological evidence has been used to support the monophyly of both the Articulata and the Ecdysozoa. Although most recent computer-assisted cladistic analyses appear to support Ecdysozoa, several zoologists remain loyal to the classic Articulata concept. We address this phylogenetic debate from two perspectives. First, we discuss the striking differences in historical epistemology adopted by different workers, and how this inevitably leads to disagreement. Second, in order to provide a bridge of sorts between the different phylogenetic epistemologies we perform a set of morphological sensitivity experiments on various published morphological data sets to explore the robustness of the Articulata and Ecdysozoa hypotheses. We vary both the relative weight of characters, as well as the selection, coding, and scoring of characters. This approach allows a better insight into the relationship between character evidence and phylogenetic hypothesis for the different data sets. Depending on the data set, support for the Ecdysozoa varied from being weak or absent (Zrzavý et al. 1998; Nielsen 2001), somewhat stronger and moderately robust to changes in the data set (Zrzavý et al. 2001; Zrzavý 2003), to quite strong and robust to introduction of conflicting characters (Peterson & Eernisse 2001). However, by excluding problematic characters, correcting character coding and scoring errors, and introducing new potential articulatan synapomorphies the modified data set of Peterson & Eernisse (2001) yields support for a monophyletic Articulata. Ultimately, whether a given analysis supports Articulata or Ecdysozoa depends to a large degree on the phylogenetic philosophy that one adopts, and an unambiguous choice between these competing hypotheses, which will be accepted by all workers, seems therefore elusive.

We dedicate this chapter to Fred Schram on the occasion of his retirement. Fred's work on animal phylogeny and evolution at various levels was often very inspiring and never boring. We trust that our respectful simultaneous consideration of multiple alternative hypotheses is directly in line with Fred's spirit of what good science is all about.

1 ARTICULATA VERSUS ECDYSOZOA: A MICROCOSM FOR CONTEMPORARY DEBATES IN METAZOAN PHYLOGENETICS

"What about all those conflicting opinions and the 'imperfect' nature of the data that so many scientific workers seem to be concerned about. The solution to this is clear. We have to realize that unanimity of agreement over the perfect and complete data set is a type of scientific 'holy grail'. Moreover, just like that other grail, unanimity and perfection are myths!" Schram (1997: 149).

For most of the 20th century, writings about animal phylogeny have not generally been regarded as a paragon of exciting literature reporting on cutting edge scientific developments. However, all this changed when in the late 1980s the wider scientific community was alerted by the appearance of rather unexpected molecular metazoan phylogenies that were significantly at odds with many of the received wisdoms ensconced in textbooks. Currently available molecular and morphological evidence suggests a number of conspicuous phylogenetic discrepancies that are in need of explanation (see also Jenner 2004d).

Among these discrepancies, the conflicting placement of the Arthropoda as either a member of the Articulata (together with Annelida) or the Ecdysozoa (together with such molting, non-coelomate taxa as Nematoda and Priapulida) has risen to become a prominent emblem of the scope of current debates in metazoan phylogenetics. The debate that immediately ensued after the publication of the seminal paper by Aguinaldo et al. (1997) potently illustrates that the conflict between the Articulata and Ecdysozoa hypotheses provides perhaps the clearest example of the surprising depth at which current molecular evidence forces us to reconceptualize the evolution of animal body organizations, and reassess how we generate our hypotheses. The Articulata-Ecdysozoa debate draws attention to many important topics on a variety of levels of generality in comparative zoology, ranging from the use of molecules versus morphology, the adoption of different epistemologies in phylogenetic research, the use of evidence from the fossil record in the reconstruction of morphological ground patterns (for example reconstructing the primitive mouth position for Onychophora and Arthropoda: Budd 1999; Eriksson & Budd 2000; Eriksson et al. 2003), and the relative likelihood of convergence of ostensibly convincing homologies such as segmentation and cuticle molting. The Articulata-Ecdysozoa debate also necessitates a renewed look at some of the perennial problems of invertebrate zoology. For example what is the evolutionary significance of different life cycles and larvae (is the lack of primary larvae in ecdysozoans primitive or derived?), and how can we determine whether morphologies with only a limited amount of similarity are potentially true homologies that have become modified during evolution, or independently evolved features (for example, comparison of onychophoran and arthropodan body cavities and nephridia with the coeloms and metanephridia in other protostomes). Clearly, the study of these ingredients provides for an engaging intellectual adventure that has the power to illuminate many aspects of the operation of metazoan phylogenetics as a science, however, without the hope of a quick and easy resolution.

1.1 *Articulata versus Ecdysozoa: morphology versus molecules*

On the broadest level this conflict once again pits molecules versus morphology, and in the minds of many this dichotomy in supporting evidence is a central feature of the Articulata-

Ecdysozoa debate. For example, in his wide-ranging exposition of life's diversity for a general readership, Colin Tudge (2000: 208) labels the traditional union of the annelids and arthropods as a "cosy, commonsensical appraisal" that has now become upset by molecular insights. Tudge (2000: 200) judges this change of ideas about animal phylogeny as "somewhat shocking," but nevertheless feels confident enough to base the scheme of animal classification adopted in his book on the molecular rather than the morphological evidence. Tudge's embrace of the new molecular view of animal phylogeny certainly seems to be shared by an increasing majority of biologists and paleontologists as invertebrate zoology textbooks have started to incorporate molecular phylogenies (Brusca & Brusca 2003), and newly described fossils of the famous Cambrian arthropod *Canadaspis perfecta* are now unambiguously assigned to the Ecdysozoa (Lieberman 2003).

The principal molecular support for the Ecdysozoa derives from phylogenetic analyses of 18S and 28S rDNA sequences (e.g., Aguinaldo et al. 1997; Eernisse 1997; Zrzavý et al. 1998, 2001; Giribet et al. 2000; Peterson & Eernisse 2001; Mallatt & Winchell 2002; Mallatt et al. 2004), although some workers have criticized the 18S rDNA evidence (Wägele et al. 1999; Wägele & Misof 2001), and several other molecules appear to provide support for the Ecdysozoa as well (see Giribet 2003a for discussion and analysis). In contrast, the previously reported support for Ecdysozoa based on the presence of multimeric β-thymosin (Manuel et al. 2000) has now been disproved (Telford 2004), and a phylogenetic analysis of the amino acid sequences of several nuclear genes for a few taxa also did not find support for Ecdysozoa (Hausdorf 2000). Moreover, several recent phylogenomic analyses based on large numbers of orthologous genes or homologous exons failed to provide support for Ecdysozoa as well (Blair et al. 2002; Dopazo et al. 2004; Wolf et al. 2004). However, Copley et al. (2004) pointed out that results of phylogenetic analyses of large numbers of genes from just a few taxa need to be interpreted with caution, because disproportionate loss of genes in taxa such as *Caenorhabditis elegans*, may lead to biased results. This indicates that molecular phylogenetics still has a task in testing the monophyly of Ecdysozoa in future studies. However, it should be noted that molecular support for Articulata has never been found (Giribet 2003a). The Articulata-Ecdysozoa debate cannot solely be conceived of as a clash between molecules and morphology. The debate is also visible when just morphological evidence is considered.

1.2 *Articulata and morphology: almost two centuries of unanimity*

In the context of this paper it is particularly noteworthy that with the current Articulata-Ecdysozoa debate we have for the first time in the history of metazoan phylogenetics a deep divide within the community of invertebrate morphologists at large with respect to the phylogenetic position of the Arthropoda. Very different traditions of comparative zoology were in place in different parts of the world before the advent of molecular systematics. Conspicuously among these, the Anglo-Saxon and German traditions have long maintained quite different views on animal phylogeny (for example, see Westheide 2004: 172 for a lament about the absence of what he calls "European views and theories" on metazoan phylogeny in many Anglo-Saxon textbooks) that were strongly influenced by the views of a relatively small number of vociferous zoologists. Until very recently, Libbie Hyman's ideas were widely endorsed as the typical Anglo-Saxon view of animal phylogeny (Jenner

2004f), from American textbooks on invertebrate zoology to recent review papers on metazoan phylogenetics. In contrast, the German literature featured an alternative scheme of animal phylogeny, which drew heavily on the views of influential zoologists such as Adolf Remane, Rolf Siewing, and Werner Ulrich.

Yet, considering the existence of these different traditions of comparative invertebrate zoology, it is striking to see the widespread unanimity about the validity of the Articulata until very recently. Ever since Georges Cuvier christened the Articulata in pre-Darwinian times, a close relationship between annelids and arthropods has been a consensus view among zoologists, from the first generation of evolutionary morphologists in the last quarter of the 19th century in Germany and Britain (Bowler 1996), to the views of most contemporary zoologists in the 1990s. Certainly, the arthropods have been proposed to be closely related to what we would now consider to be members of the Ecdysozoa at various points in the history of our discipline in both the German and English literature, for example Bütschli (1876), Rauther (1909), and Kristensen (1991), but these views have never attained the status of canonical textbook knowledge. However, this unanimity now seems to have dissolved, and the Articulata-Ecdysozoa debate is now firmly established within morphological phylogenetics.

1.3 *Articulata versus Ecdysozoa: morphology and a clash of irreconcilable epistemologies*

One might expect that the almost universal adoption of cladistic methods would have facilitated the development of one well-supported hypothesis on the position of the arthropods within the animal kingdom. This, however, did not happen. Although the first published phylogenetic-systematic analyses of the Metazoa (Hennig 1972, 1979, 1983; Gruner 1980; Ax 1984; Dohle 1986) as well as the first computer-assisted cladistic analyses (Brusca & Brusca 1990; Meglitsch & Schram 1991; Schram 1991) supported the Articulata hypothesis, Eernisse et al.'s (1992) analysis united the arthropods more closely with some nemathelminth representatives than with the annelids. After the Ecdysozoa concept received molecular support from Aguinaldo et al. and Eernisse in 1997, most subsequently published comprehensive morphological cladistic analyses came to support Ecdysozoa as well (Zrzavý et al. 1998, 2001; Giribet et al. 2000; Peterson & Eernisse 2001; Zrzavý 2003). Nevertheless, several invertebrate morphologists and phylogeneticists upheld the Articulata concept in various major books, and papers (Ax 1999; Wägele et al. 1999; Sørensen et al. 2000; Nielsen 2001; Wägele & Misof 2001; Scholtz 2002, 2003; Brusca & Brusca 2003). This contradiction is in need of an explanation.

In order to understand this conflict, we have to distinguish between two different kinds of support that have recently been advanced for the Articulata hypothesis. First, several morphological cladistic analyses have supported Articulata (Sørensen et al. 2000; Nielsen 2001; Brusca & Brusca 2003), but these studies are contradicted by others that yielded support for Ecdysozoa (Zrzavý et al. 1998, 2001; Giribet et al. 2000; Peterson & Eernisse 2001; Zrzavý 2003). To determine whether this disagreement is due to ambiguity of available evidence, or idiosyncrasies of the cladistic analyses, we will take a closer look in the sections below.

Second, several workers have also argued strongly that the potential synapomorphies of annelids and arthropods are much more convincing than those supporting the monophyly of Ecdysozoa (Ax 1999; Wägele et al. 1999; Nielsen 2001; Wägele & Misof 2001; Scholtz 1997; 2002, 2003; Brusca & Brusca 2003), independent of whether cladistic analyses may support this conclusion or not. For example, Brusca & Brusca (2003: 499) talk about "numerous" and "powerful" synapomorphies shared between annelids and arthropods, even though at that point in their discussion these anatomical and developmental similarities are at most potential synapomorphies. And Wägele et al. (1999: 220, 221) state that the greater complexity of articulatan similarities indicate that the "probability of homology of these characters is much higher than that of the Ecdysozoa pattern" and that when "weighting the characters according to their complexity the balance is clearly heavier on the side of the Articulata hypothesis". So far, almost no one seems to have argued for the quality of potential ecdysozoan synapomorphies prior to a phylogenetic analysis (Schmidt-Rhaesa et al. 1998 is a notable exception).

The reason for these differences in the judgment of ecdysozoan and articulatan characters is their perceived difference in complexity. Prior to performing a cladistic analysis, Wägele et al. (1999), Wägele & Misof (2001), and Scholtz (2002, 2003) use the criterion of complexity in an attempt to separate more and less reliable characters. Because complex characters are composed of a greater number of potentially independent details than simpler ones, a complex character that is shared between different taxa may be more likely to be homologous than a simple character with less comparable details. This principle is widely recognized and applied, at least since Adolf Remane (1952) credited more complex characters with a greater probability of being homologous (see Riedl 1975; Dohle 1989; Donoghue & Sanderson 1994). In fact, in a paper on the epistemology of phylogenetic inference, Grant & Kluge (2003) consider evaluations of character complexity to be at the heart of homology-testing in cladistics. It has been argued that the many anatomical and developmental similarities of segmentation shared between annelids and arthropods build a convincing case for homology (see Scholtz 2002).

In contrast, the ecdysozoan synapomorphies are considered less convincing by those workers (Wägele et al. 1999; Nielsen 2001; Wägele & Misof 2001; Scholtz 2002, 2003), even if previous morphological cladistic analyses apparently support Ecdysozoa, because of the lesser complexity of those characters (they include many character losses, for example). In addition, several subsidiary criteria have been employed to argue that potential ecdysozoan synapomorphies are unconvincing, for example, that some characters are convergent (loss of motile epidermal cilia, molting), or not yet sufficiently studied (hormonal control of molting) (Nielsen 2001; Wägele & Misof 2001). Others have also applied such criteria in an attempt to discern reliable homologies for other taxa prior to a phylogenetic analysis, however, until now with little success (Jenner 2004a).

Wägele et al. (1999) and Wägele & Misof (2001) also adduce arguments derived from functional morphology in support of the Articulata hypothesis. They construct an evolutionary scenario that connects the disparate body plans of annelids and arthropods by taking explicitly into account functional changes in morphology in the context of the environment and hypothesized selection pressures. Their scenario links the origin of the arthropods to an increased efficiency of locomotion as annelid parapodia are transformed into more efficient segmented limbs. Furthermore, these authors argue that a similar scenario cannot be constructed to link the arthropods to ecdysozoans (however, see Budd 1999, 2001 for a functional scenario in line with the Ecdysozoa hypothesis). In addition, Wägele & Misof (2001) also incorporated information from the fossil record into their functional morphological scenario to link the annelids and arthropods, principally by pointing to fossils with morphologies that look intermediate between the organizations of annelids and arthropods (again, for a different view see Budd 1999, 2001).

Such approaches to phylogenetics, with a strong emphasis on data quality evaluation prior to a phylogenetic analysis, a priori separation of good and bad characters, a concern for functional hypothetical ancestors implied by the combination of characters at internal nodes of cladograms, and a consideration of presumed selection pressures and the functional correlates of changes in morphology are summarily rejected by other workers who abide by a very different phylogenetic epistemology (Zrzavý 2001; Giribet 2003a). It should be emphasized that not all workers necessarily follow all these epistemological precepts, so that accepting the criterion of complexity in evaluating character homologies does not necessarily imply the acceptance of the use of functional evolutionary scenarios in the evaluation of phylogenetic hypotheses. However, to let a priori ideas about data quality influence the selection of characters prior to a phylogenetic analysis is an important decision that divides the opinions of phylogeneticists.

Because the judging of data quality is considered as a subjective matter that may be biased by the expectations of the researcher, most contemporary phylogeneticists who use parsimony or standard cladistic analysis would initially collect as many pertinent characters as possible, which will all be treated equally. These workers purposely adopt a stance of agnosticism with regards to the presumed phylogenetic value of characters prior to parsimony analysis. They keep considerations of the evolutionary process to an absolute minimum, while solely focusing on the pattern of character distribution. The most parsimonious tree will then allow one to distinguish provisionally accepted and rejected homology proposals. In no instance can the choice of a less parsimonious solution be allowed, neither on the basis of a functional scenario, nor on the basis of differential character weighting (Kluge 1998). To other workers such a dressed-down approach to phylogenetics is seriously impoverished without the consideration of what Ghiselin (1991: 290) has called "contextual information" pertaining to ecology, niches, habitats, functional morphology, and adaptive significance of characters. Ghiselin (1984: 220) concluded that "phylogenetics without evolutionary biology is like astronomy without mechanics".

When workers adhere to such different phylogenetic epistemologies (see also Jenner 2004e), the Articulata-Ecdysozoa debate could be in danger of stalling, especially when proponents of the Articulata and Ecdysozoa hypotheses largely fall on opposite sides of this epistemological division. To prevent this from happening we need to approach epistemology with an open mind. According to Bowler (2000) scientists typically engage in methodological debate only when challenged to defend their views against those of others. Philosopher Stephen Asma (1996: 168) opined that the general unwillingness of biologists to actively engage philosophical history has often resulted in "an inflexible ossification of one sort of explanatory account".

Among scientists, phylogeneticists may be exceptional in that they frequently engage in philosophical debate. The history of cladistics is a rich repository of such debates, with phenetics pitted against cladistics at the cradle of our discipline, and supermatrices versus supertrees in more modern times. Although such debates in general are a healthy sign of critical scrutiny of our conceptual instruments, according to some workers our enthusiasm for philosophical debate should be labeled as one of the "pointless and suicidal tendencies in systematics" (Schram 2004). We believe that the problem here is not the existence of epistemological debates per se, but the fundamentalist adherence to a single epistemological framework, without the willingness to seriously consider any results generated within different epistemological frameworks.

In view of such deep-rooted differences in outlook, we are naïve to expect an imminent resolution. Thus, there is some justification in Schram's negative characterization of methodological debates in phylogenetics, because history shows that these often result in the establishment of opposing schools of thought characterized by unbridgeable differences. This effectively kills debate as people are fundamentally convinced of their right on the basis of first principles, and as a consequence people will just talk past each other in pursuit of the hopeless goal of converting one's opponents. This is all too clear in the morphological Articulata-Ecdysozoa debate. It is noteworthy that one person's central epistemological commitments can be an opponent's gravest logical fallacies. This is forcefully illustrated in the last paragraph of a paper by Zrzavý (2001), in which he lists as the greatest shortcomings of the views of his opponents Wägele et al. exactly those principles most central to their reasoning (the reverse is true as well). For example, Zrzavý (2001: 162) labels as "shortcomings" several central components of Wägele et al.'s method of phylogeny reconstruction, including "hypothetical scenario building" and the use of "speculative arguments on the genetic complexity of characters". Similarly, Giribet (2003a: 315) urges the phylogeneticist "to leave plausibility and complexity arguments aside".

Perhaps, instead of upholding epistemological dogmatism we should heed the words of those who have seen it all before: "why argue, why not do both?" (Schram 2004). The phylogenetic community, with its patchwork of carefully staked-out epistemological territories may seem an especially unreceptive substrate for an appeal to epistemological pluralism. We acknowledge that defending epistemological pluralism may be misguided on a certain level (Giribet et al. 2002 and Grant & Kluge 2003). However, the differences between the advocates of complexity and plausibility arguments on the one hand, and proponents of standard cladistic analysis on the other, are perhaps more profitably seen as differences of degree rather than sharp qualitative breaks. These workers all aim to base clades only on shared derived characters, but they differ in their convictions about the relative importance of different kinds of characters. In fact, the distinction is likely to be even less than that, namely just a practical difference. Most systematists think that some characters are more informative for phylogeny reconstruction than others, because convergence of some characters may seem more likely for some than for others. The problem, however, is that it is very difficult, if not impossible, to separate more from less informative characters at a certain level. Because of this difficulty, standard cladistic analysis maximizes its power to test competing hypotheses and objectivity by treating all characters equally, and by minimizing process assumptions. However, the results may not represent the true phylogeny if in fact different characters have different susceptibilities to convergence. In contrast, other workers may want to incorporate their a priori assumptions about the evolutionary process, however conjectural they may be, into the reconstruction of phylogeny. But if these assumptions turn out to be wrong, the true phylogeny may not be reconstructed. However, if one is of the opinion that "systematics as a realist, truth-seeking activity is doomed" (Schram 2004) and that the true tree is in principle unknowable anyway (Siddall & Kluge 1997), then the differences between these two methodologies just boil down to what type of uncertainty one prefers, uncertainty of a priori assumptions, or of results.

We want to make a general appeal to phylogeneticists to pay respectful attention to results generated within different epistemological frameworks. In that way, we can at least minimize dogmatic myopia, maximize exciting dialogue, and broaden our horizons. In a perceptive paper, Lee & Doughty (1997) broadly divided phylogenetic approaches into pattern- and process-oriented approaches. They conclude that both may have value, while none has absolute logical priority. When initially kept separate, the value of either of these methods lies in their ability to reciprocally illuminate each other's results. We wholeheartedly agree. Both approaches have their strengths and weaknesses, and it is by no means obvious that standard cladistic analysis without any consideration of the evolutionary process will necessarily lead to the best answer. A possible example is provided by metazoan phylogenetics. An increasingly robust molecular phylogenetic framework of the Metazoa is emerging, which allows us to test whether a cladistic or functional approach to comparative morphology yields the most congruent results. Interestingly, in many cases of conflict between morphological cladistic analyses and molecular evidence, a more functional approach to comparative morphology may provide for a better alignment of molecules and morphology (Jenner 2004d). This may especially be the case for taxa that have likely undergone substantial morphological modifications during their evolutionary origin, such as secondarily simplified taxa and parasites.

In this paper we want to build a bridge of sorts between the two approaches by performing a set of sensitivity analyses under different assumptions about the relative importance of characters, the choice of characters, and the coding of characters. These experiments are intended to add to our current understanding of the Articulata and Ecdysozoa hypotheses and their relative support based on morphological evidence.

2 EXPLORING ARTICULATA VERSUS ECDYSOZOA WITH MORPHOLOGICAL SENSITIVITY ANALYSES

Sensitivity analysis in the context of phylogenetics can be described as the analysis of how different assumptions differentially affect the outcomes of a phylogenetic analysis. An increasingly common use of sensitivity analysis is to explore how robust molecular phylogenetic conclusions are against changes in sequence alignment parameters (Wägele & Stanjek 1995; Wheeler; 1995; Giribet 2003b). Within the wider context of comparative biology, sensitivity analysis has been used to assess the effect of various types of phylogenetic uncertainties on the robustness of conclusions (Donoghue & Ackerly 1996). Although the scientific merit of sensitivity analysis in systematics is not entirely uncontested (Grant & Kluge 2003), we believe it can provide valuable insights into the link between evidence (synapomorphies) and favored hypothesis (cladogram) that are otherwise difficult to obtain. Yet, the application of sensitivity analysis to morphological phylogenetics is distinctly less widespread than to molecular analyses (but see Simmons & Geisler 2002; Prendini 2003; Bivort & Giribet 2004), notably to studies of metazoan phylogeny (Jenner 2004b). This is noteworthy because experiments on the selection, coding, and scoring of characters clearly show that relatively minor changes in data sets may have far-reaching consequences for the placement of individual taxa, clades, or even for the overall topology and resolution of cladograms (Hawkins et al. 1997; Jenner & Schram 1999; Hawkins 2000; Forey & Kitching 2000; Donoghue et al. 2000; Rouse & Fauchald 1997; Rouse 2001; Jenner 2002, 2003,

2004b; Simmons & Geisler 2002; Turbeville 2002).

Despite the fact that most of the most comprehensive recent morphological cladistic analyses support Ecdysozoa, critics are reluctant to accept these results because they consider the potential synapomorphies of Articulata to be more convincing (Scholtz 1997, 2002, 2003; Wägele et al. 1999; Wägele & Misof 2001). To explicitly flesh out these intuitions, in the following section on '*Stability of phylogenetic hypotheses and weight of evidence*' we re-analyzed the major morphological data sets supporting Ecdysozoa. We increased the weight of the Articulata characters to see exactly when the monophyly of the Ecdysozoa collapses.

The second section below on '*Peterson & Eernisse (2001) and the stability of Ecdysozoa in the context of character selection, coding, and scoring*' has a related but somewhat different purpose. Although most recent comprehensive morphological cladistic analyses support Ecdysozoa (Zrzavý et al. 1998, 2001; Peterson & Eernisse 2001; Zrzavý 2003), those of Sørensen et al. (2000) and Nielsen (2001) instead support Articulata (the recent analysis by Brusca & Brusca 2003 also supports Articulata, but we were unable to duplicate their results). In order to understand this disagreement, we will focus on the adopted strategies of character selection in these studies to determine whether the analyses can truly be regarded as effective tests of competing hypotheses (see Jenner 2003 for another example of this approach). The analysis of Sørensen et al. (2000) will not be further discussed because the data set is very similar to that of Nielsen (2001), which will be discussed below. Although the selection of hypotheses to be tested may be based on non-scientific reasons, one obvious choice would be to focus on the most strongly corroborated hypothesis (Kluge 1997; Grant & Kluge 2003), which currently seems to be Ecdysozoa. We will pay particular attention to the analysis by Peterson & Eernisse (2001: 187), who claimed to show "using morphological data that the monophyly of Ecdysozoa is much more parsimonious than the monophyly of Articulata," despite their consideration of most of the Articulata characters mentioned in Wägele et al. (1999). Specifically, we evaluate whether all known similarities between annelids and arthropods have been included in Peterson & Eernisse (2001), in particular the various independent subcomponents of segmentation discussed in Scholtz (2002, 2003).

2.1 *Stability of phylogenetic hypotheses and weight of evidence*

Scholtz (2002, 2003), Wägele et al. (1999), and Wägele & Misof (2001) argue that the complexity of segmentation similarities found between annelids and arthropods are a reason for attaching more weight to these characters than is done in published morphological cladistic analyses. Here we test how much more weight has to be assigned to segmentation characters in order to collapse support for the Ecdysozoa in recent studies based on different data sets.

One guideline for increasing the weight of segmentation characters included in the data sets is the number of distinct levels at which segmentation is manifested. Scholtz (2002, 2003) argues, for example, that anatomical segmentation is manifested at least in similarities on 6 distinct levels: 1) the nervous system, 2) the epidermis, 3) coelomic cavities, 4) metanephridia, 5) muscles, and 6) limbs. Consequently, in view of these similarities on 6 levels we may assign a weight of 6 to a simple 'segmentation absent/present' character.

Developmental similarities include 1) a preanal proliferation zone, 2) neurogenesis, 3) coelomogenesis, 4) dorsal blood vessel formation, 5) the presence of parasegments, and 6) expression of a number of developmental genes involved in segmentation. To capture these 6 levels of developmental complexity, we may again assign a weight of 6 to a character describing the developmental similarity of segments (this approach reflects the assumed independence of the single characters), making up a total of 12 for the weight of a single segmentation character to reflect both the anatomical and developmental similarity.

A different but logically equivalent way of testing the stability of Ecdysozoa is to introduce new Articulata synapomorphies scored as present only in the annelids and arthropods. If, for example, one has to introduce 25 synapomorphies for the Articulata to collapse the Ecdysozoa hypothesis, then it could be concluded that the monophyly of Ecdysozoa is relatively robust because this would require a number of similarities exceeding the number of segmentation similarities (6 anatomical and 6 developmental similarities as determined above) currently known to be shared between annelids and arthropods.

However, it should be realized that several ambiguities attend the interpretation of such experiments. First, comparing the stability of the Ecdysozoa hypothesis through experiments in character weights or numbers is complicated in some cases by the inclusion of different sets of characters with different distributions that may support Ecdysozoa and the more inclusive clades in which Ecdysozoa is nested. Second, it should be noted that several of the above mentioned similarities are not uniformly distributed throughout both arthropods and annelids, creating uncertainty about the ground pattern states of the Annelida and Arthropoda. However, in the context of this sensitivity analysis, we want to devise the strongest possible test of the stability of the Ecdysozoa; i.e., we perform the most severe test of the Ecdysozoa hypothesis by assuming that these similarities are all present in the ground patterns of the Annelida and Arthropoda.

In all the following experiments we started with the original morphological data sets of Zrzavý et al. (1998, 2001), Zrzavý (2003), Peterson & Eernisse (2001), and Nielsen (2001). In the different sensitivity analyses we modified these data sets by changing character selection, character weighting, and character coding. We have adopted ground pattern character coding (in some cases multistate) and scoring for all analyses. Unless stated otherwise, all phylogenetic analyses are parsimony analyses performed with PAUP* (Swofford 2002), employing heuristic searches with 1000 random addition replicates with TBR branch swapping (Tree Bisection and Reconnection, which is the most effective method of branch swapping). The results of the analyses of the original and modified data sets are then compared in terms of numbers of equally most parsimonious trees (MPTs), and the topology of the corresponding strict consensus trees (in terms of the absence or presence of the clades of interest).

2.1.1 *Zrzavý et al. (1998)*

The morphological matrix in Zrzavý et al. (1998) is the largest compiled for the Metazoa to date, including 276 characters. Analysis of this matrix (heuristic search with 100 random addition replicates and TBR branch swapping) yields 258 MPTs with a strict consensus tree that supports neither Articulata nor Ecdysozoa. When we assigned a weight of 15 to character 18, which codes for a body segmented with serially repeated organs developed from 4d-mesoderm or ectomesoderm, and which comes as close to a potential synapomorphy of Articulata as any character in the matrix of Zrzavý et al. (1998), then Pan-Arthropoda becomes the sister group to Mollusca, and together they form the sister group to Annelida. In subsequent studies Zrzavý modified his data set (Zrzavý et al. 2001; Zrzavý 2003), and these later matrices seemed to provide more support for Ecdysozoa.

2.1.2 Zrzavý et al. (2001)

The data set in Zrzavý et al. (2001) includes two characters coding for segmentation shared between annelids and arthropods (character 3: segmentation; character 4: teloblastic growth). Analysis of the original data set produced 234 MPTs, with Ecdysozoa supported in the strict consensus tree. When we added 10 or less potential Articulata synapomorphies (uniquely scored present for annelids, arthropods, onychophorans, and tardigrades) to the matrix of Zrzavý et al. (2001), the strict consensus tree continued to support a monophyletic Ecdysozoa. However, when 11 potential Articulata synapomorphies were added to the matrix, the phylogenetic position of the pan-arthropods remained unresolved in the strict consensus tree. Only when 13 or more potential synapomorphies of the Articulata are added to the matrix, a monophyletic Articulata emerges in the strict consensus tree. It thus appears that the morphological matrix of Zrzavý et al. (2001) quite robustly supports Ecdysozoa. However, it should be noted that this matrix only includes two segmentation characters, one of which codes simply for segmentation absent/present, and the other for teloblastic growth absent/present. If the added potential Articulata synapomorphies are interpreted as representing the different levels of structural and developmental complexity uniquely shared between the segmentation patterns of annelids and (pan-)arthropods as identified in Scholtz (2002, 2003) then the support for Ecdysozoa in Zrzavý et al. (2001) may be less compelling than appears at first sight. It can at least be concluded that annelids and arthropods share more details of segmentation than captured by the two characters present in the matrix of Zrzavý et al. (2001), the scoring of which is problematic as well (teloblastic growth is scored present for molluscs and sipunculans which is unsupported; Scholtz 2002, 2003).

2.1.3 Zrzavý (2003)

Only two characters are coded to represent the detailed structural and developmental similarities of annelid and arthropod segmentation in the matrix of Zrzavý (2003). These two characters, 10 and 13, are the same characters as characters 3 and 4 in Zrzavý et al. (2001), albeit not with identical scoring. Analysis of the original data set of Zrzavý (2003) yields 15 MPTs with Ecdysozoa supported in the strict consensus tree. When one potential Articulata synapomorphy (uniquely scored present for annelids, arthropods, onychophorans, and tardigrades) is added to this matrix, and its weight is increased from 1 to 17, support for a monophyletic Ecdysozoa is maintained. When this character is given a weight of 18, however, the phylogenetic position of the pan-arthropods is unresolved, and with a weight of 20 this character supports a monophyletic Articulata. Because the weight of potential articulatan synapomorphies needs to be increased more than for the data set of Zrzavý et al. (2001) to collapse Ecdysozoa and yield support for a monophyletic Articulata it thus appears that morphological support for Ecdysozoa in the matrix of Zrzavý (2003) is more compelling than that in the data set of Zrzavý et al. (2001). This is mostly due to the inclusion of additional potential ecdysozoan synapomorphies in the matrix of Zrzavý (2003), while no additional characters were included that could be synapomorphies for Articulata.

2.1.4 *Nielsen (2001)*

Analysis of Nielsen's original matrix (excluding character 64, as Nielsen did as well) results in 13 MPTs with a monophyletic Articulata. However, when character 14, which codes for cuticle moulting, is upweighted to 5, then the analysis yields 12 MPTs, while neither Articulata nor Ecdysozoa is supported in the strict consensus. When character 14 is given a weight of 6 or higher, a monophyletic Ecdysozoa is supported. This is significant, because it is easier to collapse Articulata for Nielsen's data set than it is to collapse Ecdysozoa for the other data sets discussed above. This is partly explained by the fact that Nielsen's data set did not include all potential ecdysozoan synapomorphies, such as characters on cuticle layers and the presence of intestinal cilia. However, neither did Nielsen include all potential Articulata synapomorphies.

2.1.5 *Peterson & Eernisse (2001)*

Analysis of the original data set of Peterson & Eernisse (2001) resulted in 14 MPTs, with a strict consensus tree supporting the Ecdysozoa. Based on this result the authors claim that their morphological data provide strong evidence in favor of the Ecdysozoa. Our results after differential character weighting seem to support this conclusion. For example, character 42 codes for teloblastic segmentation and is scored present for arthropods, onychophorans, and annelids. When we gradually increased the weight of this character from 1 to 26 the monophyly of Ecdysozoa remains supported. Ecdysozoa collapses when character 42 is give a weight of 27, and Articulata (excluding tardigrades) is supported only when character 42 is given a weight of 28. These experiments indicate that Ecdysozoa seems to be quite robustly supported. This may not have been expected on the basis of clade support values in Peterson & Eernisse (2001). For example, the clade Gastrotricha + Ecdysozoa has a bootstrap value of 64. However, even when the weight of character 42, which conflicts with this clade, is increased to 26, the clade is still supported in the strict consensus tree. This nicely illustrates the value of Giribet's (2003b) recommendations that clade support and clade stability should be analyzed independently because there is no simple relationship between these measures, and both may yield different insights about the confidence we may have in the results. Cases of low clade support measures and high stability, such as for the clade Gastrotricha + Ecdysozoa in Peterson & Eernisse (2001), may be explained by the presence of few supporting characters but little conflict in the matrix. Although Peterson & Eernisse (2001: p. 187) stated that most of the potential articulatan synapomorphies discussed in Wägele et al. (1999) were considered in their study, in fact only character 42, which codes for teloblastic segmentation, is included in the matrix of Peterson & Eernisse (2001). In addition, we identified problems with the coding and scoring of other characters that are relevant to the issue of Articulata versus Ecdysozoa, for example character 40. This character codes for a lateral coelom derived from mesodermal bands, a character at least shared between arthropods, onychophorans, and annelids, although it was scored as present only for the latter taxon. These problems led us to re-analyze the data set of Peterson & Eernisse (2001) in more detail in the next section in order to determine whether all evidence critical to resolving this phylogenetic conflict was properly dealt with.

2.2 Peterson & Eernisse (2001) and the stability of Ecdysozoa in the context of character selection, coding, and scoring

In this section we focus on the morphological data set of Peterson & Eernisse (2001), because it supports the monophyly of the Ecdysozoa despite their claim to have considered most of the potential articulatan synapomorphies discussed in Wägele et al. (1999). However, as discussed above, Peterson & Eernisse only included a single character for teloblastic segmentation. Therefore we studied this data set in more detail with the aim of identifying the quality of included evidence in terms of character coding and scoring, and how rigorously available data, potentially in favor of Articulata, have been treated. Several problems with the data set of Peterson & Eernisse (2001) are discussed in previous papers (Jenner 2002, 2004b: p. 300, 2004c), but here we focus specifically on those characters most relevant to Ecdysozoa versus Articulata. We made no attempt to check the entire data matrix. Most of the relevant morphological evidence has previously been discussed (see Schmidt-Rhaesa et al. 1998; Wägele et al. 1999; Nielsen 2001; Wägele & Misof 2001; Zrzavý 2001; Scholtz 2002, 2003; Giribet 2003a), but not yet included in an explicit cladistic analysis. The complete modified data set of Peterson & Eernisse (2001) is available upon request from the authors.

We performed a number of sensitivity analyses of the matrix of Peterson & Eernisse (2001) incorporating different combinations of two types of changes: 1) we identified and corrected errors in character coding and scoring; 2) we included additional characters not considered by Peterson & Eernisse (2001). A summary of all changes and their justifications that have been variously included in the different sensitivity analyses is given in Appendix 1. An overview of the different sensitivity experiments is given in Table 1 (Appendix 2). The results are discussed in the following sections.

2.2.1 The effect of errors in character scoring

The first thing we wanted to know is whether character scoring errors could have had an effect on the efficacy to test the Ecdysozoa and Articulata hypotheses using the data set of Peterson & Eernisse (2001). Therefore we principally checked two types of characters. First, characters underlying the clades Ecdysozoa, Ecdysozoa + Gastrotricha, and [Ecdysozoa + Gastrotricha] + Chaetognatha in the analysis of Peterson & Eernisse (2001). Second, characters dependent on the possession of spiral cleavage, or trochophore or ciliated larvae. Since pan-arthropods do not have spiral cleavage or ciliated larvae, all characters pertaining to spiral cleavage or ciliated larvae should be scored as 'inapplicable' for pan-arthropods to prevent the phylogenetic separation of the neotrochozoans from the pan-arthropods on the basis of multiple logically related 'absence characters'. The changes we made to the data set are summarized under 'Modified scoring' in Appendix 1.

When we ran the corrected matrix (Experiment 1; hereafter abbreviated as Exp.; see Table 1 in Appendix 2), the monophyly of Ecdysozoa remains supported, identical to the original analysis (Fig. 1). This indicates that the Ecdysozoa hypothesis is robust to these changes for the data of Peterson & Eernisse (2001). However, by upweighting character 42 (teloblastic segmentation), just as we did in the previous section on '*Stability of phylogenetic hypotheses and weight of evidence*', we were able to show that the character scoring errors we identified have contributed spurious clade support to the Ecdysozoa (Exp. 2-4). Whereas in the original data set we need to give character 42 a weight of 28 to collapse

Ecdysozoa and support monophyly of Articulata (minus Tardigrada), we only needed to increase the weight of this character to 15 to achieve the same effect for the corrected data set. This indicates that the correction of these scoring errors significantly reduces spurious support for the Ecdysozoa hypothesis.

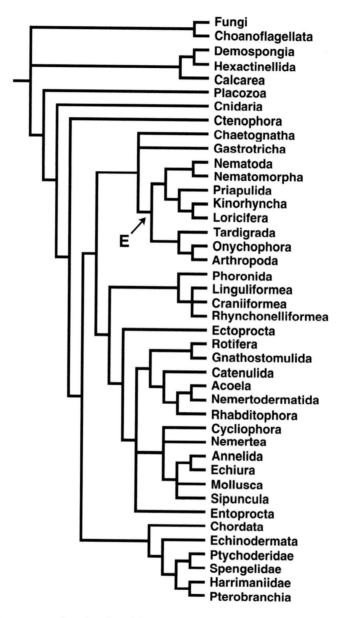

Figure 1. Strict consensus of results of sensitivity experiment 1 (see Table 1, Appendix 2). Ecdysozoa is monophyletic (indicated by clade E in the tree). See text for discussion.

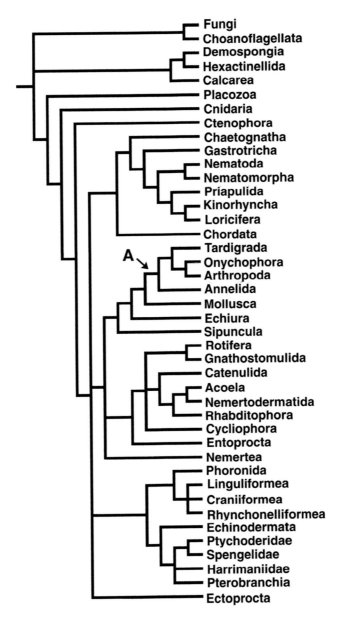

Figure 2. Strict consensus of results of sensitivity experiment 5 (see Table 1, Appendix 2). Articulata is monophyletic (indicated by clade A in the tree). See text for discussion.

2.2.2 *The effect of adding new characters and deleting old ones*

All of the following experiments start with the corrected data set of Peterson & Eernisse (2001). We have added a number of new characters to the data set (139-156) that have previously been discussed in the literature, and which have been variously included in different published cladistic data sets (see section 2.2 for references). These characters are listed under 'New characters' in Appendix 1. Several of these new characters are replacements of old characters: characters 149, 146-147, and 155 replace characters 36, 42, and 48-50, respectively. In addition, several characters had very problematic codings and scorings, and were therefore excluded from the analysis (see listing under 'Deletions' in Appendix 1). Making these additions and deletions to the data set, and modifying the coding of character 101 to incorporate shared similarities between pan-arthropods and annelids (Exp. 5), Ecdysozoa is no longer supported. Instead Articulata is now supported, including Tardigrada (Fig. 2). The same result is obtained when the original characters 36, 42, and 48-50 are left in the analysis, while their replacements are excluded (Exp. 6). When experiment 5 is repeated, but then with the original coding of character 101 (Exp. 45) neither Articulata nor Ecdysozoa is supported. These results indicate that the modification of the coding of character 101 provides important support for the monophyly of Articulata.

Differential inclusion of the deleted characters has different effects upon the outcome of the analysis. This is explored by starting with the data set used for Exp. 5, which includes all original and newly added characters, and which excludes all problematic characters (see under 'Deletions' in Appendix 1 for these characters). For example, when the excluded problematic character 28 is again added to the data set Articulata remains supported (Exp. 7; compare results with Exp. 5). In contrast, when the excluded problematic character 114 is again included in the data set (Exp. 8; compare results with Exp. 5), neither Articulata nor Ecdysozoa is supported (Fig. 3). A similar bilaterian polytomy is generated when in addition to character 28 also the excluded problematic characters 30 and 31 are again included (Exp. 9). However, when only characters 130-133 are again included (Exp. 10; compare results with Exp. 5) Ecdysozoa is supported (Fig. 4). These experiments show that the results are quite sensitive to character selection.

When we start with the corrected data set of characters 1-138, we can explore the effect of the gradual addition of the new characters. For this data set, support for Ecdysozoa appears to be very robust (Exps. 11-19). Even when all new characters are added to the corrected data set Ecdysozoa remains supported. Similarly, when all problematic characters (28, 30, 31, 33, 37, 47, 54, 114, 130-133) are deleted from the corrected data set of 138 characters, Ecdysozoa remains supported (Exps. 20-26). A shift in the support for the competing hypotheses occurs only when we combine the addition of the new characters and the deletion of problematic ones.

When we include all characters, except 36, 42, and 48-50 (these are replaced by the new characters 149, 146-147 and 155, respectively), and exclude all problematic characters, then Articulata is supported (Exp. 5). When we include variable numbers of problematic characters, Articulata may be supported (Exps. 32, 33, 48, 49), Ecdysozoa may be supported (Exps. 27, 28, 34-41, 46), or neither hypothesis may be supported (Exps. 29-31, 42-45). Note that these results partly depend on including the modified coding of character 101, where the pairs of repeated ventral ganglia or swellings in the pan-arthropods and annelids are considered homologous (Scholtz 2002, 2003). When the original coding of this character is adopted (nerve cells organized in ganglia or not) support for Articulata emerges later than in the modified coding option (compare Exp. 43 and 44 with 48 and 49).

Interestingly, none of our experiments supports the hypothesis of Nielsen (2003) and Almeida et al. (2003) that an ecdysozoan clade forms the sister group of the Annelida or a subclade thereof.

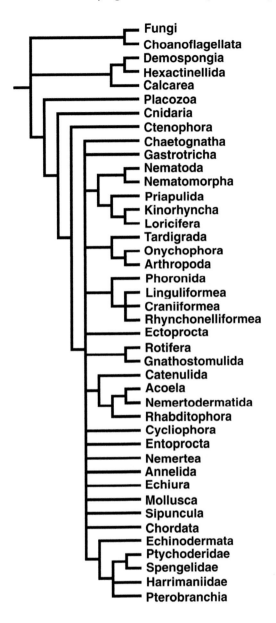

Figure 3. Strict consensus of results of sensitivity experiment 8 (see Table 1, Appendix 2). Neither Articulata nor Ecdysozoa is supported. See text for discussion.

3 CONCLUSIONS

Claims that morphological data sets strongly support the Ecdysozoa (Peterson & Eernisse 2001) are clearly premature and not justified. Our study shows a bias in most morphological cladistic studies towards finding support for Ecdysozoa, which is caused by several factors.

First, it might sound trivial but the choice of characters is very important. If one excludes characters that support a particular grouping, the testing power of the phylogenetic analysis is crippled, and the corresponding clade may not appear in the results. So far, the characters supporting Articulata were never adequately included in the analyses.

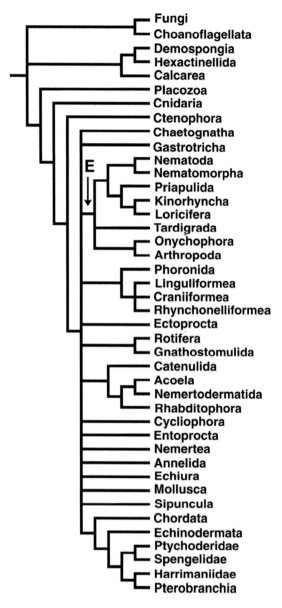

Figure 4. Strict consensus of results of sensitivity experiment 10 (see Table 1, Appendix 2). Ecdysozoa is monophyletic (indicated by clade E in the tree). See text for discussion.

Second, it is also important to stress once again that the definition, coding, and scoring of characters is important in determining the outcome of the phylogenetic analysis. We show that errors in the coding and scoring of characters in the data set of Peterson & Eernisse (2001) have caused spurious support for Ecdysozoa.

In contrast, our sensitivity experiments clearly reveal that a careful evaluation of the morphological data at hand lends support to the Articulata hypothesis. However, we believe that available morphological evidence might not allow the unambiguous choice between the Ecdysozoa and Articulata hypotheses. This unambiguity can be seen as a paradigm for the general problems of reconstructing the phylogeny of Metazoa using morphological characters. There are two reasons for this.

First, the morphological evidence is often uncertain. The precise distribution of many of the character states within the higher-level taxa is not well understood. This creates inevitable uncertainty of ground patterns. Moreover, homologies may in several cases be very uncertain, irrespective of whether the character is uniformly distributed in the higher taxa. Our new character 139, which homologizes parapodia, lobopodia, and arthropodia, is a good example of a contentious character. This uncertainty can only be overcome when more attention is paid to character quality and to character treatment in data matrices instead of stressing just data quantity and focusing solely on the improvement of the analytical methods of the given data.

Second, different workers may adopt radically different phylogenetic philosophies. These differences concern, for instance, the choice and quality of characters, the a priori and independent test of homology based on character complexity, the inclusion of assumptions of evolutionary processes, functionality and plausibility criteria, or the kind of terminal taxa chosen for the analysis. In some cases this will inevitably lead to conflicting phylogenies, because the available evidence is interpreted very differently. It is obvious that many workers strictly adhere to a single phylogenetic epistemology, and they will often tend to dismiss results generated on the basis of conflicting philosophies. Phylogenetic inference is an inexorably and intensely interpretative activity, and, therefore, such disagreements will continue to exist.

As one example of the difficulties encountered, some workers are intent on purging phylogenetic reconstruction from all traces of "bias and subjectivity"; what is more, in order to resolve this phylogenetic debate "we will have to rely on observation rather than inference, by coding exemplars instead of using inferred ground-patterns" (Giribet 2003a: 315). How one could reconstruct ancient divergence events while strictly avoiding bias in perspective, subjectivity, and inference, remains a mystery to us. In contrast, the substitution of exemplar species for ground patterns does not correspond to the substitution of observation for inference in the reconstruction of higher level phylogeny. As Prendini (2001: 290) noted, both the exemplar and the ground pattern approaches have the same aim, namely "to estimate the groundplan, or plesiomorphic states of the higher taxa concerned". This task cannot simply be performed by 'observation', and inference is unavoidable for both approaches. For the exemplar approach, selection of the exemplar species is of central importance, while for the ground pattern approach it is the choice of plesiomorphic character states. As noted by Prendini (2001: 295), "the selection of exemplar species scarcely differs epistemologically from the estimation of plesiomorphic states for a supraspecific terminal taxon". Judicious choice of exemplar species is critical for avoiding basing a phylogenetic analysis on phylogenetically misleading derived character states. Thus, studies of comparative morphology that underlie all morphological cladistic analyses can simply not be performed in any meaningful theory-free manner. However, workers may differ in the allowed amount of assumptions fed into the analysis.

Much the same holds true for the use of molecular data - contrary to the widespread opinion that molecular data offer a direct and more objective approach to reconstructing phylogenetic relationships. In molecular analyses there are many subjective inferences involved attending gene choice, gene homologies, sequence alignment, and the performance of sensitivity analyses under different sets of assumptions (e.g., Giribet 2003a, b).

Consequently, the results reported in this paper should be interpreted as very tentative and preliminary. Phylogenetic research is composed of repeated cycles of data analysis, and it is in this context that an open-minded and experimental approach to phylogenetic analysis (sensitivity analysis) will prove useful. Previous morphological cladistic analyses focusing on the problem of Ecdysozoa versus Articulata have typically not adopted such an experimental approach. This approach allowed us to show that support for Articulata or Ecdysozoa depends critically upon which characters are included in the analysis, and how these characters are coded. For example, the data set of Peterson & Eernisse (2001) provided spurious support for Ecdysozoa due to scoring errors. Nevertheless, the remaining support for Ecdysozoa was quite robust. However, when most problematic characters were excluded, and our new potential articulatan synapomorphies were included, Ecdysozoa gave way to Articulata.

ACKNOWLEDGEMENTS

RAJ gratefully acknowledges support from a Marie Curie Individual Fellowship of the European Community program Improving Human Potential under contract number HPMF-CT-2002-01712. GS' studies on segmentation are supported by grants of the Deutsche Forschungsgemeinschaft. Stefan Richter, Gonzalo Giribet, Stefan Koenemann, and an anonymous reviewer provided valuable comments on the manuscript. We are grateful for the invitation to contribute a chapter to this book.

APPENDIX 1

Summary of new characters added and character selection, coding and scoring changes. The following changes are made to the morphological data set of Peterson & Eernisse (2001). The numbers identify the characters: 1-138 are the original characters of Peterson & Eernisse (2001), several of which may be modified or deleted in different sensitivity analyses, and characters 139-156 are newly added. If no specific reference is cited the information is discussed in one or more recent papers discussing Ecdysozoa versus Articulata (Schmidt-Rhaesa et al. 1998; Wägele et al. 1999; Nielsen 2001; Wägele & Misof 2001; Zrzavý 2001; Scholtz 2002, 2003; Giribet 2003a).

New characters

139. Iterated limb pairs absent (0) / present (1); Although the homology between polychaete parapodia, and the lobopodia and arthropodia of pan-arthropods remains contentious we scored them as homologous.
140. Mushroom bodies (corpora pedunculata) in anterior brain absent (0) / present (1); These mushroom-shaped neuropil regions have been identified in arthropods, onychophorans and annelids. Tardigrades appear to possess similar structures (Dewel & Dewel 1996).
141. Metanephridia: absent (0) / present (1); Peterson & Eernisse (2001) do not accept the potential homology of metanephridia in neotrochozoans (Annelida, Mollusca, Echiura, Sipuncula), brachiopods, phoronids, arthropods, and onychophorans. We disagree, and consider the ultrastructural similarities sufficient to warrant a proposal of primary homology.
142. Anterior boundaries of *Hox* gene expression: two segments distance between subsequent *Hox* genes (0) (annelids, arthropods) / three segments distance (1) (chordates). Remaining non-segmented taxa not applicable. See Scholtz (2002, 2003) for discussion.
143. *Engrailed* expression: not iterated (0) (*Saccoglossus*) / iterated dorsal stripes (1) (molluscs 0/1) / iterated ventral stripes (2) (annelids, onychophorans arthropods) / in mesodermal blocks (3) (chordates 0/3). Most taxa unknown. See Scholtz (2002, 2003) for discussion.
144. *Wingless* expression not in stripes (0) (nematodes, cnidarians, chordates) / iterated stripes anterior to engrailed (1) (annelids, arthropods). Most taxa unknown.
145. Germ elongation: scattered cell divisions (0) (molluscs, chordates, basically unknown for most taxa) / from a preanal growth zone (1) (annelids, onychophorans, arthropods).
146. Teloblasts in ectoderm (new version of character 42) : absent (0) (all taxa except annelids and arthropods) / present (1) (annelids and arthropods 0/1). See Scholtz (2002, 2003).
147. Teloblasts in mesoderm (new version of character 42): absent (0) / present (1) (annelids, echiurids, molluscs, arthropods 0/1). See Scholtz (2002, 2003).
148. Circulatory system: absent (0) / present (1). Present in all coelomates, including ectoprocts (funiculus) and chaetognaths (associated with gut), except sipunculans, absent in remaining bilaterians, and inapplicable in non-bilaterians, which lack mesoderm.
149. Mode of coelomogenesis (character 36 modified): enterocoely (0) (echinoderms, enteropneusts, chaetognaths) / schizocoely (1) (neotrochozoans, arthropods, onychophorans, nemerteans, pterobranchs). We regard the primitive mode of coelomogenesis in phoronids, ectoprocts, brachiopods, and chordates as uncertain. Taxa lacking coeloms are scored as inapplicable. See Nielsen (2001) and Jenner (2004c) for discussion.

150. Dorsal heart: absent (0) (ectoprocts, chaetognaths, echiurans, sipunculans, nemerteans) / present (1) (molluscs, brachiopods, echinoderms, enteropneusts, pterobranchs) / long tube (2) (annelids, onychophorans, arthropods, enteropneusts [the dorsal vessel is contractile in addition to the pericardial heart in the posterior part of the prosoma]). Inapplicable for non-coelomates, and phoronids and chordates (due to uncertainty of what is dorsal with respect to other phyla) scored '?'.
151. Heart with ostia: absent (0) / present (1) (onychophorans, arthropods). Not applicable to taxa without circulatory system.
152. Metanephridia serially repeated: (0) for taxa with one pair / (1) present (annelids, onychophorans, arthropods; molluscs, chordates and sipunculans scored 0/1). Not applicable to taxa without metanephridia.
153. Coelom absent (0) / present (1). Inapplicable if mesoderm is absent.
154. Thin visceral coelomic walls and thick somatic coelomic walls in the embryo absent (0) (all taxa except annelids, onychophorans, and arthropods) / present (1) (annelids, onychophorans, arthropods). Inapplicable in non-coelomates. Phoronids, sipunculans, echiurans, ectoprocts scored as '?'.
155. Trochophore larva absent (0) / present with opposed band system (prototroch + metatroch) (1) / with prototroch only (2) (This character replaces characters 48-50).
156. More or less regular commissures between ventral connectives in the adult nervous system (inapplicable when multiple ventral nerve cords are absent) absent (0) / present (1).

Modified scoring
13. Ciliated epidermis absent (0) / present with monociliated cells (1) / present with multi- or multi- + monociliated cells (2). Scored 1 for onychophorans (based upon mono-ciliated cells in ectodermally derived hypocerebral organ) (Eriksson et al. 2003).
23. Gonads present with gametes passing through coelom and metanephridia absent (0) / present (1). Scored 1 for brachiopods and phoronids. Both groups shed gametes through coelom and metanephridia.
38. 4d endomesoderm absent (0) / present (1). Scored inapplicable for all non-spiral cleaving taxa.
39. Mesodermal germ bands derived from 4d absent (0) / present (1). Scored inapplicable for all non-spiral cleaving taxa.
40. Lateral coelom derived from mesodermal bands absent (0) / present (1). Scored 1 for onychophorans and arthropods. Both have coeloms developed in lateral mesodermal bands.
43. Somatoblast absent (0) / present (1). Scored inapplicable for taxa without spiral cleavage.
45. Apical organ / tuft absent (0) / present (1). Scored inapplicable for taxa lacking ciliated larvae/juveniles.
46. Apical organ with muscles extending to the hyposphere absent (0) / present (1). Scored inapplicable for taxa lacking apical organs.
47. Pretrochal *Anlagen* absent (0) / present (1). Scored inapplicable for taxa without trochophore larva.
48. Prototroch absent (0) / present (1). Scored inapplicable for taxa without trochophore larva; scored '?' for ectoprocts, rotifers, cycliophorans, nemerteans. See Jenner (2004c) for discussion.

49. Metatroch absent (0) / present (1). Scored inapplicable for taxa without trochophore larva; scored '?' in rotifers and cycliophorans, molluscs, and sipunculans. See Jenner (2004c) for discussion.
50. Adoral ciliary band absent (0) / present (1). Scored inapplicable for taxa without trochophore larva; scored '?' for rotifers. See Jenner (2004c) for discussion.
51. Telotroch absent (0) / present (1). Scored inapplicable for taxa without ciliated larva.
52. Neurotroch absent (0) / present (1). Scored inapplicable for taxa without trochophore larva.
53. Neotroch absent (0) / present (1). Scored inapplicable for taxa without ciliated larva.
54. Nonmuscular peritoneal cells in lateral regions of coelom absent (0) / present (1). Scored annelids 0/1.
78. Cuticle with chitin absent (0) / present (1). Brachiopods and echiurans scored 1, rotifers, ectoprocts, and cnidarians scored 0/1 (Ax 1999, Nielsen 2001).
79. Trilaminate epicuticle absent (0) / present (1). Scored 0/1 for brachiopods, molluscs, and ectoprocts (Schmidt-Rhaesa et al. 1998).
80. Trilayered cuticle absent (0) / present (1). Scored 1 also in at least molluscs and brachiopods.
83. Ecdysis absent (0) / present (1). Annelids 0/1 (general mode in leeches), sipunculans 0/1 (Giribet 2003a).
86. Head divided into 3 segments absent (0) / present (1). Scored inapplicable for taxa without segmentation.
87. Terminal mouth absent (0) / present (1). Scored 1 for onychophorans and arthropods because they primitively may have had a terminal mouth.
93. Digestive gut without cilia absent (0) / present (1). Intestinal cilia present in rhabditophorans and nemerteans. See Jenner (2004c) for discussion.
110. Gliointerstitial cell system absent (0) / present (1). Scored 1 for echiurans, cycliophorans, brachiopods, and 0/1 for chordates. See Jenner (2004c) for discussion.

Modified coding
101. New version: Ganglia: absent (0), present (1) serially repeated pairs (2) (annelids, arthropods, onychophorans, tardigrades)

Deletions
The following characters in the data set of Peterson & Eernisse (2001) had problems regarding definition, coding, or scoring, and we decided to exclude them from several of the analyses.
28. Stereotypical cleavage pattern absent (0) / present (1). Without a specification of different cleavage types we find this character not meaningful.
30-31. Annelid cross (30) and molluscan cross (31) absent (0) / present (1). There is no support for these characters (Jenner 2003; Maslakova et al. 2004).
33. Blastopore associated with larval / adult mouth absent (0) / present (1). This character is greatly variable across taxa such as arthropods, onychophorans, and nemerteans, in which the blastopore may contribute to the mouth or not. In addition taxa such as rhabditophorans and cnidarians are scored differently although the blastopore may form the mouth in both. These concerns lead us to discard this character.
37. Ectomesenchyme absent (0) / present (1). This character is not well-defined and cell lineage data are unknown for many taxa (see Jenner 2004c).
47. Pretrochal *Anlagen* absent (0) / present (1). A vaguely defined character, with no clear distinction between absence and presence.

54.   Nonmuscular peritoneal cells in lateral regions of coelom absent (0) / present (1). Other taxa that also possess peritoneal cells, such as chaetognaths and echinoderms, should be examined for this character as well.

114.   Closed circulatory system with dorsal and ventral blood vessels absent (0) / present (1). Closed ventral and dorsal blood vessels are more widespread in the Metazoa, for example in hemichordates, chordates, and phoronids, and should be rescored.

The scoring of the following characters should undergo major adjustments.

130.   *Antp* absent (0) / present (1).
131.   *Ubx/abd-A* absent (0) / present (1).
132.   *Lox2/4* absent (0) / present (1).
133.   *Hox 6-8* absent (0) / present (1).

APPENDIX 2

Table 1. Summary of sensitivity analyses performed on the morphological data set of Peterson & Eernisse (2001). In the left 'Exp.' column all sensitivity experiments are numbered. The 'Description' column gives a summary of the modifications to the data set used for each sensitivity experiment. The 'Results' column shows whether the monophyly of Articulata, Ecdysozoa, or neither is supported in the strict consensus of each sensitivity experiment. The reported numbers and length of the MPTs may differ between repeated heuristic searches. They should therefore be considered as an indication, not as fixed. See text for discussion of the results. Abbreviations: chr(s). = character(s); Exp. = sensitivity experiment; MPTs = most parsimonious trees.

Exp.	Description	Results
1	Corrected matrix chrs. 1-138	Ecdysozoa; 48 MPTs of 242 steps
2	As Exp. 1 with chr. 42 upweighted to 10 or 12	Ecdysozoa
3	As Exp. 1 with chr. 42 upweighted to 14	Neither
4	As Exp. 1 with chr. 42 upweighted to 15 or higher	Articulata
5	Corrected matrix chrs. 1-138, deletion of problematic chrs. (see 'Deletions' in Appendix 1), modified chr. 101, included all new chrs., deletion of 36, 42, 48-50 (replaced by new chrs.)	Articulata; 42 MPTs of 250 steps
6	As Exp. 5, with replaced chrs. 36, 42, 48-50 included, and their new replacements (146, 147, 149, 155) excluded	Articulata; 90 MPTs of 252 steps
7	As Exp. 5 including problematic chr. 28	Articulata; 42 MPTs of 253 steps
8	As Exp. 5 including problematic chr. 114	Neither; 355 MPTS of 252 steps
9	As Exp. 7 including problematic chrs. 30, 31	Neither; 338 MPTs of 257 steps
10	As Exp. 5 including problematic chrs. 130-133	Ecdysozoa; 510 MPTs of 258 steps
11	As Exp. 1 including new chrs. 139-141	Ecdysozoa; 48 MPTs of 249 steps
12	As Exp. 11 including new chrs. 148, 149, excluding 36 which is replaced by 149	Ecdysozoa; 86 MPTs of 254 steps
13	As Exp. 12 including new chrs. 150-154	Ecdysozoa; 42 MPTs of 267 steps
14	As Exp. 13 including new chrs. 155, 156, excluding 48-50 which are replaced by 155	Ecdysozoa; 32 MPTs of 270 steps
15	As Exp. 14 including new chrs. 146, 147, excluding 42 which is replaced by them	Ecdysozoa; 32 MPTs of 268 steps
16	As Exp. 15 including new chr. 145	Ecdysozoa; 33 MPTs of 270 steps

Table 1 continued.

Exp.	Description	Results
17	As Exp. 16 including new chr. 142	Ecdysozoa; 32 MPTs of 271 steps
18	As Exp. 17 including new chr. 143	Ecdysozoa; 32 MPTs of 273 steps
19	As Exp. 18 including new chr. 144	Ecdysozoa; 60 MPTs of 275 steps
20	As Exp. 1 excluding problematic chrs. 28, 30, 31	Ecdysozoa; 24 MPTs of 236 steps
21	As Exp. 20 excluding problematic chr. 33	Ecdysozoa; 60 MPTs of 232 steps
22	As Exp. 21 excluding problematic chr. 37	Ecdysozoa; 61 MPTs of 230 steps
23	As Exp. 22 excluding problematic chr. 47	Ecdysozoa; 329 MPTs of 229 steps
24	As Exp. 23 excluding problematic chr. 54	Ecdysozoa; 224 MPTs of 227 steps
25	As Exp. 24 excluding problematic chrs. 130-133	Ecdysozoa; 86 MPTs of 220 steps
26	As Exp. 25 excluding problematic chr. 114	Ecdysozoa; 183 MPTs of 219 steps
27	All old and new chrs. included, excluding the replaced chrs. 36, 42, 48-50, and problematic chr. 114, original coding for 101	Ecdysozoa; 32 MPTs of 273 steps
28	As Exp. 27 including 114, excluding problematic chrs. 130-133	Ecdysozoa; 36 MPTs of 260 steps
29	As Exp. 27 excluding problematic chrs. 130-133	Neither; 57 MPTs of 265 steps
30	As Exp. 28 excluding problematic chr. 54	Neither; 110 MPTs of 263 steps
31	As Exp. 30 excluding problematic chr. 47	Neither; 50 MPTs of 261 steps
32	As Exp. 31 excluding problematic chr. 37	Articulata (minus Tardigrada); 10 MPTs of 250 steps
33	As Exp. 32 excluding problematic chr. 33	Articulata (minus Tardigrada); 10 MPTs of 258 steps
34	As Exp. 27 including 114, excluding problematic chr. 28	Ecdysozoa; 32 MPTs of 269 steps
35	As Exp. 34 excluding problematic chr. 30	Ecdysozoa; 32 MPTs of 268 steps
36	As Exp. 35 excluding problematic chr. 31	Ecdysozoa; 28 MPTs of 266 steps
37	As Exp. 36 excluding problematic chr. 33	Ecdysozoa; 225 MPTs of 265 steps
38	As Exp. 37 excluding problematic chr. 37	Ecdysozoa; 628 MPTs of 260 steps
39	As Exp. 38 excluding problematic chr. 47	Ecdysozoa; 490 MPTs of 257 steps

Table 1 continued.

Exp.	Description	Results
40	As Exp. 39 excluding problematic chr. 54	Ecdysozoa; 490 MPTs of 256 steps
41	As Exp. 40 excluding problematic chr. 114	Ecdysozoa; 510 MPTs of 255 steps
42	As Exp. 41 excluding problematic chr. 130	Neither; 641 MPTs of 253 steps
43	As Exp. 42 excluding problematic chr. 131	Neither; 896 MPTs of 251 steps
44	As Exp. 43 excluding problematic chr. 132	Neither; 895 MPTs of 250 steps
45	As Exp. 44 excluding problematic chr. 133	Neither; 486 MPTs of 248 steps
46	As Exps. 34-41, but chr. 101 newly coded	Ecdysozoa
47	As Exp. 42 but chr. 101 newly coded	Neither; 642 MPTs of 255 steps
48	As Exp. 43 but chr. 101 newly coded	Articulata; 33 MPTs of 252 steps
49	As Exp. 44 but chr. 101 newly coded	Articulata; 32 MPTs of 251 steps

REFERENCES

Aguinaldo, A.M.A., Turbeville, J.M., Linford, L.S., Rivera, M.C., Garey, J.R., Raff, R.A. & Lake, J.A. 1997. Evidence for a clade of nematodes, arthropods and other moulting animals. *Nature* 387: 489-493.

Almeida, W.O., Christoffersen, M.L., Amorim, D.S., Garraffoni, A.R.S. & Silva, G.S. 2003. Polychaeta, Annelida, and Articulata are not monophyletic: articulating the Metameria (Metazoa: Coelomata). *Rev. Brasil. Zool.* 20: 23-57.

Asma, S.T. 1996. *Following Form and Function. A Philosophical Archaeology of Life Science.* Evanston: Northwestern Univ. Press.

Ax, P. 1984. *Das phylogenetische System*. Stuttgart: Gustav Fischer Verlag.

Ax, P. 1999. *Das System der Metazoa II*. Stuttgart: Gustav Fischer Verlag.

Bivort, B.L. de & Giribet, G. 2004. A new genus of cyphophtalmid from the Iberian Peninsula with a phylogenetic analysis of the Sironidae (Arachnida: Opiliones, Cyphophtalmi) and a SEM database of external morphology. *Invert. Syst.* 18: 7-52.

Blair, J.E., Ikeo, K., Gojobori, T., Hedges, S.B. 2002. The evolutionary position of nematodes. *BMC Evol. Biol.* 2: 7.

Bowler, P.J. 1996. *Life's Splendid Drama; Evolutionary Biology and the Reconstruction of Life's Ancestry, 1860-1940.* Chicago: University of Chicago Press.

Bowler, P.J. 2000. Philosophy, instinct, intuition: what motivates the scientist in search of a theory? *Biol. Phil.* 15: 93-101.

Brusca, R.C. & Brusca, G.J. 1990. *Invertebrates*. Massachusetts: Sinauer Assoc,.

Brusca, R.C. & Brusca, G.J. 2003. *Invertebrates; Second edition*. Massachusetts: Sinauer Assoc,.

Budd, G.E. 1999. The morphology and phylogenetic significance of *Kerygmachela kierkegaardi* Budd (Buen Formation, Lower Cambrian, N. Greenland). *Trans. R. Soc. Edinburgh, Earth Sci.,* 89: 249-290.

Budd, G.E. 2001. Why are arthropods segmented? *Evol. Dev.* 3: 332-342.

Bütschli, O. 1876. Untersuchungen über freilebende Nematoden und die Gattung *Chaetonotus*. *Z. wiss. Zool.* 26: 363-413.

Copley, R.R., Alor, P., Russell, R.B. & Telford, M.J. 2004. Systematic searches for molecular synapomorphies in model metazoan genomes give some support for Ecdysozoa after accounting for the idiosyncrasies of *Caenorhabditis elegans*. *Evol. Devel.* 6: 164-169.

Dewel, R.A. & Dewel, W.C. 1996. The brain of *Echiniscus viridissimus* Peterfi, 1956 (Heterotardigrada): A key to understanding the phylogenetic position of tardigrades and the evolution of the arthropod head. *Zool. J. Linn. Soc.* 116: 35-49.

Dohle, W. 1986. Die Evolution der Wirbellosen. *Sitzungsber. Ges. Naturf. Freunde Berlin, N.F.,* 26: 67-90.

Dohle, W. 1989. Zur Frage der Homologie ontogenetischer Muster. *Zool. Beitr., N.F.,* 32: 67-90.

Donoghue, M.J. & Ackerly, D.D. 1996. Phylogenetic uncertainties and sensitivity analyses in comparative biology. *Phil. Trans. R. Soc. London, Ser. B,* 351: 1241-1249.

Donoghue, M.J. & Sanderson, M.J. 1994. Complexity and homology in plants. In: Hall, B.K. (ed.), *Homology; The Hierarchical Basis of Comparative Biology*: 393-421. San Diego: Academic Press.

Donoghue, P.C.J., Forey, P.L. & Aldridge, R.J. 2000. Conodont affinity and chordate phylogeny. *Biol. Rev.* 75: 191-251.

Dopazo, H., Santoyo, J. & Dopazo, J. 2004. Phylogenomics and the number of characters required for obtaining an accurate phylogeny of eukaryote model species. *Bioinformatics, Suppl. 1,* 20: i116-i121.

Eernisse, D.J. 1997. Arthropod and annelid relationships re-examined. In: Fortey, R.A. & Thomas, R.H. (eds.), *Arthropod Relationships*: 43-56. London: Chapman & Hall.

Eernisse, D.J., Albert, J.S. & Anderson, F.E. 1992. Annelida and Arthropoda are not sister taxa: a phylogenetic analysis of spiralian metazoan morphology. *Syst. Biol.* 41: 305-330.

Eriksson, B.J. & Budd, G.E. 2000. Onychophoran cephalic nerves and their bearing on our understanding of head segmentation and stem-group evolution of Arthropoda. *Arthr. Struc. Devel.* 29: 197-209.

Eriksson, B.J., Tait, N.N. & Budd, G.E. 2003. Head development in the onychophoran *Euperipatoides kanangrensis* with particular reference to the central nervous system. *J. Morph.* 255: 1-23.

Forey, P.L. & Kitching, I.J. 2000. Experiments in coding multistate characters. In: Scotland, R. & Pennington, R.T. (eds.), *Homology and Systematics; Coding Characters for Phylogenetic Analysis*: 54-80. London: Taylor & Francis.

Ghiselin, M.T. 1984. Narrow approaches to phylogeny: a review of nine books of cladism. *Oxford Surv. Evol. Biol.* 1: 209-222.

Ghiselin, M.T. 1991. Classical and molecular phylogenetics. *Boll. Zool.* 58: 289-294.

Giribet, G. 2003a. Molecules, development and fossils in the study of metazoan evolution; Articulata versus Ecdysozoa revisited. *Zoology* 106: 303-326.

Giribet, G. 2003b. Stability in phylogenetic formulations and its relationship to nodal support. *Syst. Biol.* 52: 554-564.

Giribet, G., DeSalle, R. & Wheeler, W.C. 2002. 'Pluralism' and the aims of phylogenetic research. In: DeSalle, R., Giribet, G. & Wheeler, W. (eds.), *Molecular Systematics and Evolution: Theory and Practice*: 141-146. Basel: Birkhäuser Verlag.

Giribet, G., Distel, D.L., Polz, M., Sterrer, W. & Wheeler, W.C. 2000. Triploblastic relationships with emphasis on the acoelomates and the position of Gnathostomulida, Cycliophora, Plathelminthes, and Chaetognatha: a combined approach of 18S rDNA sequences and morphology. *Syst. Biol.* 49: 539-562.

Grant, T. & Kluge, A.G. 2003. Data exploration in phylogenetic inference: scientific, heuristic, or neither. *Cladistics* 19: 379-418.

Gruner, H.-E. 1980. Einführung. In: Gruner, H.-E. (ed.), *Lehrbuch der Speziellen Zoologie, Band I, 1. Teil*: 15-156. Stuttgart: Gustav Fischer Verlag.

Hausdorf, B. 2000. Early evolution of the Bilateria. *Syst. Biol.* 49: 130-142.

Hawkins, J.A. 2000. A survey of primary homology assessment: different botanists perceive and define characters in different ways. In: Scotland, R. & Pennington, R.T. (eds.), *Homology and Systematics; Coding Characters for Phylogenetic Analysis*: 22-53. London: Taylor & Francis.

Hawkins, J.A., Hughes, C.E. & Scotland, R.W. 1997. Primary homology assessment, characters and character states. *Cladistics* 13: 275-283.

Hennig, W. 1972. *Wirbellose II, Gliedertiere; Taschenbuch der Speziellen Zoologie, Teil 2, 3. Auflage*. Frankfurt: H. Deutsch.

Hennig, W. 1979. *Wirbellose I, Ausgenommen Gliedertiere; Taschenbuch der Speziellen Zoologie, Teil 1, 4. Auflage*. Frankfurt: H. Deutsch.

Hennig, W. 1983. *Die Stammesgeschichte der Chordaten*. Hamburg: Parey Verlag.

Jenner, R.A. 2002. Boolean logic and character state identity: pitfalls of character coding in metazoan cladistics. *Contr. Zool.* 71: 67-91.

Jenner, R.A. 2003. Unleashing the force of cladistics? Metazoan phylogenetics and hypothesis testing. *Integ. Comp. Biol.* 43: 207-218.

Jenner, R.A. 2004a. Accepting partnership by submission? Morphological phylogenetics in a molecular millennium. *Syst. Biol.* 53: 333-342.

Jenner, R.A. 2004b. The scientific status of metazoan cladistics: why current research practice must change. *Zool. Scr.* 33: 293-310.

Jenner, R.A. 2004c. Towards a phylogeny of the Metazoa: evaluating alternative phylogenetic positions of Platyhelminthes, Nemertea, and Gnathostomulida, with a critical reappraisal of cladistic characters. *Contr. Zool.* 73: 3-163.

Jenner, R.A. 2004d. When molecules and morphology clash: reconciling conflicting phylogenies of the Metazoa by considering secondary character loss. *Evol. Devel.* 6: 372-378.

Jenner, R.A. 2004e. Historical imagination, colonial theories and phylogenetic fashion. *Pal. Ass. Newsl.* 56: 50-59.

Jenner, R.A. 2004f. Libbie Henrietta Hyman (1888-1969): from developmental mechanics to the evolution of animal body plans. *J. Exp. Zool. (Mol. Dev. Evol.)* 5: 413-423.

Jenner, R.A. & Schram, F.R. 1999. The grand game of metazoan phylogeny: rules and strategies. *Biol. Rev.* 74: 121-142.

Kluge, A.G. 1997. Testability and the refutation and corroboration of cladistic hypotheses. *Cladistics* 13: 81-96.

Kluge, A.G. 1998. Sophisticated falsification and research cycles: consequences for differential character weighting in phylogenetic systematics. *Zool. Scr.* 26: 349-360.

Kristensen, R.M. 1991. Loricifera. In: Harrison, F.W. & Ruppert, E.E (eds.), *Microscopic Anatomy of Invertebrates; Vol. 4., Aschelminthes*: 351-375. New York: Wiley-Liss.

Lee, M.S.Y. & Doughty, P. 1997. The relationship between evolutionary theory and phylogenetic analysis. *Biol. Rev.* 72: 471-495.

Lieberman, B.S. 2003. A new soft-bodied fauna: the Pioche formation of Nevada. *J. Paleontol.* 77: 674-690.

Mallatt, J., Garey, J.R. & Shultz, J.W. 2004. Ecdysozoan phylogeny and Bayesian inference: first use of nearly complete 28S and 18S rRNA gene sequences to classify the arthropods and their kin. *Mol. Phyl. Evol.* 31: 178-191.

Mallatt, J. & Winchell, C.J. 2002. Testing the new animal phylogeny: first use of combined large-subunit and small-subunit rRNA gene sequences to classify the protostomes. *Mol. Biol. Evol.* 19: 289-301.

Manuel, M., Kruse, M., Müller, W.E.G. & Le Parco, Y. 2000. The comparison of b-thymosin homologues among Metazoa supports an arthropod-nematode clade. *J. Mol. Evol.* 51: 378-381.

Meglitsch, P.A. & Schram, F.R. 1991. *Invertebrate Zoology*. Oxford: Oxford Univ. Press.

Nielsen, C. 2001. *Animal evolution; Interrelationships of the Living Phyla*. Oxford: Oxford Univ. Press.

Nielsen, C. 2003. Proposing a solution to the Articulata-Ecdysozoa controversy. *Zool. Scr.* 32: 475-482.

Peterson, K.J. & Eernisse, D.J. 2001. Animal phylogeny and the ancestry of bilaterians: inferences from morphology and 18S rDNA gene sequences. *Evol. Dev.* 3: 170-205.

Prendini, L. 2001. Species or supraspecific taxa as terminals in cladistic analysis? Groundplans versus exemplars revisited. *Syst. Biol.* 50: 290-300.

Prendini, L. 2003. A new genus and species of bothriurid scorpion from the Brandberg Massif, Namibia, with a reanalysis of bothriurid phylogeny and a discussion of the phylogenetic position of *Lisposoma* Lawrence. *Syst. Entomol.* 28: 149-172.

Rauther, M. 1909. Morphologie und Verwandschaftsbeziehungen der Nematoden. *Ergeb. Fortschr. Zool.* 1: 491-596.

Remane, A. 1952. *Die Grundlagen des natürlichen Systems der vergleichenden Anatomie und der Phylogenetik*. Leipzig: Geest & Portig.

Riedl, R. 1975. *Die Ordnung des Lebendigen*. Hamburg: Parey.

Rouse, G.W. 2001. A cladistic analysis of Siboglinidae Caullery, 1914 (Polychaeta, Annelida): formerly the phyla Pogonophora and Vestimentifera. *Zool. J. Linn. Soc.* 132: 55-80.

Rouse, G.W. & Fauchald, K. 1997. Cladistics and polychaetes. *Zool. Scr.* 26: 139-204.

Schmidt-Rhaesa, A., Bartolomaeus, T., Lemburg, C., Ehlers, U. & Garey, J.R. 1998. The position of the Arthropoda in the phylogenetic system. *J. Morph.* 238: 263-285.

Scholtz, G. 1997. Cleavage, germ band formation and head segmentation: the ground pattern of the Euarthropoda. In: Fortey, R.A. & Thomas, R.H. (eds.), *Arthropod Relationships*: 317-332. London: Chapman & Hall.

Scholtz, G. 2002. The Articulata hypothesis - or what is a segment? *Org. Divers. Evol.* 2: 197-215.

Scholtz, G. 2003. Is the taxon Articulata obsolete? Arguments in favour of a close relationship between annelids and arthropods. In: Legakis, A., Sfentourakis, S., Polymeni, R. & Thessalou-Legaki, M. (eds.), *The new panorama of animal evolution*: 489-501. Sofia: Pensoft Publishers.

Schram, F.R. 1991. Cladistic analysis of metazoan phyla and the placement of fossil problematica. In: Simonetta, A.M. & Conway Morris, S. (eds.), *The Early Evolution of Metazoa and the Significance of Problematic Taxa*: 35-46. Cambridge: Cambridge Univ. Press.

Schram, F.R. 1997. Of cavities - and kings. *Contrib. Zool.* 62: 143-150.

Schram, F.R. 2004. The truly new systematics - megascience in the information age. *Hydrobiologia* 519: 1-7.

Siddall, M.E. & Kluge, A.G. 1997. Probabilism and phylogenetic inference. *Cladistics* 13: 313-336.

Simmons, N.B. & Geisler, J.H. 2002. Sensitivity analysis of different methods of coding taxonomic polymorphism: an example from higher-level bat phylogeny. *Cladistics* 18: 571-584.

Sørensen, M.V., Funch, P., Willerslev, E., Hansen, A.J. & Olesen, J. 2000. On the phylogeny of the Metazoa in the light of Cycliophora and Micrognathozoa. *Zool. Anz.* 239: 297-318.

Swofford, D.L. 2002. *PAUP**. Sunderland: Sinauer Assoc.

Telford, M.J. 2004. The multimeric *b-thymosin* found in nematodes and arthropods is not a synapomorphy of the Ecdysozoa. *Evol. Dev.* 6: 90-94.

Tudge, C. 2000. *The Variety of Life.* Oxford: Oxford Univ. Press.

Turbeville, J.M. 2002. Progress in nemertean biology: development and phylogeny. *Integ. Comp. Biol.* 42: 692-703.

Wägele, J.W., Erikson, T., Lockart, P. & Misof, B. 1999. The Ecdysozoa: artifact or monophylum? *J. Zool. Syst. Evol. Res.* 37: 211-223.

Wägele, J.W. & Misof, B. 2001. On quality of evidence in phylogeny reconstruction: a reply to Zrzavý's defence of the 'Ecdysozoa' hypothesis. *J. Zool. Syst. Evol. Res.* 39: 165-176.

Wägele, J.W. & Stanjek, G. 1995. Arthropod phylogeny inferred from partial 12S rRNA revisited: monophyly of the Tracheata depends on sequence alignments. *J. Zool. Syst. Evol. Research* 33: 75-80.

Westheide, W. 2004. Review of "Anderson, D.T. (ed.), Invertebrate Zoology". *J. Zool. Syst. Evol. Res.* 42: 171-172.

Wheeler, W.C. 1995. Sequence alignment, parameter sensitivity, and the phylogenetic analysis of molecular data. *Syst. Biol.* 44: 321-331.

Wolf, Y.I., Rogozin, I.B. & Koonin, E.V. 2004. Coelomata and not Ecdysozoa: evidence from genome-wide phylogenetic analysis. *Genome Res.* 14: 29-36.

Zrzavý, J. 2001. Ecdysozoa versus Articulata: clades, artifacts, prejudices. *J. Zool. Syst. Evol. Res.* 39: 159-163.

Zrzavý, J. 2003. Gastrotricha and metazoan phylogeny. *Zool. Scr.* 32: 61-81.

Zrzavý, J., Hypsa, V. & Tietz, D.F. 2001. Myzostomida are not annelids. Molecular and morphological support for a clade of animals with anterior sperm flagella. *Cladistics* 17: 170-198.

Zrzavý, J., Mihulka, S., Kepka, P., Bezdek, A. & Tietz, D. 1998. Phylogeny of the Metazoa based on morphological and 18S ribosomal DNA evidence. *Cladistics* 14: 249-285.

Appendix A

PUBLICATIONS OF FREDERICK R. SCHRAM

References are given in alphabetical order for each year.

1967
Schram, F.R. An important Middle Pennsylvanian crustacean fauna from the Mazon Creek area of Illinois. *Abstracts with Program, Annual Meeting 1967, Geol. Soc. Amer.*: p. 197.

1968
Schram, F.R. *Paleosquilla* gen. nov., a stomatopod crustacean from the Cretaceous of Colombia. *J. Paleontol.* 42: 1297-1301.

1969
Schram, F.R. Polyphyly in the Eumalacostraca? *Crustaceana* 16: 243-250.
Schram, F.R. Stratigraphic distribution of the Paleozoic Eumalacostraca. *Fieldiana, Geol.* 12: 213-234.
Schram, F.R. Some Middle Pennsylvanian Hoplocarida and their phylogenetic significance. *Fieldiana, Geol.* 12: 235-289.
Schram, F.R. Insights into the evolution of the Syncarida. *Abstracts with Program, Annual Meeting 1969, Geol. Soc. Amer.*: p. 201.

1970
Schram, F.R. Isopod from the Pennsylvanian of Illinois. *Science* 169: 854-855.
Schram, F.R. & Turnbull, W.D. Structural composition and dental variations in the murids of the Broom Cave fauna, Late Pleistocene, Wombeyan Caves area, N.S.W., Australia. *Rec. Austr. Mus.* 28: 1-24, pls.1-3.

1971
Schram, F.R. A strange arthropod from the Mazon Creek area of Illinois; and the trans-Permo-Triassic Merostomoidea. *Fieldiana, Geol.* 20: 85-102.
Schram, F.R. *Litogaster turnbullensis*, a Lower Triassic glypheid decapod crustacean from Idaho. *J. Paleontol.* 45: 534-537.

1973
Schram, F.R. On some phyllocarids and the origin of the Hoplocarida. *Fieldiana, Geol.* 26: 77-94.
Schram, F.R. Pseudocoelomates and a nemertine from the Illinois Pennsylvanian. *J. Paleontol.* 47: 985-989.
Turnbull, W.D. & Schram, F.R. Broom Cave *Cercartetus*, with observations on pygmy possum dental morphology, variation, and taxonomy. *Rec. Austr. Mus.* 28: 427-464, pls. 25-31.

1974
Schram, F.R. Mazon Creek caridoid Crustacea. *Fieldiana, Geol.* 30: 9-65.
Schram, F.R. Crustacean chronofaunas of the Late Paleozoic. *Abstracts with Program, Annual Meeting 1974, Geol. Soc. Amer.*: pp. 1061-1062.
Schram, F.R. Late Paleozoic Peracarida of North America. *Fieldiana, Geol.* 33: 95-124.
Schram, F.R. Convergences between Late Paleozoic and modern caridoid Malacostraca. *Syst. Zool.* 23: 323-332.

Schram, J.M. & Schram, F.R. *Squillites spinosus* Scott 1938 (Syncarida: Malacostraca) from the Mississippian Heath Shale of central Montana. *J. Paleontol.* 48: 95-104.

1975

Gilliam, J.K. & Schram, F.R. The fauna and paleoecology of a Pennsylvanian shale. *Trans. Illin. Acad. Sci.* 68: 136-144.

O'Neill, T.C. & Schram, F.R. Conodonts of the Mecca Quarry Shale. *Trans. Illin. Acad. Sci.* 68: 18-23.

Schram, F.R. Tutmosis III: History's first male chauvinist? *Field Mus. Nat. Hist. Bull.* 46 (2): 8-12.

Schram, F.R. Lepadomorph barnacle from Mazon Creek. *J. Paleontol.* 49: 928-930.

Schram, F.R. & Nitecki, M.H. 1975. Hydra from the Illinois Pennsylvanian. *J. Paleontol.* 49: 549-551.

1976

Nitecki, M.H. & Schram, F.R. *Etacystis communis*, a fossil of uncertain affinities from the Mazon Creek fauna (Pennsylvanian of Illinois). *J. Paleontol.* 50: 1157-1161.

Schram, F.R. Crustaceans from the Pennsylvanian Linton vertebrate beds of Ohio. *Palaeontology* 19: 411-412.

Schram, F.R. *Peachocaris*, a new name for *Peachella* Schram, 1974, non Walcott, 1910. *J. Paleontol.* 50: 994.

Schram, F.R. Some notes on Pennsylvanian crustaceans of the Illinois basin. *Fieldiana, Geol.* 35: 21-28.

1977

Feldmann, R.M., Schram, F.R. & Copeland, M.J. *Palaeopalaemon newberryi*, the earliest known (Late Devonian) Decapod crustacean. *Abstracts with Program, Annual Meeting 1977, Geol. Soc. Amer.*: p. 973.

Schram, F.R. Reenter the scribes. *Eastern Education Journal* 10: 9-12.

Schram, F.R. Ralph Gordon Johnson, Research Associate. *Field Mus. Nat. Hist. Bull.* 48 (7): 15.

Schram, F.R. Paleozoogeography of Late Paleozoic and Triassic Malacostraca. *Syst. Zool.* 26: 367-379.

1978

Felgenhauer, B.E. & Schram, F.R. Differential epibiont fouling in relation to grooming behavior in *Palaemonetes kadiakensis*. *Fieldiana, Zool.* 72: 83-100.

Felgenhauer, B.E. & Schram, F.R. Grooming behavior and functional morphology of the freshwater prawn, *Palaemonetes kadiakensis*. *Trans. Illin. State Acad. Sci.* 70: 237.

Herrick, E.M. & Schram, F.R. Malacostracan crustacean fauna from the Sundance Formation (Jurassic) of Wyoming. *Amer. Mus. Novitates* 2652: 1-12.

Schram, F.R. Arthropods, a convergent phenomenon. *Fieldiana, Geol.* 39: 61-108.

Schram, F.R. Carboniferous faunal setting of the Mazon Creek biotas. *Abstracts with Program, North-Central Section Geol. Soc. Amer.*: p. 284.

Schram, F.R. *Jerometichenoria grandis* n. gen., n. sp. (Crustacea: Mysidacea) from the Lower Permian of the Soviet Union. *J. Paleontol.* 52: 605-607.

Schram, F.R., Feldmann, R.M. & Copeland, M.J. The Late Devonian Palaeopalaemonidae Brooks, 1962, and the earliest decapod crustaceans. *J. Paleontol.* 52: 1375-1387.

Schram, F.R. & Hedgpeth, J.W. Locomotory mechanisms in some Antarctic pycnogonids. In: Fry, W.D. (ed.), *Sea Spiders (Pycnogonida). Zool. J. Linn. Soc.* 63: 145-169.

Schram, F.R. & Horner, J. Crustacea of the Mississippian Bear Gulch Limestone of central Montana. *J. Paleontol.* 52: 394-406.

1979

Deméré, T.A., Sundberg, F.A. & Schram, F.R. Paleoecology of a protected biotope from the Eocene Mission Valley Formation, San Diego Co., CA. In: Abbott, P.L. (ed.), *Eocene Depositional Systems*: pp. 97-102. Los Angeles: Pacific Section, SEPM.

Felgenhauer, B.E. & Schram, F.R. The functional morphology of the grooming appendages of *Palaemonetes kadiakensis*. *Fieldiana, Zool.* 2: 1-17.

Schram, F.R. Book review: "Arthropod Phylogeny with Special Reference to Insects". *Syst. Zool.* 28: 635-638.

Schram, F.R. British Carboniferous Malacostraca. *Fieldiana, Geol.* 40: 129 pp.

Schram, F.R. Crustacea. In: Fairbridge, R.W. & Jablonski, D. (eds.), *The Encyclopedia of Paleontology*: pp. 238-244. Stroudsberg, Pennsylvania: Dowden, Hutchinson & Ross, Inc.

Schram, F.R. Fossil shrimp: An adventure of the mind. *Environm. Southwest* 487: 8-11.

Schram, F.R. Limulines of the Mississippian Bear Gulch Limestone of central Montana. *Trans. San Diego Soc. Nat. Hist.* 19: 67-74.

Schram, F.R. Manton on arthropods. *Paleobiology* 5: 63-66.

Schram, F.R. Shrimping in the Coal Age. *Environm. Southwest* 486: 24-27.

Schram, F.R. The genus *Archaeocaris* and a general review of the Palaeostomatopoda. *Trans. San Diego Soc. Nat. Hist.* 19: 57-66.

Schram, F.R. The Mazon Creek biotas in the context of a Carboniferous faunal continuum. In: Nitecki, M.H. (ed.), *Mazon Creek Fossils*: pp. 159-190. New York: Academic Press.

Schram, F.R. *The Myth of Science; or, the Fantasy of Truth*. New York: Vantage Press.

Schram, F.R. Worms of the Bear Gulch Limestone of central Montana. *Trans. San Diego Soc. Nat. Hist.* 19: 107-120.

Schram, F.R. & Schram, J.M. Some shrimp of the Madera Formation (Pennsylvanian) Manzanita Mountains, New Mexico. *J. Paleontol.* 53: 169-174.

Schram, J.M. & Schram, F.R. *Joanellia lundi* n. sp. from the Mississippian Heath Shale of central Montana. *Trans. San Diego Soc. Nat. Hist.* 19: 53-56.

1980

Schram, F.R. *Pygocephalus* from the Upper Carboniferous of the Soviet Union. *J. Paleontol.* 54: 50-56.

Schram, F.R. Notes on miscellaneous crustaceans from the Late Paleozoic of the Soviet Union. *J. Paleontol.* 54: 542-547.

Schram, F.R. O relyavistsko-kvanto-mekhanicheskom podkhodye k evolyutsii. *Zhurnal Obshchey Biologii* 41: 557-573.

Schram, F.R. & Newman, W.A. *Verruca withersi* sp. nov. from the middle of the Cretaceous of Colombia. *J. Paleontol.* 54: 229-233.

1981

Pacaud, G., Rolfe, W.D.I., Schram, F.R., Secrétan, S. & Sotty, D. Quelques invertébrés nouveaux du Stéphanien de Montceau-les-Mines. *Bull. Soc. Hist. Nat. Autun* 97: 37-43.

Schram, F.R. On the classification of the Eumalacostraca. *J. Crust. Biol.* 1: 1-10.

Schram, F.R. Stalking wild mountain shrimp 'round the world, Part I. *Environm. Southwest* 493: 8-11.

Schram, F.R. Late Paleozoic crustacean communities. *J. Paleontol.* 55: 126-137.

Schram, F.R. Stalking wild mountain shrimp 'round the world, Part II. *Environm. Southwest* 494: 20-23.

1982

Rolfe, W.D.I., Schram, F.R., Pacaud, G., Sotty, D. & Secrétan, S. A remarkable Stephanian biota from Montceau-les-Mines, France. *J. Paleontol.* 56: 426-428.

Schram, F.R. Medalists of the Paleontological Society. *J. Paleontol.* 56: 469-476.

Schram, F.R. Book review: "The Evolutionary Timetable". *Palaeontol. Assc. Circ.* 108: 5.

Schram, F.R. The fossil record and evolution of Crustacea. In: Abele, L.G. (ed.), *The Biology of Crustacea, Vol. I*: pp. 93-147. New York: Academic Press.

Schram, F.R. & Felgenhauer, B.E. Collecting *Procaris* from the Hawaiian Islands. *Environm. Southwest* 498: 18-21.

Schram, F.R. & Rolfe, W.D.I. New euthycarcinoid arthropods from the Upper Pennsylvanian of France and Illinois. *J. Paleontol.* 56: 1434-1450.

1983

Dyer, J.C. & Schram, F.R. *A Manual of Invertebrate Paleontology*. Champaign, Illinois: Stipes Publ. Co.

Schram, F.R. Charles Robert Darwin. *Environm. Southwest* 500: 4-9.

Schram, F.R. Lower Carboniferous biota of Glencartholm, Eskdale, Dumfriesshire, Scotland. *Scot. J. Geol.* 19: 1-15.

Schram, F.R. Preface and Introduction. *Crust. Issues* 1: ix-xiii.

Schram, F.R. Remipedia and crustacean phylogeny. *Crust. Issues* 1: 23-28.

Schram, F.R. Method and madness in phylogeny. *Crust. Issues* 1: 331-350.

Schram, F.R. Living crustacean with trilobite similarities. *Abstracts with Program, Geol. Soc. Amer.* 16 (6): 681.

1984

Hessler, R.R. & Schram, F.R. Leptostraca as living fossils. In: Eldredge, N. & Stanley, S.M. (eds.), *Living Fossils*: pp. 189-191. New York: Springer-Verlag.

Schram, F.R. Charles Darwin and evolutionary theory. *Space Reflections* 1: 6-10.

Schram, F.R. Fossil Syncarida. *Trans. San Diego Soc. Nat. Hist.* 20: 189-246.

Schram, F.R. Relationships within eumalacostracan crustaceans. *Trans. San Diego Soc. Nat. Hist.* 20: 301-312.

Schram, F.R. People and rocks: Geologists at the museum. *Environm. Southwest* 507: 14-18.

Schram, F.R. Upper Pennsylvanian arthropods from black shales of Iowa and Nebraska. *J. Paleontol.* 58: 197-209.

Schram, F.R. Russell Patterson MacFall 1903-1983. *Environm. Southwest* 504: 9-10.

Schram, F.R. & Hessler, R.R. Anaspidid Syncarida. In: Eldredge, N. & Stanley, S.M. (eds.), *Living Fossils*: pp. 192-195. New York: Springer-Verlag.

Schram, F.R. & Malzahn, E. The fossil leptostracan *Rhabdouraea bentzi* (Malzahn, 1958). *Trans. San Diego Soc. Nat. Hist.* 20: 95-98.

Schram, F.R. & Mapes, R.H. *Imocaris tuberculata*, n. gen., n. sp. (Crustacea: Decapoda) from the Upper Mississippian Imo Formation, Arkansas. *Trans. San Diego Soc. Nat. Hist.* 20: 165-168.

1985

Schram, F.R. The mystique of dinosaurs. *Environm. Southwest* 509: 6-8.

Schram, F.R. Book review: "Towards a Monograph of *Euchirella*". *Syst. Zool.* 34: 246-247.

Schram, F.R. The Bear Gulch crustaceans and their bearing on Late Paleozoic diversity and Permo-Triassic evolution of Malacostraca. *C.R. IX Congr. Int. Strat. Geol. Carbon.* 5: 468-472.

Schram, F.R. West Indies cave divers discover new crustaceans. *Skin Diver* 1985: 106-107.

1986

Burkenroad, M.D. & Schram, F.R. Heinrich Balss (1886-1957). *J. Crust. Biol.* 6: 300-301.

Schram, F.R. *Crustacea.* New York: Oxford Univ. Press.

Schram, F.R. Martin David Burkenroad (20 March 1910 - 12 January 1986). *J. Crust. Biol.* 6: 302-307.

Schram, F.R. & Emerson, M.J. Review and redescription of *Tesnusocaris goldichi* Brooks, 1955, a lower Pennsylvanian crustacean (Remipedia: Enantiopoda). *Program and Abstracts, IV North American Paleo. Conv.*: p. A41. Boulder: Univ. Colorado.

Schram, F.R. & Emerson, M.J. The great Tesnus Fossil Expedition of 1985. *Environm. Southwest* 515: 16-21.

Schram, F.R., Sieg, J. & Malzahn, E. Fossil Tanaidacea. *Trans. San Diego Soc. Nat. Hist.* 21: 127-144.

Schram, F.R., Yager, J. & Emerson, M.J. Remipedia; Part 1, Systematics. *Mem. San Diego Soc. Nat. Hist.* 15: 1-60.

Yager, J. & Schram, F.R. *Lasionectes entrichoma* n. gen., n. sp. (Crustacea: Remipedia) from anchialine caves in the Turks and Caicos, B.W.I. *Proc. Biol. Soc. Wash.* 99: 65-70.

1988

Itô, T. & Schram, F.R. Gonopores and the reproductive system of nectiopodan Remipedia. *J. Crust. Biol.* 8: 250-253.

Schram, F.R. *Pseudotealliocaris palincsari* n. sp., a pygocephalomorph from the Pocono Formation, Mississippian of Pennsylvania. *Trans. San Diego Soc. Nat. Hist.* 21: 221-225.

Schram, F.R. Beyond collecting. *Field Notes San Diego Soc. Nat. Hist.* 1 (3): 7.

1989

Schram, F.R. Eumalacostracan crustaceans from the quarries at Hamilton. *Kansas Geol. Surv. Guidebook* 6: 105-107.

Schram, F.R. Darwin, evolution, and pseudoscience. *Environm. Southwest* 524: 8-11.

Schram, F.R. Designation of a new name and type for the Mazon Creek (Pennsylvanian, Francis Creek Shale) Tanaidacean. *J. Paleontol.* 63: 536.

Schram, F.R. Correspondence - Cry from the past. *Nature* 340: 94.

Schram, F.R. & Lewis, C.A. Functional morphology of feeding in Nectiopoda. *Crust. Issues* 6: 115-122.

1990

Emerson, M.J. & Schram, F.R. A new view of arthropod evolution. *Amer. Zool.* 30 (4): 19A.

Emerson, M.J. & Schram, F.R. A novel hypothesis for the origin of biramous limbs in arthropods. In: Mikulic, D.G. (ed.), *Arthropod Paleobiology, Short Courses in Paleontology, No. 3*: pp. 157-176. Knoxville: Univ. Tennessee.

Emerson, M.J. & Schram, F.R. Phylogenetic implications of some heretofore enigmatic Paleozoic and Mesozoic arthropods. *Abstracts and Program, 3rd Int. Crust. Conf., Brisbane*: p. 38.

Emerson, M.J. & Schram, F.R. The origin of crustacean biramous appendages and the evolution of Arthropoda. *Science* 250: 667-669.

Schram, F.R. Book review: "S.J. Gould, 1989, Wonderful Life; Norton: N.Y.". *J. Crust. Biol.* 10: 571-572.

Schram, F.R. Crustacean phylogeny. In: Mikulic, D.G. (ed.), *Arthropod Paleobiology, Short Courses in Paleontology, No. 3*: pp. 285-302. Knoxville: Univ. Tennessee.

Schram, F.R. Michael James Emerson (8 May 1954 - 22 March 1990). *J. Crust. Biol.* 10: 563-564.

Schram, F.R. On Mazon Creek Thylacocephala. *Proc. San Diego Soc. Nat. Hist.* 3: 1-16.

Schram, F.R. & Deméré, T.A. Fossils: handle with care. *Fieldnotes S. D. Soc. Nat. Hist.* 3 (5): 6.

1991

Emerson, M.J. & Schram, F.R. Remipedia; Part 2, Paleontology. *Proc. San Diego Soc. Nat. Hist.* 7: 1-52.

Meglitsch, P.A. & Schram, F.R. *Invertebrate Zoology, 3rd. Edit.* New York: Oxford Univ. Press.

Schram, F.R. Are natural history museums becoming dinosaurs? *San Diego Union*, July 21: p. C-3.

Schram, F.R. Book review: "Dall, W., B.J. Hill, P.C. Rothlisberg & D.J. Staples. 1990. The Biology of the Penaeidae; Advances In Marine Biology 27: i-xiii, 1-489". *J. Crust. Biol.* 11: 653.

Schram, F.R. Cladistic analysis of metazoan phyla and the placement of fossil problematica. In: Simonetta, A. & Conway Morris, S. (eds.), *The Early Evolution of Metazoa and the Significance of Problematic Taxa*: pp. 35-46. Cambridge Univ. Press.

Schram, F.R. & Emerson, M.J. Arthropod pattern theory: a new approach to arthropod phylogeny. *Mem. Queensland Mus.* 31: 1-18.

1992

Schram, F.R. Book review: "Anatomy of a Controversy; A.M. Wenner & P.H. Wells; Columbia Univ. Press, New York, 1990, 399 pp.". *Amer. Zool.* 32: 357.

Schram, F.R. Museum collections: Why are they there? *Science* 256: 1502.

Schram, F.R. & Dively, D. *Legacy Resource Management Programs Outside of the Department of Defense, Final Report.* Los Angeles: Nat. Hist. Mus. Los Angeles County.

Schram, F.R. Exposition - *Wiwaxia. Terra* 31 (2): 43.

1993

Schram, F.R. A correspondence between Martin Burkenroad and Libbie Hyman: or, whatever did happen to Libbie Hyman's lingerie. *Crust. Issues* 8: 119-142.

Schram, F.R. The British School: Calman, Cannon, and Manton and their effect on carcinology in the English speaking world. *Crust. Issues* 8: 321-348.

Schram, F.R. Book review: "Boxshall, G.A., J.-O. Strömberg & E. Dahl, 1992, The Crustacea: origin and evolution; Acta Zoologica 73: 271-392". *J. Crust. Biol.* 13: 440-442.

1994

Minelli, A. & Schram, F.R. Owen revisited: a reappraisal of homology and morphology in evolutionary biology. *Bijd. Dierk.* 64: 65-74.

Schram, F.R. Book review: "Watling, L. & P.G. Moore, eds., 1993, Amphipods, a noble obsession: essays in memory of J. Laurens Barnard (1928-1991); Journal of Natural History 27: 723-988". *J. Crust. Biol.* 14: 612-613.

Schram, F.R. Book review: "Biological Systematics: the State of the Art; A. Minelli; Chapman Hall, London, 1993, 387 pp.". *Amer. Zool.* 34: 476.

Schram, F.R. & Rolfe, W.D.I. The Stephanian (Late Carboniferous) Euthycarcinoidea from the Montceau-les-Mines basin (Massif Central, France). In: Poplin, C. & Heyler, D. (eds.), *Quand le Massif Central était sous l'equateur: un écosystème carbonifère à Montceau-les-Mines*: pp. 139-144. Mémoires de la Section des Sciences; CTHS Ed. no. 12.

Schram, F.R. & Secrétan, S. The Stephanian (Late Carboniferous) Syncarida from the Montceau-les-Mines basin (Massif Central, France). In: Poplin, C. & Heyler, D. (eds.), *Quand le Massif Central était sous l'equateur: un écosystème carbonifère à Montceau-les-Mines*: pp. 155-169. Mémoires de la Section des Sciences; CTHS Ed. no. 12.

1995

Glenner, H., Grygier, M.J., Høeg, J.T., Jensen, P.G. & Schram, F.R. Cladistic analysis of the Cirripedia Thoracica. *Zool. J. Linn. Soc., London* 114: 365-404.

Schram, F.R. An update on ETI. *ASC Newsletter* 23 (5): 63.

Schram, F.R. Institutional activities and the biodiversity crisis. In: de Cininck, E., Dall'Asta, U. & Fermon, H. (eds.), *Proceedings 2nd EuroLOOP Workshop, Amsterdam, 27-30 March 1995*: pp. 3-6. Tervuren, Belgium: Royal African Museum.

Schram, F.R. & Ellis, W.N. Metazoan relationships: a rebuttal. *Cladistics* 10: 331-337.

Schram, F.R. & Høeg, J.T. (eds.). New Frontiers in barnacle evolution. *Crust. Issues* 10: 297-312.

Sluys, R., Schalk, P.H. & Schram, F.R. 1995. Systematic cirripedology in the 21st century: multimedia biodiversity information systems. *Crust. Issues* 10: 43-48.

1996

Schram, F.R. Book review: "Sponges in Time and Space; R.W.M. van Soest, T.M.G. van Kempen, J.-C. Braekman; A.A. Balkema, Rotterdam, 1994, 515 pp.". *Amer. Zool.* 36: 60.

Schram,. F.R. Recent advances in understanding the phylogeny of phyla. In: Backup, L. (ed.), *Resumos XXI Congr. Brasil. Zool.*: p. 265. Puerto Allegre: Univ. Fed. Rio Grande do Sul.

Schram, F.R. Animal phylogeny at last in the mainstream. *Cladistics* 11: 219-222.

Schram, F.R. Book review: "Das System der Metazoa I: ein Lehrbuch der phylogenetischen Systematik, Peter Ax, 1995; Stuttgart, Gustav Fischer Verlag, 226 pages". *J. Comp. Biol.* 1:73-74.

Schram, F.R. William Anderson Newman: Recipient of Award for Excellence in Research. *J. Crust. Biol.* 16: 814-816.

Schram, F.R. & Los, W. Training systematists for the 21st century. In: Blackmore, S. (ed.), *Systematics Agenda 2000, a Challenge for Europe*: pp. 89-101. London: Linnean Soc.

1997

Emerson, M.J. & Schram, F.R. Theories, patterns, and reality: game plan for arthropod phylogeny. In: Fortey, R.A. & Thomas, R.H. (eds.), *Arthropod Relationships*: pp. 67-86. London: Chapman Hall.

Schram, F.R. Of cavities - and kings. *Contrib. Zool.* 67: 143-150.

Schram, F.R. & Hay, A.A. Introduction to arthropods. In: Hay, A.A. & Shabica, C.W. (eds.), *Richardson's Guide to the Fossil Fauna of Mazon Creek*: pp.131-133. Chicago: Northeastern Illin. Univ.

Schram, F.R. & Rolfe, W.D.I. Euthycarcinoids and thylacocephalans. In: Hay, A.A. & Shabica, C.W. (eds.), *Richardson's Guide to the Fossil Fauna of Mazon Creek*: pp.211-214. Chicago: Northeastern Illin. Univ.

Schram, F.R., Rolfe, W.D.I. & Hay, A.A. Crustacea. In: Hay, A.A. & Shabica, C.W. (eds.), *Richardson's Guide to the Fossil Fauna of Mazon Creek*: pp.155-171. Chicago: Northeastern Illin. Univ.

Schram, F.R. & Vonk, R. Jan Hendrik Stock: 22 February 1931 - 17 February 1997. *J. Crust. Biol.* 17: 562-564.

Schram, F.R. & Vonk, R. Jan Hendrik Stock: 22 February 1931 - 17 February 1997. *Contrib. Zool.* 66: 3-7.

Schram, F.R., Vonk, R. & Hof, C.H.J Mazon Creek Cycloidea. *J. Paleontol.* 71: 261-284.

Taylor, R.S., Schram, F.R. & Shen, Y.-B. New species of Spelaeogriphacea (Crustacea: Peracarida) from the Upper Jurassic of the People's Republic of China. *Abstracts, Annual Meeting 1997, Geol. Soc. Amer.*: p. A99.

van der Brugghen, W, Schram, F.R. & Martill, D.M. The fossil *Ainiktozoon* is an arthropod. *Nature* 385: 589-590.

1998

Hof, C.H.J. & Schram, F.R. Stomatopods (Crustacea: Malacostraca) from the Miocene of California. *J. Paleontol.* 72: 317-331.

Jenner, R.A., Hof, C.H.J. & Schram, F.R. Palaeo- and archaeostomatopods (Hoplocarida: Crustacea) from the Bear Gulch Limestone, Mississippian (Namurian), of central Montana. Contrib. Zool. 67: 155-185.

Jenner, R.A. & Schram, F.R. Reconstructing the phylogeny of the animal phyla. *Abstract Book, 5th Benelux Congr. Zool., Univ. Ghent*: p. 31.

Koenemann, S., Vonk, R. & Schram, F.R. A cladistic analysis of 37 Mediterranean Bogidiellidae (Crustacea, Amphipoda), including *Bogidiella arista* n. sp. from Turkey. *J. Crust. Biol.* 18: 383-404.

Schram, F.R. Crustaceans in the art of Amsterdam. *Proceedings and Abstracts, 4th Int. Crust. Congr., Amsterdam 1998*: p. 204-206.

Schram, F.R. & Hof, C.H.J. Fossil taxa and the relationships of major crustacean groups. In: Edgecombe, G. (ed.), *Arthropod Fossils and Phylogeny*: pp. 273-302. New York: Columbia Univ. Press.

Taylor, R.S., Schram, F.R. & Shen, Y.-B. Crayfish from the Upper Jurassic of Liaoning Province, China. *Proceedings and Abstracts, 4th Int. Crust. Congr., Amsterdam 1998*: p. 76.

Taylor, R.S., Shen, Y.-B. & Schram, F.R. New pygocephalomorph crustaceans from the Permian of China and their phylogenetic relationships. *Palaeontol.* 41: 815-834.

Vonk, R. & Schram, F.R. On the distribution and phylogeny of ingolfiellid amphipods. *Proceedings and Abstracts, 4th Int. Crust. Congr., Amsterdam 1998*: p. 58.

Whyte, M.A., Glenner, H., Høeg, J.T. & Schram, F.R. New evidence on the basic phylogeny of the Cirripedia Thoracica. *Proceedings and Abstracts, 4th Int. Crust. Congr., Amsterdam 1998*: p. 122.

1999

Høeg, J.T., Whyte, M.A., Glenner, H. & Schram, F.R. New evidence on the basic phylogeny of the Cirripedia Thoracica. In: Schram, F.R. & von Vaupel Klein, J.C. (eds.), *Crustaceans and the Biodiversity Crisis*: pp. 101-114. Leiden: Koninkl. Brill Publ.

Jenner, R.A. & Schram, F.R. The grand game of metazoan phylogeny: rules and strategies. *Biol. Rev.* 74: 121-142.

Lange, S. & Schram, F.R. Crustacean evolution and phylogeny. In: Melic, A., de Haro, J.J., Mendez, M. & Ribera, I. (eds.), *Evolucion y Filogenia de Arthropoda*: pp. 235-254. Entomol. Soc. Aragon.

Schram, F.R. Jan Stock: a personal reminiscence. *Crustaceana* 72: 723-724.

Schram, F.R. & von Vaupel Klein, J.C. (eds.). *Crustaceans and the Biodiversity Crisis*: 1021 pp. Leiden: Koninkl. Brill Publ.

Schram, F.R., Hof, C.H.J. & Steeman, F. Thylacocephala (Arthropoda: ?Crustacea) from the Cretaceous of Lebanon. *Palaeontology* 42: 769-797.

Shen, Y.-B., Schram, F.R. & Taylor, R.S. *Liaoningogriphus quadripartitus* (Crustacea: Malacostraca: Spelaeogriphacea) from the Jehol Biota and notes on its paleoecology. In: Chen Pei-ji & Jin Fan (ed.), *The Jehul Biota*: pp. 175-185, pls. 1-2. Beijing: Press Univ. Sci. Techn. China.

Shen, Y.B., Taylor, R.S. & Schram, F.R. New spelaeogriphacean from the Jurassic of China and the biogeographic implications for the evolution of the group. *Contrib. Zool.* 68: 19-35.

Taylor, R.S. & Schram, F.R. Meiura (anomalan and brachyuran crabs). In: Savazzi, E. (ed.), *Functional Morphology of the Invertebrate Skeleton*: pp. 517-528. Chichester: John Wiley.

Taylor, R.S., Schram, F.R. & Shen Y.-B. A new crayfish Family (Decapoda: Astacida) from the Upper Jurassic of China, with a reinterpretation of other Chinese crayfish taxa. *Palaeontol. Res.* 3: 121-136.

2000

Lange, S. & Schram, F.R. Advances in thylacocephalan morphology. *Stud. Ric., Assc. Amici Mus. Civ. "G. Zannato", Mont. Maggiore* 2000: 51.

Schram, F.R. Phylogeny of decapods. *Stud. Ric., Assc. Amici Mus. Civ. "G. Zannato", Mont. Maggiore* 2000: 61.

Schram, F.R. & Shen, Y.-B. An unusual specimen of crayfish molt. *Acta Palaeontol. Sinica* 39: 416-418.

Schram, F.R., Shen, Y.-B., Vonk, R. & Taylor, R.S. The first fossil stenopodidean. *Crustaceana* 73: 235-242.

von Vaupel Klein, J.C. & Schram, F.R. (eds.). *The Biodiversity Crisis and Crustacea*: 848 pp. Rotterdam: A.A. Balkema.

Watling, L., Hof, C.H.J. & Schram, F.R. The place of the Hoplocarida in the malacostracan pantheon. *J. Crust. Bio.* 20: 1-11.

2001

Belk, D. & Schram, F.R. A new species of anostracan from the Miocene of California. *J. Crust. Biol.* 21: 49-55.

Lange, S., Hof, C.H.J., Schram, F.R. & F.A. Steeman. New genus and species from the Cretaceous of Lebanon links the Thylacocephala with the Crustacea. *Palaeontology* 44: 905-912.

Schram, F.R. Phylogeny of decapods: moving towards a consensus. Hydrobiologia 449: 1-20.

Schram, F.R. Review: "Rachel Wood, 1999. Reef Evolution: i-xi, 1-414, ill. (Oxford University Press, Oxford)". *Crustaceana* 74: 703-704.

Schram, F.R. & Jenner, R.A. The origin of Hexapoda: the crustacean perspective. In: Deuve, T. (ed.), *Origin of the Hexapoda*: 243-264. *Ann. Soc. Entomol. France* 37.

Schram, F.R. & Koenemann, S. Developmental genetics and arthropod evolution: Part I, On Legs. *Evolution & Development* 3: 343-354.

Shen, Y.-B., Schram, F.R. & Taylor, R.S. Morphological variation in fossil crayfish of the Jehol biota, Liaoning Province, China and its taxonomic discrimination. *Chin. Sci. Bull.* 46: 26-33.

Taylor, R.S., Schram, F.R. & Shen, Y.-B. A new Upper Middle Triassic shrimp (Crustacea: Lophogastrida) from Guizhou, China, with discussion regarding other fossil "mysidaceans". *J. Paleontol.* 75: 310-318.

2002

Garassino, A., Shen Y.-B., Schram, F.R. & Taylor, R.S. *Yongjiacaris zhejiangensis*, n. gen, n. sp. (Crustacean: Decapoda: Caridea) from the Lower Cretaceous of Zhenjiang Province, China. *Bull. Mizunami Fossil Mus.* 29: 73-80.

Jenner, R.A. & Schram, F.R. Systematic Zoology: Invertebrates. In: *Knowledge for Sustainable Development - An Insight into the Encyclopedia of Life Support Systems*. Paris: UNESCO Publ.; Eolss Publ.

Koenemann, S. & Schram, F.R. *Hox* gene expression patterns and the location of gonopores: a re-examination of crustacean body plans. *Abstracts, 6th Int. Congr. Syst. Evol. Biol.*: p. 22. Greece: Univ. Patras.

Koenemann, S. & Schram, F.R. The limitations of ontogenetic data in phylogenetic analyses. *Contrib. Zool.* 71: 47-66.

Lange, S. & Schram, F.R. Possible lattice organs in Cretaceous Thylacocephala. *Contrib. Zool.* 71: 159-169.

Schram, F.R. (ed.). *Metazoan Deep History: Evaluating Alternative Hypotheses about the Macroevolution of Animal Body Plans*: pp. 1-113. The Hague: SPB Acad. Publ.

Schram, F.R. Evolution and developmental biology in the Netherlands. *Contrib. Zool.* 71: 3-8.

Schram, F.R. Stomatopod systematics in the 21st century. *Abstracts, 6th Int. Congr. Syst. Evol. Biol.*: p. 126. Greece: Univ. Patras.

Schram, F.R. Trails and tribulations of assembling and working with international taxonomic teams. *Abstracts, 6th Int. Congr. Syst. Evol. Biol.*: p. 156. Greece: Univ. Patras.

Schram, F.R. & Ahyong, S.T. The higher affinities of *Neoglyphea inopinata* in particular and the Glypheoidea (Reptantia: Decapoda) in general. *Crustaceana* 75: 629-635.

Schram, F.R. & Koenemann, S. Shrimp Cocktail: are the crustaceans monophyletic? *Abstracts, 6th Int. Congr. Syst. Evol. Biol.*: p. 13. Greece: Univ. Patras.

Schram, F.R., Minelli, A. & Lange, S. Arthropods Other than Insects. In: *Knowledge for Sustainable Development - An Insight into the Encyclopedia of Life Support Systems*. Paris: UNESCO Publ.; Eolss Publ.

van den Biggelaar, J.A.M., Edsinger-Gonzales, E. & Schram, F.R. The improbability of dorso-ventral axis inversion during animal evolution, as presumed by Geoffroy Saint Hilaire. *Contrib. Zool.* 71: 29-36.

2003

Dixon, C.J., Ahyong, S.T. & Schram, F.R. A new hypothesis of decapod phylogeny. *Crustaceana* 76: 935-975.

Fraaije, R.H.B., Schram, F.R. & Vonk, R. *Maastrichtiocaris rostratus*, n. gen., n. species, the first Cretaceous cycloid. *J. Paleontol.* 77: 386-388.

Schram, F.R. Insect History. *Contrib. Zool.* 72: 73-76.

Schram, F.R. Our evolving understanding of biodiversity through history and its impact on the recognition of higher taxa in Metazoa. In: A. Legakis et al. (eds.), *The New Panorama of Animal Evolution, Proc. XVIIIth Int. Congr. Zool.* pp. 359-368. Sofia: Pensoft Publ.

Schram, F.R. Review: "Peter J.F. Davie, Zoological Catalogue of Australia: vol. 19.3A, Crustacea: Malacostraca - Phyllocarida, Hoplocarida, Eucarida (Part 1), i-xii, 1-551; vol. 19.3B, Crustacea: Malacostraca - Eucarida (Part 2), Decapoda - Anomura, Brachyura, i-xiv, 1-641; CSIRO Publ., Collingwood, Victoria". *Crustaceana* 76: 125-126.

Schram, F.R. The barnacle years. *Nature* 422: 472.

Schram, F.R. & Dixon, C. Fossils and decapod phylogeny. *Contrib. Zool.* 72: 169-172.

Schram, F.R., Hof, C.H.J., Mapes, R.H. & Snowdon, P. Paleozoic cumaceans (Crustacea: Malacostraca: Peracarida) from North America. *Contrib. Zool.*

Schram, F.R. & Koenemann, S. Developmental genetics and arthropod evolution: On body regions of crustaceans. In: G. Scholtz (ed.), *Crustacean Issues 15, Evolutionary Developmental Biology of Crustacea*: 75-92. Balkema: Lisse.

Vonk, R. & Schram, F.R. Ingolfiellidea (Crustacea, Malacostraca, Amphipoda): a phylogenetic and biogeographic analysis. *Contrib. Zool.* 72: 39-72.

2004

Lange, S. & Schram, F.R. Evolución y filogenia de los crustáceos. In: Bousquets, J.L. Morrone, J.J., Ordóñes, O.Y., & Fernández, I.V. (eds.), *Biodiversidad, Taxonomía y Biogeografía de Arthrópodes de México: Hacia una sítesis de su conocimiento, Vol. IV*: pp. 93-111. Mexico: UNAM.

Marijnissen, S.A.E., Schram, F.R., Cumberlidge, N. & Michel, E. Two new species of *Platythelphusa* A. Milne Edwards, 1887 (Decapoda, Potamopidea, Platythelphusidae) and comments on the taxonomic position of *P. denticulata* Capaart, 1952 from Lake Tanganyika, East Africa. *Crustaceana* 77: 513-532.

Schram, F.R. Review: "Escobar-Briones, E. & F. Alvarez 2002, Modern Approaches to the Study of Crustacea: i-xix, 1-355 (Kluwer Acad. Publ., New York)". *Crustaceana* 77: 253-254.

Schram, F.R. Review: "Jayachandran, K.V. 2001, Palaemonid Prawns: Biodiversity, Taxonomy, Biology and Management (Science Publ., Inc., Enfield)". *Hydrobiology* 515: 247-246.

Schram, F.R. The systematics of sponges. *Contrib. Zool.* 73: 253-254.

Schram, F.R. The truly new systematics - megascience in the information age. *Hydrobiology* 519: 1-7.

Schram, F.R. & Koenemann, S. Are crustaceans monophyletic? In: Cracraft, J. & Donaghue, M.J. (eds.), *Assembling the Tree of Life*: pp. 319-329. Oxford; New York: Oxford Univ. Press.

Schram, F.R. & Müller, H.-G. *Catalog and Bibliography of the Fossil and Recent Stomatopoda.* Leiden: Backhuys Publ.

Sluys, R., Martens, K. & Schram, F.R. The PhyloCode: naming of biodiversity at a crossroads. *TREE* 19: 280-281.

In Press

Guinot, D., Wilson, G.D.F. & Schram, F.R. Jurassic isopod (Malacostraca: Peracarida) from Ranville, Normandy, France. *J. Paleontol.*

Schram, F.R. The place of Collembola amongst the arthropods. *Crustaceana.*

Schram, F.R. The evolution of fossil ecosystems. *Contrib. Zool.*

Schram, F.R. Brine shrimps and other worthy creatures. *Contrib. Zool.*

Schram, F.R. Review: "Gary C.B. Poore, 2002. Zoological Catalogue of Australia: Tanaidacea, Mictacea, Thermosbaenacea, Spelaeogriphacea, i-xii, 1-433; and James K. Lowry & Helen E. Stoddart, 2003. Zoological Catalogue of Australia: vol. 19.2B, Crustacea: Malacostraca: Peracarida, Amphipoda, Cumacea, Mysidacea, i-xii, 1-531". *Crustaceana.*

Schram, F.R., Boere, A.C. & Thomas, N. *Halicyne montanaensis,* a new species of Cycloidea from the Mississippian Bear Gulch Limestone of Central Montana. *Contrib. Sci.*

Schram, F.R. & Dixon, C.J. Decapod phylogeny: addition of fossil evidence to a robust morphological cladistic analysis. *Bull. Mizunami Fossil Mus.*

Appendix B

TAXA ERECTED BY OR IN COLLABORATION WITH F.R. SCHRAM

See *Publications of Frederick R. Schram* (Appendix A) for complete references.

PHYLUM ARTHROPODA
Subphylum Crustacea

   Class Remipedia
     Order Enantiopoda
        Cryptocaris Schram, 1974b (sensu Emerson & Schram, 1991)
          Cryptocaris hootchi Schram, 1974b (sensu Emerson & Schram, 1991)
     Order Nectiopoda Schram, 1986
       Speleonectidae
        Lasionectes Yager & Schram, 1986
          Lasionectes entrichoma Yager & Schram, 1986
       Godzilliidae Schram, Yager & Emerson, 1986
        Godzillius Schram, Yager & Emerson, 1986
          Godzillius robustus Schram, Yager & Emerson, 1986

   Class Maxillopoda
   Subclass Halicyna
     Cyclidae
        Cyclus obesus Schram, Vonk & Hof, 1997
        Halicyne max Schram, Vonk & Hof, 1997
        Halicyne montanaensis Schram, Boere & Thomas, in press
      Apionicon Schram, Vonk & Hof, 1997
        Apionicon apioides Schram, Vonk & Hof, 1997
      Maastrichtiocaris Fraaije, Schram & Vonk, 2003
        Maastrichtiocaris rostratus Fraaije, Schram & Vonk, 2003
   Subclass Thecostraca
     Order Thylacocephala
        Concavicaris georgeorum Schram, 1990
        Concavicaris remipes Schram, 1990
       Convexicaris Schram, 1990
        Convexicaris mazonensis Schram, 1990
       Thylacocephalus Lange et al., 2001
        Thylacocephalus cymolopos Lange, Hof, Schram & Steeman, 2001

Order Thoracica
    Iblidae
        Illilepas Schram, 1986
            Illilepas damrowi (Schram, 1975)
        Verrucidae
            Verruca withersi Schram & Newman, 1980

Class Branchiopoda
    Order Anostraca
        Branchinectidae
            Branchinecta barstowensis Belk & Schram, 2001

Class Malacostraca
Subclass Phyllocarida
    Order Archaeostraca
        Rhinocarididae
            Dithyrocaris rolfei Schram & Horner, 1978
    Order Hoplostraca Schram, 1973
        Sairocarididae Schram, 1973
            Sairocaris centurion Schram & Horner, 1978
            Kellibrooksia Schram 1973
                Kellibrooksia macrogaster Schram, 1973a
    Order Leptostraca
        Rhabdouraeidae Schram & Malzahn, 1984
            Rhabdouraea Schram & Malzahn, 1984
Subclass Hoplocarida
    Order Aeschronectida Schram, 1969
        Kallidecthidae Schram, 1969
            Kallidecthes Schram, 1969
                Kallidecthes richardsoni Schram, 1969
                Kallidecthes eagari Schram, 1979a
        AratidechtidaeSchram, 1979a
            Aractidecthes Schram, 1969
                Aractidecthes johnsoni Schram, 1969
        Aenigmacarididae Schram & Horner, 1978
            Aenigmacaris Schram & Horner, 1978
                Aenigmacaris cornigerum Schram & Horner, 1978
                Aenigmacaris minima F.R. Schram & J.M. Schram, 1979
            Joanellia Schram, 1979
                Joanellia elegans Schram, 1979a
                Joanellia lundi J.M. Schram & F.R. Schram, 1979

Order Palaeostomatopoda
   Perimecturidae
      Perimecturus rapax Schram & Horner, 1978
     Bairdops Schram, 1979a
      Bairdops beargulchensis Schram & Horner, 1978
Order Stomatopoda
  Suborder Archaeostomatopodea Schram, 1969
   Tyrannophontidae Schram, 1969
     Tyrannophontes Schram, 1969
      Tyrannophontes theridion Schram, 1969
      Tyrannophontes acanthocercus Jenner, Hof & Schram, 1998
     Gorgonophontes Schram, 1984a
      Gorgonophontes peleron Schram, 1984a
  Suborder Opisterostomatopodea Schram, 1969 [= Unipeltata Letreille, 1825]
   Superfamily Gonodactyloidea
   Family uncertain
    Paleosquilla Schram, 1968
    Paleosquilla brevicoxa Schram, 1968
   Superfamily Squilloidea
   Squillidae
     Squilla laingae Hof & Schram, 1998
    Angelosquilla Hof & Schram, 1998
     Angelosquilla altamirensis Hof & Schram, 1998
   Superfamily Lysiosquilloidea
   Lysiosquillidae
    Topangosquilla Hof & Schram, 1998
    Topangosquilla gravesi Hof & Schram, 1998
Subclass Eumalacostraca
  Order Syncarida
   Acantotelsonidae
    Acanthotelson kentuckiensis Schram & Chestnut, 1984 (in Schram, 1984b)
   Uronectidae
    Uronectes kinniensis F.R. Schram & J.M. Schram, 1979
    Palaeosyncaris micra Schram, 1984b
   Squillitidae J.M. Schram & F.R. Schram, 1974
   Palaeocarididae
    Palaeocaris secretanae Schram, 1984b
   Minicarididae Schram, 1984b
    Minicaris Schram, 1979a
     Minicaris brandi Schram, 1979a
    Erythrogaulos Schram, 1984b
     Erythrogaulos carrizoensis Schram, 1984b
   Family uncertain
    Williamocalmania Schram, 1984b
    Brooksyncaris Schram, 1984b

[Cohort Mysoida Schram, 1981]
    Order Lophogastrida
        Peachocarididae Schram, 1986
            Peachocaris Schram, 1976 [= Peachella Schram 1974a]
                Peachocaris acanthouraea Schram, 1984
        Eucopiidae
            Schimperella acanthocercus Taylor, Schram & Shen, 2001
    Order Pygocephalomorpha
        Pygocephalidae
            Mamayocaris jaskoskii Schram, 1974a
            Pseudotealiocaris palincsari Schram, 1988
        Jerometichenoriidae Schram, 1978a
            Jerometichenoria Schram, 1978a
                Jerometichenoria grandis Schram, 1978a
        Tylocarididae Taylor, Shen & Schram, 1998
            Tylocaris Taylor, Shen & Schram, 1998
                Tylocaris asiaticus Taylor, Shen & Schram, 1998
            Fujianocaris Taylor, Shen & Schram, 1998
                Fujianocaris bifurcatus Taylor, Shen & Schram, 1998
    Order Belotelsonidea Schram, 1981
        Belotelsonidae Schram, 1974a
    Order Waterstonellidea Schram, 1981
        Waterstonellidae Schram, 1979a
            Waterstonella Schram, 1979a
                Waterstonella grantonensis Schram, 1979a
["Order" Acaridea Schram, 1981 (Edriophthalma Leach 1815)]
    Order Isopoda
        Palaeophreatoicidae
            Hesslerella Schram, 1970
                Hesslerella shermani Schram, 1970
        Sphaeromatoidea
            Reboursia Guinot, Wilson & Schram, in press
                Reboursia ranvillensis Guinot, Wilson & Schram, in press
    Order Amphipoda
        Ingolfiellidae
            Proleleupia Vonk & Schram, 2003
            Paraleleupia Vonk & Schram, 2003
        Bogidiellidae
            Bogidiella arista Koenemann, Vonk & Schram, 1998
[Cohort Brachycarida Schram, 1981]
    Order Hemicaridea Schram, 1981
    Suborder Tanaidacea
    Infraorder Anthracocaridomorpha
        Anthracocarididae Schram, 1979b
            Eucryptocaris Schram, 1989 [not *Cryptocaris* Schram 1974b]
                Eucryptocaris asherorum Schram 1989 [not *Cryptocaris hootchi*]

Infraorder Apseudomorpha
    Superfamily Jurapseudoidea Schram, Sieg & Malzahn, 1986
    Jurapseudidae Schram, Sieg & Malzahn, 1986
      Jurapseudes Schram, Sieg & Malzahn, 1986
      Carlclausus Schram, Sieg & Malzahn, 1986
        Carlclausus emersoni Schram, Sieg & Malzahn, 1986
Infraorder Tanaidomorpha
    Superfamily Cretitanaoidea Schram, Sieg & Malzahn, 1986
    Cretitanaidae Schram, Sieg & Malzahn, 1986
      Cretitanais Schram, Sieg & Malzahn, 1986
Suborder Cumacea
Infraorder Neocumacea Schram, Hof, Mapes & Snowdon, 2003
Infraorder Ophthalcumacea
    Ophthalmdiastylidae
      Ophthaldiastylus parvulorostrum Schram, Hof, Mapes & Snowdon, 2003
      Carbocuma Schram, Hof, Mapes & Snowdon, 2003
        Carbocuma imoensis Schram, Hof, Mapes & Snowdon, 2003
      Securicaris Schram, Hof, Mapes & Snowdon, 2003
        Securicaris spinosus Schram, Hof, Mapes & Snowdon, 2003
Suborder Spelaeogriphacea
    Acadiocarididae Schram, 1974b
    Spelaeogriphidae
      Liaoningogriphus Shen, Taylor & Schram, 1998
        Liaoningogriphus quadripartitus Shen, Taylor & Schram, 1998
Order? Euphausiacea
      Anthracophausia ingelsorum Schram, 1976a
Order Decapoda
Suborder Caridea
    Palaemonidae
      Yongjicaris Garassino, Shen, Schram & Taylor
        Yongjicaris zhejiangensis Garassino, Shen, Schram & Taylor
Suborder Stenopodidea [= Euzygida]
    Spongicolidae Schram, 1986
      Jilinocaris Schram, Shen, Vonk & Taylor, 2000
        Jilinocaris chinensis Schram, Shen, Vonk & Taylor, 2000
Suborder Reptantia
    Glypheoidea
    Glypheidae
      Litogaster turnbullensis Schram, 1971b
    Astacoidea
    Cambaridae
      Palaeocambarus Taylor, Schram & Shen, 1999
    Cricoidoscelosidae Taylor, Schram & Shen, 1999
      Cricoidoscelosus Taylor, Schram & Shen, 1999
        Cricoidoscelosus aethus Taylor, Schram & Shen, 1999

Dromioidea
   Dromiidae?
      Imocaris Schram & Mapes, 1984
         Imocaris tuberculata Schram & Mapes, 1984
Potomoidea
   Platythelphusidae
      Platythelphusa immaculata Marijnissen, Schram, Cumberlidge & Michel, 2004
      Platythelphusa praelongata Marijnissen, Schram, Cumberlidge & Michel, 2004
Order uncertain
   Essoidiidae Schram, 1974a
     Essoidea Schram, 1974a
        Essoidea epiceron Schram, 1974a

Subphylum Atelocerata

Class Euthycarcinoidea
   Euthycarcinidae
      Kottixerxes Schram, 1971a
         Kottixerxes gloriosus Schram, 1971a
         Schramixerxes gerem (Schram & Rolfe, 1982)
      Smithixerxes Schram & Rolfe, 1982
         Smithixerxes juliarum Schram & Rolfe, 1982
   Sottixerxidae Schram & Rolfe, 1982
      Sottixerxes Schram & Rolfe, 1982
         Sottixerxes multiplex Schram & Rolfe, 1982
         Pieckoxerxes pieckoae (Schram & Rolfe, 1982)

Subphylum Cheliceriformes Schram, 1978b

Superclass Chelicerata
Class Merostomata
   Paleolimulidae
      Rolfeia longispinus (Schram, 1979b)

Class Eurypterida
   Mycteropidae
      Mycterops whitei Schram, 1984a

PHYLUM CNIDARIA

Class Hydrozoa
   Order Hydroida
      Mazohydra Schram & Nitecki, 1975
         Mazohydra megabertha Schram & Nitecki, 1975

PHYLUM PRIAPULIDA

>*Priapulites* Schram, 1973b
>>*Priapulites konecniorum* Schram, 1973b

PHYLUM NEMATODA

>*Nemavermes* Schram, 1973b
>>*Nemavermes mackeei* Schram, 1973b

PHYLUM CHAETOGNATHA

>*Paucijaculum* Schram, 1973b
>>*Paucijaculum samamithion* Schram, 1973b

PHYLUM NEMERTINA

>*Archisymplectes* Schram, 1973b
>>*Archisymplectes rhothon* Schram, 1973b

PHYLUM ANNELIDA

>Goniadidae
>>*Carbosesostris* Schram, 1979c
>>>*Carbosesostris megaliphagon* Schram, 1979c
>>Lumbrinereidae
>>>*Phiops* Schram, 1979c
>>>>*Phiops aciculorum* Schram, 1979c
>>Order and Family uncertain
>>>*Soris* Schram, 1979c
>>>>*Soris labiosus* Schram, 1979c
>>>*Ramesses* Schram, 1979c
>>>>*Ramesses magnus* Schram, 1979c

PHYLUM HEMICHORDATA

>*Etacystis* Nitecki & Schram, 1976
>>*Etacystis communis* Nitecki & Schram, 1976

PHYLUM UNCERTAIN

>*Deuteronectanebos* Schram, 1979c
>>*Deuteronectanebos papillorum* Schram, 1979c

Contributors

Babbitt, Courtney C.: Committee on Evolutionary Biology, Division of Biological Sciences, University of Chicago, 5841 S. Maryland Ave., Chicago, IL 60637, U.S.A.
E-mail: cbabbitt@uclink.berkeley.edu

Baron, Christian: Biological Institute, University of Copenhagen, Universitetsparken 15, DK-2100 Copenhagen, Denmark.
E-mail: cbaron@zi.ku.dk

Bergström, Jan: Department of Palaeozoology, Swedish Museum of Natural History, P.O. Box 50007, SE-104 05 Stockholm, Sweden.
E-mail: jan.bergstrom@nrm.se

Bitsch, Colette and Jacques: UMR 5174 Evolution et Diversité Biologique, Université Paul Sabatier, 118 route de Narbonne, Bat IVR3-b1, 31062 Toulouse cedex 4, France.
E-mail: cbitsch@cict.fr

Bonato, Lucio: Department of Biology, University of Padova, Via Ugo Bassi 58 B, 35131 Padova, Italy.
E-mail: luciob@bio.unipd.it

Boore, Jeffrey L.: U.S. Department of Energy Joint Genome Institute and Lawrence Berkeley National Laboratory, and University of California, 2800 Mitchell Drive Walnut Creek, Berkeley, CA 94598, U.S.A.
E-mail: JLBoore@lbl.gov

Brena, Carlo: Department of Biology, University of Padova, Via Ugo Bassi 58 B, 35131 Padova, Italy.
E-mail: brena@bio.unipd.it

Carapelli, Antonio: Department of Evolutionary Biology, University of Siena, via A. Moro 2, 53100 Siena, Italy
E-mail: carapelli@unisi.it

Cotton, Trevor: 1 Colewood Drive, Higham, Rochester, Kent ME2 3UE, U.K.
E-mail: cotton@hotmail.com

Dallai, Romano: Department of Evolutionary Biology, University of Siena, via A. Moro 2, 53100 Siena, Italy
E-mail: dallai@unisi.it

Edgecombe, Gregory D.: Australian Museum, 6 College Street, Sydney, NSW 2010, Australia.
E-mail: greged@austmus.gov.au

Feldmann, Rodney M.: Department of Geology, Kent State University, Kent, OH 44242, U.S.A.
E-mail: rfeldman@kent.edu

Fortey, Richard A.: Department of Palaeontology, The Natural History Museum, Cromwell Rd., London SW7 5BD, U.K.
E-mail: R.Fortey@nhm.ac.uk

Frati, Francesco: Department of Evolutionary Biology, University of Siena, via A. Moro 2, 53100 Siena, Italy.
E-mail: frati@unisi.it

Fusco, Giuseppe: Department of Biology, University of Padova, Via Ugo Bassi 58 B, 35131 Padova, Italy.
E-mail: giuseppe.fusco@unipd.it

Giribet, Gonzalo: Department of Organismic and Evolutionary Biology, Museum of Comparative Zoology, Harvard University, 16 Divinity Avenue, Cambridge, MA 02138, U.S.A.
E-mail: ggiribet@oeb.harvard.edu

Høeg, Jens T.: Biological Institute, University of Copenhagen, Universitetsparken 15, DK-2100 Copenhagen, Denmark.
E-mail: jthoeg@bi.ku.dk

Horne, David J.: Department of Geography, Queen Mary, University of London, Mile End Road, London E1 4NS, U.K.
E-mail: d.j.horne@qmul.ac.uk

Hou, Xian-Guang: Yunnan Research Center for Chengjiang Biota, Yunnan University, Kunming 650091, People's Republic of China.
E-mail: xghou@ynu.edu.cn

Hrycaj, Steven: Department of Biological Sciences, 5047 Gullen Mall, Wayne State University, Detroit, MI 48202, U.S.A.

Jenner, Ronald A.: Section of Evolution & Ecology, University of California, 1 Shields Avenue, Davis, CA 95616, U.S.A.
E-mail: rajenner@ucdavis.edu

Liò, Pietro: Computer Laboratory, University of Cambridge, 15 JJ Thomson Avenue, CB3 0FD Cambridge, U.K.
E-mail: Pietro.Lio@cl.cam.ac.uk

Maas, Andreas: Section for Biosystematic Documentation, University of Ulm, Helmholtzstr. 20, D-89081 Ulm, Germany.
E-mail: andreas.maas@biologie.uni-ulm.de

Martens, Koen: Freshwater Biology, Royal Belgian Institute of Natural Sciences, Vautierstraat 29, 1000 Brussels, Belgium.
E-mail: martens@naturalsciences.be

Maruzzo, Diego: Department of Biology, University of Padova, Via Ugo Bassi 58 B, 35131 Padova, Italy.
E-mail: maruzzo@bio.unipd.it

Minelli, Alessandro: Department of Biology, University of Padova, Via Ugo Bassi 58 B, 35131 Padova, Italy.
E-mail: alessandro.minelli@unipd.it

Nardi, Francesco: Department of Evolutionary Biology, University of Siena, via A. Moro 2, 53100 Siena, Italy
E-mail: nardifra@unisi.it

Patel, Nipam H.: Department of Integrative Biology, University of California, 3060 VLSB #3140, Berkeley, CA 94720-3140, U.S.A.
E-mail: nipam@calmail.berkeley.edu

Popadić, Aleksandar: Department of Biological Sciences, 5047 Gullen Mall, Wayne State University, Detroit, MI 48202, U.S.A.
E-mail: apopadic@biology.biosci.wayne.edu

Richter, Stefan: Institut für Spezielle Zoologie und Evolutionsbiologie, Friedrich-Schiller-Universität Jena, Erbertstr. 1, D-07743 Jena, Germany.
E-mail: richter.stefan@uni-jena.de

Scholtz, Gerhard: Humboldt-Universität zu Berlin, Institut für Biologie/Vergleichende Zoologie, Philippstr. 13, D-10115 Berlin, Germany.
E-mail: gerhard.scholtz@rz.hu-berlin.de

Schön, Isa: Freshwater Biology Section, Royal Belgian Institute of Natural Sciences, Vautierstraat 29, 1000 Brussels, Belgium.
E-mail: schoen@naturalsciences.be

Smith, Robin J.: Department of Earth Sciences, Kanazawa University, Kakuma, Kanazawa 920-1192, Japan.
E-mail: smith@earth.s.kanazawa-u.ac.jp

Schweitzer, Carrie E.: Department of Geology, Kent State University Stark Campus, Canton, OH 44720, U.S.A.
E-mail: cschweit@kent.edu

Stein, Martin: Department of Earth Sciences, Palaeobiology, Uppsala University, Norbyvägen 22, SE-752 36 Uppsala, Sweden.
E-mail: martin.stein@geo.uu.se

Waloszek, Dieter: Head of the Section for Biosystematic Documentation, University of Ulm, Helmholtzstr. 20, D-89081 Ulm, Germany.
E-mail: dieter.waloszek@biologie.uni-ulm.de

Wheeler, Ward C.: Division of Invertebrate Zoology, American Museum of Natural History, Central Park West at 79th Street, New York, NY 10024-5192, U.S.A.
E-mail: wheeler@amnh.org

Index

A

Abakolia · 113, 123, 126
abdomen · 59, 74, 76, 77, 87, 88, 223, 301, 302, 339
abdominal-A (abd-A) · 149, 150, 378
Acari · 219, 323
Achaeranea tepidariorum 171
Achaeta · 229
Acheta · 173, 229, 230
  domestica · 176
Acidiscus · 102, 106, 113, 123, 126
Acimetopus · 102, 106, 109, 113, 118, 123, 126
acron · 75, 144, 147, 152
adaptive radiation · 27
adductor muscle · 266, 321
adult structure · 168
Aeglidae · 23, 29, 38
aglaspidids · 74
Agnostida · 69, 70, 73, 81-83, 85-87, 90, 95-101, 103-113, 115-117, 126
Agnostina · 95-97, 100, 101, 110, 115, 116, 123
Agnostoidea · 96-98, 103, 105, 108, 110, 113
Agnostus · 55-57, 65-68, 79, 81-84, 86, 90, 96, 97, 101, 105, 106, 109
  pisiformis · 60, 62, 64, 69
Alacomenaeus · 158
Alaskadiscus · 113, 123, 126
alimentary caeca · 104, 119, 337
Amblypygi · 152, 154, 220, 237
amino acid · 313, 321
  sequences · 296, 298-300, 357
Amphipoda · 189, 223, 286, 287, 289, 325

Analox · 113, 119, 121, 123, 126
anamorphic development · 236
Anaspides · 289, 325, 328, 343-347
Anlage · 150, 202, 336, 376, 377
Annelida · 152, 153, 197, 201, 233, 356-359, 363-366, 370, 375-377
anomalocaridids · 74, 79, 90
Anomura · 299
Anopheles gambiae · 299
Anoplura · 196, 202
Anostraca · 56, 154, 189, 222, 277, 289, 299, 319, 324
antenna · 57-62, 64, 67-69, 73-77, 79-83, 85-90, 139, 140, 142, 144-147, 150-158, 168, 217, 221-223, 225-228, 230-234, 237, 256, 276, 301, 332-334
  1st (~ antennule) · 55, 57-61, 64, 67-69, 73, 76, 83, 153, 154, 223, 231, 234, 332, 334
  2nd · 77, 79, 85-87, 90, 144, 147, 148, 153, 155, 168, 331, 334, 341
  primary · 139, 146, 152-155
  secondary · 139, 146, 151-155, 157
antennal · 77, 80-82, 85, 144, 147, 151, 156, 172, 223, 236, 255, 269, 331, 334, 341
  segment · 62, 64, 69, 88, 89, 148, 157, 227, 231
Antennapedia (Antp) · 147, 341, 378
antennomere · 222
antennulary · 59, 62, 64, 68, 69, 333
anterior

head · 68, 87
limb · 55, 64, 84
antibody · 169, 170, 172
apomorphy · 67, 86, 143, 155, 259, 321
appendage · 55-57, 59, 60, 64, 65, 67-69, 73-77, 79-82, 84-90, 96, 97, 109, 140, 142-144, 146-148, 151-158, 167-173, 175-177, 180, 204, 215-227, 229-239, 251, 258, 259, 276, 320, 332, 333, 335-337, 340
  flagellar · 236
  frontal · 90, 146, 154
  great · 76, 79, 80, 89, 90, 146, 154, 155, 157
  preferred breakage point (PBP) · 215, 217-235, 237, 238
  regeneration · 217, 223, 239
appendotomy · 217-222, 225-229, 231, 235, 237
apposition eye · 333
Apterygota · 206
apterygotans · 295-297, 301
apterygote · 191, 200
Arachnata · 139, 141-144, 157
Arachnida · 83, 139, 142-144, 148, 149, 178, 185, 201, 203, 204, 207, 220, 313, 319, 320
Arachnomorpha · 139, 140, 143, 157, 204
Araneae · 220, 279, 323
Archaeognatha · 190, 192, 206, 227, 319, 324
Archegozetes · 148
arhabdomeric · 188, 195, 198, 203
Armadillidium vulgare · 170, 171, 325
Artemia · 328, 342-346

franciscana · 289, 299, 301, 324
salina · 195, 324
arthrodial · 61, 62, 67, 334, 338, 339
Arthropod Pattern Theory (APT) · 9
Arthropoda
   appendages · 172, 173, 215, 216, 233, 234, 237, 320
   evolution · 3-8, 10, 89, 95
   monophyly · 142, 276, 313, 315, 317, 320
Arthropodisation · 74
Articulata · 152, 276, 355-375, 379-381
   monophyly · 355, 363, 368, 370, 379
Ascothoracida · 189
Atelocerata · 142, 168, 169, 172, 173, 179, 186, 201, 204-207, 276, 277, 280, 287, 295, 296, 303, 307, 308, 319
Atelocerata-Pancrustacea controversy · 303
atrium oris · 60, 67, 68, 85
autapomorphy · 60, 67-69, 95-97, 143, 197, 203, 207
autospasy · 217, 220-222, 226, 229, 238
autotilly · 217, 220, 221, 229, 238
autotomy · 215-217, 219-221, 223-231, 233, 235, 237, 238
   plane (see also 'preferred breakage point') · 217
Axiidae · 29, 36
axon · 158, 187, 189, 279, 333
axonogenesis · 278

B

Baconian · 9
Bairdiocopina · 265
Bairdioidea · 257, 258, 270
Balanus · 223, 325, 328, 342-344, 346, 347
barnacle · 18
basipod · 55, 59, 61, 62, 64, 67-69
basis · 4, 8, 56, 61, 79, 81, 83, 84, 97, 99, 100, 107, 110-112, 156, 169, 173, 176-178, 193, 223-225, 238, 252, 255-257, 266, 268, 269, 303, 360, 361, 366, 367, 373
Bathydiscus · 103, 105, 113, 121, 123, 126
Bayesian · 283, 285-287, 300
   analysis · 281, 287, 298
   inference · 303
Bellonci Organ · 259, 269
Bilateria · 152
biramous · 74, 76, 81, 255, 277
   appendage · 85, 140, 156, 158
blastopore · 331, 377
Blattodea · 228, 324
blindness · 100
body
   axis · 152, 233, 236
   organization · 356
   plan · 86, 96, 167, 171, 179, 276, 359
   region · 152
   wall · 332, 336
Bolboparia · 106, 109, 113, 118, 123, 126
bootstrap · 112, 115, 249, 252, 260-263, 266, 366
Brachyura · 20, 21, 143, 217, 224-226
Bradoriida · 74, 86, 259
brain · 83, 140, 142, 147-152, 180, 186, 187, 278, 309, 332, 375
branchial plate · 257, 258, 269
Branchiocaris · 154
Branchiopoda · 9, 77, 85, 88, 189, 222, 275, 277-280, 287-289, 299, 307, 309, 311, 315-321, 324
   monophyly · 286, 287, 288, 321
Branchipus · 222
Branchiura · 189, 223, 267, 307, 317-319, 321, 325
branchless (bnl) · 178
breathless (btl) · 178
Bredocaris admirabilis · 56
Bremer support · 112, 115, 281, 282, 284
Burgess Shale · 6-9, 27, 74, 77, 79-81, 83, 87
Burgessia · 77, 79, 81, 83, 87, 88, 143
burgessiids · 74, 86
Bythocytheridae · 257

C

Caenorhabditis · 357
Callianassidae · 21, 29, 37
Calodiscidae · 100, 113, 116
Cambrian · 7, 10, 27, 62, 69, 86-90, 95-98, 110, 116, 153-155, 252, 259, 278, 320, 357
   arthropods · 65, 67, 73, 77, 79, 81, 86-88, 90, 154, 155, 158
   explosion · 74, 89
   Lower · 7, 74, 77, 88, 108, 111, 140, 141, 197
   Upper · 55, 57, 65, 67, 73, 74, 96, 97
Cambrocaris · 83
Campanian · 20, 25, 42
canadaspidids · 74, 86
Canadaspis · 75-77, 79, 81, 90, 357
Candonidae · 255
carapace · 57, 75-77, 79-81, 85, 89, 145, 250, 255, 259, 265, 277, 333, 341
   adductor muscle · 266, 321
   bivalved · 251, 321
Carausius · 230
Carboniferous · 10, 197, 250
Carcineretidae · 26, 30
Caribbean · 18
Caridea · 275, 281, 286, 287
caudal · 80, 339
   furca · 83
caudal (cad) · 80, 83, 339
cell
   division · 228, 375

lineage · 139, 377
Cenomanian · 20, 25, 42
Cenozoic · 23, 24
central
　body · 146, 147, 148, 150
　nervous system (CNS) · 150, 205, 278, 279
Central Americas · 17, 18, 20, 24-27, 29
cephalic
　outline · 100, 101, 119
　segment · 157, 259, 269, 331
cephalization · 145
Cephalocarida · 143, 145, 154, 252, 255, 258-260, 264, 270, 276, 278, 309, 311, 315, 318, 319, 321, 325
cephalon · 69, 82, 84, 87-89, 99-103, 105, 108, 119, 120, 145, 156
　larva · 88
Cephalopyge · 113, 121, 123, 126
cephalosoma · 148
cephalothoracic articulation · 108
cephalothorax · 76
Ceraurus · 79, 82
cerci · 227, 229, 231, 339
character, morphological · 205, 252, 255, 277, 301-303, 308-311, 373
chelae · 77, 226, 334
Chelediscus · 97, 100, 102, 105-107, 109, 113, 115, 123, 126
chelicerae · 149, 151, 153, 154, 155, 218, 220, 237, 308, 320
Chelicerata · 19, 64, 74, 90, 139-144, 146, 148-155, 158, 167-171, 173-176, 178, 179, 185, 187, 188, 195, 197, 199, 201-204, 207, 215, 218, 237, 238, 278, 279, 281, 287, 290, 307, 308, 310, 313, 315, 317-319, 323, 335
Cheliceriformes · 276
chelifores · 308, 320

chemosensory · 147, 152
Chengjiang · 87, 90, 250
　fauna · 74-77, 79-81, 89
Cherax · 289
chiasmata · 309, 339
Chicxulub impact   17, 20, 24-26
Chilenophoberidae · 30
Chilopoda · 178, 186, 188, 193, 194, 199, 200, 205-207, 308, 319, 320, 323
Chirostylidae · 18, 23, 29, 37
Cindarella · 143
circumesophageal · 149, 150
cirri · 223
Cirripedia · 153, 189, 223, 267, 307, 317, 319, 321, 325
cladism · 9
cladist · 3, 8-10
cladistic · 6-10, 98, 99, 186, 322, 355, 358, 369, 371, 374
　analysis · 8, 95-97, 110-112, 115, 123, 151, 252, 265, 359-363, 367
cladistics · 7-10, 97, 359, 360
Cladocera · 189, 222, 321, 325
Cladocopoidea · 252, 257, 270
cladogram · 3, 143, 156, 252, 307, 312, 316, 319, 360, 362
class · 10, 206, 278
classification · 8, 9, 87, 96, 110, 249, 250, 252, 277, 357
claws · 75, 76, 82, 85, 217, 221, 227, 229, 269, 338, 339
cleavage
　intralecithal · 331
　pattern · 276, 377
　spiral · 367, 376
　total · 331
Clypecaris · 75, 90
Cobboldites · 113, 123, 126
Coccoidea · 196, 202
coelom · 363, 366, 376, 377, 378
Coleiidae · 30
Coleoptera · 190, 192, 193, 196, 200, 202, 232, 299

Collembola · 191-194, 200, 206, 227, 280, 295-303, 324
Colymbosathon · 250
commissure · 146-150, 376
common ancestor · 77, 88, 97, 100, 177, 202, 204, 320
comparative
　anatomy · 204
　approach · 216
　morphology · 96, 98, 110, 111, 116, 215, 362, 374
compound eye · 60, 83, 84, 141, 142, 148, 185, 187-191, 193, 195, 197-207, 259, 265, 269, 278, 309
Conchostraca · 290
Condylopyge · 99, 108, 111, 113, 123, 126
Condylopygoidea · 96-98, 103, 105, 108, 109
cone cell · 142, 189, 190, 191, 194, 205, 207, 333
constraint · 217, 299
contingency · 3, 6, 7, 9, 17, 27
Copepoda · 195, 223-235, 286, 289, 317, 319, 321, 325
corm/cormus · 86
Cormogonida · 307, 308, 315, 320
corneagenous cell · 188, 189, 191, 192, 194, 195, 197, 333
coxa · 67, 68, 84-87, 90, 173, 175, 176, 219-222, 224, 225, 228, 230, 231, 238, 256, 257, 269, 335, 336, 338
coxal endite · 84, 336
coxopodite · 157, 158, 169, 172
crab · 20, 21, 148, 217, 225, 226, 233, 235, 237
cranidial spine · 105, 120
Cretaceous · 5, 18, 31
　Lower · 29, 30, 36-41, 250
　Upper · 17, 19-27, 29, 30, 36-42
Crioceris duodecimpunctata · 299
crown group · 73, 150, 252, 257, 276

Crustacea
  crown group · 73, 252
  monophyly · 275, 276, 287, 288, 296, 307, 309
  paraphyly · 280, 301, 309
  stem group · 55, 57, 64, 83, 86, 87, 95, 97, 252, 258
  stem lineage · 82, 83, 86, 87, 90
Crustaceomorpha · 85
crustaceomorphs · 73, 87, 88
crystalline · 340
  cone · 142, 185, 187-193, 195-198, 200, 201, 205, 207, 333
Ctenochelidae · 21, 23, 29, 37
Ctenolepisma · 192, 324
Cumacea · 189, 287, 289
cuticle · 60, 69, 141, 149, 188, 189, 196, 224, 234, 235, 331, 336, 356, 366, 377
cuticular · 60, 187-189, 193, 196-198, 201, 203, 223, 224, 226, 333, 335, 336, 338
Cuvier · 358
Cyclestheria · 321
Cyclestherida · 321
Cylindroleberidoidea · 264, 267, 270
Cypridinoidea · 252, 264, 267, 270
Cypridocopina · 255, 265
Cypridoidea · 254, 255, 257-259, 270
Cytherelloidea · 255-258, 260, 264, 267, 270
Cytherocopina · 265
Cytheroidea · 255, 257-259, 264, 265, 270
cytochrome b (cob) · 298, 299
cytochrome oxidase I-III (cox 1-3) · 297, 298, 300

D

dachshund (dac) · 142, 173-175, 177
Dakoticancridae · 26, 30
Dala peilerta · 56

Danian · 19, 21, 24, 41
Daphnia · 222, 289, 297, 325, 328, 342-345, 347
Daphniidae · 289, 325
Darwin, Charles · 4, 223
Darwinulocopina · 266
Darwinuloidea · 255-257, 261, 270
Dawsonia · 113, 123, 126
decapentaplegic (dpp) · 173, 178
Decapoda · 17-27, 189, 190, 215-217, 224-226, 230, 234-237, 280, 289, 299
Deformed (Dfd) · 147, 341
Delgadella · 110, 113, 121, 123, 126
dendritic · 188, 339
Dermaptera · 190, 229
determination · 159, 258, 265
deuterostome · 170
deutocerebral · 139, 334
  commissure · 147, 149
  segment · 146, 147, 149-153
deutocerebrum · 145-150, 152, 153, 332
development ·
  anamorphic · 236
  evolutionary · 167, 169, 216
  mode · 236
developmental ·
  genetics · 140, 216, 295
  pathway · 180
  stage · 171, 172, 176, 224, 225
Devonian · 20
Diaphanosoma · 290
diaphragm · 215, 221, 223, 227-230, 238
Dicerodiscus · 113, 121, 123, 126
Dicondylia · 297
differentiation · 89, 147, 148, 152, 157, 189, 197, 200, 202, 215, 217, 224-226, 234-236
Diplopoda · 178, 187, 193, 194, 206, 222, 299, 324
Diplostraca · 289
Diplura · 192, 227, 295-297, 299-303, 319, 324

Diptera · 190, 193, 196, 299, 324
distal pigment cell · 333
Distal-less (Dll) · 142, 151, 170-175, 177, 279, 335
diversity · 56, 95, 167, 173, 179, 186, 190, 195, 199, 201, 216, 233, 275, 276, 278, 307, 357
DNA · 266, 280, 281, 287, 311, 312, 322
  mtDNA · 297, 298
Dorippidae · 29, 39
dorsal shield · 269
doublure · 69, 83, 121
Drosophila · 178, 297, 328, 342-346
  melanogaster · 170, 173, 174, 176, 179, 189, 324
Dynomenidae · 20-22, 29, 38, 41

E

ecdysis · 224
Ecdysozoa · 355-368, 370-375, 379-381
  monophyly · 357, 359, 363-367
echinoderms · 375, 376, 378
Eciton · 197, 198, 202
ectoderm · 178, 179, 276, 331, 332, 335, 340, 375
  cells · 178
egg · 172, 331, 336, 339, 341
  nauplius · 331
Egyngolia · 113, 123, 126
Ekwipagetia · 113, 123, 126
embryo · 147, 149, 150, 157, 170-178, 202, 236, 331, 341, 376
embryogenesis · 149, 172, 175, 228, 233
embryology · 139, 216
embryonic · 149, 200, 216, 218, 222, 233, 235, 236, 331, 333, 336, 337
emeraldellids · 74, 81
endite · 62, 63, 67-69, 79, 81, 83, 85, 86, 88, 90, 144, 153,

157, 222, 256, 257, 269, 335, 336, 338
endopod · 55, 61-64, 67, 68, 73, 75-85, 88, 156, 269
endopodite · 255, 257, 258, 269
Endopterygota · 232, 238, 299
engrailed (en) · 79, 145-149, 178, 279, 331, 375
Entocytheridae · 257
Entognatha · 295, 296, 301-303, 319
Entomostraca · 307, 309, 310, 318
Eocene · 17, 19-27, 30, 31, 36-41
Eodiscidae · 86, 97, 98, 100, 101, 105-109, 113, 116
Eodiscina · 95, 96, 97, 103, 106, 110, 115, 123
eodiscinids · 95-118
Eodiscus · 103, 113, 123, 124, 126
Ephemeroptera · 227, 324
Ephestia · 190
epidermal · 178, 221, 359
  cell · 194, 202, 235
epidermis · 215, 218, 234, 235, 331, 363, 376
epipod (epipodite) · 88, 257, 258, 338
Erymidae · 21, 30, 36
Ethmostigmus · 279, 323
Etyidae · 20, 30
Euarthropoda · 60, 62, 64, 68, 69, 87, 97, 139, 142, 146, 149-154, 157, 158, 197-203, 207
Euchelicerata · 142, 143, 313
Eucrustacea · 55-57, 60, 64, 69, 85, 252
Eumalacostraca · 8, 277, 286, 287, 320
Euphausiacea · 189, 278, 286-289, 325
Eurypterida · 143, 144, 148, 185, 187, 197, 203
eurytypy · 23
evo-devo · 139, 167, 169, 177, 179, 180
evolution ·

GTR model · 298-300
Γ model   299-302
evolutionary ·
  developmental biology · 167, 169, 216
  rate · 249, 260, 261, 267
exite · 257
exopod · 55, 59, 62-64, 67-69, 73, 74, 76, 77, 79-82, 85, 87, 88, 97, 140, 144, 156, 158
exopodite · 255-258, 269
exoskeleton · 26, 73, 83, 85, 97, 104, 111, 167, 224, 228, 331
expression domain · 147, 175, 177, 337, 341
extinction · 5, 10, 17-27, 320
extradenticle (exd) · 173-177
eye
  apposition · 333
  compound · 60, 83, 84, 141, 142, 148, 185, 187-193, 195, 197-207, 259, 265, 269, 278, 309
  dispersed · 185, 190, 191, 196, 199, 200, 202, 203
  lateral · 60, 143, 185, 187, 191-197, 200-202, 204, 206, 207
  median · 59, 60, 187, 188, 195, 201, 333
  naupliar · 195, 198, 201, 334
  stalk · 226, 333
  superposition · 333
eye-slit · 86

F

facial suture · 81, 83, 99, 100
feeding · 18, 22, 55, 57, 64, 65, 67, 68, 73, 76, 77, 82-86, 88, 89, 145, 148, 149, 151, 152, 157, 169, 171, 259, 277, 334

  organ · 82, 85, 172
feeler · 64, 69, 335
filament · 223, 339

filiform · 79, 80, 85, 341
filter feeding · 334
flattened setae · 74, 79
Folsomia · 227
food · 55, 57, 61, 62, 64, 67-69, 73, 86, 88, 90, 148, 151, 153, 169, 222
  groove · 334
Fortiforceps · 76, 77, 79-81, 90, 154, 158
fossil taxa · 140, 146, 155, 309
frontal
  appendage · 90, 146, 154
  organ · 334
  processes · 153
furca · 76, 83, 85, 251, 259
furcal rami · 223, 339
furrow · 82, 87, 100, 103-110, 112, 117-122, 156, 157, 226, 229, 333
Fuxianhuia · 74, 76, 90, 146, 153-155
  protensa · 64, 68, 77, 79
fuxianhuiids · 74

G

Galatheidae · 18, 21, 22, 29, 37, 41
Gammarus · 223, 224
ganglia · 145, 146, 149, 150, 152, 153, 226, 235, 309, 332, 370
  mother cells · 279
genal · 117
  caecae · 104
  spines · 100, 101, 103, 143
GenBank · 281, 297, 299, 326
gene
  expression patterns · 142, 144, 148-151, 167, 169, 171-177, 179, 200, 216, 279
  protein-encoding · 295, 296, 298, 312
  regulation · 177
generative zone · 157
genetic basis · 173, 178
Geophilomorpha · 194, 222, 319

germ band · 376
gill · 83, 140, 144, 178, 227, 228, 237, 337
glabellar · 103-106, 119, 120, 156, 157
  node · 106
  segmentation · 103
gland, maxillary · 259
Glypheidae · 21, 29, 36
gnathal · 335
  segments · 147
gnathobase · 67, 151, 167, 169, 172, 179, 269, 279
gnathobasic mandible · 279
Gnathomorpha · 140
Gomphiocephalus hodgsoni · 297, 299
Goneplacidae · 29, 40, 41
gonopore · 340
Goticaris longispinosa · 57
Gould, Stephen J. · 3, 5-7, 10, 11
Gouldian · 17, 27
'great appendage' · 68, 76, 79, 80, 89, 90, 146, 154, 155, 157
grooming · 62
ground pattern · 60, 62, 64, 68, 69, 87, 146, 147, 151, 156, 158, 186, 190, 201, 207, 356, 364, 373
growth · 87, 89, 101, 157, 178, 215, 217, 224-226, 228-230, 232-235, 251, 259, 341, 365
  zone · 331, 375

H

Habelia · 80
Haikoucaris · 158
Halocyprida · 261, 262
Halocypridoidea · 252, 256, 258, 270
head · 55-61, 64, 67-69, 75, 76, 80-90, 139-158, 167, 171, 187, 192-196, 200, 234, 257, 276, 301, 331-336, 339
  arthropod · 155
  head/trunk boundary · 141, 147, 152, 157, 158
  segmentation · 64, 140, 144, 151, 155, 157, 158
  shield · 56-58, 60, 76, 80, 89, 143, 145, 146, 152, 156-158, 331
Hebediscidae · 106, 113, 116
Hebediscina · 113, 121, 124, 126
Helepagetia · 100, 106, 113
helmetiids · 74, 81, 158
Hemisphere, Southern · 17, 18, 25
Henningsmoenicaris · 55, 57, 60, 62, 64, 65, 68, 82, 83, 84
hermit crabs · 20
Hesslandona · 79, 250
heterochely · 217, 225, 226, 235
Heteroptera · 190, 232, 234, 299
Hexapoda · 85, 87, 142-154, 180, 185-193, 196, 200-207, 215, 227, 238, 275-280, 286-290, 295-303, 307-310, 313, 315, 324, 339
  monophyly · 279, 296, 299-303, 319, 320
  paraphyly · 303
Hexapodidae · 18, 21, 29, 40, 41
holometabolous · 176, 185, 187, 189, 191-193, 195, 196, 200, 202, 206, 207, 217
Homarus · 24, 36, 189, 226, 289, 325, 328, 343-347
homeotic · 169
Homolidae · 18, 21, 22, 29, 38
Homolodromiidae · 18, 21-23, 29, 38
homologous · 103-106, 108, 109, 117, 118, 121, 139, 144, 148, 150, 151, 155, 168, 191, 192, 199, 201, 202, 238, 251, 255, 257-259, 279, 312, 357, 359, 370, 375
homology · 8, 10, 96, 98, 100, 101, 104, 106, 107, 109, 118, 121, 140, 143, 151, 154, 155, 167, 169, 177, 180, 194, 198, 199, 201, 205, 249, 251, 257, 277, 279, 298, 356, 359, 360, 373-375
  dynamic · 312, 313
  serial · 144, 237
homonomy · 150, 278
Homoptera · 196, 232
homothorax (hth) · 173, 174, 176
horseshoe crab · 148
Hox · 169, 378
  cluster · 279
  gene expression · 148-151, 375
  genes · 89, 142
Hutchinsoniella · 325, 329, 342-346
hypopharynx · 334, 335
hypostome · 57-61, 64, 67, 69, 83, 85, 88, 90, 97, 106, 139, 154-158

I

illite · 75, 90
in-group · 69, 113, 260
innervation · 153
insect · 19, 167-180, 185-196, 200, 202, 205, 207, 215, 217, 232, 235, 237, 238, 278-280
Insecta · 168, 188, 191, 195, 199, 201, 295, 296, 299-303
instar · 56, 69, 218, 219, 221, 227, 229, 230-232, 250, 251, 257
intercalary segment · 145, 147, 148, 334, 341
inter-segmental · 338
intralecithal cleavage · 331
Isopoda · 143, 171, 172, 175, 190, 191, 215, 223, 234, 235, 238, 286, 287, 289, 325
Isoptera · 229

J

Japyx solifugus · 297, 299
jaw · 86, 139, 146, 150-154, 169, 334
Jianfengia · 80, 90
Jinghediscus · 100, 113, 115, 117, 119, 124
joint · 61, 176, 220, 223-226, 228-230, 258, 269, 312, 338, 339
Jurassic · 20, 29, 30, 36-41

K

K/P
  boundary · 19-22, 24, 26, 31, 36
  event · 21, 23, 24
  extinction · 17, 19
Katharina tunicata · 298-300
key gene · 178, 180
Kirkbyocopina · 266
Kiskinella · 113, 124, 127
Korobovia · 104, 113, 124, 127
Kunmingella · 250

L

labial (lab) · 145, 147-149, 216, 268, 336, 341
Labrophora · 55, 60, 65, 67, 68, 85
labrum · 57, 58, 60, 67, 68, 85, 86, 88, 90, 144, 145, 148, 154, 334
Laevicaudata · 324
Lagerstätte · 74
lamellipedian setae · 85, 86
lamellipedians · 74, 79, 81-83, 85, 86, 89
lamina · 191, 333, 335
larva · 56, 57, 67, 80, 82, 84-90, 153, 154, 172, 175, 191-196, 206, 218, 219, 220, 221, 232, 236, 331, 356, 367, 376, 377
insect · 185-187, 191, 193, 194, 196, 200, 202, 206, 207
lateral furrows · 103, 104, 106
law · 4-6
leanchoiliids · 74, 80
leg · 74, 76, 80, 82, 84, 85, 87-89, 153, 168, 177, 216-238, 257, 258, 269, 301, 332, 334, 335, 337-341
  patterning · 173-176
  segments · 87, 175, 176
  walking · 149, 171, 216, 217, 220, 226, 237, 338
Lenadiscus · 113, 119, 121, 122, 124, 127
Lepas · 223
Leperditicopida · 250, 252
Lepidothrichidae · 192, 206
Lepidurus · 189, 190, 222, 289
Lepismatidae · 192, 206, 311, 324, 327, 342-346
Leptochilodiscus · 105, 113, 117-119, 121, 124, 127
Leptodora · 325, 328, 342-344, 346, 347
Leptostraca · 287, 320, 325
life
  cycle · 331, 356
  style · 64, 153
Ligia · 190
limb
  5th · 60, 257-259, 269, 336
  6th · 257, 258, 269
  8th · 251, 258, 269, 337
  bud · 56, 172, 224, 226, 233, 337, 338
  homologies · 140
  morphology · 81, 89, 252, 255, 307
  phyllopodous · 337
  rod-like · 58-60, 335
  stem · 67
  thoracic · 153, 257, 277
  types · 151
  walking · 148-150, 152
Limnadia · 290, 324, 328, 342-346
Limnadiidae · 290, 324
limulavids · 74
Limulus · 151, 153, 195, 198, 203, 218, 326, 342-346
  polyphemus · 188, 290, 297, 299, 323
Lithobiomorpha · 221
Lithobius · 194, 199, 221, 323, 327, 342-346
Litometopus · 99, 103, 113, 121, 124, 127
lobster · 20
locomotion · 55, 57, 62, 64, 67, 68, 82, 85, 86, 90, 153, 333, 334, 337, 359
Locusta migratoria · 290, 299, 324
Lophogastrida · 189
lower lip · 58
Luvsanodiscus · 113, 124, 127
Lynceus · 324, 328, 342-346

M

Maastrichtian · 17, 20, 21, 24-27, 30, 31, 42, 43
Macannaia · 113, 120, 124, 127
Macrocypridoidea · 255, 257, 258, 261, 270
Malacostraca · 8, 76, 85, 180, 189, 190, 215, 238, 275, 277, 278-280, 284-289, 299, 301, 307, 309, 311, 315, 317, 318-321, 325, 334
  monophyly · 301, 320
Mallagnostus · 113, 115, 117, 119, 124, 127
Mallophaga · 196
Manawa · 250-252
mandible · 57, 58, 62-64, 67, 68, 77, 85, 87, 90, 142, 145, 147, 151, 167-172, 179, 255, 276, 279, 309, 334, 335
mandibular · 62-68, 144, 167, 169, 170-172, 179, 255, 335, 341
  ganglion · 145
  palp · 256
  segment · 58, 276

Mandibulata · 139-158, 168, 185, 190, 194, 198-207, 276, 277, 279, 307-309, 313, 315, 317-318, 320
 stem-group · 56
Manson impact · 24
Mantodea · 229
Manton, Sidnie M. · 8, 168, 169, 172, 173, 276, 308
 Mantonian · 3, 8-10
marrellomorphs · 74
Martinssonia · 60, 67, 79, 82, 83, 85, 252, 255, 257, 258, 260, 264, 270
mass extinction · 5, 17-19, 23, 26, 27
masticatory · 86
mating · 149
maxilla · 60, 142, 151, 257, 276, 334
 1st (~ maxillule) · 69, 85, 145-147, 257-259, 335, 341
 2nd · 69, 145, 147, 335, 336
maxillary
 ganglion · 145
 glands · 259
 segment · 145, 147, 172, 258, 331, 341
maxilliped · 145, 336
Maxillopoda · 85, 250, 277, 278, 307, 309, 311, 315, 317, 321, 325
maxillule (1st maxilla) · 69, 85, 145-147, 257-259, 335, 341
Mecochiridae · 20, 30
median eye · 59, 60, 187, 188, 195, 201, 333
medulla · 333
megacheirans · 80
Meniscuchus · 102, 113, 118, 121, 124, 127
meraspis · 84
mesoderm · 144, 331, 340, 364, 375, 376
Mesozoic · 20, 26, 27
metamorphosis · 193, 197, 198
Metazoa · 3, 8, 89, 233, 297, 356-358, 364, 373, 378
 phylogeny · 357, 362
Microcoryphia · 296

Miocene · 17, 27, 31, 36, 38-41
mite · 148, 195, 219, 235, 237
mitochondrial · 204, 312, 340
 DNA · 297, 298
 genes · 169, 280, 295, 299, 301, 307, 321, 322
 genome · 168, 295, 297, 299, 301, 303, 308
mitosis · 202, 219, 220, 235
model
 GTR model of evolution · 298, 300
 Γ model   298-300, 302
 species · 216
model-based · 3, 5, 6, 8, 10
Molaria · 80
molecular
 analysis · 142, 266, 276, 278, 303, 308, 309, 320-322, 362, 374
 phylogenetics · 357
Mollisonia · 81
Mollusca · 19, 24, 27, 153, 197, 201, 298, 365, 375-377
molt · 157, 215, 217-236, 331, 339, 356, 359, 366
morphogenesis · 148, 179, 234, 235
morphological phylogenetics · 358, 362
mosaic · 77, 88, 89, 234
mouth · 57, 60-62, 64, 65, 67, 68, 83, 84, 88, 146, 154, 155, 219, 259, 332, 334, 336, 341, 356, 377
mouthparts · 74, 139, 151-154, 157, 171, 217, 237, 279, 296, 302, 336
Müller, Klaus J. · 7, 9, 27, 37, 55-67, 73, 79, 81-87, 90, 96, 97, 99, 101, 103, 105, 106, 109, 115, 142, 145, 158, 188, 190, 199, 207, 238, 250, 259, 276, 309, 322
mushroom body · 146-148, 150, 152, 153
Myodocopa · 249, 250, 252, 253, 255, 259-261, 263, 264, 266-268
Myodocopida · 199, 258, 259, 261, 262

myodocopids · 74
Myriapoda · 19, 142, 143, 145-148, 151, 167-169, 171-180, 185, 187, 190, 193, 194, 197, 199-207, 215, 221, 238, 275, 276, 278-280, 286, 287, 290, 295, 308, 310, 313, 315, 319, 323
 monophyly · 206, 320
 myriapod-like · 77, 81
Myriochelata · 142, 204, 207, 215, 238, 309
Mysida · 189, 286, 287
mysids · 190, 275, 278, 286, 287
Mystacocarida · 321

N

NADH dehydrogenase (nad4L) · 297
Naraoia · 74, 87, 88, 97, 98
naraoiids · 62, 69, 81, 83, 156, 158
Narceus annularus · 299
Natalina · 106, 113, 124, 127
naupliar eye · 334
nauplius · 64, 82, 85, 86, 90, 172, 195, 198, 201, 331
Nebalia · 223, 325, 328, 342, 344-347
Necrocarcinidae · 18, 29, 30, 39
nectaspidids · 74
Neighbor Joining · 260, 263
Nematoda · 279, 298, 321, 356, 375
Neocobboldia · 113, 115, 124, 127
Neopagetina · 113, 124, 127
Nephropidae · 21, 29, 36, 289, 325
Nereis · 152
nervous system · 146, 147, 149, 150, 152, 205, 278, 309, 332, 363, 376
neural stem cell · 279
neurite · 332
neuroblast · 147, 279, 309, 332

neurogenesis · 204, 278, 279, 364
neuromere · 144, 149, 150
neuron · 142, 188, 278, 279, 309, 332
neuropil · 139, 142, 146-150, 152, 187, 189, 332, 333, 375
Ninadiscus · 113, 118, 124, 127
North Pacific Ocean · 17, 18, 27
Notostraca · 143, 153, 222, 223, 289, 324

O

occacaridids · 74
Occacaris · 77, 79, 80, 90, 146, 153, 154
occipital ring · 103-105, 107, 108, 120
ocelli · 185, 187, 193-199, 201-204, 207
ocular · 139, 145, 147, 148, 150-154
ocular-protocerebral · 145, 147, 150, 152
Odonata · 227, 290
Oelandocaris · 55-69

Ö

Öland, Isle of · 55-57

O

olenellids · 86
Olenoides · 79, 82, 97
olfaction · 152, 153
olfactory · 153
    lobe · 149, 332
    neuropil · 139, 146, 147, 149, 150, 152
Oligocene · 22, 30, 31, 38, 40
ommatidia · 142, 185, 187-194, 197-207
Oncopeltus · 232

Oniscidea · 311, 325, 328, 343-346, 347
ontogenetic · 97, 111, 144, 217
    stages · 256
ontogeny · 99, 101, 111, 140, 141, 149, 157, 236, 250, 333
Onychophora · 139, 146, 149-151, 154, 203, 310, 319, 323, 356
Oodiscus · 113, 124, 127
Opiliones · 218, 290, 323
opisthosoma · 148, 149, 171, 331
Opsidiscus · 113, 120, 124, 127
optic · 186, 187, 190, 192
    ganglia · 150, 309
    lobe · 205
    neuropil · 148, 189, 333
Orchestia · 223-325, 328, 343-347
Ordovician · 74, 79, 95, 141
Orithopsidae · 18, 29, 30, 40
Orsten · 7, 9, 55, 56, 64, 67, 68, 73, 74, 79, 97, 250
Orthoptera · 229, 235, 290, 299, 324
Ostracoda · 19, 83, 86, 199, 223, 249-259, 265-270, 307, 317, 318, 321, 325
    monophyly · 250
    pseudo-ostracodes · 74
    superfamilies · 250, 252, 256, 258, 268
Ostrinia furnacalis · 299
out-group · 113, 201, 252, 255, 260, 261, 263, 265, 267, 270, 281, 284, 286, 298, 300, 320, 322
Ovalicephalus · 79, 80, 90
Oxidus gracilis · 171

P

Pagetia · 102, 103, 106, 107, 113, 124, 127
Pagetides · 113, 124, 127
Paguridae · 29, 36, 226, 289
Pagurus longicarpus · 299

Palaemonetes · 286, 289
Palaeocene · 18-23, 30, 31, 36-41
Palaeocopida · 249, 250, 252, 260, 262, 266
Palaeogene · 17-19, 22, 24-26, 30-36
Palaeoxanthopsidae · 18, 20, 21, 29, 30, 40, 41
Palaeozoic · 64, 73, 74, 83, 197, 249, 250, 252
Palinuridae · 21, 29, 36, 289
palp · 152, 220, 227, 234, 237, 255-257, 269, 335, 336
palpigrades · 148
Panarthropoda · 364, 365, 367, 370, 375
Pancrustacea · 79, 185, 205, 207, 215, 238, 275-277, 279, 280, 288, 295, 296, 298, 299, 301, 303, 308
Panopeidae · 19, 41
Panulirus japonicus · 299
Paradoxides · 141
Paradoxopoda · 142, 204, 207, 307-309, 315, 317, 320
paragnaths · 65, 67, 85
Parapeytoia · 155
parasegment · 364
parasite · 362
parsimony · 8, 10, 140, 249, 251, 252, 260-262, 266, 281, 282, 284, 286, 287, 307, 314, 360, 364
passalid beetle · 190
pathway · 178, 179, 180, 198, 200, 203, 229, 233, 235, 278
pattern
    expression · 142, 144, 148, 149, 151, 167, 169, 171-177, 179, 200, 216, 279
    tagmosis · 140, 141, 157, 301
patterning · 173-175, 177, 179, 189, 234, 331
Pattersoncypris · 250
PCR · 7, 281, 297
pedipalp · 148, 149, 171, 333-335
peduncle · 223

Penaeidae · 21, 29, 36, 289
penis · 223
Pentastomida · 307, 311, 313, 315, 317-319, 321-323, 342, 344-347
Peracarida · 275, 278, 281, 286-288
Permian · 20, 140, 250
Peronopsis · 99, 107, 108, 111, 113, 122, 123, 126
Perspicaris · 76
Phacops · 155, 156
Phanerozoic · 18, 251, 252
Phasmida · 230
phenetics · 9, 121, 360
phosphatic · 74
Phosphatocopida · 74, 83, 86, 87
Phosphatocopina · 56, 57, 60, 64, 154, 250, 252, 255-261, 264, 267, 270
photoreceptors · 185-189, 191, 192, 194, 195, 197, 199-203, 331
Phyllocarida · 223, 277, 278, 287
Phyllopoda · 277
phyllopodous · 337
phylogenetic analysis · 9, 55, 83, 110, 116, 117, 126, 139, 186, 207, 251, 255, 266, 269, 275, 278, 295, 298, 303, 310, 320, 357, 359, 360, 362, 364, 372-374
phylogenetics · 3, 7-9, 310, 356-358, 360-362
phylogeny · 8, 11, 96, 97, 103, 110, 111, 116, 139, 140, 142, 167-169, 186, 201, 203, 215, 249, 250, 252, 257, 265-268, 280, 295, 298, 307, 310, 355-357, 361, 362, 373
phylum · 74, 89, 167, 185, 197, 201, 203, 207, 376
physical ideal of science, the · 4, 5, 8-10
Pilumnidae · 29, 40
Pinnotheridae · 40, 41
pioneer · 8, 10, 249, 309
Platycopida · 260, 262, 266

Platycopina · 265
Pleistocene · 31
pleon · 337
plesiomorphic character · 62, 64, 185, 201, 373
Pliocene · 31
Podocopa · 249, 250, 252, 254, 255, 257, 259-261, 263, 264, 266-268
Podocopida · 249, 260, 262, 266
podocopids · 74, 259
podomeres · 62, 67, 82, 83, 85, 86, 89, 175-177, 255, 257, 258, 269
Podotremata · 21
Polychelidae · 29, 36
Polycopoidea · 252
polymerid trilobites · 95, 112
polyphyletic · 97, 98, 100, 250, 265, 276
polyphyly · 10, 268
polytomy · 315, 370
Polyxenus · 193, 194, 200, 324
Pontocypridoidea · 255, 257, 258, 261, 270
Popperian · 9
Porcellanidae · 26
Porcellio · 191, 223, 325
Portunidae · 41, 289
post-antennal · 64, 69, 77, 81, 82, 85, 88, 89, 157
post-cephalic · 84
  segments · 331
posterior
  appendages · 55, 57, 69, 85, 153, 227
  commissure · 146
  head · 145
  probabilities · 283, 285, 287, 299, 300
  trunk · 62
post-larval · 220, 221
Poupiniidae · 29, 39
precursor · 87, 279
preferred breakage point (PBP) · 215, 217
pre-glabellar furrow · 106, 119
'primary antennae' · 139, 146, 152-155
primer · 252, 280, 297, 313

proboscipedia (pb) · 147-149, 341
Procambarus · 189, 226
proctodeum · 337
proliferation · 96, 224, 235, 279, 364
propeltidium · 148
prosoma · 74, 75, 80, 148, 149, 171, 172, 178, 185, 331, 376
Prosopidae · 21, 30, 39
prostomium · 152
protaspides · 157
protaspis · 88, 96
protein kinase B (PKB) · 178
proterosoma · 148
protocerebral · 139, 145, 147, 150-154, 332
  commissure · 148
protocerebrum · 145-147, 150, 152, 153, 186, 187, 332, 333
Protohermes · 196
protopod · 88, 269, 335
protopodite · 256-258, 269
protostomes · 170, 356
Protura · 153, 154, 192, 295, 296, 302, 324
proximal endite · 62, 63, 67-69, 84, 86, 90, 269
Pseudocobboldia · 113, 124, 127
pseudoextinction · 17, 26, 27
Pseudoiulia · 81
Pterygota · 206, 297
pterygotes · 192, 215
Ptychagnostidae · 111
Ptychagnostus · 106, 111, 113, 122, 123, 126
Ptychopyge · 141, 159
Puncioidea · 249, 252, 255, 258, 260, 261, 264, 266, 267, 270
punctuated equilibrium · 5, 7
Pycnogonida · 139, 142-144, 148, 149, 151, 153, 218, 310, 313, 315, 317-319, 320, 323
pygidium · 84, 97, 99, 101, 102, 108-111, 121, 122, 141, 156, 157

R

ramus · 55, 62, 67, 84, 223, 255, 258, 269, 334, 339
Raninidae · 21, 22, 29, 39, 41
rank · 6
raptorial · 148, 155, 237
rDNA
  18S · 277, 357
  28S · 252, 280, 288, 357
Recent · 17, 20-23, 26, 29, 36-41, 76, 86, 140, 142, 145, 149, 151, 153, 154, 173, 207, 249, 256, 275, 280
recovery
  dynamics · 21
  patterns · 19
recruitment · 189, 235
refugium · 22, 25
  taxa · 17, 19, 23
regeneration · 215-237, 239
  antennae · 223
  leg · 220, 223, 224, 226
  mechanisms · 233
regulatory · 179, 180
  mechanism · 167, 175, 177
Rehbachiella · 56, 154, 155
Remipedia · 10, 153, 180, 277, 278, 309, 311, 315, 319, 325, 329, 334, 342-346
  relationships · 318
respiratory · 144, 177-179, 337
retina · 187, 194-197, 200-202, 333
retinula cells · 188-191, 195, 200, 201
Retroplumidae · 21, 29, 40
Retrorsichelidae · 20, 30
rhabdom · 187-197, 200, 201, 204, 333
rhabdomeres · 187, 190, 192
rhabdomeric cells · 188, 192
Rhizocephala · 64
ribosomal · 168, 204, 266, 280, 297, 298, 307
  RNA · 170, 280, 282, 284, 312
    ribosomal (rRNA) · 199, 297, 310, 311, 313, 321, 322, 326, 330, 341
  rrnL · 297
  rrnS · 297
tRNA · 295, 297, 341
rostrum · 57-59
rudiment · 331
rudimentary · 197, 198, 202, 230
Runcinodiscus · 113, 124, 127

S

Sanctacaris · 80, 87
Santonian · 25, 42
Sarsielloidea · 270
Schistocerca · 189
schizomids · 148
Schizoramia · 141, 204, 207, 307
Schram, Frederick R. · 3, 5, 8-10, 20, 27, 28, 55, 73, 89, 95, 97, 139, 142, 153, 172, 180, 186, 215, 249, 250, 257, 275, 276, 278, 287, 296, 301, 303, 307, 309, 310, 320, 321, 355, 356, 358, 361, 362
Scolopendromorpha · 222
Scorpiones · 218, 290, 319, 323
scorpions · 74, 144, 195, 197, 218, 236, 237
Scutigera · 188, 190, 199, 221, 238, 239, 323, 326, 342-346
Scutigeromorpha · 190, 199, 200, 205, 206, 221
Scyllaridae · 29, 41
'secondary antenna' · 139, 146, 151-155, 157
secretion · 150
segment · 58, 60, 62, 64, 67, 69, 73, 75, 76, 79-90, 99-101, 104, 107-109, 140, 141, 144-157, 171, 172, 175, 176, 196, 215-224, 227-236, 238, 251, 258, 259, 269, 277, 301, 331, 334, 336, 337-341, 364, 375, 377
  identity · 279
  polarity · 145, 149, 279
thoracic · 84, 99, 101, 105, 107, 108, 120-122, 141, 157, 338
segmental composition · 89
segmentation · 10, 87, 97, 99, 103, 108, 109, 121, 140, 144, 149, 157, 158, 176, 219, 234, 251, 259, 278, 279, 334, 339, 356, 359, 363-367, 374, 377
  appendages · 228, 236
  patterns · 365
  reduced · 220, 229, 232
  tarsus · 227, 228, 231
segment-polarity · 279
  gene · 145, 149
SEM · 57, 61, 65
Semadiscus · 113, 125, 127
Semibalanus · 325, 328, 342, 343, 345, 346, 347
Semper cells · 187, 189, 190, 192-194, 197, 199, 202
sensilla · 219, 227, 230, 231, 234, 332, 334
sensitivity analysis · 322, 362, 364, 374
sensory · 82, 86, 139, 145, 147, 148, 192, 218, 220, 221, 229, 259, 339
  cells · 191, 196, 201, 334
  organs · 64, 149, 152-154, 230
Sepkoski, Jack J. · 10, 18, 21
serial homology · 144
serotonergic · 332
serotonin · 142
Serrodiscus · 98, 104, 106, 110, 113, 121, 125, 127
sessile · 191, 333
setae · 55, 61-64, 67, 68, 73, 74, 76, 77, 79-82, 84-90, 140, 144, 157, 158, 218, 219, 221, 222, 255, 256, 258, 269, 339
setules · 76, 79
Sex combs reduced (Scr) · 147
sexual dimorphism · 255
shield · 56-60, 64, 76, 80, 82, 87, 88, 89, 143, 145, 146, 152, 156-158, 269, 331, 333
Sigilliocopina · 266

Sigillioidea · 255, 257, 258, 270
Signor-Lipps effect 21
Silurian · 250
Simocephalus · 222, 289
Sinodiscus · 101, 104, 113, 125, 127
  changyanensis · 115
Sinopagetia · 113, 125, 127
Siphonaptera · 196, 202
sister group · 98, 103, 112, 113, 115, 141, 142, 151, 168, 169, 200, 203, 206, 252, 275, 277, 278, 280, 287, 308, 309, 315, 319, 320-322, 365, 370
Skara
  anulata · 56
  minuta · 56
slime papillae · 150
solifuges · 148
somite · 337, 340
Southern Hemisphere · 17, 18, 25
spermatozoa · 341
spider · 152, 178, 179, 195, 203, 220, 221, 235-238, 279
  embryos · 172
Spinicaudata · 319, 421, 324
stem
  cell · 279
  group · 56, 83, 86, 87, 95, 97, 252, 258, 322
  line · 82, 83, 86, 87, 90
  lineage · 55, 57, 64, 67-69, 82, 86, 87, 90, 139, 141, 143, 149, 151-155, 157, 158, 320, 322
  species · 151, 152
stemmata · 185-187, 190, 191, 193, 194, 196, 199, 200, 202, 207, 333
sternum · 64, 67, 68, 85, 143, 334, 338, 340
Stigmadiscus · 110, 113, 125, 127
stomatogastric · 146, 147, 149, 150, 152
Stomatopoda · 223, 289, 311, 325, 328, 343-347
Strepsiptera · 196, 202

subphylum · 167, 168, 176
superposition eye · 333
symplesiomorphic character · 278, 302
symplesiomorphy · 69
synapomorphic · 251
  character · 296
synapomorphy · 85, 95-97, 100, 101, 104, 105, 107, 109, 110, 115, 142, 143, 251, 255, 259, 276, 278, 288, 355, 358, 359, 362, 363-367, 374
Syncarida · 278, 287, 288
synonym · 87, 217
systematics · 10, 19, 143, 169, 179, 180, 249, 357, 361, 362

T

tagma · 75, 80-82, 87, 89, 139, 141, 145, 148, 152, 157, 158, 333
tagmatization · 149, 277
tagmosis · 76, 77, 82, 85, 139-141, 158, 278, 307
  pattern · 140, 141, 157, 301
tail · 56, 64, 80, 81, 85
Tanglangia · 90
Tannudiscus · 97, 100, 106, 110, 113, 115, 125, 127
Tardigrada · 142, 203, 310, 315, 319, 320, 322, 323, 365, 366, 368, 370, 377, 380
taxa, higher · 265, 373
taxon, name · 323
taxonomy · 6, 10, 110
TCC-group · 141, 168
Tchernyshevioides · 113, 125, 127
tegopeltids · 158
telencephalon · 152
telopodite · 169, 172, 173, 335, 336
telson · 79, 80, 88, 218, 251, 259, 339
tergite · 58, 60, 75, 333, 337, 338

terrestrialization · 295, 301
Terrestricytheroidea · 256, 257, 261, 265, 270
test
  Kishino-Hasegawa · 300, 301
  Shimodaira-Hasegawa · 300, 301
Tethys, Super scenario · 26
Tetraconata · 142, 185, 205, 207, 295, 307-309, 313, 315, 317-319, 322
Tetrodontophora bielanensis · 297, 298, 299
Thaumatocypridoidea · 252, 253, 256, 270
Thecostraca · 321
Thelxiope · 81
Thermobia · 227, 234
  domestica · 170, 171, 172, 192, 324
thoracic
  limbs · 153, 156, 222, 237, 257, 277
  segments · 84, 99, 101, 105, 107, 108, 120-122, 141, 157, 338
thoracopod · 337
Thoracopoda · 307, 309, 310, 318
Thoracotremata · 26
thorax · 59, 84, 103, 105, 107, 108, 111, 156, 301, 337
  posterior · 63
Thyropygus · 299
Torynommidae · 30
tracheal systems · 167, 177-179
trachealess (trh) · 178
Tracheata · 177, 186, 204, 205, 308
transcription factor · 178, 216
transformation · 144, 152, 154, 197-199, 201, 207, 279, 312
Triarthrus · 156
Triassic · 20, 29, 30, 36
Triatoma dimidiata · 299
Tribolium · 189, 200
  castaneum · 172, 173, 176, 299
trilobation · 69, 97, 143

Trilobita · 69, 81, 87, 96, 97, 107, 112, 115, 139-143, 146, 153, 156-158, 185, 187, 197, 203, 204
trilobites · 55, 56, 64, 68-70, 73, 74, 79, 81-89, 95-112, 116, 168, 238
Trilobitomorpha · 140, 276
trilobitomorphs · 88, 141-144, 155, 158
Triops · 189, 190, 223, 289, 324, 328, 342-346
tritocerebrum · 144-147, 152
trunk · 58-60, 80, 81, 85, 140, 141, 144, 145, 147-153, 156-158, 251, 336-341
   boundary · 147, 152
   limbs · 56, 59, 60, 62, 257, 259, 336
Tsunyidiscidae · 112, 116
Tsunyidiscus · 101, 107, 112, 115, 118, 125, 128
Turonian · 25, 42

U

Uca · 225, 235
Ultrabithorax (Ubx) · 149, 150, 279, 337, 378
Uniramia · 204
uniramous antennae · 75, 79
unsegmented · 122, 228, 233, 269
Upogebiidae · 21, 29, 37
Ur-arthropod · 87, 88
Ur-crustacean · 88
uropod · 226, 237, 251, 259

V

vascular network · 83
Västergötland · 55-57, 60
ventral nerve cord · 332, 376
ventral veinless/drifter (vvl/dfr) · 178
Vertebrata · 6, 10, 18, 152
vestigial · 153, 172, 251, 336
visual · 83, 185-188, 190, 191, 194, 196-199, 201-207, 298
   system · 186

W

walking · 62, 171, 172, 216, 217, 220, 226, 237, 297, 338, 339
   limbs · 148-150, 152
Walossekia · 56
Waloszek, Dieter · 7, 9, 55-57, 60, 62, 64, 65, 67, 69, 73, 79, 81-84, 86, 87, 90, 96, 97, 99, 101, 103, 105, 106, 109, 115, 145, 148, 149, 153-155, 158, 203, 238, 250, 255, 259, 276, 309, 310, 321, 322
   = Walossek 7
Waptia · 77, 87, 88
Weinbergina · 149
Weymouthia · 113, 125, 128
Weymouthiidae · 95, 100, 101, 103, 105-107, 109, 110, 113, 115, 116
whip spiders · 152
whole-limb · 172, 173
wing · 57, 337
wingless (wg) · 173, 178, 206
Wonderful Life · 6, 7, 10

X

Xandarella · 155
xandarellids · 74, 143, 158
Xanthidae · 20, 41
Xanthoidea · 26
Xiphosura · 142-144, 148, 149, 153, 185, 187, 188, 198, 199, 202-205, 207, 218, 237, 290, 299, 313, 319, 320, 323

Y

Yohoia · 77, 80, 87, 88, 90, 158
yohoiids · 74
yolk · 331
Yukonia · 100, 113, 121, 125, 128
Yukonides · 113, 125, 128
Yukoniidae · 105, 113, 116

Z

Zanthopsidae · 41
Zenker's Organ · 255, 265, 269
Zygentoma · 143, 191, 194, 200, 206, 227, 296, 297, 299, 324

CRUSTACEAN ISSUES

General editor
RONALD VONK
Institute for Biodiversity and Ecosystem Dynamics, University of Amsterdam, Netherlands

---

1. Crustacean Phylogeny
   Schram, F.R. (ed.)
   1983   ISBN 90 6191 231 8 Sold out

2. Crustacean Growth: Larval Growth
   Wenner, A. (ed.)
   1985   ISBN 90 6191 294 6

3. Crustacean Growth: Factors in Adult Growth
   Wenner, A. (ed.)
   1985   ISBN 90 6191 535 X

4. Crustacean Biogeography
   Gore, R.H. & Heck, K.L. (eds.)
   1986   ISBN 90 6191 593 7

5. Barnacle Biology
   Southward, A.J. (ed.)
   1987   ISBN 90 6191 628 3

6. Functional Morphology of Feeding and Grooming in Crustacea
   Felgenhauer, B.E., Thistle, A.B. & Watling, L. (eds.)
   1989   ISBN 90 6191 777 8

7. Crustacean Egg Production
   Wenner, A. & Kuris, A. (eds.)
   1991   ISBN 90 6191 098 6

8. History of Carcinology
   Truesdale, F.M. (ed.)
   1993   ISBN 90 5410 137 7

9   Terrestrial Isopod Biology
    Alikhan, A.M.
    1995   ISBN 90 5410 193 8

10 New Frontiers in Barnacle Evolution
    Schram, F.R. & Hoeg, J.T. (eds.)
    1995   ISBN 90 5410 626 3

11 Crayfish in Europe as Alien Species - How to Make the Best of a Bad Situation?
    Gherardi, F. & Holdich, D.M. (eds.)
    1999   ISBN 90 5410 469 4

12 The Biodiversity Crisis and Crustacea - Proceedings of the Fourth International
    Crustacean Congress, Amsterdam, Netherlands
    Vaupel Klein, J.C. von & Schram, F.R. (eds.)
    2000   ISBN 90 5410 478 3

13 Isopod Systematics and Evolution
    Kensley, B. & Brusca, R.C. (eds.)
    2001   ISBN 90 5809 327 1

14 The Biology of Decapod Crustacean Larvae
    Anger, K.
    2001   ISBN 90 2651 828 5

15 Evolutionary Developmental Biology of Crustacea
    Scholtz, G. (ed.)
    2004   ISBN 90 5809 637 8

16 Crustacea and Arthropod Relationships
    Koenemann, S. & Jenner, R.A. (eds.)
    2005   ISBN 0 8493 3498 5